# Environmental Biotechnology

*Edited by*
*Hans-Joachim Jördening and*
*Josef Winter*

# Related Titles from WILEY-VCH

H.-J. Rehm, G. Reed, A. Pühler, P. Stadler, J. Klein, J. Winter (Eds.)

## Biotechnology

**Second, Completely Revised Edition,**
**Volume 11a–c, Environmental Processes I–III**

2000
ISBN 3-527-30242-5

G. M. Evans, J. C. Furlong (Eds.)

## Environmental Biotechnology

**Theory and Application**

2002
ISBN 0-470-84372-1

G. Bitton (Ed.)

## Encyclopedia of Environmental Microbiology

2002
ISBN 0-471-35450-3

G. Bitton

## Wastewater Microbiology

1999
ISBN 0-471-32047-1

P. Singleton

## Bacteria in Biology, Biotechnology and Medicine

2004
ISBN 0-470-09026-X

R. D. Schmid, R. Hammelehle

## Pocket Guide to Biotechnology and Genetic Engineering

2003
ISBN 3-527-30895-4

# Environmental Biotechnology

## Concepts and Applications

*Edited by*
*Hans-Joachim Jördening and Josef Winter*

**WILEY-
VCH**

WILEY-VCH Verlag GmbH & Co. KGaA

Environmental
biotechnology : concepts
c2005.

**Edited by**

**Priv.-Doz. Dr. Hans-Joachim Jördening**
Technical University Braunschweig
Institute for Technical Chemistry
Division Technology of Carbohydrates
Langer Kamp 5
38106 Braunschweig
Germany

**Prof. Dr. Josef Winter**
University Karlsruhe
Institute of Biological Engineering
Am Fasanengarten
76131 Karlsruhe
Germany

■ This book was carefully produced. Nevertheless, editors, authors and publisher do not warrant the information contained therein to be free of errors. Readers are advised to keep in mind that statements, data, illustrations, procedural details or other items may inadvertently be inaccurate.

**Library of Congress Card No.: applied for**

**British Library Cataloguing-in-Publication data:**
A catalogue record for this book is available from the British Library.

**Bibliographic information published
by Die Deutsche Bibliothek**
Die Deutsche Bibliothek lists this publication in the Deutsche Nationalbibliografie; detailed bibliographic data is available in the Internet at <http://dnb.ddb.de>.

© 2005 Wiley-VCH Verlag GmbH & Co. KGaA, Weinheim

Printed in the Federal Republic of Germany
printed on acid-free paper

**Composition**  Detzner Fotosatz, Speyer
**Printing**  betz-druck GmbH, Darmstadt
**Bookbinding**  J. Schäffer GmbH, Grünstadt

**ISBN**  3-527-30585-8

# Preface

Josef Winter, Claudia Gallert, Universität Karlsruhe, Germany
Hans-Joachim Jördening, Technische Universität Braunschweig, Deutschland

The growing awareness of environmental problems, caused especially by the predominate use of fossil resources in connection with pure chemical pathways of production, has led the focus on those alternatives, which sounds environmentally more friendly. Here, biotechnology has the chance to influence and improve the quality of the environment and production standards by:
– introduction of renewable instead of fossil raw materials
– controlled production of very specific biocatalysts for the
– development of new and environmentally improved production technologies with less purified substrates and generation of fewer by-products
– bioproducts as non-toxic matters, mostly recyclable.

Some impressive studies on industrial applications of biotechnology are published in two OECD reports, which summarized, that biotechnology has the potential of a reduction of operational and/or capital cost for the realization of more sustainable processes (OECD1, OECD2). However, until today the sustainability of technical processes is more the exception than the rule and therefore so-called "End-of-Pipe"-technologies are absolutely necessary for the treatment of production residues.

In 1972 the Club of Rome published its study "Limits of Growth" and prognosted an upcoming shortage of energy and primary resources as a consequence of exponential growth of population and industry (Meadows et al. 1972). Although the quantitative prognoses of Dennis Meadows and his research team have not been fulfilled, the qualitative statements are today well accepted. Aside of a shortage of resources for production of commodities the limits for an ecologically and economically compatible disposal of production residues and stabilized wastes have to be more and more taken into consideration. The limits for disposal of solid and liquid pollutants in soil and water or of waste gases in the atmosphere are a major issue, since soil, water and air are no longer able to absorb/adsorb these emissions without negative consequences for ecology and life in general. The ultimate oxidation product of organic residues by incineration or – more smooth – by biological respiration in aquatic or terrestrial environment led to a significant increase of the carbon diox-

*Environmental Biotechnology. Concepts and Applications.* Edited by H.-J. Jördening and J. Winter
Copyright © 2005 WILEY-VCH Verlag GmbH & Co. KGaA, Weinheim
ISBN: 3-527-30585-8

ide content of the atmosphere in the last centuries and thus influences the overall climate. This increase is abundantly attributed to combustion of fossil fuels by traffic and of fossil fuels and coal for industrial production processes and house heating. Increasing concentrations of carbon dioxide in the atmosphere from incineration of fossil energy sources and from decomposition of organic matter are the main reason for the greenhouse effect.

Whereas the pollution of soil with waste compounds and subsequently with their (bio)conversion products generally remains a locally restricted, national problem, as long as evaporation of volatile compounds into the air or solubilization of solids in rain or groundwater can be prevented, emissions into water or the atmosphere are spreading rapidly and soon reach an international dimension. A disturbance of the equilibrium of the natural cycles of carbon, nitrogen, phosphate, sulfur or halogen compounds causes an ecological imbalance and endangers nature. In the Brundtland-report "Our common future" (Hauff 1987) a discussion was started about "sustainable development". The practical realization of this concept was suggested at the "Conference on Environment and Development" of the United Nations in Rio de Janeiro in 1992 and enforced as an action programme in the so-called Agenda 21. A sustainable development to maintain the basis for future generations is contraindicated by exploitation of non-regenerative energy and material resources and a shortening of life cycles (e.g. in information technologies).

A life cycle assessment is required to reduce or at least to bring to everybodies attention the flood of waste material. By the obligate demand for recycling of waste components, which is fixed in European Council Directive 91/156/EEC and e.g. translated to the German waste law (KrW/AbfG 1996), production and the use of commodities should minimize the amount of wastes. The practicability of this approach must be demonstrated in industrialized countries and then should be adopted by less developed or developing countries.

Environmental biotechnology initially started with wastewater treatment in urban areas at the turn of the 19/20th century (Hartmann 1999) and has been extended among others to soil remediation, off gas purification, surface and groundwater cleaning, industrial wastewater purification, deposition techniques of wastes in sanitary landfills and composting of bioorganic residues, mainly in the second half of the 20 century.

The available processes for the protection of the terrestric and aquatic environment were summarized in the first edition of "Biotechnology" still in one volume. Some ten years later in the second edition of "Biotechnology" the development in the above mentioned environmental compartments was updated and decribed by experts in the field from Europe and the United States of America. Although the description was kept very stringent, the above mentioned areas of environmental processes finally were issued in 3 volumes. Volume 11a of "Biotechnology" was subtitled "Environmental Processes I – Wastewater Treatment" (edited in 1999) and was devoted to general aspects and the process development for carbon, nitrogen and phosphate removal during wastewater treatment and anaerobic sludge stabilization. Volume 11b of "Biotechnology" was subtitled "Environmental Processes II – Soil Decontamination" (edited in 2000) and summarized microbial aspects and the processes that were applied for soil (bio-)remediation and Volume 11 c, subtitled "Environmental Processes III – Solid Waste and Waste Gas Treatment, Preparation of Drinking water" (edited in 2000) covered general aspects, microbiology and processes for solid waste treatment, waste gas purification and potable water preparation.

The new book "Environmental Biotechnology" covers what we think the most relevant topics of the previous volumes 11 a, b and c of "Biotechnology" in a comprehensive form. The invited authors were given the opportunity to update their contributions when a significant progress was achieved in their field in recent years. For instance, although many alternatives were existing in the past for domestic sewage treatment to remove nitrogenous compounds, the development of new biological processes for nitrogen removal in the laboratory and in pilot scale-dimension was reported recently. These processes work with a minimized requirement for an additional carbon source. Although these processes are not yet widely applied in praxi, they are investigated in detail in pilot- or demonstration-scale in single wastewater treatment plants. The results seem to be promising and might get importance in the future.

The authors and the editors of the new book hope that the presented comprehensive overview on processes of environmental biotechnology for liquid, solid and gaseous waste treatment will help students and professional experts to obtain a fast fundamental information and an overview over the biological background and general process alternatives. This might then be a useful basis or starting point to tackle a specific process in more detail.

Josef Winter, Claudia Gallert, Hans-Joachim Jördening
Karlsruhe and Braunschweig, September 2004

## References

L. Hartmann (1999) Historical Development of Wastewater Treatment Processes. In: Biotechnology – Environmental processes I (Volume editor J. Winter), page 5–16. WILEY-VCH, Weinheim 1999.

Hauff V. (ed) (1987) Unsere gemeinsame Zukunft. Der Brundtland-Bencht der Weltkommission für Umwelt und Entwicklung. Eggenkamp Verlag, Greyen

KrW/AbfG 1996. Kreislaufwirtschafts- und Abfallgesetz – Gesetz zur Förderung der Kreislaufwirtschaft und Sicherung der umweltverträglichen Beseitigung von Abfällen. Vom 27.9.1994. Bundesgesetzblatt BGBL I. 2705 pp.

Meadows D. H., Meadows D. L., Zahn E., Milling P. (1972) Die Grenzen des Wachstums. Bericht des Club of Rome zur Lage der Menschheit. Stuttgart

OECD (1998), Biotechnology for Clean Industrial Products and Processes: Towards Industrial Sustainability, OECD Publications, Paris

OECD (2001), The Application of Biotechnology to Industrial Sustainability, OECD Publications, Paris

# Contents

*Environmental Biotechnology. Concepts and Applications.* Edited by H.-J. Jördening and J. Winter
Copyright © 2005 WILEY-VCH Verlag GmbH & Co. KGaA, Weinheim
ISBN: 3-527-30585-8

*Karl-Heinz Rosenwinkel, Ute Austermann-Haun, and Hartmut Meyer*

# List of Contributors

Ute Austermann-Haun
Fachhochschule Lippe und Höxter
Fachbereich Bauingenieurwesen
Emilienstr. 45
32756 Detmold
Germany

Klaus Buchholz
Institut für Technische Chemie
Abt. Technologie der Kohlenhydrate
Technische Universität Braunschweig
Langer Kamp 5
38106 Braunschweig Germany

Derek E. Chitwood
University of Southern California
Civil and Environmental Engineering
Los Angeles, CA 90098-1450
USA

Joseph S. Devinny
University of Southern California Civil
and Environmental Engineering
Los Angeles, CA 90098-1450
USA

Helmut Dörr
Arcadis Consult GmbH
Berliner Allee 6
64295 Darmstadt
Germany

Wolfgang Fritsche
Institut für Mikrobiologie
Friedrich-Schiller-Universität Jena
Philosophenweg 12
07743 Jena
Germany

Claudia Gallert
Institut für Ingenieurbiologie
und Biotechnologie des Abwassers
Universität Karlsruhe (TH)
Am Fasanengarten
Postfach 6980
76128 Karlsruhe
Germany

Thomas Held
Arcadis Consult GmbH
Berliner Allee 6
64295 Darmstadt
Germany

Mogens Henze
Department of Environmental Science
and Engineering
Building 115
Technical University of Denmark
2800 Lyngby
Denmark

*Environmental Biotechnology. Concepts and Applications.* Edited by H.-J. Jördening and J. Winter
Copyright © 2005 WILEY-VCH Verlag GmbH & Co. KGaA, Weinheim
ISBN: 3-527-30585-8

Veerle Herrygers
CENTEXBEL
Rue Montoyer 24 – boîte 2
1000 Bruxelles
Belgium

Kai-Uwe Heyer
Ingenieurbüro für Abfallwirtschaft
Bleicherweg 6
21073 Hamburg
Germany

Martin Hofrichter
International Graduate School Zittau
Unit of Environmental Biotechnology
Markt 23
02763 Zittau
Germany

Karsten Hupe
Arbeitsbereich Abfallwirtschaft
TU Hamburg-Harburg
Harburger Schloßstr. 37
21079 Hamburg
Germany

Hans-Joachim Jördening
Institut für Technische Chemie
Abt. Technologie der Kohlenhydrate
Technische Universität Braunschweig
Langer Kamp 5
38106 Braunschweig
Germany

Rolf Kayser
Adolf-Bingel-Str. 2
38116 Braunschweig
Germany

René H. Kleijntjens
Partners in Milieutechnik b.v.
Mercuriusweg 4
2516 AW's-Gravenhage
The Netherlands

Michael Koning
Arbeitsbereich Abfallwirtschaft
TU Hamburg-Harburg
Harburger Schloßstr. 37
21079 Hamburg
Germany

Karel Ch. A. M. Luyben
Technical University of Delft
Julianalaan 67a
2628 Delft 8
The Netherlands

Herbert Märkl
AB Bioprozeß- und
Bioverfahrenstechnik
Technische Universität Hamburg-
Harburg
Denickestr. 15
21071 Hamburg
Germany

Hartmut Meyer
Institut für Siedlungswasser-
wirtschaft und Abfalltechnik
Universität Hannover
Welfengarten 1
30167 Hannover
Germany

Norbert Rilling
TU Hamburg-Harburg
Arbeitsbereich Abfallwirtschaft
und Stadttechnik
Harburger Schloßstr. 37
21079 Hamburg
Germany

Karl-Heinz Rosenwinkel
Institut für Siedlungswasser-
wirtschaft und Abfalltechnik
Universität Hannover
Welfengarten 1
30167 Hannover
Germany

Bernhard Schink
Institut für Mikrobielle Ökologie
Universität Konstanz
Universitätsstr. 10
78457 Konstanz
Germany

Frank Schuchardt
Bundesforschungsanstalt für Landwirt-
schaft Braunschweig-VÖlkenrode (FAL)
Institut für Technologie
und Biosystemtechnik
Bundesallee 50
38116 Braunschweig
Germany

Volker Schulz-Behrendt
Umweltschutz Nord GmbH & Co.
Industriepark 6
27767 Ganderkesee
Germany

Rainer Stegmann
Arbeitsbereich Abfallwirtschaft
TU Hamburg-Harburg
Harburger Schloßstr. 37
21079 Hamburg
Germany

Michael S. Switzenbaum
Department of Civil and Environmental
Engineering
Marquette University
PO Box 1881
Milwaukee, WI 53201-18811
USA

Herman van Langenhove
University of Gent
Coupure L. 653
9000 Gent
Belgium

Willy Verstraete
University of Gent
Coupure L. 653
9000 Gent
Belgium

Muthumbi Waweru
University of Gent
Coupure L. 653
9000 Gent
Belgium

Josef Winter
Institut für Ingenieurbiologie
und Biotechnologie des Abwassers
Universität Karlsruhe (TH)
Am Fasanengarten
Postfach 6980
76128 Karlsruhe
Germany

# 1
# Bacterial Metabolism in Wastewater Treatment Systems

Claudia Gallert and Josef Winter

## 1.1
## Introduction

Water that has been used by people and is disposed into a receiving water body with altered physical and/or chemical parameters is defined as wastewater. If only the physical parameters of the water were changed, e.g., resulting in an elevated temperature after use as a coolant, treatment before final disposal into a surface water may require only cooling close to its initial temperature. If the water, however, has been contaminated with soluble or insoluble organic or inorganic material, a combination of mechanical, chemical, and/or biological purification procedures may be required to protect the environment from periodic or permanent pollution or damage. For this reason, legislation in industrialized and in many developing countries has reinforced environmental laws that regulate the maximum allowed residual concentrations of carbon, nitrogen, and phosphorous compounds in purified wastewater, before it is disposed into a river or into any other receiving water body However, enforcement of these laws is not always very strict. Enforcement seems to be related to the economy of the country and thus differs significantly between wealthy industrialized and poor developing countries. In this chapter basic processes for biological treatment of waste or wastewater to eliminate organic and inorganic pollutants are summarized.

## 1.2
## Decomposition of Organic Carbon Compounds in Natural and Manmade Ecosystems

Catabolic processes of microorganisms, algae, yeasts, and lower fungi are the main pathways for total or at least partial mineralization/decomposition of bioorganic and organic compounds in natural or manmade environments. Most of this material is derived directly or indirectly from recent plant or animal biomass. It originates from carbon dioxide fixation via photosynthesis ($\rightarrow$ plant biomass), from plants that served as animal feed ($\rightarrow$ detritus, feces, urine, etc.), or from fossil fuels or biologi-

*Environmental Biotechnology. Concepts and Applications.* Edited by H.-J. Jördening and J. Winter
Copyright © 2005 WILEY-VCH Verlag GmbH & Co. KGaA, Weinheim
ISBN: 3-527-30585-8

cally or geochemically transformed biomass ($\rightarrow$ peat, coal, oil, natural gas). Even the carbon portion of some xenobiotics can be tracked back to a biological origin, i.e., if these substances were produced from oil, natural gas, or coal. Only because the mineralization of carbonaceous material from decaying plant and animal biomass in nature under anaerobic conditions with a shortage of water was incomplete, did the formation of fossil oil, natural gas, and coal deposits from biomass occur through biological and/or geochemical transformations. The fossil carbon of natural gas, coal, and oil enters the atmospheric $CO_2$ cycle again as soon as these compounds are incinerated as fuels or used for energy generation in industry or private households.

Biological degradation of recent biomass and of organic chemicals during solid waste or wastewater treatment proceeds either in the presence of molecular oxygen by respiration, under anoxic conditions by denitrification, or under anaerobic conditions by methanogenesis or sulfidogenesis. Respiration of soluble organic compounds or of extracellularly solubilized biopolymers such as carbohydrates, proteins, fats, or lipids in activated sludge systems leads to the formation of carbon dioxide, water, and a significant amount of surplus sludge. Some ammonia and $H_2S$ may be formed during degradation of sulfur-containing amino acids or heterocyclic compounds. Oxygen must either be supplied by aeration or by injection of pure oxygen. The two process variant for oxygen supply differ mainly in their capacity for oxygen transfer and the stripping efficiency for carbon dioxide from respiration. Stripping of carbon dioxide is necessary to prevent a drop in pH and to remove heat energy. Respiration in the denitrification process with chemically bound oxygen supplied in the form of nitrate or nitrite abundantly yields dinitrogen. However, some nitrate escapes the reduction to dinitrogen in wastewater treatment plants and contributes about 2% of the total $N_2O$ emissions in Germany (Schön et al., 1994). Denitrifiers are aerobic organisms that switch their respiratory metabolism to the utilization of nitrate or nitrite as terminal electron acceptors, if grown under anoxic conditions. Only if the nitrate in the bulk mass has been used completely does the redox potential become low enough for growth of strictly anaerobic organisms, such as methanogens or sulfate reducers. If anaerobic zones are allowed to form in sludge flocs of an activated sludge system, e.g., by limitation of the oxygen supply, methanogens and sulfate reducers may develop in the center of sludge flocs and form traces of methane and hydrogen sulfide, found in the off-gas.

Under strictly anaerobic conditions, soluble carbon compounds of wastes and wastewater are degraded stepwise to methane, $CO_2$, $NH_3$, and $H_2S$ via a syntrophic interaction of fermentative and acetogenic bacteria with methanogens or sulfate reducers. The complete methanogenic degradation of biopolymers or monomers via hydrolysis/fermentation, acetogenesis, and methanogenesis can proceed only at a low $H_2$ partial pressure, which is maintained mainly by interspecies hydrogen transfer. Interspecies hydrogen transfer is facilitated when acetogens and hydrogenolytic methanogenic bacteria are arranged in proximity in flocs or in a biofilm within short diffusion distances. The reducing equivalents for carbon dioxide reduction to methane or sulfate reduction to sulfide are derived from the fermentative metabolism, e.g., of clostridia or *Eubacterium* sp., from β oxidation of fatty acids, or the oxidation of alcohols. Methane and $CO_2$ are the main products in anaerobic environments where sulfate is absent, but sulfide and $CO_2$ are the main products if sulfate is present.

## 1.2.1
### Basic Biology, Mass, and Energy Balance of Aerobic Biopolymer Degradation

To make soluble and insoluble biopolymers – mainly carbohydrates, proteins, and lipids – accessible for respiration by bacteria, the macromolecules must be hydrolyzed by exoenzymes, which often are produced and excreted only after contact with respective inductors. The exoenzymes adsorb to the biopolymers and hydrolyze them to monomers or at least to oligomers. Only soluble, low molecular weight compounds (e.g., sugars, disaccharides, amino acids, oligopeptides, glycerol, fatty acids) can be taken up by microorganisms and be metabolized for energy production and cell multiplication.

Once taken up, degradation via glycolysis (sugars, disaccharides, glycerol), hydrolysis and deamination (amino acids, oligopeptides), or hydrolysis and $\beta$ oxidation (phospholipids, long-chain fatty acids) proceeds in the cells. Metabolism of almost all organic compounds leads to the formation of acetyl-CoA as the central intermediate, which is used for biosyntheses, excreted as acetate, or oxidized to $CO_2$ and reducing equivalents in the tricarboxylic acid (TCA) cycle. The reducing equivalents are respired with molecular oxygen in the respiration chain. The energy of a maximum of only 2 mol of anhydridic phosphate bonds of ATP is conserved during glycolysis of 1 mol of glucose through substrate chain phosphorylation. An additional 2 mol of ATP are formed during oxidation of 2 mol of acetate in the TCA cycle, whereas 34 mol ATP are formed by electron transport chain phosphorylation with oxygen as the terminal electron acceptor. During oxygen respiration, reducing equivalents react with molecular oxygen in a controlled combustion reaction.

When carbohydrates are respired by aerobic bacteria, about one third of the initial energy content is lost as heat, and two thirds are conserved biochemically in 38 phosphoanhydride bonds of ATP. In activated sludge reactors or in wastewater treatment ponds that are not loaded with highly concentrated wastewater, wall irradiation and heat losses with the off-gas stream of aeration into the atmosphere prevent self-heating. In activated sludge reactors for treatment of highly concentrated wastewater, however, self-heating up to thermophilic temperatures may occur if the wastewater is warm in the beginning, the hydraulic retention time for biological treatment is short (short aeration time), and the air or oxygen stream for aeration is restricted so as to supply just sufficient oxygen for complete oxidation of the pollutants (small aeration volume).

The conserved energy in the terminal phosphoanhydride bond of ATP, formed during substrate chain and oxidative phosphorylation by proliferating bacteria is partially used for maintenance metabolism and partially for cell multiplication. Partitioning between both is not constant, but depends on the nutritional state. In highly loaded activated sludge reactors with a surplus or at least a non-growth–limiting substrate supply, approximately 50% of the substrate is respired in the energy metabolism of the cells and 50% serves as a carbon source for cell growth (Table 1.1). The biochemically conserved energy must be dissipated to be used for the maintenance metabolism of existing cells and cell growth.

If the substrate supply is growth-limiting, e.g., in a low-loaded aerobic treatment system, a higher proportion of ATP is consumed for maintenance, representing the energy proportion that bacteria must spend for non-growth–associated cell survival metabolism, and less energy is available for growth. Overall, more of the substrate carbon is respired, and the ratio of respiration products to surplus sludge formed is higher, e.g., around 70% : 30% (Table 1.1). In a trickling filter system, an even higher proportion of the substrate seems to be respired. This might be due to protozoa grazing off part of the biofilm.

For comparison, Table 1.1 also summarizes carbon dissipation in anaerobic methanogenic degradation. Only about 5% of the fermentable substrate is used for cell growth (surplus sludge formation) in anaerobic reactors, whereas 95% is converted to methane and $CO_2$, and most of the energy of the substrates is conserved in the fermentation products.

### 1.2.1.1 Mass and Energy Balance for Aerobic Glucose Respiration and Sewage Sludge Stabilization

In most textbooks of microbiology, respiration of organic matter is explained by Eq. 1, with glucose used as a model substance. Except for an exact reaction stoichiometry of the oxidative metabolism, mass and energy dissipation, if mentioned at all, are not quantified. Both parameters are, however, very important for activated sludge treatment plants. The surplus sludge formed during wastewater stabilization requires further treatment, causes disposal costs, and – in the long run – may be an environmental risk, and heat evolution during unevenly high-loaded aerobic treatment may shift the population toward more thermotolerant or thermophilic species and thus, at least for some time, may decrease the process efficiency.

$$1 \text{ mol } C_6H_{12}O_6 + 6 \text{ mol } O_2 \rightarrow 6 \text{ mol } CO_2 + 6 \text{ mol } H_2O + \text{heat energy} \qquad (1)$$

**Table 1.1** Carbon flow during (A) aerobic degradation in an activated sludge system under (a) saturating and (b) limiting substrate supply[a] and during (B) anaerobic degradation.

| | |
|---|---|
| (A) | Aerobic degradation: |
| | (a) Saturating substrate supply = high-load conditions |
| | 1 unit substrate carbon → 0.5 units $CO_2$ carbon + 0.5 units cell carbon |
| | (b) Limiting substrate supply = low-load conditions |
| | 1 unit substrate carbon → 0.7 units $CO_2$ carbon + 0.3 units cell carbon |
| (B) | Anaerobic degradation: |
| | 1 unit substrate carbon → 0.95 units ($CO_2$ + $CH_4$) carbon + 0.05 units cell carbon |

[a] Estimated from surplus sludge formation in different wastewater treatment plants.

If 1 mol of glucose (MW = 180 g) is degraded in an activated sludge system at a high BOD loading rate (e.g., >0.6 kg $m^{-3}$ $d^{-1}$ BOD), approximately 0.5 mol (90 g) is respired to $CO_2$ and water, with consumption of 3 mol of $O_2$ (96 g), releasing 19 mol of ATP (Fig. 1.1). The other 0.5 mol of glucose (90 g) is converted to pyruvate by one of three glycolytic pathways, accompanied by the formation of 0.5–1 mol ATP. Pyruvate or its subsequent metabolic products, e.g., acetate or dicarboxylic acids, are directly used as carbon substrates for cell multiplication and surplus biomass formation. A maximum amount of 20 mol ATP is thus available for growth and maintenance (Fig. 1.1). At a pH of 7, about 44 kJ of energy is available for growth per mol of ATP hydrolyzed to ADP and inorganic phosphate (Thauer et al., 1977). For an average molar growth yield of aerobes of 4.75 g per mol ATP (Lui, 1998), 90 g biomass can be generated from 180 g glucose. If the combustion energy per g of cell dry mass is 22 kJ, about 890 kJ (2870–980 kJ) is lost as heat during respiration (Fig. 1.1). The energy loss is the sum of heat losses during respiration and cell growth.

At a low BOD loading rate, the proportion of glucose respired in relation to the proportion of glucose fixed as surplus biomass can shift. Up to 0.7 mol (126 g) of glucose can be oxidized to $CO_2$, requiring 4.2 mol of oxygen (134.4 g $O_2$). Thus, for respiration of 1 mol of glucose, different amounts of oxygen may be consumed, depending on the loading rate of the wastewater treatment system and the different amounts of carbon dioxide and of surplus sludge formed (Fig. 1.1, Table 1.1).

The energy and carbon balance deduced above can be analogously applied to aerobic stabilization of raw sewage sludge. If the initial dry matter content is around 36 g $L^{-1}$ (average organic dry matter content of sewage sludge) and if a biodegradability of 50% within the residence time in the sludge reactor is obtained, about 9 g $L^{-1}$ of new biomass is formed, and thus 27 g $L^{-1}$ (36 – 18 + 9) remains in the effluent.

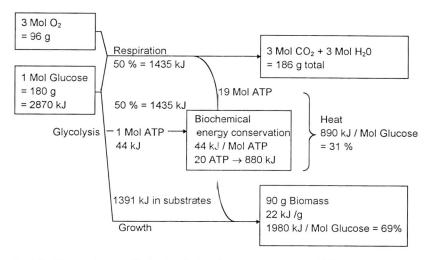

**Fig. 1.1** Mass and energy dissipation during glucose respiration at pH 7.

The released heat energy is approximately 89 kJ $L^{-1}$ of reactor content. To estimate the theoretical temperature rise, this amount of heat energy must be divided by 4.185 kJ (specific energy requirement for heating 1 L of $H_2O$ from 14.5–15.5 °C). Thus, by respiration of 18 g $L^{-1}$ organic dry matter, the reactor temperature increases by 21.3°C within the residence time required for degradation (≤16 h), provided that no heat energy is lost. A great proportion of the heat energy is, however, transferred via the liquid phase to the aeration gas and stripped out, whereas a smaller proportion is lost through irradiation from the reactor walls. Since air, containing almost 80% nitrogen, is normally used as an oxygen source in aeration ponds or activated sludge reactors, the heat transfer capacity of the off-gas is high enough to prevent a significant increase in the wastewater temperature. Thus, ambient or at least mesophilic temperatures can be maintained. An increasing temperature of several degrees Celsius would lead to a shift in the population in the reactor and – at least temporarily – would result in reduced process stability, but an only slightly increased temperature of a few degrees Celsius might simply stimulate the metabolic activity of the prevalent mesophilic population. In practice, in activated sewage sludge systems no self-heating is observed because degradability is only about 50% and complete heat transfer to the atmosphere occurs via the off-gas at a retention time of more than 0.5 d. If, however, wastewater from a dairy plant or a brewery with a similar COD concentration, but with almost 100% biodegradable constituents, is stabilized with pure oxygen, twice as much heat evolves, leading to a theoretical temperature rise of 57 °C. Self-heating is observed, since there is much less off-gas and the heat loss is thus significantly lower. In addition, due to higher reaction rates than with sewage sludge, the heat is generated during a shorter time span (shorter retention time).

### 1.2.1.2 Mass and Energy Balance for Anaerobic Glucose Degradation and Sewage Sludge Stabilization

For anaerobic wastewater or sludge treatment, oxygen must be excluded to maintain the low redox potential that is required for survival and metabolic activity of the acetogenic, sulfidogenic, and methanogenic populations. Hydrolysis of polymers, uptake of soluble or solubilized carbon sources, and the primary metabolic reactions of glycolysis up to pyruvate and acetate formation seem to proceed identically or at least analogously in aerobic and anaerobic bacteria. Whereas aerobes oxidize acetate in the TCA cycle and respire the reducing equivalents with oxygen, anaerobes, such as *Ruminococcus* sp., *Clostridium* sp., or *Eubacterium* sp., either release molecular hydrogen or transform pyruvate or acetate to highly reduced metabolites, such as lactate, succinate, ethanol, propionate, or *n*-butyrate. For further degradation within the anaerobic food chain, these reduced metabolites must be oxidized anaerobically by acetogenic bacteria. Since the anaerobic oxidation of propionate or *n*-butyrate by acetogenic bacteria is obligately accompanied by hydrogen production but is only slightly exergonic under conditions of a low $H_2$ partial pressure (Bryant, 1979), acetogens can grow only when hydrogen is consumed by hydrogen-scavenging organisms such as methanogens or sulfate reducers.

During anaerobic degradation of 1 mol glucose, approximately 95% of the glucose carbon is used for biogas formation (171 g = 127.7 L $CH_4$ + $CO_2$), and only about 5% of the substrate carbon (9 g) converted to biomass (Table 1.1). Much less heat energy is released during anaerobic metabolism than during aerobic respiration (131 kJ $mol^{-1}$ versus 890 kJ $mol^{-1}$, respectively), and the biogas contains almost 90% of the energy of the fermented substrate (Fig. 1.2). Due to the heat energy requirement to warm the wastewater and due to heat losses via irradiation from pipes and reactor walls, heat generation is not nearly sufficient to maintain a constant mesophilic fermentation temperature. For this reason, anaerobic digesters must be heated.

In sewage sludge with 36 g $L^{-1}$ organic dry matter content and 50% biodegradability, 0.9 g surplus sludge and 17.1 g biogas (equivalent to 12.75 L) are formed during anaerobic stabilization. Only 13.1 kJ of heat energy per mol of glucose is released, leading to a self-heating potential of 3.1 °C. Since the heat energy is released only during the hydraulic residence time of the wastewater in the reactor, which is usually more than 10 d (except for high-rate industrial wastewater treatment, where in special situations the HRT may be shorter than one day), much more heat energy is lost by irradiation via the reactor walls than is required to maintain the temperature. If highly concentrated wastewater streams of the food and beverage industry are stabilized anaerobically at hydraulic retention times of <1 d (high space loading), more heat energy is generated within a much shorter time. But even then, process energy from external sources has to be supplied to maintain a temperature of 30–37 °C. Proper insulation of anaerobic reactors can minimize but not replace external heating of the reactor. The methane in the biogas, generated during anaerobic sludge or wastewater stabilization processes, contains about 90% of the energy of the fermented substrate. Since methane is a climate-relevant gas it may not be emitted into the atmosphere but must be combusted to $CO_2$. Methane from anaerobic reac-

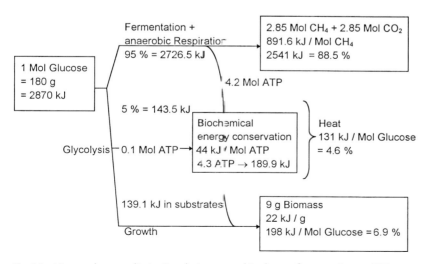

**Fig. 1.2** Mass and energy dissipation during anaerobic glucose fermentation at pH 7.

tors can be used as a fuel for gas engines to generate electricity and/or heat energy (Eq. 2):

$$CH_4 + 2\ O_2 \rightarrow CO_2 + 2\ H_2O + \text{heat energy } (\Delta G^{o\prime} = -891.6 \text{ kJ mol}^{-1}\ CH_4) \qquad (2)$$

### 1.2.2
### General Considerations for the Choice of Aerobic or Anaerobic Wastewater Treatment Systems

If a producer of wastewater has to decide whether to install an aerobic or an anaerobic waste or wastewater treatment system, several points should be considered:

- Anaerobic treatment in general does not lead to the low pollution standards of COD, BOD$_5$, or TOC that can be met with aerobic systems and which are required by environmental laws. Anaerobic treatment of wastes and wastewater is often considered a pretreatment process to minimize the oxygen demand and surplus sludge formation in a subsequent aerobic post-treatment stage. Only after a final aerobic treatment can the COD, BOD$_5$, or TOC concentration limits stated in the environmental laws be met. If limiting concentrations for nitrogen and phosphate also have to be achieved, further treatment steps such as nitrification, denitrification, and biological or chemical phosphate removal, must be considered.
- Highly concentrated wastewater should in general be treated anaerobically, because of the possibility of energy recovery in biogas and the much lower amounts of surplus sludge to be disposed of. For aerobic treatment, a high aeration rate is necessary and much surplus sludge is generated. Aeration causes aerosol formation and eventually requires off-gas purification.
- The efficiency of COD degradation for the bulk mass in concentrated wastewater or sludges (degradability of organic pollutants) generally seems to be about similar in aerobic or anaerobic bacteria. However, the degradation rates may be faster in aerobic treatment procedures than in anaerobic treatment procedures.
- Wastewater with a low concentration of organic pollutants should be treated aerobically due to its higher process stability at low pollutant concentrations, although aerobic treatment is more expensive and more sludge remains for disposal. If mineralized sludge is required, aerobic treatment at a low loading or at prolonged hydraulic retention times is necessary to reinforce respiration of all endogenous reserve material.
- Anaerobic treatment systems are more expensive to construct but less expensive to operate than aerobic treatment systems.

### 1.2.3
### Aerobic or Anaerobic Hydrolysis of Biopolymers: Kinetic Aspects

Hydrolysis of biopolymers and fermentation or respiration of monomers can be catalyzed by strictly anaerobic, facultative anaerobic, and aerobic microorganisms. With some exceptions (e.g., small protein molecules, dextran), biopolymers are insoluble and form fibers (cellulose), grains (starch) or globules (casein after enzymatic precipitation) or can be melted or emulsified (fat). Henze et al. (1997) reported hydrolysis constants $k_h$ for dissolved organic polymers of 3–20 d$^{-1}$ under aerobic conditions and of 2–20 d$^{-1}$ under anaerobic conditions, whereas for suspended solids the hydrolysis constants $k_h$ were 0.6 to 1.4 d$^{-1}$ under aerobic conditions and 0.3 to 0.7 d$^{-1}$ under anaerobic conditions. For a kinetic description of hydrolysis and fermentation, a substrate-limited first-order reaction was assumed by Buchauer (1997), who deduced that the temperature-dependent reaction rate for hydrolysis is a little lower than the reaction rate for fermentation of the hydrolysis products. The rate-limiting step is therefore hydrolysis of particles and not fermentation of solubilized material (Buchauer, 1997). Since hydrolysis is catalyzed not only by freely soluble exoenzymes, diluted in the bulk mass of liquid, but to a much higher extent by enzymes that are excreted in the neighborhood of bacterial colonies growing attached to the surface of the particles, the above description of complex fermentation processes is not always valid. Cellulases can be arranged in cellulosomes, which attach to the particles, which in turn serve as carriers until they themselves are solubilized. For this reason, Vavilin et al. (1997) included biomass in their description of the hydrolysis of cellulose, cattle manure, and sludge. Shin and Song (1995) determined the maximum rates of acidification and methanation for several substrates. For hydrolysis of particulate organic matter, the ratio of surface area to particle size is important. They found that for glucose, starch, carboxymethyl cellulose, casein, and food residues from a restaurant, hydrolysis proceeded faster than methanogenesis, whereas for newspaper and leaves hydrolysis was the rate-limiting step.

## 1.2.4
### Hydrolysis of Cellulose by Aerobic and Anaerobic Microorganisms: Biological Aspects

Cellulose and lignin are the main structural compounds of plants. Both substances are the most abundant biopolymers on earth. Cellulose fibers are formed of linear chains of 100–1400 glucose units linked together by β-1,4-glycosidic bonds. Inter- and intramolecular hydrogen bonds and van der Waals interactions arrange the highly organized fibrous regions (crystalline region), which alternate with less organized amorphous regions in the cellulose fibers. The fibers are embedded in a matrix of hemicelluloses, pectin, or lignin. The hemicelluloses consist mainly of xylans or glucomannans, which have sidechains of acetyl, gluconuryl, or arabinofuranosyl units. To make cellulose fibers accessible to microorganisms, the hemicellulose, pectin, or lignin matrix must be degraded microbiologically or solubilized chemically. Cellulose degradation in the presence of oxygen in soil or in the absence of oxygen in the rumen of ruminants, in swamps, or in anaerobic digesters is the most important step in mineralization of decaying plant material. Cellulolytic organisms are found among aerobic soil fungi, e.g., within the genera *Trichoderma* and *Phanaerochaete* and in anaerobic rumen fungi, e.g. *Neocallimastix* and *Piromyces*, and among bacteria, e.g., within the genera *Cellulomonas, Pseudomonas*, and *Thermomonospora* (aerobic cellulose degraders) and *Clostridium, Fibrobacter, Bacteroides*, and *Ruminococcus* (anaerobic cellulose degraders). For more details on cellulolytic bacteria and the mechanism of cellulose cleavage, please see Coughlan and Mayer (1991).

Glycosyl hydrolases are involved in cellulose and hemicellulose degradation by cleaving glycosidic bonds between different carbohydrates and between carbohydrates and noncarbohydrates. Endo- and exocellulases – in some organisms organized in cellulosomes – must be excreted into the medium. Cellulases are complex biocatalysts and contain a catalytic site and a substrate-binding site. The presence of a noncatalytic substrate binding site permits tight attachment to the different forms of cellulose substrate and keeps the enzyme close to its cleaving sites. Substrate binding is reversible, which allows the enzyme to 'hike' along the fibers and obtain total solubilization. Many aerobic fungi and some bacteria excrete endoglucanases that hydrolyze the amorphous region of cellulose (degradation within the chain), whereas exoglucanases hydrolyze cellulose from the ends of the glucose chains. Cellobiose is cleaved off by cellobiohydrolases from the nonreducing ends in the amorphous region, and finally the crystalline region is also hydrolyzed.

Cleavage of cellobiose to glucose units by β-glucosidases is necessary to prevent cellobiose accumulation, which inhibits cellobiohydrolases (Be'guin and Aubert, 1994).

Anaerobic bacteria such as *Clostridium thermocellum* form a stable enzyme complex, the cellulosome, at the cell surface (Lamed and Bayer, 1988). Cellulosomes are active in degrading crystalline cellulose. The catalytic subunits of a cellulosome, endoglucanases and xylanases, cleave the cellulose fiber into fragments, which are simultaneously degraded further by β-glucosidases. Cellulosome-like proteins are found also in *Ruminococcus* sp. and *Fibrobacter* sp. cultures. The cell-bound enzymes are associated with the capsule or the outer membrane. Other specific adhesions or ligand formations with the cellulose can be facilitated by fimbrial connections, glycosylated epitopes of carbohydrate binding proteins, or the glycocalyx and carbohydrate binding modules (Krause et al., 2003). Some bacteria have not developed a mechanism to adhere to cellulose fibers, but excrete cellulases into the medium. Adsorption of bacteria onto cellulose fibers via cellulosomes offers the advantage of close contact with the substrate, which is hydrolyzed mainly to glucose, which is then taken up and metabolized. Small amounts of cellobiose must be present initially to induce cellulase expression. The contact of bacteria with the solid substrate surface keeps them close to cellobiose and thus keeps cellulase activities high. However, accumulation of hydrolysis products such as glucose repress cellulase activity.

In nature most cellulose is degraded aerobically. Only about 5%–10% is thought to be degraded anaerobically, which may be an underestimate. Since most ecosystems are rich in carbonaceous substances but deficient in nitrogen compounds, many cellulolytic bacteria can also fix dinitrogen. This is advantageous to them and to syntrophic or symbiotic organisms (Leschine, 1995). Other examples of mutual interactions between organisms of an ecosystem are interspecies hydrogen transfer between anaerobes (anaerobic food chain), transfer of growth factors (mycorhizzae, Kefir), and production of fermentable substrates for the partner organisms (bacterial interactions in the rumen).

Hydrolysis of biological structural components such as cellulose, lignin, and other structural or storage polymers (Table 1.2) is difficult. The limiting step of hydrolysis seems to be liberation of the cleavage products. In contrast to the slow hydrolysis of celluloses, mainly due to lignin encrustation of naturally occurring celluloses, starch can be easily hydrolyzed. The branching and helical structure of starch facilitates hydrolysis (Warren, 1996). Whereas cellulose forms fibers with a large surface covered with lignin, starch forms grains with an unfavorable surface-to-volume ratio for enzymatic cleavage. Thus, although amylases may be present in high concentrations, the hydrolysis rate is limited by the limited access of the enzymes to the substrate.

Whereas cellulose and starch are biodegradable, other carbohydrate-derived cellular compounds are not biodegradable and – after reaction with proteins – form humic acid-like residues by the Maillard reaction.

**Table 1.2** Polysaccharides and derivatives occurring in nature.

| Compound | Bond | Unit | Occurrence |
|---|---|---|---|
| Cellulose | β-1,4- | glucose | plant cell wall |
| Chitin | β-1,4- | N-acetyl glucosamine | fungal cell and insect wall exoskeleton |
| Murein | β-1,4- | N-acetyl-glucosamine | bacterial cell wall |
| | | N-acetyl-muramic acid | |
| Chitosan | β-1,4- | N-glucosamine, substituted with acetyl residues | fungal and insect materials |
| Mannans | β-1,4- | mannose | plant material |
| Xylans | β-1,4- | xylopyranose, substituted with acetyl-arabinofuranosice residues | plant material |
| Starch | α-1,4- | glucose | storage material in plant |
| | | amylose (contains few α-1 6-branches) | |
| | | amylopectin (contains many α-1,6-branches) | |
| Glycogen | α-1,4- | highly branched glucose with α-1,6 bonds | storage material in bacteria |
| Dextran | α-1,2- | glucose | storage material in yeasts |
| | α-1,3- | glucose | exopolymer of bacteria |
| | α-1,4- | glucose | |
| | α-1,6- | glucose | |
| Laminarin | β-1,3- | glucose | reserve material in algae |

1.2.5
**Biomass Degradation in the Presence of Inorganic Electron Acceptors and by an Anaerobic Food Chain**

In ecosystems in which molecular oxygen is available, plant and animal biomass is degraded to $CO_2$ and $H_2O$, catalyzed by either single species of aerobic microorganisms or the whole population of the ecosystem, in competition for the substrates. A single organism may be able to hydrolyze the polymers and oxidize the monomers to $CO_2$ and $H_2O$ with oxygen. In ecosystems where molecular oxygen is deficient – such as swamps, wet soil, the rumen of animals, the digestive tract of humans, or in river and lake sediments – oxidation of dead biomass proceeds anoxically by reduction of electron acceptors such as nitrate and nitrite or anaerobically by reduction of sulfate, $Fe^{3+}$, $Mn^{4+}$, or $CO_2$. The oxidation of the carbon source is either complete or incomplete with acetate excretion. In the absence of inorganic electron acceptors, oxidized metabolites such as pyruvate or acetate are reduced to lactate or ethanol or biotransformed to, e.g., *n*-butyrate or *n*-butanol. In permanently anaerobic ecosystems with seasonal overfeeding, periodic accumulation of such metabolites can occur, e.g., in autumn after the non-evergreen plants drop their leaves or decay completely. The biopolymers of the leaves or the plants themselves decompose by extracellular enzymatic hydrolysis. The monomers are fermented, and the fermentation products may be degraded further to biogas by acetogenic and methanogenic bacteria. Whereas single cultures of aerobes can catalyze the whole mineralization process to finally form $CO_2$ and $H_2O$, single cultures of strictly anaerobic bacteria are not capable of degrading biopolymers to $CH_4$ and $CO_2$. Under anaerobic conditions biopolymers must be degraded by a food chain via depolymerization (hydrolysis), fermentation (acidogenesis), oxidation of fatty acids (acetogenesis), and biogas formation (methanogenesis) as the last step (McInerney, 1988). In an initial exoenzyme-catalyzed reaction the biopolymers are hydrolyzed to soluble mono-, di-, or oligomers. These are taken up by the bacteria and fermented to $CO_2$, $H_2$, formate, acetate, propionate, butyrate, lactate, etc. If fatty acid isomers are produced, they are mainly derived from degradation of amino acids after proteolysis. Fatty acids are further oxidized by acetogenic bacteria, before the cleavage products $CO_2$, $H_2$, and acetate can be taken up by methanogens and be converted to methane and $CO_2$. Lactate is oxidized to pyruvate, which is decarboxylated to yield acetate, $CO_2$, and $H_2$. If ethanol is present, it is oxidized to acetate and hydrogen, and the hydrogen is used for $CO_2$ reduction.

Table 1.3 summarizes the reactions that can be catalyzed by methanogens and that can contribute to methane emission in various ecosystems. In sewage digesters about two thirds of the methane is derived from acetate cleavage and one third from $CO_2$ reduction with $H_2$. If hexoses are the substrates and glycolysis is the main degradation pathway, then the 2 mol of pyruvate can be decarboxylated by pyruvate: ferredoxin oxidoreductase to yield 2 mol acetate and 2 mol $CO_2$. The hydrogens of the 2 mol $NADH_2$ from glycolysis and the 2 mol $FdH_2$ from pyruvate decarboxylation are then released as molecular hydrogen at low $H_2$ partial pressure (Eq. 3). Two

**Table 1.3** Reactions catalyzed by methanogens and standard changes in free energy.

| Reaction | | | $\Delta G^{o\prime}$ (kJ per mol of methane) |
|---|---|---|---|
| Substrates (mol) | | Products (mol) | |
| Acetate | $\rightarrow$ | $CH_4 - CO_2$ | −31.0 |
| $4 H_2 + CO_2$ | $\rightarrow$ | $CH_4 + 2 H_2O$ | −135.6 |
| 4 HCOOH | $\rightarrow$ | $CH_4 + 3 CO_2 + 2 H_2O$ | −130.1 |
| $4 CO + 2 H_2O$ | $\rightarrow$ | $CH_4 + 3 CO_2$ | −211.0 |
| 4 Methanol | $\rightarrow$ | $3 CH_4 + CO_2 + 2 H_2O$ | −104.9 |
| Methanol + $H_2$ | $\rightarrow$ | $CH_4 + H_2O$ | −112.5 |
| 2 Ethanol[a] + $CO_2$ | $\rightarrow$ | $CH_4 + 2$ acetate | −116.3 |
| 4 2-Propanol[b] + $CO_2$ | $\rightarrow$ | $CH_4 + 4$ acetone + $2 H_2O$ | −36.5 |
| 4 Methylamine + $2 H_2O$ | $\rightarrow$ | $3 CH_4 + CO_2 + 4 NH_3$ | −75.0 |
| 2 Dimethylamine + $2 H_2O$ | $\rightarrow$ | $3 CH_4 + CO_2 + 2 NH_3$ | −73.2 |
| 4 Trimethylamine + $6 H_2O$ | $\rightarrow$ | $9 CH_4 + 3 CO_2 + 4 NH_3$ | −74.3 |
| 2 Dimethylsulfide + $2 H_2O$ | $\rightarrow$ | $3 CH_4 + CO_2 + 2 H_2S$ | −73.8 |

[a] Other primary alcohols that are used as hydrogen donors for $CO_2$ reduction are 1-propanol and 1-butanol (in a few species).
[b] Other secondary alcohols used as hydrogen donors for $CO_2$ reduction are 2-butanol, 1,3-butanediol, cyclopentanol, and cyclohexanol (in a few species).
Compiled from Whitman et al. (1992) and Winter (1984).

moles of $CH_4$ are then formed from acetate and 1 mol of $CH_4$ by $CO_2$ reduction (reactions 1 and 2 of Table 1.3).

$$1 \text{ mol glucose} \rightarrow 2 \text{ mol acetate} + 2 \text{ mol } CO_2 + 4 \text{ mol } H_2 \text{ (at low } pH_2) \qquad (3)$$

In complex ecosystems formate is formed if high concentrations of hydrogen accumulate. Syntrophic interactions are usually associated with interspecies hydrogen transfer, but evidence for interspecies formate transfer was also reported (Thiele et al., 1988). The feasibility of the electron carrier depends on its solubility, which is much less for hydrogen than for formate, and on its diffusion coefficient in water, which favors hydrogen 30 times over formate. The efficiency of the appropriate electron transfer depends mainly on the distance between the producing and consuming bacteria. It can be expected, that formate transfer is favored when the distance between communicating bacteria is high and hydrogen transfer when the distance is small (de Bok et al., 2004). Interspecies formate transfer is thought to play a major role in degradation of syntrophic butyrate (Boone et al., 1989) and propionate (Stams, 1994; Schink, 1997). However, at an increased $H_2$ partial pressure formate is also produced by methanogens, either in pure cultures or in a sewage sludge population (Bleicher and Winter, 1994), and this may also contribute to increasing formate concentrations. Other substrates for methanogenic bacteria (Table 1.3), such as methanol (derived, e.g., from methoxy groups of lignin monomers) or methylamines and dimethylsulfide (e.g., from methylsulfonopropionate in algae; Fritsche, 1998) are relevant only in ecosystems where these substances are produced during microbial decay. A few methanogens can also use reduced products such as primary, secondary, and cyclic alcohols as a source of electrons for $CO_2$ reduction (Widdel, 1986; Zellner and Winter, 1987a; Bleicher et al., 1989; Zellner et al., 1989).

## 1.2.6
### Roles of Molecular Hydrogen and Acetate During Anaerobic Biopolymer Degradation

Molecular hydrogen is produced during different stages of anaerobic degradation. In the fermentative stage, organisms such as *Clostridium* sp. and *Eubacterium* sp. produce fatty acids, $CO_2$, and hydrogen from carbohydrates. In the acetogenic stage, acetogens such as *Syntrophobacter wolinii* and *Syntrophomonas wolfei* produce acetate, $CO_2$, and hydrogen or acetate and hydrogen by anaerobic oxidation of propionate and *n*-butyrate (McInerney, 1988). Fermentative bacteria release molecular hydrogen even at a high $H_2$ partial pressure and simultaneously excrete reduced products (e.g., clostridia, *Ruminococcus*, *Eubacterium* sp.). However, the release of molecular hydrogen during acetogenesis of fatty acids or of other reduced metabolites may occur only when hydrogen does not accumulate, for thermodynamic reasons. Molecular hydrogen is consumed by methanogens (Table 1.4, reaction 1) or, alternatively, by sulfate reducers (Table 1.4, reaction 2) via interspecies hydrogen transfer. In the rumen and in sewage sludge digesters, the hydrogen concentration can be decreased by acetate formation from $CO_2$ and $H_2$ (Table 1.4, reaction 3) by bacteria such as *Acetobacterium woodii* and *Clostridium thermoaceticum*. Some additional reactions consuming hydrogen to decrease its concentration are also listed in Table 1.4 (reactions 4–6).

To maintain a low $H_2$ partial pressure, a syntrophism of acetogenic, hydrogen-producing and methanogenic, hydrogen-utilizing bacteria is essential (Ianotti et al., 1973). Complete anaerobic degradation of fatty acids with hydrogen formation by obligate proton-reducing acetogenic bacteria is possible only at $H_2$ partial pressures $<10^{-4}$ atm (*n*-butyrate) or $10^{-5}$ atm (propionate), which cannot be maintained by

**Table 1.4** Hydrogen-consuming reactions in anaerobic ecosystems (Schink, 1997).

| Substrates (mol) | | Products (mol) | $\Delta G^{o\prime}$ (kJ per mol) |
|---|---|---|---|
| (1) | $4 H_2 + CO_2$ | $\rightarrow$ $CH_4 + 2 H_2O$ | −131.0 |
| (2) | $4 H_2 + SO_4^{2-}$ | $\rightarrow$ $S^{2-} + 4 H_2O$ | −151.0 |
| (3) | $4 H_2 + 2 CO_2$ | $\rightarrow$ $CH_3COO^- + H^+ + 2 H_2O$ | −0.9 |
| (4) | $H_2 + S^0$ | $\rightarrow$ $H_2S$ | −0.9 |
| (5) | $H_2C(NH_3^+)COO^- + H_2$ | $\rightarrow$ $CH_3COO^- + NH_4^+$ | 0.0 |
| (6) | $COOH-CH-CH-COOH + H_2$ | $\rightarrow$ $COOH-CH_2-CH_2-COOH$ | 0.0 |

methanogens or sulfate reducers. However, by reversed electron transport electrons can be shifted to a lower redox potential suitable for proton reduction (Schink, 1997). If hydrogen accumulates beyond this threshold concentration, the anaerobic oxidation of fatty acids becomes endergonic and does not proceed (for details, see Chapter 8, this volume). Whereas hydrogen prevents β oxidation of fatty acids by acetogens even at very low $H_2$ partial pressure, much higher concentrations of acetate (in the millimolar range) are required for the same effect.

The fermentative metabolism of acidogenic bacteria is exergonic even at $H_2$ partial pressures $>10^{-4}$ atm. Whereas acetogenic bacteria apparently depend mainly on ATP generation by chemiosmotic phosphorylation, fermentative bacteria produce most of their ATP by substrate chain phosphorylation. This may be why fermentative bacteria do not depend on a syntrophic interaction with electron-consuming bacteria, such as methanogens or sulfate reducers.

In addition to the possibility of anaerobic oxidation of organic compounds via synthrophic interactions between acetogenic bacteria and acetoclastic + hydrogenotrophic methanogens (see also Section 1.2.5), other synthrophic associations between acetate-oxidizing bacteria and $H_2/CO_2$-utilizing methanogens under thermophilic (Lee and Zinder 1988) and mesophilic (Schnuerer et al. 1996) growth conditions have been observed. Thermodynamic analysis of mesophilic synthrophic acetate oxidation revealed a hydrogen partial pressure of < 0.1–2.6 Pa, which is in the range found in methanogenic ecosystems (Dolfing, 2001). In thermophilic methanogenic reactors, acetate is degraded either by synthrophic acetate oxidizers (dominant process at low acetate concentrations) and acetate-degrading methanogens (acetate concentration above the threshold concentration) or by acetate-utilizing methanogens of the genera *Methanosaeta* or *Methanosarcina* (Ahring, 2003). Synthrophic acetate oxidation and methane formation from the cleavage products may explain the lack of acetoclastic methanogens (*Methanosarcina* sp. or *Methanosaeta* sp.) in anaerobic reactors.

## 1.2.7
## Anaerobic Conversion of Biopolymers to Methane and $CO_2$

The principle of the anaerobic metabolism of biopolymers was outlined by the pioneering work of Wolin (1976) and Bryant (1979). There is an essential requirement for syntrophic interaction between different metabolic groups for complete anaerobic degradation (Wolin, 1976, 1982). The most sensitive switch of the carbon flow of substrates to biogas is the $H_2$ partial pressure. The substrate supply for biomethanation processes must be limited so that the most slowly growing group in the food chain, the obligate proton-reducing acetogens, can still excrete hydrogen at a maximum rate, but at the same time hydrogen accumulation $>10^{-5}$ atm is prevented by active methanogenesis or sulfate reduction (Bryant, 1979). Whereas hydrogen seemed to be the most sensitive regulator of anaerobic degradation, formate (Bleicher and Winter, 1994), acetate, or other fatty acids accumulated to much higher concentrations (McInerney, 1988) but did not repress anaerobic degradation. If hydrogen accumulated due to an oversupply or to inhibition of methanogens, anaerobic biodegradation was disturbed successively in different stages. Initially, in the acetogenic stage β oxidation of fatty acids and alcohols failed, leading to an accumulation of these acid metabolites. Later, the spectrum of metabolites of the fermentative flora changed toward more reduced products like ethanol, lactate, propionate, and *n*-butyrate, leading to an even higher concentration of volatile fatty acids and a further decrease in pH. At pH values below 6.5, methanogenic reactions were almost completely prevented. At this stage acidification with a rich spectrum of reduced products still proceeded.

### 1.2.7.1 Anaerobic Degradation of Carbohydrates in Wastewater

Carbohydrates are homo- or heteropolymers of hexoses, pentoses, or sugar derivatives, which occur in soluble form or as particles, forming grains or fibers of various sizes. In some plants, starch forms grains up to 1 mm in diameter, which is 1000 times the diameter of bacteria. Starch metabolism by bacteria requires hydrolytic cleavage by amylases to form soluble monomers or dimers, since only soluble substrates can be taken up and metabolized.

The anaerobic degradation of biopolymers in general and of cellulose in particular can be divided into hydrolytic, fermentative, acetogenic, and methanogenic phases (Fig. 1.3). Hydrolysis and fermentation of the hydrolysis products can be catalyzed by the same trophic group of microorganisms. The distinction of the two phases is of more theoretical than practical relevance. Concerning reaction rates in a methane fermenter that is fed with a particulate substrate, the rate-limiting step is hydrolysis rather than the subsequent fermentation of the monomers, if acetogenesis and methanogenesis proceed faster. The hydrolysis rates of polymers can be very different. Hemicellulose and pectin are hydrolyzed ten times faster than lignin-encrusted cellulose (Buchholz et al., 1986, 1988). In the acidification reactor of a two-stage anaerobic process, hydrolysis of polymers to monomers is normally slower than fermentation of monomers to fatty acids and other fermentation products. For this reason, no sugar monomers can be detected during steady-state operation. In the methane reactor, β oxidation of fatty acids, especially of propionate or *n*-butyrate, is the rate-limiting step (Buchholz et al., 1986). Fatty acid degradation is the slowest reaction overall in a two-stage methane reactor fed with carbohydrate-containing wastewater from sugar production. Thus, the methane reactor has to be larger than the acidification reactor to permit longer hydraulic retention times.

The rate of cellulose degradation depends strongly on the state of the cellulose in the wastewater. If cellulose is lignin-encrusted, lignin prevents access of cellulases to the cellulose fibers. If cellulose is mainly in a crystalline form, cellulases can easily attach to it, and then hydrolysis can be a relatively fast process. At increasing loading in an anaerobic reactor fed with crystalline cellulose, acetogenesis became the rate-limiting process, leading to propionate and butyrate formation (Winter and Cooney, 1980). In decaying plant material, cellulose is very often lignin-encrusted. Due to the highly restricted access to these complexes by cellulases, hydrolysis of cellulose is the rate-limiting step in its degradation to methane and $CO_2$.

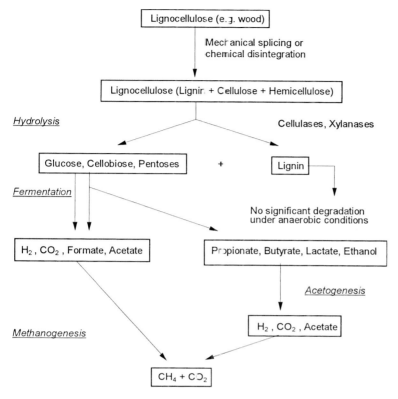

**Fig. 1.3** Anaerobic degradation of lignocellulose and cellulose to methane and $CO_2$ (according to ATV, 1994).

Whether microorganisms are capable of degrading lignin under anaerobic conditions is still under discussion. In a natural environment without time limitation, lignin was reported to be degraded anaerobically (Colberg and Young, 1985; Colberg, 1988). However, since these results were based on long-term experiments performed in situ and anaerobiosis was not controlled, it remains doubtful, whether the small amount of lignin that disappeared was really degraded under strictly anaerobic conditions. The occurrence of coal and fossil oil suggests that lignin compounds are highly resistant to microbial attack.

During anaerobic degradation of starch, hydrolysis by amylases proceeds with high velocity if good contact between starch grains and amylases is maintained. Whereas in an anaerobic reactor at low loading, starch degradation can proceed in the absence of acetogenic bacteria, as indicated in Figure 1.4 (route a), at high loading, volatile fatty acids are formed and acetogens are essential for total degradation (Fig. 1.4, route b). Figure 1.4 illustrates how the rate-limiting acetogenic reactions may be avoided by adjusting the conditions so that hydrolysis and fermentation occur no faster than methanogenesis. The rationale behind this is that many fermentative bacteria produce only acetate, formate, $CO_2$, and hydrogen when $H_2$-scavenging methanogens or sulfate reducers are able to maintain a sufficiently low $H_2$ partial pressure, but a wide spectrum of fermentation products, typical for the metabolism of the respective bacterial species in pure culture, is produced at higher $H_2$ partial pressure (Winter, 1983, 1984). Methanogenesis in continuous syntrophic methanogenic cultures can be disturbed by spike concentrations of sugars or – at low concentrations of sugars – by the presence of inhibitory substances like $NH_3$, $H_2S$, antibiotics (Hammes et al., 1979; Hilpert et al., 1981), or xenobiotics. In consequence, the $H_2$ partial pressure increases and volatile fatty acids are generated (Winter, 1984; Winter et al., 1989; Wildenauer and Winter, 1985; Zellner and Winter, 1987b). Once propionate or *n*-butyrate are produced, anaerobic degradation requires acetogens for β oxidation (Fig. 1.4, route b).

### 1.2.7.2 Anaerobic Degradation of Protein

Proteins are biological macromolecules, either soluble or solid (e.g., feathers, hair, nails). Outside the cell at an acid pH or in the presence of enzymes, soluble proteins precipitate, e.g., precipitation of casein by addition of rennet enzyme. The reaction sequences necessary for protein degradation in a methanogenic ecosystem are outlined in Figure 1.5. Hydrolysis of precipitated or soluble protein is catalyzed by sev-

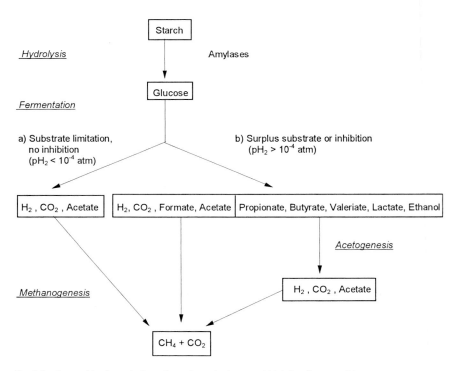

**Fig. 1.4** Anaerobic degradation of starch under low- and high-loading conditions.

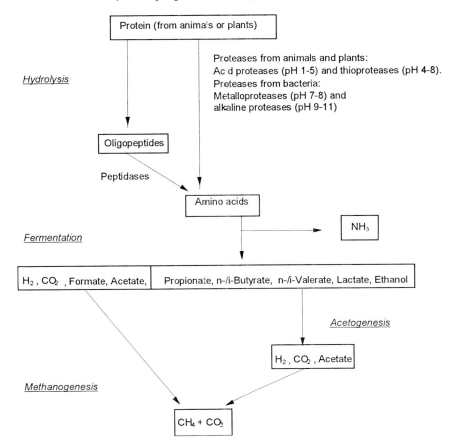

**Fig. 1.5** Anaerobic degradation of proteins.

eral types of proteases that cleave membrane-permeable amino acids, dipeptides, or oligopeptides. In contrast to the hydrolysis of carbohydrates, which proceeds favorably at a slightly acid pH, optimal hydrolysis of proteins requires a neutral or weakly alkaline pH (McInerney, 1988). In contrast to the fermentation of carbohydrates, which lowers the pH due to volatile fatty acid formation, fermentation of amino acids in wastewater reactors does not lead to a significant pH change, due to acid and ammonia formation. Acidification of protein-containing wastewater proceeds optimally at pH values of 7 or higher (Winterberg and Sahm, 1992), and ammonium ions together with the $CO_2$–bicarbonate–carbonate buffer system stabilize the pH. Acetogenesis of fatty acids from deamination of amino acids requires a low $H_2$ partial pressure for the same reasons as for carbohydrate degradation. This can be maintained by a syntrophic interaction of fermentative, protein-degrading bacteria and acetogenic and methanogenic or sulfate-reducing bacteria. Except for syntrophic interaction of amino acid-degrading bacteria with methanogens for maintenance of a low $H_2$ partial pressure, clostridia and presumably also some other sludge bacteria may couple oxidative and reductive amino acid conversion via the Stickland reaction. One amino acid, e.g., alanine (Eq. 4) is oxidatively decarboxylated and the hydrogen or reducing equivalents produced during this reaction are used to reductively convert another amino acid, e.g., glycine, to acetate and ammonia (Eq. 5).

$$CH_3–CHNH_2–COOH + 2\ H_2O \rightarrow CH_3–COOH + CO_2 + NH_3 + 2\ H_2 \tag{4}$$
$$\Delta G^{\circ\prime} = +7.5\ kJ\ mol^{-1}$$
$$2\ CH_2NH_2–COOH + 2\ H_2 \rightarrow 2\ CH_3COOH + 2\ NH_3 \tag{5}$$
$$\Delta G^{\circ\prime} = -38.9\ kJ\ mol^{-1}$$

For complete degradation of amino acids in an anaerobic system therefore, a syntrophism of amino acid-fermenting anaerobic bacteria with methanogens or sulfate reducers is required (Wildenauer and Winter, 1986; Winter et al., 1987; Örlygsson et al., 1995). If long-chain amino acids are deaminated (Eqs. 6–9), fatty acids such as propionate, *i*-butyrate, or *i*-valerate are formed directly. The fatty acids require acetogenic bacteria for their degradation.

$$valine + 2\ H_2O \rightarrow i\text{-butyrate} + CO_2 + NH_3 + 2\ H_2 \tag{6}$$
$$leucine + 2\ H_2O \rightarrow i\text{-valerate} + CO_2 + NH_3 + 2\ H_2 \tag{7}$$
$$i\text{-leucine} + 2\ H_2O \rightarrow 2\text{-methylbutyrate} + CO_2 + NH_3 + 2\ H_2 \tag{8}$$
$$glutamate + 2\ H_2O \rightarrow propionate + 2\ CO_2 + NH_3 + 2\ H_2 \tag{9}$$

In contrast to carbohydrate degradation, where the necessity for propionate- and butyrate-degrading acetogenic bacteria can be circumvented by substrate limitation (Fig. 1.4a), during protein degradation these fatty acids are a product of deamination, and their formation cannot be avoided by maintaining a low $H_2$ partial pressure. In the methanogenic phase there is no difference in methanogenic activity whether carbohydrates or proteins are fermented, except that the methanogens in a reactor fed with protein need to be more tolerant to ammonia and higher pH.

### 1.2.7.3 Anaerobic Degradation of Neutral Fats and Lipids

Fats and lipids are another group of biopolymers that contribute significantly to the COD in sewage sludge, cattle and swine manures, and wastewater from the food industry, e.g., slaughterhouses or potato chip factories (Winter et al., 1992; Broughton et al., 1998). To provide a maximum surface for hydrolytic cleavage by lipases or phospholipases, solid fats, lipids, or oils must be emulsified. Glycerol and saturated and unsaturated fatty acids (palmitic acid, linolic acid, linolenic acid, stearic acid, etc.) are formed from neutral fats. Lipolysis of phospholipids generates fatty acids, glycerol, alcohols (serine, ethanolamine, choline, inositol), and phosphate. Lipolysis of sphingolipids generates fatty acids and amino alcohols (e.g., sphingosine), and lipolysis of glycolipids generates fatty acids, amino alcohols, and hexoses (glucose, galactose). A scheme for anaerobic degradation of fats is shown in Figure 1.6. Sugar moieties and glycerol can be degraded to methane and $CO_2$ by interaction between fermentative and methanogenic bacteria in low-loaded systems or by cooperation between fermentative, acetogenic, and methanogenic bacteria in high-loaded systems. The long-chain fatty acids are degraded by acetogenic bacteria by $\beta$ oxidation to acetate and molecular hydrogen. If acetate and molecular hydrogen accumulate, the anaerobic digestion process is inhibited (Hanaki et al., 1981). Odd-numbered fatty acids are degraded to acetate, propionate, and hydrogen, and even-numbered fatty acids to acetate and hydrogen (Bryant, 1979). Only at a very low $H_2$ partial pressure, which can be maintained by hydrogen-utilizing methanogens or sulfate reducers, is $\beta$ oxidation of at least n-butyrate or propionate exergonic. Methanol, ethanol, and ammonia are formed from choline (Fig. 1.6). After hydrolysis, fermentation, and acetogenesis of the fat components in the methanogenic phase acetate, $CO_2$, and hydrogen are converted to biogas. All subsequent intracellular reactions can be influenced by syntrophic interaction via interspecies hydrogen transfer, except for the extracellular initial lipase reaction.

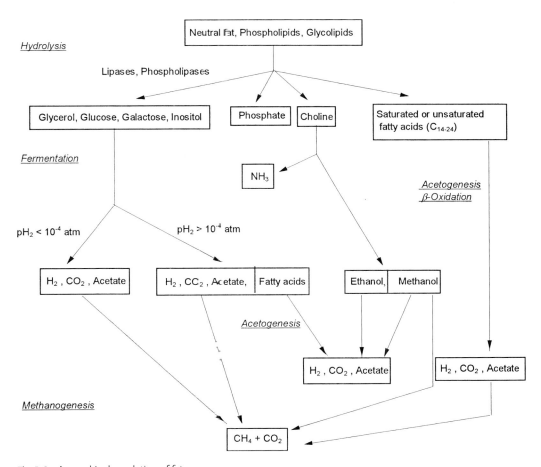

**Fig. 1.6** Anaerobic degradation of fats.

Carbohydrates, proteins, fats, and biogenic oils can also be degraded anaerobically under thermophilic conditions. The overall degradation scheme is the same, but the microorganism populations are different (e.g., Winter and Zellner, 1990). Thermophilic fat degradation is becoming more important in practice, since waste fat from fat separators and fat flotates of the food industry are often cofermented in agricultural biogas plants. Since, for hygienic reasons, the input material must be autoclaved, a thermophilic process should be used, keeping the fat in a melted, soluble form for more effective metabolism. Biogas plants with cofermentation of waste fat residues are considered waste treatment systems (Chapter 11, this volume) and must be designed to meet the hygienic demands relevant to treatment of the respective waste.

## 1.2.8
## Competition of Sulfate Reducers with Methanogens in Methane Reactors

Municipal wastewater or wastewater from sugar production, slaughterhouses, breweries, etc., normally contains less than 200 mg $L^{-1}$ of sulfate. If sulfuric acid is used, e.g., to clean stainless steel containers and pipes in the dairy industry or to maintain an acid pH in bioreactors for bakers' yeast or citric acid production, or if ammonium sulfate is used to inhibit metabolic routes in bakers' yeast for the production of biochemicals, the wastewater contains large amounts of sulfate. Sulfite-containing wastewater is also generated by the starch and cellulose industry during bleaching of the raw products.

If these wastewaters are subjected to anaerobic treatment, the methanogenic bacteria must compete with sulfate reducers for the hydrogen equivalents from COD degradation (Omil et al., 1998). In anaerobic digesters sulfate reduction is the favored reaction, due to the higher affinity of sulfate reducers for reducing equivalents or hydrogen, leading to sulfide production at the expense of reduced biogas formation. Sulfide subsequently forms heavy metal precipitates and, if still present in larger amounts, remains in solution and may be toxic to acetogens and methanogens. Some hydrogen sulfide leaves the reactor with the biogas, which may contain up to a few percent of $H_2S$. Then gas purification is necessary before the gas can be used as fuel for gas engines. Hydrogen sulfide in biogas causes not only odor problems but also corrosion of fermenters and pipes. Methanogenesis of dissolved and particulate organic material in sulfate-rich wastewater is possible, if acidogenesis and sulfate reduction (stripping of $H_2S$ together with $CO_2$) is separated from methanogenesis in a phased or a staged process (Lens et al., 2002). The higher affinity of sulfate reducers than methanogens for reducing equivalents can also be used for sulfate and heavy metal ion removal from wastewater in the first stage of a two-stage anaerobic process (Elferink et al., 1994).

In comparison to methanogens, which have a rather restricted substrate spectrum (Table 1.3), sulfate reducers are metabolically more versatile. Sulfate reducers can utilize polymers such as starch, monomers such as sugars, fatty acids, formate, aliphatic and aromatic compounds, as well as molecular hydrogen (Widdel, 1988) to generate reducing equivalents for sulfate reduction (Eq. 10):

$$8\,(H) + 2\,H^+ + SO_4^{2-} \rightarrow H_2S + 4\,H_2O \tag{10}$$

Undissociated hydrogen sulfide is toxic to both methanogens and sulfate reducers. A 50% inhibition of methanogenesis was observed at a total sulfide concentration of 270 mg $L^{-1}$ (Oleszkiewicz et al., 1989), whereas 85 mg $L^{-1}$ sulfide inhibited sulfate reducers (McCartney and Oleszkiewicz, 1991). For stable methanogenesis, no more than 150 mg $L^{-1}$ sulfide should be accumulated (Speece et al., 1986). $H_2S$ toxicity can be avoided by intensive flushing with biogas to strip the $H_2S$ for adsorption onto $Fe_2(OH)_3$. If large amounts of $H_2S$ are formed during anaerobic treatment, essential heavy metal ions for methanogens (Fe, Ni, Mo, Co, etc.) may be precipitated as sulfides, which may lead to deficiencies in heavy metal bioavailability for the wastewater population.

The metabolites of the fermentative and acetogenic phase of anaerobic wastewater treatment systems, mainly acetate and $CO_2 + H_2$, are substrates for methanogens and for sulfate reducers. If wastewater with a high sulfate concentration is treated in a methane reactor, the population may gradually shift from hydrogenotrophic methanogens toward hydrogenotrophic sulfate reducers, due to a more favorable $K_s$ value for hydrogen of the sulfate reducers. For pure cultures of *Desulfovibrio* sp. the $K_s$ value for hydrogen is 1 µM, but for hydrogenotrophic methanogens it is only 6 µM (Kristjansson et al., 1982). For acetate, such a dramatic disadvantage of methanogens was not observed. The affinities of sulfate reducers and of methanogenic bacteria (valid at least for *Methanosaeta* sp.) for hydrogen are in the same range, 0.2–0.4 µM (Elferink et al., 1994).

## 1.2.9
### Amount and Composition of Biogas During Fermentation of Carbohydrates, Proteins, and Fats

From an economic point of view, the specific biogas amounts and the biogas quality from anaerobic treatment of wastewater and sludge are important process parameters. Therefore, a basis for prediction of amount and composition of biogas was elaborated as early as the 1930s (Buswell and Neave, 1930; Buswell and Sollo 1948; Buswell and Mueller, 1952). If the elemental composition of a wastewater is known, the theoretical amount and composition of the biogas can be predicted with the Buswell equation (Eq. 11). The actual biogas amount is lower and can be calculated by including a correction factor for the degree of degradability, the pH (which influences $CO_2$ absorption), and a 5%–10% discount for biomass formation.

$$C_cH_hO_oN_nS_s + 1/4\,(4c-h-2o+3n+2s)H_2O \rightarrow 1/8\,(4c-h+2o+3n+2s)CO_2 \tag{11}$$
$$+ 1/8\,(4c+h-2o-3n-2s)CH_4 + nNH_3 + sH_2S$$

According to the Buswell equation (Eq. 11), for the anaerobic treatment of a wastewater with carbohydrates as pollutants, the gas composition should theoretically be 50% methane and 50% $CO_2$ (Eq. 12):

$$1 \text{ mol } C_6H_{12}O_6 \rightarrow 3 \text{ mol } CH_4 + 3 \text{ mol } CO_2 \tag{12}$$

Since $CO_2$ is increasingly soluble in water with decreasing temperature and increasing pH, $CO_2$ reacts to form bicarbonate/carbonate, and the biogas may contain more than 80% methane. The total amount of gas is then diminished by the amount of $CO_2$ that is absorbed and solubilized in the liquid. From a fat- and protein-containing wastewater, theoretically more than 50% methane can be generated (Table 1.5).

## 1.3
## Nitrogen Removal During Wastewater Treatment

Nitrogen-containing substances in wastewater are inorganic or organic. Together with phosphates, they represent the main source for eutrophication of surface water. For this reason they must be eliminated together with the organic carbon during wastewater treatment. Whereas phosphates form insoluble precipitates with many heavy metal ions and the precipitates can be separated by sedimentation or flotation, all nitrogen compounds, except for magnesium ammonium phosphate, are easily soluble in water and thus cannot be removed chemically by precipitation. For biological removal of amino nitrogen and of heterocyclic nitrogen compounds, their conversion to ammonia in an aerobic or anaerobic treatment process is the first step. Then ammonia must be nitrified, and the nitrate denitrified to yield nitrogen. Thus, depending on the kind of nitrogen compounds present in wastewater, nitrogen removal requires up to three processes in sequence: ammonification, nitrification, and denitrification.

The major portion of nitrogen compounds in municipal wastewater are reduced nitrogen compounds such as ammonia, urea, amines, amino acids, and proteins. Oxidized nitrogen compounds such as nitrate and nitrite normally are not present at all or not in relevant amounts. Nitrate and nitrite may, however, represent the main nitrogen load in wastewater of certain food or metal industry branches (Gensicke et al., 1998; Zayed and Winter, 1998).

**Table 1.5** Amount of biogas, biogas composition, and energy content.

| Substrate | Amount $(cm^3\,g^{-1})$ | Composition % $CH_4$ | % $CO_2$ | Energy content $(kW\,h \times m^{-3})^d$ |
|---|---|---|---|---|
| Carbohydrates[a] | 746.7 | 50 | 50 | 4.95 |
| Fats[b] | 1434 | 71 | 29 | 7.02 |
| Proteins[c] | 636 | 60 | 40 | 5.93 |

[a] Calculated for hexoses.
[b] Calculated for triglycerides containing glycerol plus 3 mol palmitic acid.
[c] Calculated for polyalanine and reaction of ammonia to $(NH_4)_2CO_3$.
[d] at standard conditions

The ammonia in raw municipal wastewater is mainly derived from urine and is formed in the sewer system by enzymatic cleavage of urea by ureases (Eq. 13):

$$NH_2CONH_2 + H_2O \rightarrow CO_2 + 2\,NH_3 \tag{13}$$

The residence time of the wastewater in the sewer system normally is not long enough for a significant contribution of ammonia from other sources, e.g., proteolysis and deamination of amino acids.

## 1.3.1
## Ammonification

The main organic nitrogen compounds in municipal wastewater are heterocyclic compounds (e.g., nucleic acids) and proteins. Proteolysis and degradation of amino acids leads to liberation of ammonia by the various mechanisms of ammonification (Rheinheimer et al., 1988), including hydrolytic, oxidative, reductive, and desaturative deamination (Eqs. 14–17, respectively)

$$R-NH_2 + H_2O \rightarrow R-OH + NH_3 \tag{14}$$
$$R-CHNH_2COOH + H_2O \rightarrow R-CO-COOH + 2\,(H) + NH_3 \tag{15}$$
$$R-CHNH_2-COOH + 2\,(H) \rightarrow R-CH_2-COOH + NH_3 \tag{16}$$
$$R-CH_2-CHNH_2-COOH \rightarrow R-CH=CH-COOH + NH_3 \tag{17}$$

A significant amount of ammonia from urea cleavage or from ammonification of amino acids is assimilated in aerobic treatment processes for growth of bacteria (surplus sludge formation). It can be estimated that bacteria consist of roughly 50% protein and that the nitrogen content of protein is about 16%. Thus, for synthesis of 1 g of bacterial biomass, about 0.08 g of ammonia-N is required. To eliminate ammonia that is not used for cell growth during wastewater treatment, it must first be nitrified and then denitrified to molecular nitrogen or anaerobically oxidized with nitrite.

## 1.3.2
## Nitrification of Ammonia

### 1.3.2.1 Autotrophic Nitrification
Autotrophic nitrifiers are aerobic microorganisms oxidizing ammonia via nitrite (Eq. 18) to nitrate (Eq. 19). Organisms catalyzing nitrification (Eq. 18) belong to the genera *Nitrosomonas*, *Nitrosococcus*, *Nitrosolobus*, *Nitrosospira*, and *Nitrosovibrio*, organisms catalyzing nitration (Eq. 19) include, e.g., members of the genera *Nitrobacter*, *Nitrococcus*, and *Nitrospira*.

$$NH_4^+ + 1.5\,O_2 \rightarrow NO_2^- + 2\,H^+ + H_2O \tag{18}$$
$$NO_2^- + 0.5\,O_2 \rightarrow NO_3^- \tag{19}$$

Ammonia oxidation to nitrite or nitrite oxidation to nitrate are energy-yielding processes for the autotrophic growth of the nitrifying bacteria. $CO_2$ is assimilated via the Calvin cycle. Since the positive redox potential of the oxidizable nitrogen compounds is not low enough to form $NADH_2$ for $CO_2$ reduction, $NADH_2$ must be formed by an energy-consuming reverse electron transport. For this reason, the growth yield of nitrifying bacteria is low. *Nitrosomonas* sp., e.g., must oxidize 30 g $NH_3$ to form 1 g of cell dry mass (Schlegel, 1992).

The oxidation of ammonia by nitrifiers is initiated by an energy-neutral monooxygenase reaction that yields hydroxylamine (Eq. 20):

$$NH_3 + XH_2 + O_2 \rightarrow NH_2OH + X + H_2O \tag{20}$$

Then hydroxylamine is further oxidized, presumably via nitroxyl (Eq. 21) to nitrite (Eq. 22), which is the energy-yielding reaction during nitrification. For microbial oxidation of 1 mg $NH_4^+$-nitrogen to nitrite, 3.42 mg $O_2$ is required, and for the oxidation of 1 mg $NO_2^-$-nitrogen to nitrate (Eq. 19), 1.14 mg $O_2$.

$$NH_2OH + X \rightarrow XH_2 + (NOH) \tag{21}$$
$$(NOH) + 0.5\ O_2 \rightarrow HNO_2 + energy \tag{22}$$

Due to their slow growth autotrophic nitrifiers cannot successfully compete with heterotrophic bacteria for oxygen. In a highly loaded activated sludge system, the autotrophic nitrifiers are overgrown by the heterotrophic sludge flora, which consume the oxygen. Ammonia oxidation starts only if the $BOD_5$ concentration in the wastewater is <110 mg $L^{-1}$ (Wild et al., 1971). During nitrification of ammonia the alkalinity of the wastewater increases slightly, due to $CO_2$ consumption for autotrophic growth (pH increase), but in a counter reaction it drastically decreases due to nitric acid formation from ammonia (pH decrease from above neutral to acidic). If the buffer capacity of the wastewater is weak, the pH drops far below 7 and thus prevents further nitrification by autotrophic nitrifiers (Rheinheimer et al., 1988).

### 1.3.2.2 Heterotrophic Nitrification

Some bacteria of the genera *Arthrobacter*, *Flavobacterium*, and *Thiosphaera* are able to catalyze heterotrophic nitrification of nitrogen-containing organic substances (Eq. 23):

$$R-NH_2 \rightarrow R-NHOH \rightarrow R-NO \rightarrow NO_3^- \tag{23}$$

Heterotrophic nitrifiers oxidize reduced nitrogen compounds, such as hydroxylamine and aliphatic and aromatic nitrogen-containing compounds, but in contrast to autotrophic nitrification, no energy is gained by nitrate formation. For this reason an organic substrate must be respired to satisfy the energy metabolism (Schlegel, 1992) .

Some heterotrophic nitrifiers can denitrify nitrate or nitrite under aerobic growth conditions. The nitrogen metabolism of *Thiosphaera pantotropha* and *Paracoccus denitrificans* are well documented examples (Stouthamer et al., 1997). These organisms express a membrane-bound nitrate reductase under anoxic growth conditions, which works only in the absence of molecular oxygen. Under aerobic growth conditions a periplasmic nitrate reductase is expressed, which catalyzes nitrate reduction at least to the state of nitrous oxide.

### 1.3.3
### Denitrification: Nitrate Removal from Wastewater

Many aerobic bacteria seem to be able to switch their oxidative metabolism to nitrate respiration. Similar to oxygen respiration, the nitrate respiration of heterotrophic bacteria requires a complex carbon source as an electron source for denitrification (e.g., Eqs. 24 and 25). Denitrification (Eq 26) starts with the reduction of nitrate to nitrite by membrane-bound nitrate reductase A (a). Then a membrane-bound nitrite reductase (b) catalyzes NO formation. Finally, NO reductase (c) and $N_2O$ reductase (d) form $N_2$. The theoretical stoichiometry of denitrification with methanol or acetate as a carbon source is shown in Eqs. 27 and 28. For practical application a surplus of carbon source must be supplied, since the wastewater is not free of oxygen, and part of the carbon source is respired until anoxic conditions are achieved.

To evaluate the stoichiometry of nitrate to organic compounds for denitrification with a complex carbon source, the oxidation–reduction state of the carbon substrates and the oxygen concentration in the wastewater should be known. In wastewater treatment plants more than 2.85 g COD is required for reduction of 1 g $NO_3^- -N$ (Bernet et al., 1996).

$$CH_3OH + H_2O \rightarrow CO_2 + 6\,H^+ + 6\,e^- \tag{24}$$
$$CH_3COOH + 2\,H_2O \rightarrow 2\,CO_2 + 8\,H^+ + 8\,e^- \tag{25}$$
$$NO_3^- \xrightarrow[a]{2\,e^-} NO_2^- \xrightarrow[b]{e^-} NO \xrightarrow[c]{e^-} 0.5\,N_2O \xrightarrow[d]{e^-} 0.5\,N_2 \tag{26}$$
$$5\,CH_3OH + 6\,HNO_3 \rightarrow 5\,CO_2 + 3\,N_2 + 13\,H_2O \tag{27}$$
$$5\,CH_3-COOH + 8\,HNO_3 \rightarrow 10\,CO_2 + 4\,N_2 + 14\,H_2O \tag{28}$$

Instead of nitrate, many denitrifying bacteria can use $NO_2^-$, NO, or $N_2O$ as terminal electron acceptors. Alternatively, they may release these intermediates during denitrification of nitrate under unfavorable conditions, as was observed in soil (Conrad, 1996). If, e.g., surplus nitrate is supplied and hydrogen donors are not sufficiently available, NO and $N_2O$ can be formed (Schön et al., 1994). Another condition for $N_2O$ formation is a pH below 7.3, at which nitrogen oxidoreductase is inhibited (Knowles, 1982).

Except for dissimilatory nitrate reduction, many aerobic and anaerobic bacteria are capable of assimilatory nitrate reduction to supply the cells with ammonia for growth (Eqs. 29 and 30). However, the enzymes for nitrate assimilation are expressed only at concentrations of ammonia <1 mM.

$$NO_3^- \xrightarrow[a]{2e^-} NO_2^- \xrightarrow[b]{2e^-} HNO \xrightarrow[b]{2e^-} NH_2OH \xrightarrow[b]{2e^-} NH_3 \qquad (29)$$

$$HNO_3 + 8\,(H) \rightarrow NH_4^+OH^- + 2\,H_2O \qquad (30)$$

Here, nitrate reductase B (enzyme *a* in Eq. 29) reduces nitrate to nitrite, which is then reduced to ammonia by the nitrite reductase complex (enzyme b in Eq. 29). Whereas nitrate reduction by the oxygen-sensitive, membrane-bound enzyme nitrate reductase A conserves energy, no energy conservation is possible in the reaction catalyzed by the soluble enzyme nitrate reductase B, which is not repressed by oxygen. For details on the cell biology and the molecular basis of denitrification, please refer to Zumft (1997).

### 1.3.4
### Combined Nitrification and Denitrification

A strict separation of the reactions participating in aerobic, autotrophic nitrification and in anoxic, heterotrophic denitrification is not required, as concluded from $N^{15}$-tracer experiments (Kuenen and Robertson, 1994). Autotrophic ammonia oxidizers seem to be able to produce NO, $N_2O$, or $N_2$ from nitrite if oxygen is limited, and ammonia as well as nitrite oxidizers can be isolated from anaerobic reactors (Kuenen and Robertson, 1994). *Nitrosomonas europaea* can use nitrite as an electron acceptor and pyruvate as an energy source under anoxic, denitrifying growth conditions (Abeliovich and Vonshak, 1992). In addition, several strains of *Nitrobacter* sp. were reported to denitrify during anoxic, heterotrophic growth (Bock et al., 1986).

Aerobic denitrification by *Thiosphaera pantotropha* was first described by Robertson and Kuenen (1984). *T. pantotropha* respires molecular oxygen and denitrifies nitrate simultaneously, provided that suitable electron acceptors are available. The conversion rate of acetate as electron donor with nitrate as electron acceptor was twice as high when the concentration of molecular oxygen was <30% of air saturation compared to 30%–80% air saturation (Robertson et al., 1988). Respiration, simultaneous nitrification, and denitrification were observed in the presence of oxygen and ammonia. During heterotrophic denitrification by *T. pantotropha*, ammonia is first oxidized to hydroxylamine by an ammonia monooxygenase, with ubiquinone serving as electron donor. Hydroxylamine is subsequently oxidized to nitrite by a hydroxylamine oxidoreductase. During coupled nitrification–denitrification, 3 of the 4 electrons from the oxidation of hydroxylamine are used for reduction of nitrite to nitrogen and are not available to the electron transport reaction catalyzed by cytochrome oxidase. Regeneration of ubiquinone is mediated by electrons that are generated during oxidation of an organic substrate (heterotrophic nitrification).

Conversion rates of ammonia by heterotrophic nitrifiers such as *T. pantotropha* (Kuenen and Robertson, 1994) are smaller than those of autotrophic nitrifiers (35.4 for *T. pantotropha* versus 130–1550 nmol $NH_3$ $min^{-1}$ $mg^{-1}$ dry weight for *Nitrosomonas* sp.). If, however, the higher population density of heterotrophic nitrifiers, resulting from higher growth rates, is taken into consideration, the specific conversion rates are in a similar range. Heterotrophic nitrifiers can, in addition to their nitrifying capability, denitrify nitrite or nitrate to molecular nitrogen. In many waste-water treatment plants autotrophic nitrifiers may exist, producing nitrite from ammonia under moderately aerobic conditions; the ammonia is then converted to nitrate and/or reduced to nitrogen by heterotrophic nitrifiers in the presence of a suitable carbon source. In practice, the only disadvantage of heterotrophic nitrification is more surplus sludge generation for final disposal.

## 1.3.5
### Anaerobic Ammonia Oxidation (Anammox®)

To date, two anaerobic ammonia-oxidizing bacteria species have been isolated and preliminarily classified in the order Planctomycetales as candidatus *Brocardia anammoxidans* and candidatus *Kueneria stuttgartiensis*. Under anaerobic conditions ammonia is oxidized to nitrite in membrane-bound intracytoplasmic anammoxosoms (Schmidt et al., 2003).

In the Anammox process ammonia is oxidized to nitrogen, with nitrite serving as electron acceptor (Van de Graaf et al., 1995). Use of $N^{15}$ isotopes indicated that nitrite is reduced to hydroxylamine (Eq. 31), which then reacts with ammonia to yield hydrazine ($N_2H_4$, Eq. 32). By oxidation of hydrazine to molecular nitrogen (Eq. 33), four reducing equivalents are generated, which are required for nitrite reduction to hydroxylamine (Van de Graaf et al., 1997).

$$2\ HNO_2 + 4\ XH_2 \rightarrow 2\ NH_2OH + 2\ H_2O + 4\ X \tag{31}$$
$$2\ NH_2OH + 2\ NH_3 \rightarrow 2\ N_2H_4 + 2\ H_2O \tag{32}$$
$$2\ N_2H_4 + 4\ X \rightarrow 2\ N_2 + 4\ XH_2 \tag{33}$$

Since the redox state is balanced in the above reactions, reducing equivalents for $CO_2$ reduction by the autotrophic microorganisms must be generated by oxidation of nitrite to nitrate (Eq. 34):

$$HNO_2 + H_2O + NAD \rightarrow HNO_3 + NADH_2 \tag{34}$$

Per mol of ammonia, 0.2 mol nitrate is generated and 20 mg biomass is produced (Van de Graaf et al., 1996). The Anammox process seems to be suitable for nitrogen removal in ammonia-rich effluents of anaerobic reactors that are fed with wastewater rich in TKN compounds. Nitrogenous compounds can be eliminated from this wastewater by a combination of nitrification for nitrite supply and anaerobic ammonia oxidation (Strous et al., 1997). The process is sensitive to high, toxic nitrite concentrations of 70–180 mg N $L^{-1}$ (depending on the kind of biomass). Furthermore, due to the low growth rate of the anammox bacteria, a startup time of 100–150 days with activated sludge as inoculum seems to be necessary (Schmidt et al., 2003).

In oxygen-limited environments (e.g. oxic–anoxic interfaces of biofilms or sludge flocs) aerobic and anaerobic ammonia oxidizers are natural partners. The ammonia is oxidized to nitrite and concomitantly the oxygen level is decreased. The anammox bacteria disproportionate the nitrite and the remaining ammonia to $N_2$. When ammonia is limited, the affinities of aerobic and anaerobic ammonia oxidizers may lead the natural partners to change to competitors (Schmidt et al., 2002). Other examples of 'uncontrolled' anammox activities were observed in different wastewater treatment plants with uncharacterized high ammonium losses.

## 1.3.6
### New N-removal Processes

The metabolic versatility of nitrogen-converting bacteria offers new concepts for nitrogen removal processes to treat wastewater. A *partial* nitrification of ammonia to nitrite (nitritation) would have the advantage of saving a significant amount of oxygen for the nitrification process, as compared to nitratation of ammonia. However, nitrite is toxic and must be kept below toxic concentrations. In a subsequent denitrification process, fewer reducing equivalents from the degradation of different carbon sources, which often are a limiting factor in wastewater treatment plants, are then required. In the SHARON process, which is a single reactor system for rapid ammonia oxidation to nitrite and subsequent denitrification of the nitrite with methanol, nitrite oxidizers that would generate nitrate are outcompeted by the slowly metabolizing ammonia oxidizers after a temperature shift toward higher temperatures, due to the higher temperature tolerance of the ammonia oxidizers, and by no sludge retention or sludge return in the reactor (Hellinga et al., 1998). Aeration was switched off periodically and denitrification of nitrite was initiated by adding methanol as a carbon source. The SHARON process in full scale was used to treat sludge liquor at the Rotterdam wastewater treatment plant (Mulder et al., 2001). The partial nitrification of ammonia to nitrite led to an ammonia/nitrite ratio that was suitable for a subsequent Anammox process in which nitrite was denitrified to $N_2$ with ammonia instead of methanol as the electron donor. This process was finally established in full-scale at the Rotterdam wastewater treatment plant (Schmidt, 2003).

Partial ammonia oxidation to nitrite and anaerobic oxidation of ammonia in one aerated reactor is the concept behind the CANON process (completely autotrophic nitrogen removal over nitrite) (Van Lossdrecht and Jetten, 1997). Aerobic and anaerobic ammonia-oxidizing bacteria cooperate as long as the nitrifying bacteria consume oxygen and create anoxic conditions for the anaerobic ammonia-oxidizing bacteria.

Another possibility for removing ammonia in one single step without using an organic carbon source is the OLAND process (oxygen-limited nitrification and denitrification) (Kuai and Verstraete, 1998). Conversion of ammonia to $N_2$ is catalyzed by aerobic nitrifiers and anaerobic granular sludge as a source of *Planctomycetes* including anaerobic ammonia-oxidizing bacteria (Pynaert et al., 2004).

Other nitrogen-removal processes also exist, such as the $NO_x$ process (simultaneous nitrification and denitrification of *Nitrosomonas*-like microorganisms under fully oxic conditions) and aerobic deammonification (conversion of ammonia to $N_2$ and nitrate), which allow ammonia removal without COD. An overview of these new concepts was presented by Schmidt et al., 2003.

## 1.4
## Enhanced Biological Phosphate Removal

Based on an early observation that microorganisms take up more phosphate than required for cell growth, it was found that single cells accumulate polyphosphate in granules containing a few to several thousand phosphate units (Egli and Zehnder, 1994). Phosphate constitutes up to 12% of the cell weight of polyphosphate-accumulating bacteria, whereas bacteria that do not accumulate polyphosphate contain only 1%–3% phosphate (van Loosdrecht et al., 1997b). The accumulated polyphosphate is assumed to be an energy source for substrate assimilation during anaerobic growth conditions and poly-β-hydroxyalkanoate (PHA) synthesis. Evidence suggests that degradation of polyphosphate may be used to regulate intracellular pH under alkaline conditions (Seviour et al., 2003). For biological phosphate removal by 'luxury phosphate uptake', the polyphosphate-accumulating bacteria in wastewater have to be subjected in sequence to an anaerobic and an aerobic environment. All previous attempts to isolate polyphosphate-accumulating bacteria (culture-dependent approach) led to recovery of *Acinetobacter* sp. which belongs to the γ-Proteobacteria and is the model organism for explanation of the polyphosphate accumulation mechanism found in textbooks. However, pure cultures of the polyphosphate-accumulating *Acinetobacter* sp. and wastewater communities seem to behave differently (van Loosdrecht et al., 1997a). A major discrepancy is that *Acinetobacter* sp. do not play a dominant role in activated sludge systems, as deduced from a survey using specific gene probes (Wagner et al., 1994). The culture-independent approach led to a mixed population for enhanced biological phosphorous removal (EBPR). The phosphorous-removing members (>10% of the total population) were assigned to the β-Proteobacteria (Rhodocylus group e.g. candidatus *Accumulibacter phosphatis*), α-Proteobacteria and Actinobacteria (*Microlunatus phosphoruvorus*) and to members of the Cytophaga–Flexibacter–Bacteroides division (Seviour et al., 2003).

In *Acinetobacter* sp. polyphosphate accumulation is closely interconnected with poly-β-hydroxybutyrate (PHB) and glycogen metabolism (van Loosdrecht et al., 1997b). Under anaerobic conditions in the absence of nitrate, acetate is taken up by *Acinetobacter* cells, metabolized to β-hydroxybutyrate, polymerized to PHB, and stored intracellularly in inclusion bodies. Reducing equivalents for PHB formation from acetate are made available by glycolysis of glucose units from stored glycogen or are alternatively formed by other microorganisms for EBPR in an anaerobically operating TCA cycle. The energy for polymerization comes from hydrolysis of the anhydride bonds of ATP and polyphosphate. To supply both the maintenance metabolism and the storage metabolism of *Acinetobacter* with sufficient energy under anaerobic conditions, anhydride bonds in polyphosphate are hydrolyzed, and inorganic phosphate is excreted (phosphate resolubilization) (Smolders et al., 1994). Stored glycogen in microorganisms for EBPR may also provide ATP for PHA synthesis. For storage of 1 mg fatty acids, about 0.6 mg orthophosphate is released (Danesh and Oleszkiewicz, 1997). Under aerobic conditions two situations may be prevalent. In wastewater that does not contain a suitable carbon source for respiration, *Acinetobacter* sp. hydrolyze PHB, which is stored in the granules, and respire the β-hydroxybutyrate to gain energy for growth, maintenance, formation of glycogen, and polymerization of phosphate, which is taken up from the wastewater. In the presence of oxygen much more $P_i$ is taken up by *Acinetobacter* sp. from the activated sludge flora than is released under anaerobic conditions ('luxury uptake'). Danesh and Oleszkiewicz (1997) reported that approximately 6 to 9 mg volatile fatty acids are required for biological removal of 1 mg phosphorous.

In wastewater that contains suitable carbon sources for *Acinetobacter* sp., these carbon sources are partially oxidized by the TCA cycle and are partially used for PHB formation (Kuba et al., 1994). Since reducing equivalents are needed for PHB formation, less ATP is formed by oxidative phosphorylation. The lower ATP yield is compensated by the energy from hydrolysis of polyphosphate to $P_i$, which the cells use to form ADP from AMP. Two moles of ADP are disproportionated to AMP and ATP by adenylate kinase, and little $P_i$ is released into the medium.

Except for obligately aerobic bacteria of the genus *Acinetobacter*, nitrate-reducing bacteria compete with microorganisms for EBPR for organic substrates, e.g., volatile fatty acids such as acetate. Nitrate-reducing bacteria also can eliminate phosphate from the aqueous environment (Kerrn-Jespersen and Henze, 1993). The simultaneous presence of obligately aerobic bacteria and nitrate-reducing bacteria in activated sludge may explain why, in the presence of nitrate in the anaerobic phosphate resolubilization phase (oxygen absent), $P_i$ does not accumulate in the medium. The denitrifying, polyphosphate-accumulating bacteria apparently do not inhibit $P_i$ excretion by *Acinetobacter,* but they take up phosphate and accumulate it as polyphosphate. This may explain why resolubilization of phosphate by degradation of stored polyphosphate is not always observed in wastewater treatment plants for biological phosphate removal. Another possibility may be the inhibitory effect of nitric oxide (Section 1.3) on adenylate kinase, which is involved in polyphosphate degradation in *Acinetobacter* sp. .

For biological phosphate accumulation, the ability of cells to store reserve material plays an essential role. Polymerization of substrates and storage in intracellular granules may offer the advantage that, at times of substrate shortage or in an environment with low concentrations of substrates (at low $k_m$ values), such organisms can survive much better than other heterotrophic bacteria that do not have effective strategies for substrate storage.

Biological phosphate elimination is a highly effective process leading to final concentrations in the wastewater of <0.1 mg $L^{-1}$ phosphate. For optimal biological phosphate removal, the COD/P ratio in the wastewater should be about 20 g COD $g^{-1}$ phosphate to permit good growth of the polyphosphate-accumulating bacteria. If biological phosphate elimination is used in combination with chemical precipitation, the minimum COD/P ratio should be 2 g COD $g^{-1}$ phosphate (Smolders et al., 1996) in processes with or without separation of primary sludge. If most of the particulate COD is separated in a primary sedimentation pond, ferrous ions should be added to supplement the biological phosphate removal with chemical precipitation. If no presedimentation of sludge is available or the sedimentation efficiency is low, no ferrous ions have to be added (van Loosdrecht et al., 1997b). A detailed description of the state of the art concerning biological phosphate elimination was given by Schön (1999).

Problems with the stability of the EBPR process in wastewater treatment plants were periodically reported. The reason for this may be the occurrence of microorganisms known as glycogen nonpolyphosphate-accumulating organisms (GAOs). These bacteria are selected under EBPR operation conditions (alternating anaerobic–aerobic phases during treatment) and compete with microorganisms for EBPR for the available volatile fatty acids (Blackall et al., 2002). Filipe et al. (2001) suggested an important role of pH during anaerobic and aerobic operation of reactors: growth of GAOs is favored at low pH, but at a pH of 7–7.5 the microorganisms for EBPR outcompete the GAOs. Further research is required to elucidate conditions under which GAOs are inhibited and growth of EBPR biomass is promoted.

## 1.5
## Biological Removal, Biotransformation, and Biosorption of Metal Ions from Contaminated Wastewater

Whereas solid organic and inorganic material in wastewater or sludge can be removed by sedimentation, soluble organic pollutants and xenobiotics should be eliminated from the aqueous environment by microbial mineralization or anaerobic degradation to gaseous products, with a varying portion (5%–50%) being used as substrates for bacterial growth. Most of the inorganic components present in wastewater are soluble and are ionized. Trace amounts of many cations (e.g., $Na^+$, $K^+$, $Ca^{2+}$, $Fe^{2+}$, $Ni^{2+}$, $Co^{2+}$, $Mn^{2+}$, $Zn^{2+}$, etc.) and anions (e.g., $PO_4^{3-}$, $Cl^-$, $S^{2-}$) are essential micronutrients for bacterial growth. Other cations such as ammonia may also be required for bacterial growth, but the surplus amount must be oxidized to nitrate, and the nitrate denitrified to gaseous nitrogen, for N elimination from wastewater. Under anaerobic conditions sulfate is reduced to sulfide, low amounts of which are required for growth of bacteria. The sulfide not required for growth is toxic for bacteria if present as $H_2S$ in high concentrations. In anaerobic reactors at a slightly alkaline pH, most of the sulfide is precipitated as heavy metal sulfides; then it is harmless to microorganisms and their environment. To support detoxification, heavy metal ions may also be precipitated chemically.

Metal ion contaminants in wastewater can be removed by microorganisms by either a direct or indirect influence on the redox state of the metal ions or through biosorption of metal ions on the cell surface (Lovley and Coates, 1997). Some microorganisms have also developed resistance mechanisms against toxic metals by changing the oxidation state without supporting anaerobic growth. Certain bacteria, yeasts, fungi, and algae can actively accumulate intracellular metal ions against a gradient. The process of bioaccumulation of metal ions depends on living, metabolically active cells, whereas biosorption is a passive, energy-independent process that can be mediated also by inactive cell material. Biosorption of metal ions includes mechanisms such as ion exchange, chelation, matrix entrapment, and surface sorption (Unz and Shuttleworth, 1996). After biosorption or active removal of metal ions from wastewater or contaminated soil, the heavy metal-containing biomass must be separated and incinerated or regenerated by desorption or remobilization of the metals. As an example of successful biosorption, wastewater from the galvanizing industry that contained 29 mg $L^{-1}$ Zn and 10.5 mg $L^{-1}$ Fe was purified in a three-stage stirred-reactor cascade. With 15 g biomass, 7.5 L of wastewater was decontaminated (Brauckmann, 1997). The removal of metal ions by biologically catalyzed changes in the redox state is an alternative. The altered speciation (valence status) of metals can lead to precipitation, solubilization, or volatilization of the metal ions.

Many bacteria use metal ions as electron acceptors for anaerobic respiration. Examples are the reduction of $Fe^{3+}$, $Cr^{6+}$, $Mn^{4+}$, $Se^{6+}$, $As^{5+}$, $Hg^{2+}$, $Pb^{2+}$, or $U^{6+}$ (Table 1.6). The dissimilatory metal-reducing bacteria can use $H_2$ or organic pollutants (e.g., xenobiotics) as electron donors and are capable of simultaneously removing organic and inorganic contaminants. Metal ions are reduced and precipitated by sulfide that is generated by sulfate-reducing bacteria.

Solubilization of most heavy metal precipitates is favored at acid pH, which is the favorable pH for soil or sludge decontamination, whereas an alkaline pH is favorable for precipitation of heavy metal ions to decontaminate wastewater.

An example of the formation of precipitates or soluble compounds at different redox states occur in ferrous and manganese compounds. Under aerobic conditions Fe(III) (ferric iron) and Mn(IV) ions form insoluble Fe(III) and Mn(IV) oxides or hydroxides, but under anaerobic conditions they form soluble Fe(II) compounds (ferrous iron) or Mn(II) compounds. Most organisms that grow with energy conserved during reduction of Fe(III) or Mn(IV) are members of the *Geobacteriaceae*. Transfer of electrons from the terminal reductase, localized in the outer membrane or at the

**Table 1.6** Metals as electron acceptors for anaerobic respiration.

| Reaction | Microorganism | Reference |
|---|---|---|
| $2\ Fe^{3+} + H_2 \leftrightarrow 2\ Fe^{2+} + 2\ H^+$ | *Geobacter metallireducens* <br> *Pelobacter carbinolicus* | Lovley and Lonergan (1990) <br> Lovley et al. (1995) |
| $Mn^{4+} + H_2 \leftrightarrow Mn^{2+} + 2\ H^+$ | *Geobacter metallireducens* <br> mixed culture | Lovley (1991) <br> Langenhoff et al. (1997) |
| $2\ Cr^{6+} + 3\ H_2 \leftrightarrow 2\ Cr^{3+} + 6\ H^+$ | *Desulfovibrio vulgaris* <br> *Bacillus* strain QC1-2 | Lovley and Phillips (1994) <br> Campos et al. (1995) |
| $Se^{6+} + H_2 \leftrightarrow Se^{4+} + 2\ H^+$ | *Thauera selenatis* <br> strains SES-1; SES-3 | Macy et al. (1993) |
| $Se^{6+} + 3\ H_2 \leftrightarrow Se^0 + 6\ H^+$ | | Oremland et al. (1989) |
| $Te^{4+} + 2\ H_2 \leftrightarrow Te^0 + 4\ H^+$ | *Schizosaccharomyces pombe* | Smith (1974) |
| $Pb^{2+} + H_2 \leftrightarrow Pb^0 + 2\ H^+$ | *Pseudomonas maltophila* | Lovley (1995) |
| $As^{5+} + H_2 \leftrightarrow As^{3+} + 2\ H^+$ | *Geospirillum arsenophilus* | Ahmann et al. (1994) |
| $Hg^{2+} + H_2 \leftrightarrow Hg^0 + 2\ H^+$ | *Escherichia coli* <br> *Thiobacillus ferrooxidans* | Robinson and Tuovinen (1984) |
| $U^{6+} + H_2 \leftrightarrow U^{4+} + 2\ H^+$ | *Shewanella putrefaciens* | Lovley et al. (1991) |

cell surface, to the insoluble Fe(III) or Mn(IV) oxides outside the cells can proceed either in a direct way (contact between the oxides and the cells) or by 'soluble electron shuttles' (e.g., by humic substances) between the metal-reducing microorganism and the mineral (Lloyd, 2003). Since Fe is an essential element for microorganisms, aerobic bacteria must excrete siderophores, which bind $Fe^{3+}$ to their phenolate or hydroxamate moiety and supply the cells with soluble $Fe^{2+}$. To accelerate $Fe^{3+}$ reduction in biotechnological processes, the chelator nitrilotriacetic acid can be added. In addition to their use in synthesis of cell components (e.g., cytochromes, ferredoxin, etc.), $Fe^{2+}$ salts can be electron donors for nitrate reduction (Straub et al., 1996).

The reduction of $Hg^{2+}$ to metallic Hg by *Escherichia coli* or *Thiobacillus ferrooxidans* facilitates Hg separation and prevents methylation reactions under aerobic or sulfate-reducing conditions. The mechanism includes nonenzymatic transfer of methyl groups from methylcobalamin to $Hg^{2+}$ to form methyl mercury or dimethyl mercury, which are both neurotoxins and become enriched in the food chain. Selenium and arsenic can be transformed microbiologically by methylation to dimethyl selenide or to di- or trimethyl arsine in a volatile form. Methylated arsenic compounds are less toxic than nonmethylated arsenic compounds (White et al., 1997).

Changing the redox state of natural and anthropogenic radionuclides by metal-reducing microorganisms offers a possibility to control their solubility and mobility by converting, e.g., $U^{6+}$ to $U^{4+}$, $Pu^{5+}$ to $Pu^{4+}$, or $Np^{5+}$ to $Np^{4+}$. The tetravalent metals can be removed by chelators (e.g., EDTA) or by immobilization onto biomass from the contaminated environment (Lloyd, 2003).

In contrast to bacteria, fungi are capable of leaching soluble as well as insoluble metal salts, because they excrete organic acids such as citric acid, fumaric acid, lactic acid, gluconic acid, oxalic acid, or malic acid, which dissolve metal salts and form complexes with the metal ions. The leaching efficiency depends on the soil microflora. Some soil microorganisms seem to be able to degrade the carbon skeleton of the metal–organic complex and thus immobilize the metal ions again (Brynhildsen and Rosswall, 1997).

Mixed bacterial cultures or *Wolinella succinogenes* use perchlorate or chlorate as electron acceptors for respiration (Wallace et al., 1996; van Ginkel et al., 1995) and thus detoxify these chemicals.

### 1.5.1
### Sulfate Reduction and Metal Ion Precipitation

Sulfate-reducing bacteria are biotechnologically relevant to sulfate removal or heavy metal precipitation in wastewater or waste and to the elimination of $SO_2$ during off-gas purification. An overview of applications of sulfate-reducing microorganisms in environmental biotechnology is given by Lens et al. (2002). Sulfate is the terminal electron acceptor and is reduced to sulfide, with reducing equivalents derived from the degradation of lactic acid or many other organic compounds (Widdel, 1988). Alternatively, some sulfate reducers can also use molecular hydrogen. Sulfate reducers gain energy in an anaerobic electron transport chain (Hansen, 1994), leading to sulfide, a weak dibasic acid, which dissociates according to Eq. 35.

$$H_2S \overset{K_1}{\longleftrightarrow} H^+ + HS^- \overset{K_2}{\longleftrightarrow} 2\,H^+ + S^{2-}$$
$$K_1 = 1.02 \times 10^{-7},\ K_2 = 1.3 \times 10^{-13}\ (25\ °C) \tag{35}$$

The total dissociation is described by Eq. 36:

$$K = \frac{[H^+][S^{2-}]}{[H_2S]} = K_1 K_2 = 1.3 \times 10^{-20} \tag{36}$$

For precipitation of heavy metal ions, sulfide ions are necessary (Eq. 37):

$$Me^{2+} + S^{2-} \leftrightarrow MeS\ (or)\ 2\,Me^+ + S^{2-} \leftrightarrow Me_2S \tag{37}$$

The concentration of sulfide is pH-dependent. At acid pH only those metal sulfides of very low solubility can be precipitated. Thus, at acid pH, HgS, $As_2S_3$, CdS, CuS, and PbS form precipitates, whereas at a more alkaline pH, ZnS, FeS, NiS, and MnS form precipitates. $Al_2S_3$ and $Cr_2S_3$ are water soluble and cannot be removed by precipitation or sedimentation. Zinc removal from zinc-contaminated groundwater by microbial sulfate reduction and zinc sulfide precipitation in a 9-$m^3$ sludge blanket reactor was demonstrated and has been transferred to a full scale reactor of 1800 $m^3$ (White and Gadd, 1996).

## 1.6
### Aerobic and Anaerobic Degradation of Xenobiotics

Except for pesticides and insecticides, which are widely used in horticulture and farming and small amounts of which are washed into the groundwater, many xenobiotics are spilled in the environment, by spot contamination or contamination of large areas. Biological remediation of soil and groundwater contaminated with gasoline (e.g., Yerushalmi and Guiot, 1998) and the volatile fractions of diesel oil (e.g., Greiff et al., 1998), even at low temperatures (Margesin and Schinner, 1998), seems to be no problem. Results of many laboratory and field studies with a variety of substances are available (e.g., Arendt et al., 1995; Hinchee et al., 1995a, b; Kreysa and Wiesner, 1996). Except for studies in complex field or wastewater environments, many xenobiotics can be degraded by aerobic or anaerobic pure or defined mixed cultures (Table 1.7). These include monoaromatic and polyaromatic substances with or without chloro substituents. Whereas the two-ring compound naphthalene is relatively easily degradable by *Pseudomonas* sp. and *Rhodococcus* sp., the four-ring compounds fluoranthene and pyrene are much less degradable by *Rhodococcus* sp. (Bouchez et al., 1996). Aromatic compounds such as phenol and benzoic acid are degradable by mixed consortia at rates up to 1g $L^{-1}$ $d^{-1}$ in either the presence or absence of oxygen (Mörsen and Rehm, 1987; Knoll and Winter, 1987; Kobayashi et al., 1989). A potent population is required, which can be obtained by preincubation. Even PCP is biodegradable by a methanogenic mixed culture from UASB granules (Juteau et al., 1995; Wu et al., 1993; Kennes et al., 1996). Some pure cultures and mixed cultures are capable of dechlorinating aromatic or aliphatic compounds

(Table 1.7, second part). As for the aerobic and anaerobic degradation of phenol, the rates of dechlorination of 2-chlorophenol under aerobic and anaerobic conditions are the same order of magnitude, 102 and 128 mg $L^{-1}$ $d^{-1}$, respectively (Kafkewitz et al., 1996; Dietrich and Winter, 1990). A pure culture of *Pseudomonas pickettii* was used for aerobic dechlorination, but for anaerobic degradation a mixed culture was used. The main problem in degradation of xenobiotic compounds in wastewater is a too short residence time in the reactors, which does not allow selection or adaptation of bacteria for dechlorination or degradation. Only if a permanent pollution is prevalent, the degradation potential may develop with time. Alternatively, biofilm reactors should be used to enrich dechlorinating and xenobiotic-degrading bacteria.

**Table 1.7** Aerobic and anaerobic degradation or dechlorination of xenobiotics.

| Reaction Substance | Rate (mg $L^{-1}$ $d^{-1}$) | Microorganisms | Reference |
|---|---|---|---|
| **Aerobic degradation** | | | |
| 2-Hydroxy-benzothiazole | 138 | *Rhodococcus rhodochrous* | De Wever et al. (1997) |
| Naphthalene | 57 | *Pseudomonas* sp. *Rhodococcus* sp. | Bouchez et al. (1996) |
| Fluoranthene | 6.6 | *Rhodococcus* sp. | Bouchez et al. (1996) |
| Pyrene | 6.6 | *Rhodococcus* sp. | Bouchez et al. (1996) |
| Pyrene | 0.56 | *Mycobacterium flavescens* | Dean-Ross and Cerniglia (1996) |
| Toluene | 57 | *Pseudomonas putida* | Heald and Jenkins (1996) |
| Phenol | 188 | *Bacillus* sp. A2 | Mutzel et al. (1996) |
| Phenol | 1000 | mixed immobilized culture | Mörsen and Rehm (1987) |
| Cresol (*o, m, p*) | 259 | mixed immobilized culture | Mörsen and Rehm (1987) |
| 2,4-Diphenoxy-acetic acid | 33 | *Alcaligenes eutrophus*[a] JMP134 (pJP4)[b] | Valenzuela et al. (1997) |
| 2,4,6-Trichlorophenol | 15 | JMP134 (pJP4)[b] | Valenzuela et al. (1997) |
| **Anaerobic degradation** | | | |
| Pentachlorophenol (PCP) | 107 | methanogenic mixed culture, fixed film reactor | Juteau et al. (1995) |
| PCP | 90 | methanogenic mixed culture UASB | Wu et al. (1993) |
| PCP | 4.4 | methanogenic granules | Kennes et al. (1996) |
| PCP | 22.7 | methanogenic mixed culture | Juteau et al. (1995) |

**Table 1.7** Aerobic and anaerobic degradation or dechlorination of xenobiotics (Continued).

| Reaction Substance | Rate (mg $L^{-1}$ $d^{-1}$) | Microorganisms | Reference |
|---|---|---|---|
| Benzene | 0.029 | sulfate-reducing mixed culture | Edwards and Gribić-Galic (1992) |
| Phenol | 1000 | methanogenic mixed culture | Knoll and Winter (1987) |
| Phenol | 31 | syntrophic mixed culture | Knoll and Winter (1989) |
| Phenol | 200 | syntrophic mixed culture | Kobayashi et al. (1989) |
| Benzoic acid | 600 | syntrophic culture | Kobayashi et al. (1989) |
| Toluene | 0.1–1.5 | sulfate-reducing mixed culture | Edwards et al. (1992) |
| Toluene | 4.6 | methanogenic mixed culture | Edwards and Gribić-Galic (1994) |
| Xylene | 0.1–1.5 | sulfate-reducing mixed culture | Edwards et al. (1992) |
| Xylene | 5.3 | methanogenic mixed culture | Edwards and Gribić-Galic (1992) |
| **Aerobic Dechlorination** | | | |
| 2-Chlorophenol | 102 | *Pseudomonas pickettii* | Kafkewitz et al. (1996) |
| 4-Chlorophenol | 41 | *Pseudomonas pickettii* | Kafkewitz et al. (1996) |
| 1,3-Dichloro-2-propanol | 671 | *Pseudomonas pickettii* | Kafkewitz et al. (1996) |
| Tetrachloroethene | 35.8 total[c] | anaerobic mixed culture | Wu et al. (1995) |
| **Anaerobic Dechlorination** | | | |
| 2,6-Dichlorophenol | 38.4 | methanogenic mixed culture | Dietrich and Winter (1990) |
| 4-Chlorophenol | 0.43 | sulfate-reducing consortium | Häggblom (1998) |
| 2-Chlorophenol | 128 | methanogenic mixed culture | Dietrich and Winter (1990) |
| 2-Chlorophenol | 1.66 | anaerobic mixed culture | Kuo and Sharak Genthner (1996) |
| 3-Chlorobenzoate | 6.08 | anaerobic mixed culture | Kuo and Sharak Genthner (1996) |
| 3-Chloro-4-hydroxy-benzoate | 29.9 | *Desulfitobacterium chlororespirans* | Sanford et al. (1996) |
| Polychlorinated biphenyls (PCBs): | | | |
| 2,3,4,5,6-CB[d] | 0.24 total | methanogenic granules | Natarajan et al. (1996) |
| 2,3,4,5-CB | 0.39 3,5-CB[c] | anaerobic sediments | Berkaw et al. (1996) |
| 2,3,4,6-CB | 13.3 tri-CB[c] | anaerobic sediment | Wu et al. (1996) |
| Tetrachloroethene | 6.13 | methanogenic granules | Christiansen et al. (1997) |

**Table 1.7** Aerobic and anaerobic degradation or dechlorination of xenobiotics (Continued).

| Reaction Substance | Rate (mg L$^{-1}$ d$^{-1}$) | Microorganisms | Reference |
|---|---|---|---|
| Tetrachloroethene | 1.64 dichloro-ethene[c] | strain TT-B | Krumholz et al. (1996) |
| Tetrachloroethene | 2.05 dichloro-ethene[c] | strain MS-1 | Sharma and McCarty (1996) |
| DCB, TCB, TeCB[e] | 1.24 | methanogenic mixed culture | Middeldorp et al. (1997) |

[a] *Alcaligenes eutrophus = Ralstonia eutropha*.
[b] pJP4 = 2,4-dichlorophenoxy acetic acid-degrading plasmid.
[c] Dehalogenation product: total dehalogenation or partial dehalogenation to the corresponding dehalogenation product.
[d] 2,3,4,5,6-CB = 2,3,4,5,6-chlorinated biphenyls.
[e] DCB = dichlorobenzene, TCB = trichlorobenzene, TeCB = tetrachlorobenzene.

## 1.7
## Bioaugmentation in Wastewater Treatment Plants for Degradation of Xenobiotics

In biotechnology and pharmacology, mutants of bacteria or fungi or genetically engineered organisms are widely used in the production of citric acid, gluconic acid, ascorbate, and pharmaceuticals such as penicillin, insulin, and blood coagulation factors. Bacteria and fungi have also been adapted or genetically transformed for soil remediation (Atlas, 1981; Margesin and Schinner, 1997, Korda et al., 1997; Megharai et al., 1997). For wastewater and sludge stabilization however, successful use of genetically modified bacteria or of bacteria that can serve as donors for plasmids encoding degradative enzymes has been rather rare (van Limbergen et al., 1998). Usually, natural selection of the most suitable microorganisms from a complex flora, simply by adapting process parameters, is used. The limited reports on successful bioengineering for wastewater treatment may result from any of several factors:

- The plasmids were unstable or the genes are not expressed in the new environment.
- Inoculated strains did not survive or, if they survived, metabolic activity was too low for successful competition with autochthonous strains.
- Inoculated strains, serving as a gene source, survived, but other strains were not competent for gene transfer.
- Wastewaters normally contain a complex spectrum of carbon sources that are better than xenobiotics, and so organisms do not express genes for degrading xenobiotics.

To increase the survival potential in wastewater, a selected flora should have the desired degradative ability, tolerance to cocontaminants, and a natural spatial and temporary abundance (van der Gast et al., 2003). Another possibility would be to add organisms containing plasmids having a broad host range, permitting conjugation and DNA exchange between different species or genera of bacteria.

To enhance degradation of organic compounds in activated sludge or by a biofilm, selected specialized bacteria, genetically modified bacteria, or bacteria as plasmid donors for degradative pathways can be added (McClure et al., 1991; Frank et al., 1996). Plasmid exchange from donor to recipient cells would be advantageous to the recipient bacteria, because the plasmids harbor genes whose products function in pathways for xenobiotic degradation, which would thus extend the substrate spectrum of the recipients. The transfer of naturally occurring mercury-resistance plasmids between *Pseudomonas* strains in biofilms was shown to occur rapidly (Bale et al., 1988).

Selvaratnam et al. (1997) added a phenol-degrading strain of *Pseudomonas putida* to an activated sludge SBR reactor, which had removed 170 mg $L^{-1}$ of phenol before being augmented with the *Pseudomonas putida* strain. Whereas the original phenol-degrading activity was partially lost in the nonaugmented reactor upon further operation, the augmented reactor almost completely degraded the phenol. A more convincing approach would have been to use non-phenol–degrading activated sludge for this experiment, although the survival of the catabolic plasmid *dmpN* of *Pseudomonas putida* and its expression in the reactor biomass was demonstrated for 44 d by molecular biology techniques under steady-state conditions in the laboratory.

Successful bioaugmentation experiments in an upflow anaerobic sludge blanket reactor were reported by Ahring et al. (1992), who introduced a suspension of a pure culture of *Desulfomonile tiedjei* or a three-member consortium into an UASB reactor. They observed a rapid increase in the dehalogenation of 3-chlorobenzoate, whereas nonamended parallel incubations had no dehalogenating activity. Even after several months at 0.5-d hydraulic residence time, which was shorter than the generation time of *Desulfomonile tiedjei*, dehalogenating activity was still observed, and *Desulfomonile tiedjei* could be found within the biofilm by the use of antibody probes. More recently, an UASB reactor was supplemented with *Dehalospirillum multivorans* to improve its dehalogenating activity (Hörber et al., 1998). In contrast, Margesin and Schinner (1998) found, for biological decontamination of fuel-contaminated wastewater, that stimulation of the autochthonous flora by adding a mineral mix enhanced biodegradation to a larger extent than bioaugmentation with a cold-adapted mixed inoculum containing *Pseudomonas* sp. and *Arthrobacter* sp.

Bioaugmentation with a complex inoculum, which naturally occurs in metalwork fluids and contains strains of *Clavibacter* sp., *Rhodococcus* sp., *Methylobacterium* sp. and *Pseudomonas* sp. is very effective in degrading COD in wastewater. The augmented consortium degraded 67% of the COD (48 g $L^{-1}$) and was therefore 50%–60% more effective than the indigenous flora. In-situ analysis showed that 100 h after the introduction, the augmented consortium constituted more than 90% of the population (van der Gast et al., 2003).

Degradation of many xenobiotics in anaerobic or aerobic pure cultures or complex ecosystems has been demonstrated. Whereas for 'intrinsic sanitation' of polluted soil, the time is not limited, so long as the pollutants are adsorbed tightly to the soil matrix, degradation of xenobiotics in wastewater must be completed during the residence time in the treatment system. Thus, merely the presence of a degradation potential within a wastewater ecosystem is not sufficient, but degradation rates must be high enough, and degradation must be faster than the residence time of bacteria in suspended systems. Supplementation with microorganisms as a potential tool to increase the degradation speed or to increase the degradation potential in wastewater cannot in general be considered a state-of-the-art procedure yet.

## References

Abeliovich, A., Vonshak, A., Anaerobic metabolism of *Nitrosomonas europaea*, *Arch. Microbiol.* **1992**, *158*, 267–270.

Ahmann, D., Roberts, A. L., Krumholz, L. R., Morel, F. M. M., Microbe grows by reducing arsenic, *Nature* **1994**, *371*, 750.

Ahring, B. K., Perspectives for anaerobic digestion. In: *Advances in Biochemical Engineering/Biotechnology*, vol 81, *Biomethanation I* (Ahring, B. K., ed.), pp. 1–30. New York **2003**: Springer-Verlag.

Ahring, B. K., Christiansen, N., Mathari, I., Hendriksen, H. V., Macario, A. J. L., De-Macario, E. C., Introduction of de novo bioremediation ability, aryl reductive dechlorination, into anaerobic granular sludge by inoculation of sludge with *Desulfomonile tiedjei*, *Appl. Environ. Microbiol.* **1992**, *58*, 3677–3682.

Arendt, F., Bosmann, R., van den Brink, W. J. (eds.), *Contaminated Soil '95*, Vol. 1. Dordrecht **1995**: Kluwer.

Atlas, R. M., Microbial degradation of petroleum hydrocarbons: an environmental perspective, *Microbiol. Rev.* **1981**, *45*, 180–209.

ATV, Arbeitsbericht Fachausschuss 7.5: Geschwindigkeitsbestimmende Schritte beim anaeroben Abbau von organischen Verbindungen in Abwässern, *Korrespondenz Abwasser* **1994**, *1*, 101–107.

Bale, M. J., Day, M. J., Fry, J. C., Novel method for studying plasmid transfer in undisturbed river epilithon, *Appl. Environ. Microbiol.* **1988**, *54*, 2756–2758.

Be'guin, P., Aubert, J.-P., The biological degradation of cellulose, *FEMS Microbiol. Rev.* **1994**, *13*, 25–58.

Berkaw, M., Sowers, K. R., May, H. D., Anaerobic ortho dechlorination of polychlorinated biphenyls by estuarine sediment from Baltimore Harbor, *Appl. Environ. Microbiol.* **1996**, *62*, 2534–2539.

Bernet, N., Habouzit, F., Moletta, R., Use of an industrial effluent as a carbon source for denitrification of a high-strength wastewater, *Appl. Microbiol. Biotechnol.* **1996**, *46*, 92–97.

Blackall, L. L., Crocetti, G. R., Saunders, A. M., Bond, P. L., A review and update of the microbiology of enhanced biological phosphorus removal in wastewater treatment plants. *Antonie van Leeuwenhoek* **2002**, *81*, 681–691.

Bleicher, K., Winter, J., Formate production and utilization by methanogens and by sludge consortia: interference with the concept of interspecies formate transfer, *Appl. Microbiol. Biotechnol.* **1994**, *40*, 910–915.

Bleicher, K., Zellner, G., Winter, J., Growth of methanogens on cyclopentanol/$CO_2$ and specificity of alcohol dehydrogenase, *FEMS Microbiol. Lett.* **1989**, *59*, 307–312.

Bock, E., Koops, H.-P., Harms, H., Cell biology of nitrifying bacteria, in: *Nitrification* (Prosser, J. I., ed.), pp. 17–38. Oxford **1986**: IRL Press.

Boone, D. R., Johnson, R. L., Liu, Y., Diffusion of the interspecies electron carriers $H_2$ and formate in methanogenic ecosystems, and applications in the measurement of $K_M$ for $H_2$ and formate uptake, *Appl. Environ. Microbiol.* **1989**, *55*, 1735–1741.

Bouchez, M., Blanchet, D., Vandecasteele, J.-F., The microbial fate of polycyclic aromatic hydrocarbons: carbon and oxygen balances for bacterial degradation of model compounds, *Appl. Microbiol. Biotechnol.* **1996**, *45*, 556–561.

Brauckmann, B., Mikrobielle Extraktion von Schwermetallen aus Industrieabwässern, *Wasser Boden* **1997**, *49*, 55–58.

Broughton, M. J., Thiele, J., Birch, E. J., Cohen, A., Anaerobic batch digestion of sheep tallow, *Water Res.* **1998**, *32*, 1323–1428.

Bryant, M. P., Microbial methane production: theoretical aspects, *J. Anim. Sci.* **1979**, *48*, 193– 201.

Brynhildsen, L., Rosswall, T., Effects of metals on the microbial mineralization of organic acids, *Water Air Soil Poll.* **1997**, *94*, 45–57.

Buchauer, K., Zur Kinetik der anaeroben Hydrolyse und Fermentation von Abwasser, *Österr. Wasser- und Abfallwirtsch.* **1997**, *49*, 69–75.

Buchholz, K., Stoppock, E., Emmerich, R., Untersuchungen zur Biogasgewinnung aus Rübenpreßschnitzeln, *Zuckerindustrie* **1986**, *111*, 873–845.

Buchholz, K., Stoppock, E., Emmerich, R., Kinetics of anaerobic hydrolysis of solid material, *5th Int. Symp. Anaerobic Digestion*, Poster papers 15–18, (Tilche, A., Rozzi, A., eds.). Bologna **1988**: Monduzzi Editore.

Buswell, A. M., Mueller, H. F., Mechanism of methane fermentation, *Ind. Eng. Chem.* **1952**, *44*, 550–552.

Buswell, A. M., Neave, S. L., Laboratory studies on sludge digestion III, *State Water Surv. Bull.* **1930**, *30*, 1–84.

Buswell, A. M., Sollo, F. W., The mechanism of methane fermentation, *J. Am. Chem. Soc.* **1948**, *7*, 1778–1780.

Campos, J., Martinez-Pacheco, M., Cervantes, C., Hexavalent-chromium reduction by a chromate-resistant *Bacillus* sp. strain, *Antonie von Leeuwenhoek* **1995**, *68*, 203–208.

Christiansen, N., Christensen, S. R., Arvin, E., Ahring, B. K., Transformation of tetrachloroethene in an upflow anaerobic sludge blanket reactor, *Appl. Microbiol. Biotechnol.* **1997**, *47*, 91–94.

Colberg, P. J., Anaerobic microbial degradation of cellulose, lignin, oligolignols and monoaromatic lignin derivatives, in: *Biology of Anaerobic Microorganisms*. Wiley Series in Ecological and Applied Microbiology (Zehnder, A. J. B., ed.). New York **1988**: Wiley.

Colberg, P. J., Young, L. J., Anaerobic degradation of soluble fractions of ($^{14}$C-lignin)-lignocellulose, *Appl. Microbiol. Biotechnol.* **1985**, *49*, 345–349.

Conrad, R., Soil microorganisms as controllers of the atmospheric trace gases ($H_2$, CO, $CH_4$, OCS, $N_2O$, and NO), *Microbiol. Rev.* **1996**, *60*, 609–640.

Coughlan, M. P., Mayer, F., The cellulose-decomposing bacteria and their enzymes, in: *The Prokaryotes – A Handbook on the Biology of Bacteria: Ecophysiology, Isolation, Identification, Applications*, (Vol. 1) (Ballows, A., Trüper, H. G., Dworkin, M., Harder, W., Schleifer, K.-H., eds.), pp. 461–516. New York **1991**: Springer-Verlag.

Danesh, S., Oleszkiewicz, J. A., Volatile fatty acid production and uptake in biological nutrient removal systems with process separation, *Water Environ. Res.* **1997**, *69*, 1106–1111.

Dean-Ross, D., Cerniglia, C. E., Degradation of pyrene by *Mycobacterium flavescens*, *Appl. Microbiol. Biotechnol.* **1996**, *46*, 307–312.

De Bok, F. A. M., Plugge, C. M., Stams, A. J. M., Interspecies electron transfer in methanogenic propionate degrading consortia, *Water Res.* **2004**, *38*, 1369–1375.

De Wever, H., De Cort, S., Noots, I., Verachtert, H., Isolation and characterization of *Rhodococcus rhodochrous* for the degradation of the wastewater component 2-hydroxybenzothiazole, *Appl. Microbiol. Biotechnol.* **1997**, *47*, 458–461.

Dietrich, G., Winter, J., Anaerobic degradation of chlorophenol by an enrichment culture, *Appl. Microbiol. Biotechnol.* **1990**, *34*, 253–258.

Dolfing, J., The microbial logic behind the prevalence of incomplete oxidation of organic compounds by acetogenic bacteria in methanogenic environments, *Microb. Ecol.* **2001**, *41*, 83–89.

Edwards, E. A., Gribić-Galic, D., Complete mineralization of benzene by aquifer microorganisms under strictly anaerobic conditions, *Appl. Environ. Microbiol.* **1992**, *58*, 2663–2666.

Edwards, E. A., Gribić-Galic, D., Anaerobic degradation of toluene and o-xylene by a methanogenic consortium, *Appl. Environ. Microbiol.* **1994**, *60*, 313–322.

Edwards, E. A., Wills, L. E., Reinhard, M., Gribić-Galic, D. Anaerobic degradation of toluene and xylene by aquifer microorganisms and sulfate-reducing conditions, *Appl. Environ. Microbiol.* **1992**, *58*, 794–800.

Egli, T., Zehnder, A. J. B., Phosphate and nitrate removal, *Curr. Opin. Biotechnol.* **1994**, *5*, 275– 284.

Elferink, S. J. W. H. O., Visser, A., Pol, L. W. H., Stams, A. J. M., Sulfate reduction in methanogenic bioreactors, *FEMS Microbiol. Rev.* **1994**, *15*, 119–136.

Filipe, C. D. M., Daigger, G. T., Grady, C. P. L., pH as a key factor in the competition between glycogen-accumulating and phosphate-accumulating organisms, *Water Environ. Res.* **2001**, *73*, 223–232.

Frank, N., Simao-Beaunoir, A. M., Dollard, M. A., Bauda, P., Recombinant plasmid DNA mobilization by activated sludge strains grown in fixed-bed or sequenced-batch reactors, *FEMS Microbiol. Ecol.* **1996**, *21*, 139–148.

Fritsche, W., *Umwelt-Mikrobiologie.* Jena **1998**: Gustav Fischer Verlag.

Gensicke, R., Merkel, K., Schuch, R., Winter J., Biologische Behandlung von Permeaten aus der Ultrafiltration zusammen mit nitrathaltigen Abwässern aus der elektrochemischen Entgratung in submersen Festbettreaktoren, *Korrespondenz Abwasser* **1998**, *1/98*, 86–91.

Greiff, K., Leidig, E., Winter, J., Biologische Aufbereitung eines BTEX-belasteten Grundwassers in einem Festbettreaktor, *Acta Hydrochim. Hydrobiol.* **1998**, *26*, 95–103.

Häggblom, M. M., Reductive dechlorination of halogenated phenols by a sulfate-reducing consortium, *FEMS Microbiol. Ecol.* **1998**, *26*, 35–41.

Hammes, W., Winter, J., Kandler, O., The sensitivity of pseudomurein-containing genus *Methanobacterium* to inhibitors of murein synthesis, *Arch. Microbiol.* **1979**, *123*, 275–279.

Hanaki, K., Matsuo, T., Nagase, M., Mechanism of inhibition caused by long-chain fatty acids in anaerobic digestion process, *Biotechnol. Bioeng.* **1981**, *23*, 1591–1610.

Hansen, T. A., Metabolism of sulfate-reducing prokaryotes, *Antonie van Leeuwenhoek* **1994**, *66*, 165–185.

Heald, S. C., Jenkins, R. O., Expression and substrate specificity of the toluene dioxygenase of *Pseudomonas putida* NCIMB 11767, *Appl. Microbiol. Biotechnol.* **1996**, *45*, 56–62.

Hellinga, C., Schellen, A. A. J. C., Mulder, J. W., Van Loosdrecht, M. C. M., Heijnen, J. J., The Sharon process: an innovative method for nitrogen removal from ammonium-rich waste water, *Water Sci. Technol.* **1998**, *37*, 135–142.

Henze, M., Harremoes, P., Jansen, J., Arvin, E., *Wastewater treatment: Biological and Chemical Processes*, 2nd edit. Heidelberg **1997**: Springer-Verlag.

Hilpert, R., Winter, J., Hammes, W., Kandler, O., The sensitivity of archaebacteria to antibiotics, *Zentralbl. Bakteriol. Mikrobiol. Hyg.* (Abt. 1 Orig. C) **1981**, *2*, 11–20.

Hinchee, R. E., Müller, R. N., Johnson, P. C. (eds ), *Bioremediation 3, Vol. 2: In-Situ Aeration, Air Sparging, Bioventing and Related Processes*. Columbus, OH **1995a**: Battelle Press.

Hinchee, R. E., Vogel, C. M., Brockman, F. J. (eds.), *Bioremediation 3, Vol. 8: Microbial Processes for Bioremediation*. Columbus, OH **1995b**: Battelle Press.

Hörber, C., Christiansen, N., Arvin, E., Ahring, B. K., Improved dechlorination performance of upflow anaerobic sludge blanket reactors by incorporation of *Dehalospirillum multivorans* into granular sludge, *Appl. Environ. Microbiol.* **1998**, *64*, 1860–1863.

Ianotti, E. L., Kafkewitz, P., Wolin, M. J., Bryant, M. P., Glucose fermentation products of *Ruminococcus albus* grown in continuous culture with *Vibrio succinogenes:* changes caused by interspecies transfer of hydrogen, *J. Bacteriol.* **1973**, *114*, 1231–1240.

Juteau, P., Beaudet, R., McSween, G., Lépine, F., Milot, S., Bisaillon, J.-G., Anaerobic biodegradation of pentachlorophenol by a methanogenic consortium, *Appl. Microbiol. Biotechnol.* **1995**, *44*, 218–224.

Kafkewitz, D., Fava, F., Armenante, P. M., Effect of vitamins on the aerobic degradation of 2-chlorophenol, 4-chlorophenol, and 4-chlorobiphenyl, *Appl. Microbiol. Biotechnol.* **1996**, *46*, 414–421.

Kennes, C., Wu, W.-M., Bhathnagar, L., Zeikus, J. G., Anaerobic dechlorination and mineralization of pentachlorophenol and 2,4,6-trichlorophenol by methanogenic pentachlorophenol degrading granules, *Appl. Microbiol. Biotechnol.* **1996**, *44*, 801–806.

Kern-Jespersen, J. P., Henze, M., Biological phosphorus uptake under anoxic and aerobic conditions, *Water Res.* **1993**, *27*, 617–624.

Knoll, G., Winter, J., Anaerobic degradation of phenol in sewage sludge: benzoate formation from phenol and $CO_2$ in the presence of hydrogen, *Appl. Microbiol. Biotechnol.* **1987**, *25*, 384–391.

Knoll, G., Winter, J., Degradation of phenol via carboxylation to benzoate by a defined, obligate syntrophic consortium of anaerobic bacteria, *Appl. Microbiol. Biotechnol.* **1989**, *30*, 318–324.

Knowles, R., Denitrification, *Microbiol. Rev.* **1982**, *46*, 43–70.

Kobayashi, T., Hashinage, T., Mikami, E., Suzuki, T., Methanogenic degradation of phenol and benzoate in acclimated sludges, *Water Sci. Technol.* **1989**, *21*, 55–65.

Korda, A., Santas, P., Tenente, A., Santas, R., Petroleum hydrocarbon bioremediation: sampling and analytical techniques, in situ treatments and commercial microorganisms currently used, *Appl. Microbiol. Biotechnol.* **1997**, *48*, 677–686.

Krause D. O., Denman S. E., Mackie R. I., Morrison M., Rae A. L., Attwood G. T., McSweeney C. S., Opportunities to improve fiber degradation in the rumen: microbiology, ecology, and genomics, *FEMS Microbiology Reviews* **2003**, *27*, 663–693.

Kreysa, G., Wiesner, J. (eds.), *In-situ-Sanierung von Böden*, Resümee und Beiträge des 11. Dechema-Fachgespräches Umweltschutz. Frankfurt am Main **1996**: Dechema, Deutsche Gesellschaft für Chemisches Apparatewesen.

Kristjansson, J. K., Schönheit, P., Thauer, R. K., Different $K_s$ values for hydrogen of methanogenic bacteria and sulfate reducing bacteria: an explanation for the apparent inhibition of methanogenesis by sulfate, *Arch. Microbiol.* **1982**, *131*, 278–282.

Krumholz, L. R., Sharp, R., Fishbain, S. S., A freshwater anaerobic coupling acetate oxidation to tetrachloroethene dehalogenation, *Appl. Environ. Microbiol.* **1996**, *62*, 4108–4113.

Kuai, L., Verstraete, W., Ammonium removal by the oxygen-limited autotrophic nitrification–denitrification system, *Appl. Environ. Microbiol.* **1998**, *64*, 4500–4506.

Kuba, T., Wachtmeister, A., Van Loosdrecht, M. C. M., Heijnen, J. J., Effect of nitrate on phosphorus release in biological phosphorus removal systems, *Water Sci. Technol.* **1994**, *30*, 263–269.

Kuenen, J. G., Robertson, L. A., Combined nitrification–denitrification processes, *FEMS Microbiol. Rev.* **1994**, *15*, 109–117.

Kuo, C.-W., Sharak Genthner, B. R., Effect of added heavy metal ions on biotransformation and biodegradation of 2-chlorophenol and 3-chlorobenzoate in anaerobic bacterial consortia, *Appl. Environ. Biotechnol.* **1996**, *62*, 2317–2323.

Lamed, R., Bayer, E. A., The cellulosome of *Clostridium thermocellum*, *Adv. Appl. Microbiol.* **1988**, *33*, 1–46.

Langenhoff, A. A. M., Brouwers-Ceiler, D. L., Engelberting, J. H. L., Quist, J. J., Wolkenfelt, J. G. P. N. et al., Microbial reduction of manganese coupled to toluene oxidation, *FEMS Microbiol. Ecol.* **1997**, *22*, 119–127.

Lee, M. J., Zinder, S. H., Isolation and characterisation of a thermophilic bacterium which oxidizes acetate in synthrophic association with a methanogen and which grows acetogenically on $H_2CO_2$, *Appl. Environ. Microbiol.* **1988**, *54*, 124–129.

Lens P., Vallero M., Esposito G., Zandvoort M., Perspectives of sulfate reducing bioreactors in environmental biotechnology, *Re/Views in Environmental Science and Technology* **2002**, *1*, 311–325.

Leschine, S. B., Cellulose degradation in anaerobic environments, *Annu. Rev. Microbiol.* **1995**, *49*, 399–426.

Lloyd, J. R., Microbial reduction of metals and radionuclides, *FEMS Microb. Rev.* **2003**, *27*, 411–425.

Lovley, D. R., Dissimilatory Fe(III) and Mn(IV) reduction, *Microbiol. Rev.* **1991**, *55*, 259–387.

Lovley, D. R., Bioremediation of organic and metal contaminants with dissimilatory metal reduction, *J. Ind. Microbiol.* **1995**, *14*, 85–93.

Lovley, D. R., Coates, J. D., Bioremediation of metal contamination, *Curr. Opin. Biotechnol.* **1997**, *8*, 285–289.

Lovley, D. R., Lonergan, D. J., Anaerobic oxidation of toluene, phenol and *p*-cresol by the dissimilatory iron-reducing organism GS-15, *Appl. Environ. Microbiol.* **1990**, *56*, 1858–1864.

Lovley, D. R., Phillips, E. J. P., Reduction of chromate by *Desulfovibrio vulgaris* (Hildenborough) and its $C_3$ cytochrome, *Appl. Environ. Microbiol.* **1994**, *60*, 726–728.

Lovley, D. R., Phillips, E. J. P., Gorby, Y. A., Landa, E. R., Microbial reduction of uranium, *Nature* **1991**, *350*, 413–416.

Lovley, D. R., Phillips, E. J. P., Lonergan, D. J., Widman, P. K., Fe(III) and $S^0$ reduction by *Pelobacter carbinolicus*, *Appl. Environ. Microbiol.* **1995**, *61*, 2132–2138.

Lui, Y., Energy uncoupling in microbial growth under substrate-sufficient conditions, *Appl. Microbiol. Biotechnol.* **1998**, *49*, 500–505.

Macy, J. M., Lawson, S., DeMoll-Decker, H., Bioremediation of selenium oxyanions in San Joaquin drainage water using *Thauera se-*

*lenatis* in a biological reactor system, *Appl. Microbiol. Biotechnol.* **1993**, *40*, 588–594.

Margesin, R., Schinner, F., Bioremediation of diesel-oil–contaminated alpine soils at low temperatures, *Appl. Microbiol. Biotechnol.* **1997**, *47*, 462–468.

Margesin, R., Schinner, F., Low-temperature bioremediation of a waste water contaminated with anionic surfactants and fuel oil, *Appl. Microbiol. Biotechnol.* **1998**, *49*, 482–486.

McCartney, D. M., Oleszkiewicz, J. A., Sulfide inhibition of anaerobic degradation of lactate and acetate, *Water Res.* **1991**, *25*, 203–209.

McClure, N. C., Weightman, A. J., Fry, J. C., Survival and catabolic activity of natural and genetically engineered bacteria in a laboratory-scale activated sludge unit, *Appl. Environ. Microbiol.* **1991**, *57*, 366–373.

McInerney, M. J., Anaerobic hydrolysis and fermentation of fats and proteins, in: *Biology of Anaerobic Microorganisms* (Zehnder, A. J. B.. ed.), pp. 373–415. New York **1988**: Wiley.

Megharaj, M., Wittich, R.-M., Blasco, R., Pieper, D. H., Timmis, K. N., Superior survival and degradation of dibenzo-*p*-dioxin and dibenzofuran in soil by soil-adapted *Sphingomonas* sp. strain RW1, *Appl. Microbiol. Biotechnol.* **1997**, *48*, 109–114.

Middeldorp, P. J. M., de Wolf, J., Zehnder, A. J. B., Schraa, G., Enrichment and properties of a 1,2,4-trichlorobenzene–dechlorinating methanogenic microbial consortium, *Appl. Environ. Microbiol.* **1997**, *63*, 1225–1229.

Mörsen, A., Rehm, H. J., Degradation of phenol by a mixed culture of *Pseudomonas putida* and *Cryptococcus elinovii* adsorbed on activated carbon, *Appl. Microbiol. Biotechnol.* **1987**, *26*, 283–288.

Mulder, J. W., Van Loosdrecht, M. C. M., Hellinga, C., Van Kempen, R., Full-scale application of the Sharon process for treatment of rejection water of digested sludge dewatering, *Water Sci. Technol.* **2001**, *43*, 127–134.

Mutzel, A., Reinscheid, U. M., Antranikian, G.. Müller, R., Isolation and characterization of a thermophilic *Bacillus* strain that degrades phenol and cresols as sole carbon source at 70 °C, *Appl. Microbiol. Biotechnol.* **1996**, *46*, 593–596.

Natarajan, M. R., Wu, W.-M., Nye, J., Wang, H., Bhatnagar, L., Jain, M. K., Dechlorination of polychlorinated biphenyl congeners by an anaerobic microbial consortium, *Appl. Microbiol. Biotechnol.* **1996**, *46*, 673–677.

Oleszkiewicz, J. A., Mastaller, T., McCartney, D. M., Effects of pH on sulfide toxicity to anaerobic processes, *Environ. Technol. Lett.* **1989**, *10*, 815–822.

Omil, F., Lens, P., Visser, A., Hulshoff, L. W., Lettinga, G., Long-term competition between sulfate reducing and methanogenic bacteria in UASB reactors treating volatile fatty acids, *Biotechnol. Bioeng.* **1998**, *57*, 676–685.

Oremland, R. W., Hollibaugh, J. T., Maest, A. S., Presser, T. S., Miller, L. G., Culbertson, C. W., Selenate reduction to elemental selenium by anaerobic bacteria in sediments and culture: biogeochemical significance of a novel, sulfate-independent respiration, *Appl. Environ. Microbiol.* **1989**, *55*, 2333–2343.

Örlygsson, J., Houwen, F. P., Svensson, B. H., Thermophilic anaerobic amino acid degradation: deamination rates and end-product formation, *Appl. Microbiol. Biotechnol.* **1995**, *43*, 235–241.

Pynaert, K., Barth, F., Smets, F., Beheydt, D., Verstraete, W., Start-up of autotrophic nitrogen removal reactors via sequential biocatalyst addition, *Environ. Sci. Technol.* **2004**, *38*, 1228–1235.

Rheinheimer, G., Hegemann, W., Raff, J., Sekoulov, I., *Stickstoffkreislauf im Wasser.* München **1988**: Oldenbourg Verlag.

Robertson, L. A., Kuenen, J. G., Aerobic denitrification: a controversy revived, *Arch. Microbiol.* **1984**, *139*, 351–354.

Robertson, L. A., van Niel, W. W. J., Torremans, R. A. M., Kuenen, J. G., Simultaneous nitrification and denitrification in aerobic chemostat cultures of *Thiosphera pantotropha*, *Appl. Environ. Microbiol.* **1988**, *54*, 2812–2818.

Robinson, J. B., Tuovinen, O. H., Mechanisms of microbial resistance and detoxification of mercury and organomercury compounds: physiological, biochemical, and genetic analyses, *Microbiol. Rev.* **1984**, *48*, 95–124.

Sanford, R. A., Cole, J. R., Löffler, F. E., Tiedje, J. M., Characterization of *Desulfitobacterium chlororespirans* sp. nov., which grows by coupling the oxidation of lactate to the reductive dechlorination of 3-chloro-4-hydroxybenzoate, *Appl. Environ. Biotechnol.* **1996**, *62*, 3800–3808.

Schink, B., Energetics of syntrophic cooperation in methanogenic degradation, *Microbiol. Mol. Biol. Rev.* **1997**, *61*, 262–280.

Schlegel, H. G., *Allgemeine Mikrobiologie*, 7th
edit. Stuttgart **1992**: Georg Thieme Verlag.

Schmidt, I., Sliekers, O., Schmid, M., Cirpus,
I., Strous, M., Bock, E., Kuenen, J. G., Jetten, M. S. M., Aerobic and anaerobic ammonia oxidizing bacteria: competitors or
natural partners? *FEMS Microbiol. Ecol.*
**2002**, *39*, 175–181.

Schmidt, I., Sliekers, O., Schmid, M., Bock,
E., Fuerst, J., Kuenen, J. G., Jetten, M. S.
M., Strous, M., New concepts of microbial
treatment processes for the nitrogen removal in wastewater, *FEMS Microbiol. Rev.*
**2003**, *27*, 481–492.

Schnuerer, A., Schink, B., Svensson, B. H.,
*Clostridium ultunense* sp. nov., a mesophilic
bacterium oxidizing acetate in syntrophic
association with a hydrogenotrophic methanogenic bacterium, *Int. J. Syst. Bacteriol.*
**1996**, *46*, 1145–1152.

Schön, G., Jardin N., Biological and chemical
phosphorus removal. In: *Biotechnology, Environmental Processes I* (H.-J. Rehm, G. Reed,
A. Pühler, P. Stadler, series eds.), Volume
11A (J. Winter ed.) **1999**, pp. 285–319 .

Schön, G., Bußmann, M., Geywitz-Hetz, S.,
Bildung von Lachgas (N$_2$O) im belebten
Schlamm aus Kläranlagen, *GWF Wasser
Abwasser* **1994**, *135*, 293–301.

Selvaratnam, C., Schoedel, B. A., McFarland,
B. L., Kulpa, C. F., Application of the poly-
merase chain reaction (PCR) and the reverse transcriptase/PCR for determining
the fate of phenol-degrading *Pseudomonas
putida* ATCC 11172 in a bioaugmented sequencing batch reactor, *Appl. Microbiol. Biotechnol.* **1997**, *47*, 236–240.

Seviour, R. J., Mino, T., Onuki, M., The microbiology of biological phosphorus removal in activated sludge systems, *FEMS Microbiol. Rev.* **2003**, *27*, 99–127.

Sharma, P. K., McCarty, P. L., Isolation and
characterization of a facultatively aerobic
bacterium that reductively dehalogenates
tetrachloroethene to *cis*-1,2-dichloroethene,
*Appl. Environ. Microbiol.* **1996**, *62*, 761–765.

Shin, H.-S., Song, Y.-C., A model for evaluation of anaerobic degradation characteristics of organic waste: focusing on kinetics,
rate-limiting step, *Environ. Technol.* **1995**,
*16*, 775–784.

Smith, D. G., Tellurite reduction in *Schizosac-
charomyces pombe*, *J. Gen. Microbiol.* **1974**, *83*,
389–392.

Smolders, G. J. F., Van der Meij, J., van Loosdrecht, M. C. M., Heijnen, J. J., Model of the
anaerobic metabolism of the biological phosphorus removal process; stoichiometry and
pH influence, *Biotechnol. Bioeng.* **1994**, *43*,
461–470.

Smolders, G. J. F., van Loosdrecht, M. C. M.,
Heijnen, J. J., Steady state analysis to evaluate the phosphate removal capacity and acetate requirement of biological phosphorus removing mainstream and sidestream process
configurations, *Water Res.* **1996**, *30*,
2748–2760.

Speece, R. E., Parkin, G. F., Bhattacharya, S.,
Takashima, S., Trace nutrient requirements
of anaerobic digestion, in: *Proc. EWPCA
Conf. Anaerobic Treatment, a Grown Up Tech-
nology*, Amsterdam, Industrial Presentations
(Europe) B.V.'s-Gravelandseweg 284–296,
Schiedam, The Netherlands **1986**, pp.
175–188.

Stams, A. J. M., Metabolic interactions
between anaerobic bacteria in methanogenic
environments, *Antonie von Leeuwenhoek*
**1994**, *66*, 271–294.

Stouthamer, A. H., de Boer, A. P. N., van der
Oost, J., van Spanning, R. J. M., Emerging
principles of inorganic nitrogen metabolism
in *Paracoccus denitrificans* and related bacteria, *Antonie von Leeuwenhoek* **1997**, *71*, 33–41.

Straub, K. L., Benz, M., Schink, B., Widdel, F.,
Anaerobic, nitrate-dependent microbial oxidation of ferrous iron, *Appl. Environ. Microbiol.* **1996**, *62*, 1458–1460.

Strous, M., Van Gerven, E., Zheng, P., Kuenen, J. G., Jetten, M. S. M., Ammonium removal from concentrated waste streams with
the anaerobic ammonium oxidation (Ana-
mox) process in different reactor configurations, *Water Res.* **1997**, *8*, 1955–1962.

Thauer, R. K., Jungermann, K., Decker, K., Energy conservation in chemotrophic anaerobic
bacteria, *Bacteriol. Rev.* **1977**, *41*, 100–180.

Thiele, J. H., Chartrain, M., Zeikus, J. G., Control of interspecies electron flow during anaerobic digestion: role of the floc formation,
*Appl. Environ. Microbiol.* **1988**, *54*, 10–19.

Unz, R. F., Shuttleworth, K. L., Microbial
mobilization and immobilization of heavy
metals, *Curr. Opin. Biotechnol.* **1996**, *7*,
307–310.

Valenzuela, J., Bumann, U., Cespedes, R., Padilla, L., Gonzalez, B., Degradation of chlorophenols by *Alcaligenes eutrophus* JMP134 (pJP4) in bleached kraft mill effluent, *Appl. Environ. Microbiol.* **1997**, *63*, 227–232.

Van de Graaf, A. A., Mulder, A., de Bruijn, P., Jetten, M. S. M., Robertson, L. A., Kuenen, J. G., Anaerobic oxidation of ammonium is a biologically mediated process, *Appl. Environ. Microbiol.* **1995**, *61*, 1246–1251.

Van de Graaf, A. A., de Bruijn, P., Robertson L. A., Jetten, M. S. M., Kuenen, J. G., Autotrophic growth of anaerobic, ammonium-oxidizing microorganisms in a fluidized bed reactor, *Microbiology* **1996**, *142*, 2187–2196.

Van de Graaf, A. A., de Bruijn, P., Robertson, L. A., Jetten, M. S. M., Kuenen, J. G., Metabolic pathway of anaerobic ammonium oxidation on the basis of $^{15}$N studies in a fluidized bed reactor, *Microbiology* **1997**, *143*, 2415–2421.

Van Der Gast, C. J., Whiteley, A. S., Starkey, M., Knowles, C. J., Thompson, I. P., Bioaugmentation strategies for remediating mixed chemical effluents, *Biotechnol. Prog.* **2003**, *19*, 1156–1161.

Van Ginkel, C. G., Plugge, C. M., Stroo, C. A., Reduction of chlorate with various energy substrates and inocula under anaerobic conditions, *Chemosphere* **1995**, *31*, 4057–4066.

Van Limbergen, H., Top, E. M., Verstraete, W., Bioaugmentation inactivated sludge: current features and future perspectives, *Appl. Microbiol. Biotechnol.* **1998**, *50*, 16–23.

Van Loosdrecht, M. C. M., Jetten, M. S. M., Method for treating ammonia-containing wastewater, Patent PCT/NL97/00482 **1997**.

Van Loosdrecht, M. C. M., Smolders, G. J., Kuba, T., Heijnen, J. J., Metabolism of microorganisms responsible for enhanced biological phosphorus removal from wastewater, *Antonie van Leeuwenhoek* **1997a**, *71*, 109–116.

Van Loosdrecht, M. C. M., Hooijmans, C. M., Brdjanovic, D., Heijnen, J. J., Biological phosphate removal processes, *Appl. Microbiol. Biotechnol.* **1997b**, *48*, 289–296.

Vavilin, V. A., Rytov, S. V., Lokshina, L. Y., Two-phase model of hydrolysis kinetics and its application to anaerobic degradation of particulate organic matter, *Appl. Biochem. Biotechnol.* **1997**, *63–65*, 45–58.

Wagner, M., Erhart, R., Manz, W., Amman, R., Lemmer, H. et al., Development of an rRNA-targeted oligonucleotide probe specific for the genus *Acinetobacter* and its application for in situ monitoring in activated sludge, *Appl. Environ. Microbiol.* **1994**, *60*, 792–800.

Wallace, W., Ward, T., Breen, A., Attaway, H., Identification of an anaerobic bacterium which reduces perchlorate and chlorate as *Wolinella succinogenes*, *J. Ind. Microbiol.* **1996**, *16*, 68–72.

Warren, R. A. J., Microbial hydrolysis of polysaccharides, *Annu. Rev. Microbiol.* **1996**, *50*, 183–212.

White, C., Gadd, G. M., Mixed sulphate-reducing bacterial cultures for bioprecipitation of toxic metals: factorial and response-surface analysis of the effects of dilution rate, sulphate and substrate concentration, *Microbiology* **1996**, *142*, 2197–2205.

White, C., Sayer, J. A., Gadd, G. M., Microbial solubilization and immobilization of toxic metals: key biogeochemical processes for treatment of contamination, *FEMS Microbiol. Rev.* **1997**, *20*, 503–516.

Whitman, W. B., Bowen, T. L., Boone, D. R., The methanogenic bacteria, in: *The Prokaryotes* (Balows, A., Trüper, H. G., Dworkin, M., Harder, W., Schleifer, K.-H. (eds.), pp. 719–767. New York **1992**: Springer-Verlag.

Widdel, F., Growth of methanogenic bacteria in pure culture with 2-propanol and other alcohols as hydrogen donors, *Appl. Environ. Microbiol.* **1986**, *51*, 1056–1062.

Widdel, F., Microbiology and ecology of sulfate- and sulfur-reducing bacteria, in: *Biology of Anaerobic Microorganisms* (Zehnder, A. J. B., ed.), pp. 469–585. New York **1988**: Wiley.

Wild, H. E., Sanyer, C. N., McMahon, T. C., Factors affecting nitrification kinetics, *J. Water Pollut. Control Fed.* **1971**, *43*, 1845–1854.

Wildenauer, F. X., Winter, J., Anaerobic digestion of high-strength acidic whey in a pH-controlled up-flow fixed film loop reactor, *Appl. Microbiol. Biotechnol.* **1985**, *22*, 367–372.

Wildenauer, F. X., Winter, J., Fermentation of isoleucine and arginine by pure and syntrophic cultures of *Clostridium sporogenes*, *FEMS Microbiol. Ecol.* **1986**, *38*, 373–379.

Winter, J., Energie aus Biomasse, *Umschau* **1983**, *25*, 26, 774–779.

Winter, J., Anaerobic waste stabilization, *Biotechnol. Adv.* **1984**, *2*, 75–99.

Winter, J. U., Cooney, C. L., Fermentation of cellulose and fatty acids with enrichments from sewage sludge, *Eur. J. Appl. Microbiol. Biotechnol.* **1980**, *11*, 60–66.

Winter, J., Zellner, G., Thermophilic anaerobic degradation of carbohydrates: metabolic properties of microorganisms from the different phases, *FEMS Microbiol. Rev.* **1990**, *75*, 139–154.

Winter, J., Schindler, F., Wildenauer, F., Fermentation of alanine and glycine by pure and syntrophic cultures of *Clostridium sporogenes*, *FEMS Microbiol. Ecol.* **1987**, *45*, 153–161.

Winter, J., Knoll, G., Sembiring, T., Vogel, P., Dietrich, G., Mikrobiologie des anaeroben Abbaus von Biopolymeren und von aromatischen und halogenaromatischen Verbindungen, in: *Biogas: Anaerobtechnik in der Abfallwirtschaft* (Thomé-Kozmiensky, K. J., ed.). Berlin **1989**: EF-Verlag für Energie und Umweltschutz.

Winter, J., Hilpert, H., Schmitz, H., Treatment of animal manures and wastes for ultimate disposal: review, *Asian Aust. J. Anim. Sci.* **1992**, *5*, 199–215.

Winterberg, R., Sahm, H., Untersuchungen zum anaeroben Proteinabbau bei der zweistufigen anaeroben Abwasserreinigung, *Lecture:* DECHEMA Arbeitsausschuß Umweltbiotechnologie, January **1992**.

Wolin, M. J., Interactions between $H_2$-producing and methane-producing species, in: *Microbial Formation and Utilization of Gases ($H_2$, $CH_4$, $CO$)* (Schlegel, H. G., Gottschalk, G., Pfennig, N., eds.), pp. 14–15. Göttingen **1976**: Göltze.

Wolin, M. J., Hydrogen transfer in microbial communities, in: *Microbial Interactions and Communities*, Vol. 1 (Bull, A. T., Slater, J. H., eds.), pp. 323–356. London **1982**: Academic.

Wu, M.-W., Nye, J., Hickey, R. F., Jain, M. K., Zeikus, J. G., Dechlorination of PCE and TCE to ethene using anaerobic microbial consortium, in: *Bioremediation of Chlorinated Solvents* (Hinchee, R. E., Leeson, A., Semprini, L., eds.), pp. 45–52. Columbus, OH **1995**: Battelle Press.

Wu, Q., Bedard, D. L., Wiegel, J., Influence of incubation temperature on the microbial reductive dechlorination of 2,3,4,6-tetrachlorobiphenyl in two freshwater sediments, *Appl. Environ. Microbiol.* **1996**, *62*, 4174–4179.

Wu, W. M., Bhatnagar, L., Zeikus, J. G., Performance of anaerobic granules for degradation of pentachlorophenol, *Appl. Environ. Microbiol.* **1993**, *59*, 389–397.

Yerushalmi, L., Guiot, S. R., Kinetics of biodegradation of gasoline and its hydrocarbon constituents, *Appl. Microbiol. Biotechnol.* **1998**, *49*, 475–481.

Zayed, G., Winter, J., Removal of organic pollutants and of nitrate from wastewater from the dairy industry by denitrification, *Appl. Microbiol. Biotechnol.* **1998**, *49*, 469–474.

Zellner, G., Winter, J., Secondary alcohols as hydrogen donors for $CO_2$ reduction by methanogens, *FEMS Microbiol. Lett.* **1987a**, *44*, 323–328.

Zellner, G., Winter, J., Analysis of a highly efficient methanogenic consortium producing biogas from whey, *Syst. Appl. Microbiol.* **1987b**, *9*, 284–292.

Zellner, G., Bleicher, K., Braun, E., Kneifel, H., Tindall, B. J. et al., Characterization of a new mesophilic, secondary alcohol-utilizing methanogen, *Methanobacterium palustre* spec. nov. from a peat bog, *Arch. Microbiol.* **1989**, *151*, 1–9.

Zumft, W. G., Cell biology and molecular basis of denitrifications, *Microbiol. Mol. Biol. Rev.* **1997**, *61*, 533–616.

# 2
# Industrial Wastewater Sources and Treatment Strategies

Karl-Heinz Rosenwinkel, Ute Austermann-Haun, and Hartmut Meyer

## 2.1
## Introduction and Targets

This chapter deals with the wastewater flows discharged from industrial plants and offers a synopsis of the applicable treatment methods.

The central topic of this book being biotechnology, the main emphasis of this chapter is on industries emitting wastewater with organic pollutants, since these can be treated biologically. Next, the possible wastewater flow fractions occurring in industrial plants are listed. According to each type of industry and individual local conditions, certain wastewater flow fractions do not occur, are disposed of, or are treated in a different manner. Then, various wastewater pollutants are investigated with regard to their direct importance for certain industries. This is followed by a typical treatment sequence for wastewater. In the main section, wastewater composition and possible treatment strategies for the individual industries are examined, with particular emphasis on the most important branches of the food processing industry.

We should note that there are not only substantial differences between the various industrial branches, but that even within one branch wastewater composition and appropriate treatment methods are determined by the following factors:
- production methods
- water supply and water processing
- technical condition and age of the production site
- training and motivation of employees
- use certain additives and cleaning agents, etc.
- number of shifts, seasonal differences (campaign operation)
- effluent requirements (direct or indirect discharge)
- extent of production-integrated environmental protection means
- number of wastewater treatment facilities

*Environmental Biotechnology. Concepts and Applications.* Edited by H.-J. Jördening and J. Winter
Copyright © 2005 WILEY-VCH Verlag GmbH & Co. KGaA, Weinheim
ISBN: 3-527-30585-8

## 2.2
**Wastewater Flow Fractions from Industrial Plants**

### 2.2.1
**Synopsis**

Due to the multitude of products and production methods in industrial plants, there is also a wide range of different wastewater flow fractions for which the respective industrial branches have coined their own particular terms (e.g., 'singlings' is a term used only in the distillery business). The wastewater flow fractions listed below represent the most important wastewater sources, without each fraction being necessarily produced by each single branch:
- rainwater
- wastewater from sanitary and employee facilities
- cooling water
- wastewater from in-plant water preparation
- production wastewater (the following flow fractions are in some branches considered part of the production water):
  - wash and flume water
  - fruit water
  - condensates
  - cleaning water

### 2.2.2
**Rainwater**

Rainwater can periodically constitute a substantial proportion of wastewater, depending on the amount of sealed surface. It is important to differentiate between nonpolluted to slightly polluted rainwater, such as that coming from roofs, which should be percolated or can be discharged directly into the storm sewer or waterways, and rainwater that is collected from sealed areas where product handling or vehicles contribute to contamination of rainwater, making treatment of rainwater necessary.

The amount of rainwater $Q_R$ (in L s$^{-1}$) can be calculated by the following formula:

$$Q_R = r_{D,n} \cdot \Psi_S \cdot A_E \tag{1}$$

where $A_E$ (in ha) is the catchment area, i.e., all areas that channel runoff into the rainwater canal in the event of rainfall. $\Psi_S$ (dimensionless) is the runoff coefficient, which indicates the percentage of fallen rainwater that actually increases the runoff flow (i.e., the portion that does not percolate or evaporate or is not held back by low-lying areas). This amount varies greatly, depending on the surface conditions of the catchment area. For sealed industrial areas it is safe to use a runoff coefficient $\Psi_S = 1$. $r_{D,n}$ represents the rainfall intensity in L s$^{-1}$ ha$^{-1}$. For a 10-min rainstorm (with a sewer flow time of >10 min, the standard design rainfall intensity), which in this intensity occurs only once a year, this value amounts in Germany to 100–200 L s$^{-1}$ ha$^{-1}$. If sewer overflow is not permissible once a year, but only once every five years, the given amounts need to be increased by a factor of 1.8. A detailed design capacity for the rainfall amount can be found in ATV (1992).

The nature of the pollutants and the resulting pollutant load in the rainwater depend mainly on the prevailing degree of air and surface pollution. According to ATV (1992), the average COD value in rainwater runoff is 107 mg L$^{-1}$. Peak values, which occur especially after long dry periods at the onset of a rainfall, are substantially higher.

### 2.2.3
### Wastewater from Sanitary and Employee Facilities

Wastewater from sanitary and employee facilities consists of the water used by the employees for washing and for flushing toilets and the wastewater from other employee facilities. If there is a large cafeteria, the wastewater from this source should be evaluated separately.

Wastewater from sanitary and employee facilities has the same basic composition as domestic wastewater and should therefore be kept separate from the effluent emerging from the production line. This flow fraction should be discharged into the municipal sewage system, but treatment in the factory's own treatment plant is also possible.

The amount of wastewater per employee and shift is approx. 40–80 L, but the condition of the plumbing fixtures and the degree of dirt involved in the operation (e.g., use of showers at the end of a shift) can have a major impact on the amount of wastewater produced. The load per employee and shift can be assumed to be 20–40 g BOD$_5$, 3–8 g N, and 0.8–1.6 g P.

### 2.2.4
### Cooling Water

If cooling is necessary and done with water, one must differentiate between continuous flow (high water consumption) and recirculation (increase in salt concentrations).

No general amounts can be given for the quantity of cooling water considered necessary, because the need for cooling water varies greatly from one industry to the next. As a first approach, one can say that, for industries using recirculation cooling towers, approx. 3%–5% of the water capacity (depending on the salt concentration in the cooling water, this ratio may even be higher) needs to be extracted to prevent excessive increase in pollutant concentrations.

If there are no leaks in the a continuous-flow cooling tower and no chemical additives are used in the cooling water, the composition of the water does not change, but its temperature increases. In recirculation cooling towers the concentration of pollutants increases with evaporation of the cooling water.

2.2.5
**Wastewater from In-plant Water Preparation**

Wastewater from in-plant water preparation can result from the preparation of drinking water from well water or from water softening, decarbonation, or desalination systems. The contents of this water depend on the characteristics of the raw water, of the treatment method(s) used, and of the chemicals used.

2.2.6
**Production Wastewater**

Production wastewater is a generic term, which includes a variety of flow fractions. Exact specifications can be made only in relation to a particular industry. Certain flow fractions arising in specific industries have been designated with a particular name (e.g., 'singlings') and are not listed here. The following wastewater flow fractions are either investigated as a separate flow fraction or as part of the production wastewater:
- *Produce washing water* is also used in many food processing industries as *flume water* for transport of the produce.
- *Fruit water* is the term used in the food processing industry for that water that is extracted from the processed fruits and vegetables, such as the water extracted from potatoes during starch production.
- *Condensates* are the exhaust vapors that have been condensed after having been removed during an evaporation or drying process.
- *Cleaning water* results from washing the production lines (pipes, containers, etc.), cleaning the production facilities, and cleaning the transportation containers (bottle washing).

2.3
**Kinds and Impacts of Wastewater Components**

Below, the most important wastewater components are listed, with their influence on the value retention of the production facilities and their impact on the necessary wastewater treatment measures.

2.3.1
**Temperature**

The temperature has a major influence on the construction material. High temperatures are not suitable for synthetic materials; e.g., they may cause damage to gaskets. Thus, the temperature of wastewater that is discharged into municipal sewage systems is restricted (in Germany the standard discharge temperature is <35 °C). With regard to the corrosion of metallic materials, the temperature is a major factor, next to the chloride and oxygen contents. Another aspect is that volatile components are easily removed from the wastewater and volatilized at higher temperatures.

An increased wastewater temperature has a positive influence on biological wastewater treatment methods, since an increase in temperature also increases the activity of microorganisms. On the other hand, any temperature increase results in a lower oxygen input capacity of the aeration system.

## 2.3.2
## pH

Similar to the temperature, the pH value has an impact on determining the suitable construction materials and the activity of microorganisms. To avoid damage to the sewage systems and the connected treatment plants, in Germany the pH value of the discharged effluent should be between 6.5 and 10. In treating wastewater, it is of importance which factors have influenced the pH. A low pH, for example, might result from organic or inorganic acids. Inorganic acids should be neutralized; organic acids should be biologically removed.

For many microorganisms the ideal living conditions are at a relatively neutral pH value. Anaerobic wastewater treatment processes are especially sensitive to fluctuations in pH. It is important to keep in mind that organic acids can, at suitable organic loads, be degraded by biological plants without having to be neutralized. This, however, does not apply to mineral acids.

## 2.3.3
## Obstructing Components

With regard to wastewater treatment, obstructing components are large inorganic particles, such as glass shards, plastic parts, cigarette butts, sand, etc., which are biologically inert and need to be removed mechanically to avoid damage, clogging, or caking in the subsequent treatment process.

## 2.3.4
## Total Solids, Suspended Solids, Filterable Solids, Settleable Solids

The content of solid particles has a substantial effect on the amount of organic matter. One differentiates between total solids (TS), which consist of suspended solids (SS, all particles that do not pass through a membrane filter with a pore size of 0.45 µm), and filterable solids (FS). Settleable solids (unit: mL L$^{-1}$) are the solids that will settle to the bottom of a cone-shaped container; therefore, they do not include all of the suspended and floating solids.

## 2.3.5
## Organic Substances

Organic substances constitute the main pollutant fraction in most industrial plants. To prevent direct oxygen consumption in the waterways into which the effluent is discharged, organic substances need to be eliminated as far as possible. A variety of parameters can be used to determine the content of organic matter in a given wastewater sample: BOD, COD, TOC, DOC, etc. These sum parameters, however, do not include any indication of the kind of organic substances measured.

The COD has developed into a major parameter, because the results of COD analysis are available much more quickly than those of BOD analysis. The use of cuvette tests gives an excellent cost–benefit ratio and requires less effort, space, and time to obtain results.

The COD/BOD ratio is an important value in determining the biodegradability of the pollutants in a particular wastewater. If the ratio is <2, the load is considered easily biodegradable.

### 2.3.6
### Nutrient Salts (Nitrogen, Phosphorus, Sulfur)

Nutrient salts are inorganic salts such as $NH_4$, $PO_4$, $SO_4$, which are considered vital for the growth of plants and microorganisms. Since nitrogen and phosphorous can cause massive growth of biomass and may, therefore, lead to eutrophication of waterways into which the effluent is discharged, in the European Union nitrogen and phosphorous must be eliminated before wastewater is discharged into sensitive waterways.

To eliminate nitrogen and phosphorous biologically, sufficient organic pollutants must be present. Therefore, not only is the concentration of these substances in the industrial wastewater important, but also the ratio of their concentrations to the COD or BOD.

Some kinds of industrial wastewater have such low nitrogen and/or phosphorous concentrations that nitrogen (in the form of urea) and/or phosphorous (as phosphoric acid) must be added to obtain the necessary minimum nutrient ratio for growth of the microorganisms needed for biodegradation.

### 2.3.7
### Hazardous Substances

*Hazardous substances* are a generic term used for those substances or substance groups contained in wastewater which must be regarded as dangerous because they are toxic, long-lived, bioaccumulative, or have a carcinogenic, teratogenic, or mutagenic impact. In industrial wastewater, the following substances are of major importance:
- absorbable organic halogen compounds (AOX)
- chlorinated hydrocarbons and halogenated hydrocarbons
- hydrocarbons (benzene, phenol, and other derivatives)
- heavy metals, in particular mercury, cadmium, chromium, copper, nickel, and zinc
- cyanides

Whether a substance is regarded as toxic or hazardous (according to the definition given above) is primarily a matter of concentration. Many heavy metals are vital as trace elements for the growth of microorganisms, but toxic in higher concentrations.

In Germany, the regulations for hazardous substances are so extensive that certain industrial branches not only have to comply with the required discharge quality of the effluent of the entire plant, but also must meet further requirements for the locations where certain flow fractions emerge before they are mixed with the eventual effluent (fractional flow treatment). Some substances have been banned entirely, which requires a specific design and schedule of particular production steps.

## 2.3.8
### Corrosion-inducing Substances

The evaluation of which substances are corrosion-inducing in what concentrations depends, not only on the substances themselves, but also on the choice of material.

For cement-bound materials three kinds of corrosion are distinguished: swelling impact through sulfates, dissolving impact through acids, and dissolving impact through exchange reactions, e.g., with chloride, ammonium, or magnesium. A sufficient concrete resistance is given when the following limits are not exceeded with quality concrete and long-term exposure: $SO_4$ <600 mg $L^{-1}$ (for HS concrete <3000 mg $L^{-1}$; inorganic and organic acids pH >6,5; lime-dissolving carbonic acid <15 mg $L^{-1}$; magnesium <1000 mg $L^{-1}$; ammonium nitrogen <300 mg $L^{-1}$). Biocorrosion may occur when organically highly polluted wastewater with a neutral pH becomes anaerobic. Then $H_2S$ and, in consequence, $H_2SO_4$ (which is aggressive to concrete) are produced in the gas phase.

Corrosion of metals is an electrochemical process for which a conductive liquid must be present, e.g., water. For metallic materials problems arise particularly with high concentrations of chloride, which can lead to corrosion even when high-grade steel is used. It is not possible to determine any general chloride concentrations, because the potential for corrosion depends on a number of parameters, such as the material quality, redox potential, gap width, flow velocity, temperature, and manufacturing quality.

Synthetic materials are commonly regarded as largely corrosion-proof. Some synthetics, however, are not stable toward organic or inorganic acids, alkalis, solvents, or oils, so that when dealing with aggressive media one has to consult resistance tables and check the producers' advice.

## 2.3.9
### Cleaning Agents, Disinfectants, and Lubricants

In industrial factories a great number of cleaning agents, disinfectants, lubricants, dyes, etc., are used. If they occur in high concentrations, many of these materials have an inhibiting or even toxic impact on biological treatment methods. It is also possible that they may contain components that cannot be eliminated in the wastewater treatment plants, e.g., AOX.

Although unrestricted production flow is the guiding principle for industrial companies (e.g., in food industry factories becoming unsterile must be entirely ruled out), samples tested in practice have shown that consumption control systems or replacement of particularly polluting substances have allowed for useful and cost-effective improvements.

## 2.4
## General Processes in Industrial Wastewater Treatment Concepts

### 2.4.1
### General Information

To an increasing extent, wastewater treatment plants have changed from being pure 'end-of-pipe' units to being modules that are fully integrated into the production process, which is referred to as production-integrated environmental protection. The technological basis of this tendency is that production residues can often be used in other ways than disposal, that the wastewater flow often presents some 'product at the wrong place', and that it is simpler and more cost-effective to clean the single flow fractions individually and at higher concentrations. Thus, the general process of wastewater treatment by industrial companies can be roughly divided into production-integrated environmental protection and post-positioned wastewater purification. The borderline between these areas is not clear-cut.

The first step in any procedure of developing wastewater treatment concepts is a detailed stock-taking of the situation of the company with regard to production methods, water supply, and wastewater production. For instance, the production-specific amounts of water, wastewater, and pollutant loads should be ascertained, as well as the characteristics of the different flow fractions of the company and the places where alkalis, acids, detergents, etc. are used.

The next step includes the gathering of various propositions for production-integrated measures and the examination of the different options for the post-positioned wastewater purification unit. The following sections provide further in-depth information on this point.

The last step is the comparison and evaluation of the various propositions and options, the most important criteria for the evaluation being operational safety, economic viability, possibilities for sustainable production, and consideration of the overall concept (combination of industrial pretreatment and post-positioned municipal wastewater treatment plant).

### 2.4.2
### Production-integrated Environmental Protection

The following methods should be examined with regard to their suitability as production-integrated measures. The main principle should be that avoidance should take precedence over utilization and utilization over disposal.

- careful treatment of raw materials (short storage periods, careful handling)
- changes in the transport facilities (dry conveyance, establishment of conveyance circuits)
- changes in the production methods (reduction of water demand, substitution of water, improvements in the production organization)
- avoidance of surplus batches and production losses
- product recovery, for example, by cleaning pipes
- production circuits and multiple use of water (flume and washing water circuits, reverse-flow cleaning, lye recirculation in CIP (cleaning in place) plants, cooling water recirculation)
- utilization of raw materials from production residues (protein coagulation, valuable substance recycling, byproduct yield, forage production)
- separate collection of residues
- extensive retention of production losses in collection containers and utilization separate from the wastewater flow
- cleaning in different steps: (1) dry cleaning, e.g., with high-pressure air; (2) washing; (3) rinsing
- useful treatment of the flow fractions
- general operation and organization (training of staff, control of water and wastewater amounts, installation of water-saving devices, use of high-pressure cleaning tools, etc.)

### 2.4.3
### Typical Treatment Sequence in a Wastewater Treatment Plant

A typical treatment sequence in a wastewater treatment plant consists of the stages listed below. Most industrial factories, however, have only a few of these operation stages, either because they do not have to cope with the respective pollutants in their wastewater or because they are exempted from particular treatment steps due to specific regulations, which occurs, e.g., with indirect dischargers.
- removal of obstructing substances (screens, grit chamber)
- solids removal (strainers, settling tank, flotation)
- storing equalization cooling
- neutralization or adjustment of the pH
- special treatment (detoxification, precipitation/flocculation, emulsion cracking, ion exchange)
- biological treatment or concentration increase (evaporation) or separation (membrane methods)

## 2.5
## Wastewater Composition and Treatment Strategies in the Food Processing Industry

### 2.5.1
### General Information

In view of the great variety of industrial branches, this section is by no means comprehensive in examining all areas of industry, but concentrates on the most important branches of the food processing industry.

When examining specific wastewater amounts and pollutant concentrations, it is important to consider the points already mentioned in Section 2.1 (e.g., age and technical state of the equipment used at the plants). One crucial aspect mentioned in the introduction as well is training and motivation of the employees, which can have a substantial impact on the amount of product loss (spillage, etc.) during the production process, since these losses have a considerable impact on the concentration of pollutants in the wastewater. Apart from the hard statistical facts about production processes, a conscientious plant designer should, when planning the layout of the wastewater treatment units, also try to consider all operational factors of the particular plant.

The general pressure to remain cost-effective is prodding the industry to find ever new solutions, which makes for ongoing optimization. This has a considerable impact on the amount and composition of the wastewater produced and thus directly on the treatment methods. Two basic tendencies have resulted from this: the first is the attempt to reduce the amount of water used, e.g., by recycling or reusing water in other processes. The result is a decreased amount of wastewater, but with a higher concentration of pollutants. The second tendency is to separate the fractions and to reduce the specific load (kg COD $t^{-1}$ of product), e.g., by disposing of the dried solid matter (e.g., as nutrients for agricultural use). A detailed description of what is generally referred to as *production-integrated environmental protection* is presented in the following section.

To make precise statements about wastewater amount and composition for any particular industry, one needs to obtain the latest data pertaining to the particular plant. Apart from the relevant periodicals dealing with this topic, the most important data sources in Germany are the *Handbücher zur Industrieabwasserreinigung* (manuals on industrial wastewater treatment) published by the Abwassertechnische Vereinigung ATV (Association for Wastewater Technology, Germany, today ATV-DVWK) (ATV, 1999, 2000, 2001). Furthermore, relevant data can be found in the reports, codes of practice, and leaflets developed and issued by the single ATV-DVWK commissions and ATV-DVWK working groups.

### 2.5.2
### Sugar Factories

Sugar can be produced from sugarcane or sugar beets. In Europe sugar is extracted from sugar beets. The average sugar content of sugar beets is 18%. The processing of sugar beets is seasonal (campaign operation) and generally takes place from September to mid-December.

In recent years, the amount of wastewater has been greatly reduced from 18 m³ to approx. 0.8 m³ t⁻¹ of sugar beets by water-saving methods and by extensive sealing of the water circuits. In modern plants the water derives almost exclusively from the sugar beets themselves (sugar beets have approx. 0.78 m³ of fruit water per ton).

According to ATV (2004), a typical sugar factory produces the following wastewater flow fractions (Table 2.1).

The most important wastewater components are the high organic loads derived from the sugar. In contrast, the amounts of nitrogen and phosphorous are comparatively low; because growing ecological awareness has led to a more controlled fertilization of sugar beet fields, the amounts of these pollutants are ever decreasing. In Germany, lime milk is added to the flume water and at the juice purification step to prevent becoming unsterile. Therefore, the wastewater has an increased calcium content. Table 2.2 shows the concentrations of the most important components. Due to the recirculation of water in partially closed cycles, the concentrations increase steadily during the course of the campaign operation.

For the treatment of sugar factory wastewater, the following processes are common:
- soil treatment
- long-term batch processing
- small-scale technical processes, usually consisting of anaerobic pretreatment and aerobic secondary treatment

**Table 2.1** Wastewater amounts in m³ t⁻¹ of sugar beets (ATV, 1990, ATV-DVWK, 2004).

| Wastewater Fraction | Specific Wastewater Amount (m³ t⁻¹) | Specific COD Load (kg t⁻¹) |
|---|---|---|
| Surplus condensates | 0.4–0.6 | 0.1–0.15 |
| Soil sludge transport water | 0.15–0.25 | 2–4 |
| Cleaning water | ~0.02 | <0.1 |
| Water from ion-exchange unit | 0.05–0.13 | 0.15–0.3 |
| Surplus water from wet dust collection of the pulp-drying vapors | ~0.02 | ~0.1 |
| Pump sealing water and cooling water | varying (0.4–0.7 (ATV, 1990)) | – |

**Table 2.2** Concentrations of pollutants in sugar factory wastewater (Jördening, 2000).

| Kind of wastewater | Amount (m³ t⁻¹) | pH | COD (mg L⁻¹) | N_inorganic (mg L⁻¹) | Calcium (mg L⁻¹) |
|---|---|---|---|---|---|
| Condensate | 0.3 | 8.5–9.5 | 200–400 | | 100–300 |
| Soil transport water, washing and flume water | 0.3 | 6–11 | 6000–14 000 | 20–40 | 80–2500 |

Agricultural soil treatment is primarily geared toward utilization of the contained nutrients and water. The simultaneous degradation of the organic substances is of less importance; it does not form the basis of dimensioning. If large quantities of ion-exchange waters occur, the salt tolerance of the irrigated plants has to be considered: 90–120 mm of total annual wastewater irrigation can be regarded as a standard value, but the maximum single occurrence should not exceed 30 mm. In Germany, the total amount of wastewater irrigation is partly restricted to <50 mm per year, depending on factors such as the water consumption of the plants and the water storage capacity of the soil, with the result that, because of the high area demand, these methods are no longer economically viable. In Germany only anaerobically pretreated wastewater or condensate are used in irrigation.

With long-term batch processing, the wastewater is discharged into sealed ponds. The degradation period lasts until the summer of the following year and consists of an anaerobic phase succeeded by an aerobic phase (if necessary, aeration equipment is used). If the batch depth does not exceed 1.2 m the COD concentration can be reduced to less than 300 mg L$^{-1}$. The disadvantages of this process are the great area demand and the emergence of noxious odors in the spring time.

In recent years, mainly anaerobic wastewater treatment processes have become established in the sugar industry. In comparison to the processes described above and to the aerobic biological treatment, anaerobic methods have a number of distinct advantages: less area demand, lower sludge production, low energy consumption, energy generation from biogas, no noxious odors. In Germany, mainly anaerobic contact sludge processes are used due to the high calcium contents, whereas in the Netherlands, e.g., where caustic soda is used instead of lime milk, many UASB reactors are used. Anaerobic fixed-film reactors have not been tried yet. The successful operation of a fluidized bed reactor has been described by Jördening (1996, 2000). In properly working reactors, the desired pH range of 6.8–7.5 automatically remains stable, the operating temperature normally is 37 °C. With a volumetric load of 5–10 kg COD m$^{-3}$ d$^{-1}$ (for UASB up to 15 kg), a COD reduction efficiency of approx. 90% is achieved. Because of the lime, nothing needs to be done to decrease the phosphorous concentrations to the limits required for effluent quality. Nitrogen is decreased only to a small extent by integration into the excess sludge; thus, further treatment is necessary before direct discharge of the effluent is possible. As a rule of thumb, nutrient salts are eliminated in anaerobic reactors in the ratio COD:N:P = 800:5:1. To prevent operational difficulties with wastewater having high calcium concentrations, for contact sludge processes decanter centrifuges have proved to be most suitable for separating the calcium carbonate particles from the bacterial sludge.

If direct discharge into a waterway is planned, aerobic secondary purification is necessary, ideally by an activated-sludge system. The condensates should be directly fed into the aerobic stage.

## 2.5.3
## Starch Factories

Starch is produced from potatoes, corn, and wheat. Depending on the original raw material the wastewater fractions, amounts, and loads emerging during starch production vary considerably.

**Corn starch:** fractional flow sources: maceration station, germ washing, starch milk dewatering, gluten thickener, glue of gluten dewatering, and chaff dehydration. The various fractions of wastewater are mainly recycled and used as processing water. The wastewater fractions that have to be treated are

- processing water
- the condensates resulting from evaporation of the maceration water

**Wheat starch:** during wheat starch production, wastewater is derived from the separation step and from thickening of the secondary starch. All other flow fractions are returned into the process water. The wastewater flow fraction that has to be treated is referred to as process water.

**Potato starch:** during potato starch production, washing water (flume and transport water), fruit water (sometimes also condensates from a fruit water evaporation plant), and production wastewater are produced. By recycling it has been possible to reduce the washing water demand from 5–9 m³ t⁻¹ to 0.3–0.6 m³ t⁻¹ (ATV-DVWK, 2002). Fruit water separated from the potatoes is usually fed into the protein recovery unit. Approx. 50% of the proteins contained in the fruit water are coagulated and separated subsequently. Protein production from fruit water is always economically viable, since the price for the protein produced is approx. 0.5 € kg⁻¹, whereas biological treatment, particularly for the elimination of nitrogen (10 g of protein contains 1.7 g N), is quite expensive (approx. 2–5 € kg⁻¹ of N).

According to ATV-DVWK (2002), the specific wastewater amounts of starch production are as shown in Table 2.3.

Wastewater from starch factories has a high organic load and usually consists of easily degradable matter. The undissolved organic contents are mainly carbohydrates and proteins. The fat content is normally ~10%. Usually, this kind of wastewater does not contain any toxic substances. The substrate ratio COD:N:P is satisfactory; actually, there may even occur an excess of N and P, so that it is not necessary to add nutrient salts. Table 2.4 shows standard wastewater values for corn, wheat, and potato starch wastewater, which have distinct differences.

The following processes are used for treatment of starch factory wastewater:

- soil treatment
- pond processing
- small-scale technical processes (anaerobic, aerobic, evaporation)

Until recently, the wastewater from most starch factories in Germany was used to irrigate fields. A summary of the results of this practice can be found in Seyfried

**Table 2.3** Specific wastewater amounts in the starch industry (ATV, 2002).

|  | Specific Total Wastewater Amount ($m^3\ t^{-1}$) | Specific Amount of Certain Flow Fractions ($m^3\ t^{-1}$) | COD ($mg\ L^{-1}$) |
|---|---|---|---|
| Corn starch production | 1.5–3.0 | condensates: 0.4–0.7 | 1500–2500 |
| Wheat starch production | 1.5–2.5 |  |  |
| Potato starch production | 1.5–2.3 | – flume and washing water: 0.3–0.6 | 25 000–30 000 |
|  | without fruit water | – process water (refinement + fruit water separation): 0.6–1.0 | 1500–2000 |
|  |  | – condensates: 0.5–1.2 | 300–2000 |
|  |  | – potato fruit water: approx. 0.6–1.0 | 30 000–60 000 |

**Table 2.4** Wastewater concentrations in the starch industry (Austermann-Haun, 1997).

|  | pH | COD ($mg\ L^{-1}$) | $BOD_5$ ($mg\ L^{-1}$) | N ($mg\ L^{-1}$) | P ($mg\ L^{-1}$) | S ($mg\ L^{-1}$) |
|---|---|---|---|---|---|---|
| Corn | – | 2500– 3500 | 1700– 2500 | 100 | 60 | 110 |
| Wheat | 3.5–4.6 | 18750–48280 | 11600–24900 | 610–1440 | 115–240 | 120–410 |
| Potato |  |  |  |  |  |  |
| – washing water | 6–7 | 2 000– 4000 |  | 200– 300 | 20–40 |  |
| – process water | 6–7 | 4000– 8000 |  | 300– 600 | 20–40 |  |
| – potato fruit water | 5–6 | 50000–60000 |  | 3600–4400 | 800–1000 |  |

and Saake (1985). However, a lack of sufficient agricultural areas and the low amounts permitted to be used put an end to this practice.

Pond processing in batch operation is possible only for production of potato starch, since this is a campaign operation. Continuously operating ponds are rare, since they require additional aeration, due to the high load concentration. They can only be recommended as a secondary treatment method, because the demands for area and aeration are too high.

Anaerobic wastewater treatment has become established alongside conventional activated sludge processes. Fixed-film reactors and UASB reactors are mainly used, which at loads between 7–30 kg COD m$^{-3}$ d$^{-1}$ can achieve degradation rates of 70%–90%. To maintain operational stability, a separator for solids is usually placed upstream of the reactor. Furthermore, a separate pre-acidification stage with a sludge removal system is recommended. The amounts of available phosphorous and nitrogen are sufficient for the anaerobic treatment. Sometimes, trace amounts of cobalt and nickel need to be added. The pH should not exceed 7.0, since at higher pH values MAP (magnesium ammonium phosphate) might precipitate.

In the potato starch industry a further possibility for fruit water treatment is reduction of the wastewater by evaporation to a solids content of 70%. The dried solids can then be utilized, e.g., as fertilizer. To save evaporation energy, a membrane process can be used to increase the concentration prior to evaporation, which would be done as a cascade operation under vacuum conditions.

Aerobic secondary treatment is always required if the company intents to discharge its wastewater directly into waterways.

## 2.5.4
### Vegetable Oil and Shortening Production

In the vegetable oil industry, one differentiates between the following branches: extraction plants for extraction of raw fat and oil, refining plants for refining of raw fat and oil, plants for further processing of the nutrient fats, e.g., into margarine, which consists of approx. 80% fat, 20% water, plus some other ingredients.

The process technology for nutrient fat and oil production largely depends on the raw materials, depending on whether oil is extracted from seeds (soy, sunflower, coconut, rapeseed, etc.), animal fats from tallow and fat liquefiers (tallow or lard), and fruit flesh fats (palm oil, olive oil, which due to its poor storage stability is usually extracted near the site of production), and fats from marine animals.

The processing of seed oils consists of the following stages: cleaning, peeling/shelling if required, masticating, conditioning (heating and moistening), pre-pressing with ensuing extraction or direct extraction without prepressing. Extraction is accomplished by means of a solvent–oil mixture (emulsion), followed by removal of the solvent (usually hexane) by steam. The vapors are condensed and the hexane is separated from the water in static separators. In a further distillation step, the remaining hexane is removed from the wastewater. The wastewater amount produced in seed oil extraction processes is less than 10 m$^3$ t$^{-1}$ of seeds. Data for the wastewater pollutant concentrations are available only for COD, which is in the range of 700–2000 mg L$^{-1}$.

Extraction of animal fats is accomplished by masticating fatty tissues and heating the macerate, followed by a separation process. Wet and dry melting processes are distinguished. The dry melting process does not produce any wastewater, whereas the wet process produces approx. 0.35 m$^3$ t$^{-1}$ raw material, which mainly consists of glue water and pump sealing water from the separators. Depending on the extraction method and the cleaning procedures used in the production plant, the concentration of pollutants in the wastewater from tallow melting plants can vary considerably. For wastewater that has passed the fat separators with topped sludge catcher implements, one can assume the values shown in Table 2.5 (the ratio of emulsified fats, however, may be dramatically higher than given here).

In refining plants, the raw oil is purified (refined) by removing undesired ingredients. The basic process steps consist of desliming, neutralizing, bleaching, and deodorizing (steaming). Depending on the raw material and the desired end product, further processes such as winterizing, fractionating, transesterification, or hydrogenation may be added. In addition to this, other waste products such as saponification water (soapstock) are processed. Wastewater fractions occur during desliming/neutralizing, oil drying and cooling, steaming (condensates), as well as during soap cracking (acidic water). The amount and composition of the wastewater depends mainly on the characteristics of the fresh water, with regard to temperature and degree of hardness, the mode of process operation (continuous or discontinuous methods, process temperature and pressure), the type and quality of raw materials, and the frequency with which the raw materials are changed. The data given in the reference literature vary greatly, due to the multitude of factors involved. A standard value for wastewater of a refining plant which recirculates the falling water is $<10$ m$^3$ t$^{-1}$ of raw material.

Provided that no falling water recirculation or static fat separators are used in the wet chemical refining process, the total wastewater of a refinery has the following characteristics (Table 2.6).

For margarine production, wastewater results only from the cleaning circuits of the CIP plants and amounts to approx. 1–3 m$^3$ t$^{-1}$ margarine. After the rinsing and cleaning water has passed the fat separator, the wastewater has the following characteristic parameters (Table 2.7).

Production-integrated environmental protection must be the central part of the wastewater treatment concept, with the major factors being the efficiency of the production process and prevention of product loss during the process.

**Table 2.5** Wastewater concentrations in animal fat processing (Rüffer and Rosenwinkel, 1991).

| Settleable Solids (mL L$^{-1}$) | COD (mL L$^{-1}$) | Temperature (after cooling) (°C) | pH | Lipophilic Substances (mg L$^{-1}$) |
|---|---|---|---|---|
| ~1 | 5000–10 000 | 30–35 | 6–7 | <100 |

**Table 2.6** Wastewater concentrations in refinement plants (Rüffer and Rosenwinkel, 1991; ATV, 1981).

| Settleable Solids (mL L$^{-1}$) | Lipophilic Substances (mL L$^{-1}$) | COD (mL L$^{-1}$) | COD/ BOD$_5$ | Temperature (°C) | pH | Sulfate (when sulfuric acid is used to crack soapstock) (mL L$^{-1}$) |
|---|---|---|---|---|---|---|
| <1 | <150 | <600 | 1.5–2.0 | <35 | 5–9 | 500–1000 |

**Table 2.7**  Wastewater concentrations in margarine production (Rüffer and Rosenwinkel, 1991).

| COD (mg L$^{-1}$) | COD/BOD$_5$ | Lipophilic Substances (mg L$^{-1}$) | Temperature (°C) | pH |
|---|---|---|---|---|
| 1000–2000 | 1.5–2.0 | <250 | <35 | 5–9 |

Wastewater treatment processes usually consist of a physical-chemical pretreatment by fat separators or flotation systems to decrease the amounts of undissolved solids and lipophilic substances. Whereas installation of fat separators is only a minimal pretreatment concept, because the efficiency of these systems may be completely reduced by a hot water surge, a more expensive flotation system using additives also allows the elimination of emulsified, mainly lipophilic, substances. Particularly effective are pressure flotation systems designed for a surface load of 4–5 m$^3$ m$^{-2}$ h$^{-1}$ with a recirculation flow of 10%–40% and operating at 4–7 bar.

Often a mixing and equalizing (M+E) tank is installed behind the flotation system to equalize pH value and temperature peaks and to allow a more constant loading of the subsequent biological treatment stage.

In biological wastewater treatment, aerobic activated sludge treatment with a sludge load of approx. 0.1 kg BOD kg MLSS$^{-1}$ d$^{-1}$ has proved to be effective. To prevent flotation of sludge and development of scum layers, the fat contents of the wastewater should be <200 mg L$^{-1}$.

## 2.5.5
### Potato Processing Industry

Potatoes are used as fodder or feed potatoes, as industry potatoes (starch production, distilleries), or as market potatoes.

Potato processing can be divided into three main production methods: fresh products (peeled potatoes, precooked potatoes, potato salads, sterilized potatoes, potato products), dried products (dehydrated potatoes, instant mashed potatoes, raw dehydrated potatoes), and fried products (French fries, chips, sticks).

Prior to the final processing the potatoes must be dry cleaned and sorted, sometimes stored for some time and then transported to the processing plant. The conveyance to the processing plant can be wet or dry. Then the potatoes are washed and peeled (mechanically and/or with steam). In the fried production process, peeling is followed by cutting, sorting/washing, blanching, steaming (drying), dewatering, frying and, if necessary, cooling.

For each of the different production methods, wastewater occurs during washing, peeling, cutting, sorting, blanching and steam drying. The specific wastewater amounts and the contents of the fractions of each process step are shown in Table 2.8. The most significant wastewater fraction results from the potato fruit water, owing to the contained starch: a starch concentration of 1 g L$^{-1}$ is equivalent to approx. 1230 mg BOD$_5$ L$^{-1}$. The wastewater load depends mainly on the type of potato (cell size), the growth and storage conditions, as well as on the care in handling (cut and bruised potato cells increase the loads). Usually, the COD/BOD ratio is between 1.6 and 2.0, which is favorable for biodegradation. The C/N ratio is also advantageous for denitrification. Wastewater from the potato processing industry contains comparatively much nitrogen and phosphorous. Nutrient salt limitations occur only in exceptional cases.

Wastewater treatment systems often consist of the following sections:
- grit chamber, screening system, settling tanks for purification of the flume, and washing water recirculation
- production-integrated screening systems and separators to recover organic solids that have been separated from the production wastewater and dewatered, to recycle valuable substances (e.g., cattle forage)
- fat separators for wastewater containing fat, when deep-fat fryers are used in the production process
- biological treatment for pretreatment and full treatment

**Table 2.8** Wastewater volume and loads in various potato processing steps (Scheffel, 1994, ATV-DVWK, 2004).

| Process Step | Specific WW Volume $(m^3\,t^{-1})$ | Settleable Solids $(ml\,L^{-1})$ | COD $(g\,L^{-1})$ | BOD$_5$ $(g\,L^{-1})$ | N $(mg\,L^{-1})$ | P $(mg\,L^{-1})$ |
|---|---|---|---|---|---|---|
| Washing | 0.3–0.5 | | | <1.5 | | |
| Peeling | | | | | | |
| – mechanical process | 0.8–1.5 | peel loss: 15%–25%; settleable solids: 80%–90% of peel loss | | | | |
| – steam process | 0.2–0.3 | peel loss: 8%–15%; settleable solids: 75%–80% of peel loss | 2.9–13.0 | 0.9 | | 1.1 |
| Cutting/sorting/washing | <0.1 | | | | | |
| Blanching | 0.05–0.15 | | | 12.0–18.0 | 1.05 | 2.3 |
| Steaming (drying) | 0.15–0.25 | | | 6.0–8.0 | | |
| Wet potato dough products | 2–3 | 4.5–8.0 | 7.0–12.0 | 3.5–6.0 | 300–400 | 30–50 |
| Dehydration products | | | | | | |
| – dried potatoes | 5–7 | | 6.0–8.0 | 3.0–4.0 | 150–300 | |
| – instant mashed potatoes | 3–6 | | | 2.0–4.0 | 100–140 | 15–30 |
| Fried products | 2.2–3.0 | | 3.6–7.5 | 2.0–5.0 | 120–600 | 25–250 |
| Potato chips | 3.9 | | 2.0–6.0 | 1.0–3.5 | 90–500 | 6–50 |

Although there are large scale applications for the entire range of aerobic and anaerobic treatment methods, good results have been achieved using activated sludge systems, which are designed as either one-stage or cascade units and which can cope with a sludge load of 0.1 kg BOD kg MLSS$^{-1}$ d$^{-1}$. A problem that can arise is bulking sludge, which can be reduced by using the cascade design. In recent years, more anaerobic plants have been built, with the UASB reactor and the EGSB reactor being the most common type in the potato processing industry. To guarantee stable operation of these reactors, it is imperative to limit the amount of filterable solids in the wastewater by installing separators or settling tanks preceding the anaerobic treatment. A relatively stable and neutral pH value is also essential. Because of the large amount of starch particles, the reactors should be preceded by an acidification unit having a retention period of more than 6 h.

## 2.5.6
## Slaughterhouses

Slaughterhouses and meat processing plants can be divided into four categories according to their different production processes.
- slaughterhouses for hogs and cattle
- slaughterhouses for poultry
- meat-cutting plants
- meat-processing industry

Local butcher shops are not considered, due to their small turnover volume.

The following statements refer only to wholesale slaughterhouses that process up to 8000 hogs per day. In the slaughtering process, not only meat from muscle tissue, but also byproducts and residues are produced. The meat and the organ meats (such as liver, etc.) are fit for human consumption and can be traded freely. The residues produced in the slaughtering process are divided into two categories:
- Residues that have commercial value and are tradable, such as fat and bones, which find use as raw materials in feed plants or in the pharmaceutical industry.
- Nontradeable residues, such as meat unfit for human consumption and other byproducts, which need to be treated in animal carcass disposal plants. Additional waste products include primary waste (stomach and intestine contents) and secondary waste (solids from wastewater screening and flotation sludge).

The primary wastewater and waste product sources are divided into three production areas:
- truck washing and animal sheds (green line)
- slaughter and cutting (red line)
- stomach, intestine, and entrails cleaning (yellow line) – the latter, however, are often not done on the slaughterhouse grounds

The waste occurring in trucks and animal sheds consists of bedding material, feces, and urine, which amount to 2.0 kg per hog and 10 kg per cow, if the stabling time is kept short. These materials should be removed without the use of water and spread on agricultural land. Sparing water consumption during truck cleaning can keep the washing water down to 100 L per truck. This wastewater should be screened to remove large particles and should then be fed into the main wastewater stream.

The specific amounts and contents of wastewater (red line) are shown in Table 2.9. However, the presented data can be achieved only so long as blood retention is approx. 90% (the COD of blood is 375 000 mg $L^{-1}$. The total blood volume is approx. 5 L per hog and approx. 30 L per cow).

In slaughterhouses, wastewater pretreatment is mainly done with mechanical procedures. Up to now, the number of plants for which physicochemical or biological operational steps have been added is comparatively small. The main cleaning effect of mechanical and physical procedures consists of retention and separation of solids with the help of stationary strainers, rotating screening drums, separators, fine rakes, screening catchers, or fat separators with a preceding sludge catcher.

In wastewater from slaughterhouses, part of the organic matter consists of oils and fats in emulsified form. Thus, it may be wise to add a physical-chemical stage to the mechanical pretreatment, which would consist of a precipitation/flocculation stage and a flotation unit.

Although most of the slaughterhouses operating in Germany do not have their own biological treatment stages, the wastewater from the slaughterhouses is suitable for biological methods, because of its composition. The following biological methods have until now been used successfully on an industrial scale:
- large space biological methods (oxidation ponds, frequent formerly)
- various activated sludge systems (single-stage, cascade, two-stage)
- anaerobic biological methods

For activated sludge systems the sludge load should not exceed 0.15 kg $BOD_5$ kg $MLSS^{-1}$ $d^{-1}$. Direct anaerobic treatment in contact sludge reactors or joint treatment in municipal digestion tanks is particularly suitable for fats, floating materials, stomach and gut contents, and the liquid phase from the dewatering of rumen contents.

**Table 2.9** Specific amounts and concentrations of slaughterhouse wastewater (red line) (ATV, 1995).

| | Amount (L per animal) | COD (g $L^{-1}$) | $BOD_5$ (g $L^{-1}$) | N (mg $L^{-1}$) | P (mg $L^{-1}$) | fat (mg $L^{-1}$) | AOX ($\mu g$ $L^{-1}$) | Settleable Solids (mL $L^{-1}$) |
|---|---|---|---|---|---|---|---|---|
| Hogs | 100–300 | 2.0–8.0 | 1.0–4.0 | 150–500 | 15–50 | 500–2500 | 20–100 | 10–60 |
| Cattle | 500–1000 | | 1.5–3.0 | | | | | |
| Chicken | approx. 25 | 2.2–4.0 | 1.0–2.5 | 150–350 | 5–30 | 300–1200 | | |

## 2.5.7
## Dairy Industry

Dairy products are classified into drinking milk, cream products, sour milk and milk mix drinks, butter, curd products, hard and soft cheeses, condensed milk, and dried milk. The difference between the listed products and the entire delivery amount is mainly due to the return of skimmed milk and whey and to the extraction of exhaust vapors during coagulation and drying of milk and whey.

For milk processing, the delivered raw milk is first – regardless of the production schedule – cleaned with separators, separated into cream and skimmed milk, and pasteurized. Cream is mainly used for the production of butter. Skimmed milk is processed into drinking milk, fresh milk products, and cheese, partly by adding cream or bacterial cultures. Skimmed milk and whey are the basic materials for the production of dried milk, lactose, and casein.

The wastewater produced in milk processing plants consists of cooling water, condensation water, sanitation water, and process water. The process water consists of the wastewater from the pretreatment, water losses during production, those residues that can no longer be used economically, washing water, detergents, rinsing and cleansing water, and water processing. In dairies the wastewater derives almost exclusively from cleaning of the conveyance and production implements. More than 90% of the organic solids in the wastewater result from milk and production residues. For the milk processing industry, wastewater discharge is almost always identical with loss of products that could otherwise be utilized or sold. This is a great incentive to decrease the production of wastewater by production-integrated measures. For untreated dairy wastewater, the following data are valid; peak values can even exceed these values (Table 2.10).

Dairy wastewater has only a small ratio of settleable solids. Thus, conventional mechanical procedures, such as settling tanks, are ineffective. Inevitable losses of fats can be retained in fat separators, which, however, have only a limited efficiency if the wastewater temperatures are comparatively high.

One major problem with dairies is the considerable variation in wastewater volume and concentration. Thus, as a first step after straining and a sand trap, it is recommended to install a mixing and equalizing tank (M+E tank). In the M+E tank wastewaters of different concentrations and pH are mixed, the wastewater flow is equalized, and partial biological degradation occurs, which can also result in a bio-

Table 2.10 Amounts and concentrations of dairy wastewater (Bertsch, 1997).

| Wastewater Amount ($m^3$ $t^{-1}$ of milk) | $BOD_5$ (g $L^{-1}$) | COD (g $L^{-1}$) | $NO_3$-N (mg $L^{-1}$) | N (mg $L^{-1}$) | P (mg $L^{-1}$) | Settleable Solids (mL $L^{-1}$) | pH | Lipophilic Substances (mg $L^{-1}$) |
|---|---|---|---|---|---|---|---|---|
| 1–2 | 0.5–2.0 | 0.5–4.5 | 10–100 | 30–250 | 10–100 | 1–2 | 6–11 | 20–250 |

logical neutralization. As a minimum volume, approx. 25% of the daily water flow has proved to be a favorable value, but it is also possible to adjust the facilities for daily or weekly equalization. With unaerated M+E tanks there is no biological degradation worth mentioning; instead, substance conversion (acidification) happens, which leads to the emergence of noxious odors. Aerated M+E tanks are mostly operated as washing-off reactors. They are able to reach $BOD_5$ efficiency rates of 20%–60% for homogenized samples.

As a further pretreatment stage, a flotation implement is recommended, which can either replace the M+E tank or be downstream from it; this facility allows for the removal of fats and proteins, i.e., the major part of the organic pollutants.

Although more than 90% of the milk processing companies in Germany discharge their wastewater indirectly, direct discharge can under certain conditions be economically viable. Direct dischargers mostly have an activated sludge system, which should, if strong variations occur, be preceded by an M+E or a calimity tank. Since dairy wastewater has a tendency to develop bulking sludge, it is recommended to design the plant so that the microorganisms in the activated sludge are intermittently subjected to high loads, which can be achieved by installing an activated sludge system in plug flow design, by a preceding contact tank (selector), or by a SBR method (sequencing batch reactor: the process steps of filling, denitrification, aeration, sedimentation happen one after the other, but in the same tank).

### 2.5.8
### Fruit Juice and Beverage Industry

Natural mineral waters and spring waters are collected and bottled at the location of the source spring. Table water consists of drinking water or natural mineral water to which salts are added. Refreshment beverages are produced from water, flavoring substances, sugar or sweetener, and carbon dioxide. The technology of fruit juice production can, in a simplified manner, be divided into the production stages of washing, grinding, refining, filtering, heating, recooling, and bottling.

The wastewater produced in these three industrial branches consists of the following streams (some do not occur in every branch): wastewater from cleaning bottles and containers and from bottling, rinsing and washing water, exhaust vapor condensate, wastewater from the production facilities, wastewater from surface cleaning (floors of the production sheds and the parts of the yards where production takes place), and wastewater from cleaning the conveyor facilities.

Wastewater produced in the mineral water industries contains the following components: adhesive materials and fibrous substances, cleaning alkalis and acids, and soiling from the deposit bottles. For soft drinks, one has to consider the fact that the wastewater additionally contains organic pollutants (with a high ratio of carbohydrates, a large part of it being sugar), which derive from residues and product losses. For the fruit juice industry, product losses – in particular the loss of fruit concentrates – and the sugar, which is often added, are a considerable part of the wastewater pollution. The COD of fruit juices ranges from about 50 g $L^{-1}$ (tomato juice) to about 200 g $L^{-1}$ (apricot juice); 1 kg of glucose (or of fructose) is equivalent to

1066 g COD. Besides the wastewater, the fruit juice industry also produces cooler sludge, filtration residues, sludge from clarifying agents, pomace, and kieselguhr. Because of their high pollution potential, these substances should be the focus of production integrated measures. If possible, they should be utilized or disposed of separately.

The specific wastewater amounts and pollutant concentrations of the three industry branches are presented in Table 2.11. For companies in the fruit juice industry, one has to differentiate between those that do only bottling, only processing, and both processing and bottling. It is apparent that the wastewater from the beverage industry has low nitrogen and phosphorous values in relation to the BOD. The pH can range from 3.5–11.5. At the time of contact with the product the pH is mostly within the acidic range.

To meet the discharge limits of the municipal sewer system, it is often necessary to equalize the pH peaks and sometimes to reduce the temperature (discharge limit in most cases: <35 °C). Thus, a wastewater pretreatment plant could – in addition to a straining station – consist of only a neutralization stage. Because of the high cost of chemicals, biological neutralization is usually recommended, e.g., in an aerated mixing and equalizing tank, which, used as a washing-off reactor, achieves $BOD_5$ elimination rates between approx. 35% (daily equalization) and >50% (weekly equalization). Particular heed, however, should be paid to the alkaline water from the bottle washing machines, as it might be necessary to collect this water in a separate container and to discharge it in controlled doses.

Extensive wastewater pretreatment can be successfully done with anaerobic reactors in the fruit juice industry. Two-stage implements (first stage: acidification reactor with mixing and equalization function; second stage: methane reactor) have proved to be advantageous. At volumetric loads in the methane reactor of up to >10 kg COD $m^{-3}$ $d^{-1}$ the COD elimination rate amounts to about 80%.

For direct discharge into waterways, the activated sludge system with cascade design has proved to be viable; it is operated at sludge loads of <0.1 kg $BOD_5$ $kg^{-1}$

Table 2.11 Specific wastewater amounts and concentrations in mineral water, refreshment beverage, and fruit juice industry (ATV, 1999).

| | Specific Demand ($m^3$ 1000 $L^{-1}$ drink) | $BOD_5$ (mg $L^{-1}$) | COD (mg $L^{-1}$) | N (mg $L^{-1}$) | P (mg $L^{-1}$) |
|---|---|---|---|---|---|
| Mineral water and beverages | 0.9–1.3 (mineral) 1.1–3.3 (beverages) | 110– 800 | 200– 1600 | 2–35 | 0–18 |
| Fruit juice (bottling only) | | 250–1000 | 1500– 3000 | 1.2–10 | 1.5–12 |
| Fruit juice (production only) | | 1700–4000 | 2500–45 000 | 5–30 | 3–15 |
| Fruit juice (production + bottling) | | 400–2000 | 400– 3000 | 9–25 | 2–14 |

MLSS d$^{-1}$. Because of the low nitrogen and phosphorous amounts in the raw wastewater, it is generally necessary to add these substances. For reuse of deposit bottles, denitrification may be necessary because, as part of the label glue, nitrogen is added to the water. Moreover, one has to deal with the danger of bulking sludge.

## 2.5.9
## Breweries

Beer of course contains both alcohol and carbonic acid. In Germany, legal regulations prescribe that it may be produced only from malt (germinated and 'oasted' barley), hops, yeast, and water. The different kinds of beer (lager, stout, top-fermented, bottom-fermented) are produced mainly by varying the original wort concentrations and by using different kinds of malt and yeast.

After malting, the main operation steps of beer production are wort production, fermentation, storing, filtration, and bottling. Prior to its fermentation into alcohol, the starch contained in the malt (which in Germany is obtained from barley) has to be converted into fermentable sugar.

Residues and wastewater flow fractions occur in the brewing room, in the fermentation and storage cellars, in the filter and pressure tank cellar, during dealcoholization, and during bottling (bottle, barrel, other containers). Brewery wastewater is prone to heavy variation with regard to volume and concentration in the single flow fractions. Where production-integrated measures have already been applied, one can assume the following characteristics for the entire wastewater flow of an average brewery (Table 2.12).

Brewery wastewater has a comparatively high temperature (25–35 °C). With decreasing wastewater volume the temperature tends to rise to 40 °C. It is likely that the pH values vary strongly. In companies that reuse deposit bottles, the wastewater from the bottle washing generally has alkaline pH values. Acidic wastewater may at times result from cleaning processes and from regeneration by ion exchange techniques (for water processing). The nitrogen consists mainly of organic nitrogen (albumen, yeast) and to some extent of nitrate (nitric acid). Furthermore, the wastewater is likely to be contaminated by cleaning and detergent agents, as well as by kieselguhr and by particles arising from abrasion of bottles and shards.

Through production-integrated measures, considerable contributions to the reduction of amounts and loads, temperature, solids content, and pH can be achieved. Some very important measures are the retention and separate disposal of cooler sludge, kieselguhr, and yeast and the addition of lye.

**Table 2.12** Brewery wastewater amounts and concentrations (Rüffer and Rosenwinkel, 1991).

| Specific Wastewater Amount (m³ 100 L$^{-1}$ beer) | BOD$_5$ (mg L$^{-1}$) | COD (mg L$^{-1}$) | N (mg L$^{-1}$) | P (mg L$^{-1}$) | Settleable Solids (mL L$^{-1}$) |
|---|---|---|---|---|---|
| 0.25–0.60 | 1100–1500 | 1800–3000 | 30–100 | 10–30 | 10–60 |

For wastewater pretreatment, the first step should be the removal of settleable solids, such as shards, labels, spent hops, bottle caps, etc., by suitable screens and strainers. To neutralize the mainly alkaline wastewater it is common to use carbonic acid from the fermentation or flue gas. It is also possible to biologically neutralize the alkalis with the carbon dioxide that is produced during the BOD degradation. Because of the heavy variations, it is always recommended to use an equalization tank, which can be run as an aerated mixing and equalizing tank with a biological partial purification (the elimination rates of these washing-off reactors range from approx. 35% with daily equalization and >50% with weekly equalization) or which may serve as an unaerated pretreatment tank or acidification reactor for the anaerobic plant. For the anaerobic pretreatment of brewery wastewater the most common implements are UASB and EGSB reactors, which usually are run without heating the wastewater (reactor temperatures are in the range of 24–36 °C).

For the full-scale purification of brewery wastewater to direct discharge quality, aerobic activated sludge systems have proved to be best, because of the need to eliminate nitrogen and phosphorous. The activated sludge system can either be the sole treatment stage or be post-positioned to an anaerobic plant or an aerobic trickling filter unit. Another reasonable solution is the use of SBR methods (sequencing batch reactor), in which all treatment steps are run one after the other, but in the same tank.

## 2.5.10
## Distilleries

Distilleries produce alcohol for human consumption by fermentation and distillation of agricultural products that contain sugar or starch. Some of this alcohol is also used for vinegar production and in the pharmaceutical and cosmetics industries. In companies that produce spirits, the alcohol is diluted to make it potable and is enhanced with flavor additives. Quite a large number of rather small fruit schnapps distilleries exist.

For the production process, it is important that raw materials containing starch be turned into sugar by enzymes, fermented into ethanol, and then distilled, whereas raw materials containing sugar are only fermented and then distilled. Wine is only distilled. Normally, the first steps are mechanical disintegration of the fruit and mashing with water. In some instances, the saccharification must be artificially boosted. After fermentation is finished, the raw spirit (approx. 80 vol.% alcohol) is cleaned of its distillation residues (slops) by a first distillation. In a further refining step, either a so-called fine spirit (approx. 86 vol.% alcohol) is produced by a second discontinuous distillation, or a fine spirit or neutral alcohol (approx. 96 vol.% alcohol) is produced by continuous rectification. The residues of this second refinement step are called singlings.

Depending on the raw product and production methods, distilleries produce washing water, steaming water or fruit water, slops, and cleaning water. The specific wastewater amounts, as well as the concentrations of the major components, are listed in Table 2.13.

**Table 2.13** Specific wastewater and slops amounts and concentrations in distilleries (ATV, 1999, 2003).

|  | Amount (m³ 100 L⁻¹) | COD (g L⁻¹) | BOD₅ (g L⁻¹) | TKN (g L⁻¹) | P (mg L⁻¹) |
|---|---|---|---|---|---|
| Washing water (potatoes) | 0.2–0.5 m³ t⁻¹ |  | 0.3–1.7 |  |  |
| Slops |  |  |  |  |  |
| – wine | 0.81–0.88 | 10–39 | 6–25 | 0.24–0.45 | 0.044–0.092 |
| – potatoes | 0.5–1.0 | 72 | 44 | 2.5 |  |
| – grain (wheat) | 0.79–0.97 | 71 | 32 | 2.8 | 0.19 |

Washing water is produced only during the cleaning of potatoes or roots, which is mostly done by the suppliers. During steaming of the potatoes a mixture of condensate and fruit water emerges, which as a rule is added to the mash. Modern methods of unpressurized starch breakdown (DSA methods), however, do not produce any steaming water. Since singlings derive from the distillate of the first distillation stage, they contain absolutely no solids and are hardly polluted. For the bottling of spirits, almost exclusively new bottles are used, so that cleaning water derives only from washing the implements, containers, and factory sheds and is not highly polluted.

Slops contain a high amount of organic acids, proteins, minerals, trace elements, unfermentable carbohydrates, or – especially with fruit slops – high solids ratios (cores, stalks, stones, skins), which result in very high COD and BOD₅ values as well as in low pH.

In wastewater treatment the slops are of particular importance, since the other fractions of the wastewater can normally be discharged into the wastewater sewage without further treatment. Thus, slops should, if possible, be collected and utilized separately from the wastewater flow. The most common utilization method for slops from grain, potatoes, or fruit is direct feeding in cattle farming (if necessary, after thickening with decanters). If direct feeding is not possible, one should consider spreading on agricultural fields (if necessary, after anaerobic treatment of the slops in a factory-owned biogas plant or after injection as cosubstrate into a municipal digestion tank). Only if these two utilization methods are not possible, may the slops wastewater be mixed with the other production wastewater flow. Direct discharge, however, is then possible only with sufficiently powerful municipal wastewater treatment plants. In any case, the wastewater should be neutralized as a major pretreatment step. Another point one has to consider is the danger of bulking sludge development and hydrogen sulfide emission (corrosion, odors).

Slops from molasses distilleries often retain very high residual COD ratios after biological treatment. Here, evaporation of slops to a dry solids content of approx. 75% with ensuing separation of potassium sulfate (fertilizer) and utilization of the evaporated slops as an additive for cattle feed has proved to be a suitable utilization method. The condensed exhaust vapors from the evaporation unit are often subjected to anaerobic secondary purification.

# References

## General Literature

Andreadakis, A. (Ed.), Pretreatment of industrial wastewaters II, *Water Sci. Technol.* **1997**, *36*.

Andreadakis, A., Christoulas, D. G. (Eds.), Pretreatment of industrial wastewaters, *Water Sci. Technol.* **1994**, *29*.

ATV, *Industrieabwasser – Grundlagen*, 4th Edn., Berlin **1999**: Ernst & Sohn.

ATV, *Industrieabwasser – Lebensmittel-industrie*, 4th Edn., Berlin **2000**: Ernst & Sohn.

ATV, *Industrieabwasser – Dienstleistungs und Veredelungsindustrie*, 4th Edn., Berlin **2001**: Ernst & Sohn.

ATV-DVWK, Empfehlungen zum Korrosionsschutz von Stahlteilen in Abwasserbehandlungsanlagen durch Beschichtungen und Überzüge, *ATV Merkblatt-M 263* **(2003)**.

ATV, Richtlinien für die Bemessung und Gestaltung von Regenentlastungsanlagen in Mischwasserkanälen, *ATV Arbeitsblatt-A 128* (**1992**).

Ballay, D., IAWQ Programme Committee (Ed.) Water Quality International '96, Part 7: Agro-industries waste management, appropriate technologies, *Water Sci. Technol.* **1996**, *34*.

Britz, T. J., Pohland, F. G. (Eds.), Anaerobic digestion VII, *Water Sci. Technol.* **1994**, *30*.

Brauer, H., Produktions und produktintegrierter Umweltschutz, In: *Handbuch des Umweltschutzes und der Umweltschutztechnik* (Brauer, H., Ed.) Vol. 2. Heidelberg **1996**: Springer-Verlag.

Cecci, F., Mata-Alvarez, J., Pohland, F. G. (Eds.), Anaerobic digestion of solid waste, *Water Sci. Technol.* **1993**, *27*.

EPA, *Handbook for Monitoring Industrial Wastewater 1973*, EPA Number 625673002

Malina, J. F., Pohland, F. G., *Design of Anaerobic Processes for the Treatment of Industrial and Municipal Wastes*. Lancaster, PA **1992**: Technomic Publishing.

Metcalf, E. (Ed.), *Wastewater Engineering*. New York **1991**: McGraw-Hill.

Noike, T., Tilche, A., Hanaki, K., Anaerobic digestion VIII, *Water Sci. Technol.* **1997**, *36*.

Nyns, E.-J., *A Guide to Successful Industrial Implementation of Biomethanisation Technol-ogies in the European Union, Report Prepared for the European Commission Directorated General for Energy Thermie Programme.* Namur, Belgium **1994**: Institut Wallon.

Rüffer, H., Rosenwinkel, K.-H. (Eds.), *Taschenbuch der Industrieabwasserreinigung.* München, Wien **1991**: R. Oldenbourg Verlag.

Speece, R. E., *Anaerobic Biotechnology for Industrial Wastewater*. Nashville, TN **1996**: Archae Press.

## Special Literature

### Sugar Factories

ATV, Reinigung organisch verschmutzten Abwassers, *Korrespondenz Abwasser* **1979**, *3*, 156–161.

ATV, Abwasser aus Zuckerfabriken, *Korrespondenz Abwasser* **1990**, *3*, 285–290.

ATV-DVWK, *Abwasser in der Zuckerindustrie*, ATV-DVWK Merkblatt-M 713 (**2004**).

Jördening, H.-J., Produktionsintegrierter Umweltschutz in der Zuckerindustrie, *Handbuch des Umweltschutzes und der Umweltschutztechnik* (Brauer, H., Ed.) Vol. 2, pp. 616–635. Heidelberg **1996**: Springer-Verlag.

Jördening, H.-J., Abwasserreinigung in Zuckerfabriken, *ATV-Seminar: Abwasserbehandlung in der Ernährungs und Getränkeindustrie*, Essen **1997**.

Jördening, H.-J., Zuckerfabriken, In: *Lehr- und Handbuch der Abwassertechnik* (ATV, Ed.), 4th Edn., pp. 43–64. Berlin **2000**: Ernst & Sohn.

Nyns, E.-J., *The Anaerobic Treatment of the Wastewater of the Sugar Refinery at Tienen*, Report prepared for the European Commission Directorated General for Energy Thermic Programme. Namur, Belgium **1994**: Institut Wallon.

### Starch Factories

Althoff, F., Betriebserfahrungen mit einer anaerob/aerob-Betriebskläranlage in der Weizenstärkeindustrie, *ATV-Seminar: Anaerobtechnik in der Abwasserbehandlung*, Magdeburg **1995**.

ATV, Abwasser der Stärkeindustrie, *Korrespondenz Abwasser* **1992**, *8*, 1177–1203.

ATV, Abwasser der Stärkeindustrie, Gewinnung nativer Stärke, Herstellung von Stärkeprodukten durch Hydrolyse und Modifikation, *Korrespondenz Abwasser* **1994**, *7*, 1147–1174.

ATV-DVWK, Abwasser der Stärke-Industrie: Gewinnung nativer Stärke, Herstellung von Stärkeprodukten durch Hydrolyse und Modifikation. *ATV-DVWK Merkblatt-M 776* (**2002**).

Austermann-Haun, U., Stärkefabriken, *ATV-Seminar: Abwasserbehandlung in der Ernährungs und Getränkeindustrie*, Essen **1997**.

Austermann-Haun, U., Seyfried, C.F., Stärkefabriken, In: *Lehr- und Handbuch der Abwassertechnik* (ATV, Ed.), 4th Edn., pp. 65–91. Berlin **2000**: Ernst & Sohn.

Seyfried, C. F., Saake, M., Stärkefabriken, Stärkezucker und Stärkesirupherstellung, In: *Lehr- und Handbuch der Abwassertechnik* (ATV, Ed.), Vol. V, pp. 182–219. Berlin **1985**: Ernst & Sohn.

### Vegetable Oil and Shortening Production

ATV, Organisch verschmutzte Industrieabwässer, *Korrespondenz Abwasser* **1979**, *11*, 664–667.

ATV, Organisch verschmutzte Industrieabwässer, *Korrespondenz Abwasser* **1981**, *9*, 651–656.

Heinrich, D., Speisefett/Speiseölfabriken, ATV-Seminar: *Abwasserbehandlung in der Ernährungs und Getränkeindustrie*, Essen **1997**.

Krause, A., Fabriken zur Gewinnung und Verarbeitung von Nahrungsfetten und -ölen, In: *Lehr- und Handbuch der Abwassertechnik* (ATV, Ed.), 4th Edn., pp. 101–153. Berlin **2000**: Ernst & Sohn.

### Potato Processing Industry

ATV, *Abwässer der Kartoffelindustrie*, ATV Merkblat-M 753 (**1985**).

ATV-DVWK, *Abwasser aus der Kartoffelverarbeitung*, ATV Merkblatt-M 753 (**2004**).

Neumann, H., Kartoffelveredelungsindustrie, In: *Lehr- und Handbuch der Abwassertechnik* (ATV, Ed.), Vol. V, pp. 252–276. Berlin **1985**: Ernst & Sohn.

Nyns, E.-J., Anaerobic wastewater treatment of the potato chips factory, Convention at Frankenthal, Germany. Namur, Belgium **1994**: Institut Wallon.

Rosenwinkel, K.-H., Austermann-Haun, U., Abwasserreinigung in der Gemüse- und Kartoffelindustrie, *ATV-Seminar in der Ernährungs- und Getränkeindustrie*, Essen (**1997**).

Scheffel, W., Kartoffelverarbeitung, *ATV-Seminar in der Ernährungs- und Getränkeindustrie*, Essen (**1994**).

Scheffel, W., Kartoffelveredelungsindustrie, In: *Lehr- und Handbuch der Abwassertechnik* (ATV, Ed.), 4th Edn., pp. 155–171. Berlin **2000**: Ernst & Sohn.

### Slaughterhouses

ATV, Abwasser aus Schlacht- und Fleischverarbeitungsbetrieben, In: *ATV-Regelwerk Abwasser: Abfall.* ATV Merkblatt-M 767 (**1992**).

ATV, Behandlung und Verwertung von Reststoffen aus Schlacht- und Fleischverarbeitungsbetrieben, *ATV Regelwerk Abwasser: Abfall*, ATV Merkblatt-M 770 (**1995**).

Blaha, M.-L., Schlachthofabfälle; Mengenanfall und Inhaltsstoffe, *Die Fleischwirtschaft* **1995**, *75*, 648–654.

Fries, D., Schrewe, N., Schlacht- und Fleischverarbeitungsbetriebe, In: *Lehr- und Handbuch der Abwassertechnik* (ATV, Ed.), 4th Edn., pp. 199–238. Berlin **2000**: Ernst & Sohn.

Jäppelt, W., Neumann, H., Verarbeitung tierischer Produkte; Schlacht- und Fleischverarbeitungsbetriebe, In: *Lehr- und Handbuch der Abwassertechnik* (ATV, Ed.), Vol. V, pp. 320–382. Berlin **1985**: Ernst & Sohn.

Sayed, S. K. I., Anaerobic treatment of slaughterhouse wastewater using the UASB process, Thesis. Wageningen, NL **1987**: Eigenverlag.

Tritt, W. P., Anaerobe Behandlung von flüssigen und festen Abfällen aus Schlacht- und Fleischverarbeitungsbetrieben, Thesis, Vol. 83, Hannover: Institut für Siedlungswasserwirtschaft und Abfalltechnik der Universität Hannover (**1992**).

### Dairy Industry

ATV, Reinigung organisch verschmutzten Abwassers, 1.5 Molkereien und Milchindustriebetriebe, *Korrespondenz Abwasser* **1978**, *4*, 114–117.

ATV, Abwasser bei der Milchverarbeitung, *ATV Merkblatt-M 708* (**1994**).

Bertsch, R., *Molkereien, ATV-Seminar:* Abwasserbehandlung in der Ernährungs- und Getränkeindustrie, Essen (**1997**).

Doedens, H., Molkereien (Verarbeitung von Milch und Milchprodukten), In: *Lehr- und Handbuch der Abwassertechnik* (ATV, Ed.), Vol. V, pp. 410–456. Berlin **1985**: Ernst & Sohn.

Doeders, H., *Molkereien, ATV-Seminar:* Abwasserbehandlung in der Ernährungs- und Getränkeindustrie, Essen.

Doedens, H. (**2000**), Verarbeitung von Milch und Milchprodukten, In: *Lehr- und Handbuch der Abwassertechnik* (ATV, Ed.), 4th Edn., pp. 259–277. Berlin **1996**: Ernst & Sohn.

### Fruit Juice and Beverage Industry

ATV, Abwasser aus Erfrischungsgetränke- und Fruchtsaft-Industrie und der Mineralbrunnen, *ATV-Regelwerk Abwasser: Abfall, ATV Merkblatt-M 766* (**1999**).

Austermann-Haun, U., Rosenwinkel, K.-H., Fruchtsaftfabriken, Erfrischungsgetränkeherstellung und Mineralbrunnen, In: *Lehr und Handbuch der Abwassertechnik* (ATV, Ed.), 4th Edn., pp. 279–315. Berlin **2000**: Ernst & Sohn.

Rosenwinkel, K.-H., *Getränkeindustrie, ATV-Seminar:* Abwasserbehandlung in der Ernährungs- und Getränkeindustrie, Essen (**1997**).

Rüffer, H. M., Rosenwinkel, K.-H., *The Treatment of Wastewater from the Beverage Industry*, in: Food and Allied Industries (Barnes, D., Forster, C. F., Hrudey, S. E., Eds.). Vol. 1, Boston (**1984**), Pitman Advanced Publishing.

Rüffer, H. M., Rosenwinkel, K.-H., Getränkeindustrie und Gärungsgewerbe, In: *Lehr- und Handbuch der Abwassertechnik* (ATV, Ed.), 3rd Edn., Vol. V, pp. 457–509. Berlin **1985**: Ernst & Sohn.

### Breweries

ATV, Reinigung hochverschmutzten Abwassers, 1.18 Brauereien, *Korrespondenz Abwasser* **1977**, 6, 182–186.

Rosenwinkel, K.-H., *Entwicklungstendenzen bei der Brauereiabwasserbehandlung, Brauereiabwasser-Seminar*, Institut für Siedlungswasserwirtschaft und Abfalltechnik der Universität Hannover (26.03.1996).

Rosenwinkel, K.-H., Schrewe, N., Brauereien und Malzfabriken, In: *Lehr- und Handbuch*

der Abwassertechnik (ATV, Ed.), 4th Edn., pp. 317–351. Berlin **2000**: Ernst & Sohn.

Seyfried, C. F., *Verfahrenstechnische Grundlagen der Brauereiabwasserbehandlung, Brauereiabwasser-Seminar*, Institut für Siedlungswasserwirtschaft und Abfalltechnik der Universität Hannover (26.03.1996).

Seyfried, C. F., Rosenwinkel, K.-H., Brauereien und Malzfabriken, In: *Lehr- und Handbuch der Abwassertechnik* (ATV, Ed.), 3rd Edn., Vol. V, pp. 509–549. Berlin **1985**: Ernst & Sohn.

### Distilleries

ATV, Abwässer aus Brennereien und der Spirituosenherstellung, *ATV Merkblatt-M 772* (**1999**).

ATV, Abwasser aus Hefefabriken und Melassebrennereien, *ATV Merkblatt-M 778* (**2003**).

Craveiro, A. M., Soares, H. M., Schmidell, W., Technical aspects and cost estimation for anaerobic systems treating vinasse and brewery/soft drink wastewaters, *Water Sci. Technol.* **1996**, 18, 123–134.

Eggert, W., Borghans, A. J. M. L. Jr., et al., Upflow Fluidized-Bed-Reaktor: In der Praxis bewährt, *Umwelt & Technik* **1990**, 7–8, 10–16.

Haberl, R., Atanasoff, K., Braun, R., Anaerobic–aerobic treatment of organic high strength industrial wastewaters, *Water Sci. Technol.* **1991**, 23, 1909–1918.

Kenkel, K.-H., Brennereien und Spirituosenbereitung, In: *Lehr- und Handbuch der Abwassertechnik* (ATV, Ed.), 4th Edn., pp. 423–439. Berlin **2000**: Ernst & Sohn.

Nyns, E.-J., *Anaerobic Treatment of Distillery Wastewater: Distercoop at Faenza, Report* prepared for the European Commission Directed General for Energy Thermic Programme. Namur, Belgium **1994**: Institut Wallon.

Recault, Y., Treatment of distillery waste waters using an anaerobic downflow stationary fixed film reactor, *Water Sci. Technol.* **1990**, 22, 361–372.

Vial, D., Malnou, D., Heyard, A., Faup, G. M., Use of yeast in pollutant removal from concentrated agricultural food effluents: application to the treatment of vinasse from molasses, *LEBEDEAU* **1987**, 40, 27–36.

Weller, G., Brennereien und Spirituosenbereitung, In: *Lehr- und Handbuch der Abwassertechnik* (ATV, Ed.), 3rd Edn., Vol. V, pp. 591–616. Berlin **1985**: Ernst & Sohn.

# 3
# Activated Sludge Process

Rolf Kayser

## 3.1
## Process description and historical development

### 3.1.1
### Single-stage process

About 1910, investigations started on the treatment of wastewater simply by aeration (e.g., Fowler and Mumford in Manchester [1]). Ardern and Lockett [2] in Manchester conducted similar experiments but after a certain aeration period they stopped aeration, let the flocs settle, decanted the supernatant, added more wastewater, and repeated the cycle again and again. After buildup of a certain amount of biomass they obtained a fully nitrified effluent at an aeration period of 6 h. The settled sludge they called 'activated sludge'. The first technical scale plant was a fill-and-draw activated sludge plant, which today is called the SBR process. Since at that time the process had to be operated manually, they had a lot of operational problems, and therefore, the next plant was built in what is called today the conventional mode (Fig. 3.1).

An activated sludge plant is characterized by four elements:

- An aeration tank equipped with appropriate aeration equipment, in which the biomass is mixed with wastewater and supplied with oxygen.

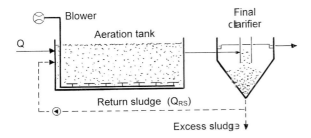

**Fig. 3.1** Flow diagram of an activated sludge plant.

*Environmental Biotechnology. Concepts and Applications.* Edited by H.-J. Jördening and J. Winter
Copyright © 2005 WILEY-VCH Verlag GmbH & Co. KGaA, Weinheim
ISBN: 3-527-30585-8

- A final clarifier, in which the biomass is removed from the treated wastewater by settling or other means.
- Continuous collection of return sludge and pumping it back into the aeration tank.
- Withdrawal of excess sludge to maintain the appropriate concentration of mixed liquor.

If one of the elements fails, the whole process fails.

Primary sedimentation is not required for activated sludge plants. For economic reasons, however, primary tanks are usually operated.

In the early times activated sludge tanks were aerated with fine bubble diffused air. Because of clogging problems in the ceramic diffusers, surface aerators were developed. In 1921 Bolton invented the vertical shaft cone surface aerator at the treatment plant of Bury. Beginning about 1965, cone surface aerators were installed in numerous plants in Germany, the largest one being the Emscher River treatment plant. Another large one is the second stage of the main treatment plant of the city of Hamburg. In the Netherlands in 1925, Kessener constructed the horizontal axis brush aerator, which was installed in spiral flow aeration tanks as they were used for diffused air aeration [1]. In the 1960s brush aerators were frequently used in Germany in high-rate activated sludge plants. Pasveer [3] installed a brush aerator in the oxidation ditch for aeration and circulation of mixed liquor. Starting about 1965, the horizontal axis mammoth rotor, diameter 1.00 m, was installed in closed-loop aeration tanks. The carrousel tank with cone surface aerators represents another closed-loop aeration tank [4]. Both systems are still used in small as well as large plants.

After membrane diffusers were developed around 1970, diffused air aeration became popular again. Usually, aeration should create sufficient turbulence to prevent mixed liquor from settling. To minimize the power requirements for aeration at plants with a low oxygen uptake rate, Imhoff installed a horizontal axis paddle in a tank with diffused air aeration as early as 1924. After Pasveer and Sweeris [5] had postulated that oxygen transfer was considerably increased if the air bubbles rise in a horizontal flow, Danjes developed a system in which diffusers were fixed on a moving bridge [6]. This system was marketed as the Schreiber Countercurrent Aeration System. As an answer to the Schreiber system, around 1970 the firm Menzel installed slow propellers in circular tanks with fine bubble diffused aeration. Today the combination of membrane diffusers in circular or closed-loop tanks with propellers creating a circulating flow is favored for intermittent aeration to remove nitrogen. Recently in some plants with EPDM (ethylene-propylene dimer) membranes after short operating periods of 1–2 years, a sharp increase in the pressure drop was observed. This is attributed to biodegradation of the plasticizer of the EPDM membrane [7, 8].

Industrial wastewater may contain substances that precipitate on ceramic diffusers. They may also contain grease or substances that can destroy membrane material. Surface aeration or coarse bubble aeration with static mixers is therefore a good choice. For deeper tanks the rotating turbine aerator, which is a combination of

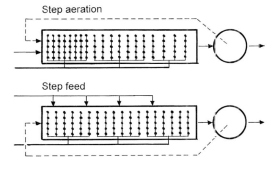

Step aeration

Step feed

**Fig. 3.2** Step aeration and step-feed activated sludge plants.

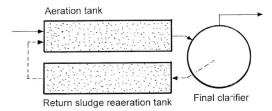

Aeration tank

Return sludge reaeration tank    Final clarifier

**Fig. 3.3** Contact stabilization process.

coarse bubble aeration and a mixer to break up the large bubbles into smaller ones, has been used successfully. In a jet or ejector diffuser, mixed liquor is pumped through a venturi nozzle into which air is introduced [9]. Very fine bubbles are released by the shear stress.

Early aeration tanks were rectangular with evenly distributed diffusers along the tank. Wastewater and return sludge entered the tank at one end, and mixed liquor left the tank at the other end. Because of the high oxygen uptake rate, the dissolved oxygen concentration in the inlet zone was almost zero. To overcome this problem the diffusers at the inlet zone were arranged in a higher density than at the outlet zone. This was called 'step aeration'. 'Step-feed , in which the return sludge is introduced at one end and wastewater is distributed along the tank, has been used in tanks having evenly distributed diffusers (Fig. 3.2).

Since organics are removed after a short period of contact with activated sludge, in the USA the 'biosorption' or 'contact stabilization' process was implemented in some full-scale plants [10]. In this process return sludge is aerated for 2–4 h to oxidize adsorbed organics before it enters the aeration tank, in which the retention period may be in 0.5–2 h (Fig. 3.3).

### 3.1.2
### Two-stage process

In Germany, Imhoff [11] implemented the first two-stage activated sludge plant, which consists of two independent activated sludge plants in series. The first stage is characterized by a high sludge loading rate (F/M) and consequently the second stage has a rather low $F/M$. The excess sludge of the second stage usually is trans-ferred to the first stage (Fig. 3.4). The AB process invented and patented by Böhnke [12] is a two-stage activated sludge plant without primary sedimentation. In this pro-cess the excess sludge of the second stage is not transferred to the first stage.

The two-stage process has several advantages. Harmful substances can be removed in the first stage, which is important for the treatment of industrial waste-water; and in the low-load second stage, due to the high sludge age microorganisms can be maintained that are able to remove slowly biodegradable organics or to oxi-dize ammonia. Furthermore, bulking sludge is only rarely observed in either stage. The disadvantages are that about twice as many clarifiers are needed as in the one-stage process and that nitrogen removal, as well as enhanced biological phosphate removal, may be inhibited owing to missing organics, which are removed in the first stage.

### 3.1.3
**Single sludge carbon, nitrogen, and phosphorous removal**

In the early 1960s three different methods for nitrogen removal were demonstrated (Fig. 3.5):
- post-denitrification [13, 14]
- pre-anoxic zone denitrification [15]
- simultaneous denitrification [16]

Post-denitrification was not successful without the addition of external organic car-bon. Bringmann [13] tried a bypass of wastewater to enhance denitrification, but then some ammonia remained in the final effluent.

The first technical scale pre-anoxic zone denitrification process in Germany was implemented at the research wastewater treatment plant of the University of Stuttgart [17]. After Barnard's [18, 19] successful experiments, the first full-scale plant with pre-anoxic zone denitrification and enhanced biological phosphate removal was built in 1974 at Klerksdorp, South Africa. In Germany the first full-scale plant with pre-anoxic zone denitrification was constructed at Biet [20].

In 1969 the wastewater treatment plant of Vienna Blumental, designed for 300 000 p.e., was put into operation [21]. Two 6000-m³ closed-loop aeration tanks operated in series, each equipped with six twin mammoth rotors, were operated in simultaneous denitrification mode [22]. At that time this was the largest single-stage

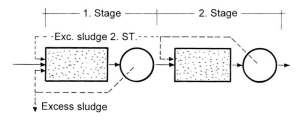

**Fig. 3.4** Flow diagram of a two-stage activated sludge plant.

Post denitrification    External carbon?    **Fig. 3.5** Processes for nitrogen removal.

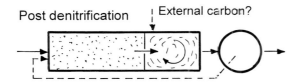

Pre anoxic zone denitrification

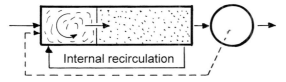

Internal recirculation

Simultaneous denitrification

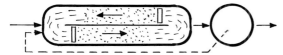

activated sludge plant in the world for nitrogen removal without the addition of external carbon.

These and other process developments for nitrogen removal are discussed in detail in Section 3.3.3.

Based on the work of Thomas [23], phosphorous is easily removed by simultaneous precipitation. Barnard [18] put an anaerobic tank upstream of the biological reactor for enhanced biological phosphate removal. Today most newer plants are built with means for enhanced biological phosphate removal and/or equipment for simultaneous precipitation.

### 3.1.4
### Sequencing batch reactor (SBR) process

Because of the development of reliable automatic process control and aeration systems, the SBR process today is a perfect alternative to the conventional activated sludge process. The reactor is usually equipped with an aeration system, a mixing device, a decanter to withdraw treated wastewater, an excess sludge removal device, and the process control system. Wastewater treatment is performed by a time series of process phases (fill, react, settle, decant). As in the conventional activated sludge process, the SBR process is capable of carbon, nitrogen, and phosphorous removal. It is used for both industrial and municipal wastewater (for further information see, e.g., [24, 25]).

### 3.1.5
### Special developments

#### 3.1.5.1 Pure oxygen-activated sludge process

Owing to poor aeration systems in many activated sludge plants in the USA, the plants were operated with a low mixed-liquor suspended-solids concentration of about 1–2 kg m$^{-3}$ MLSS and for long aeration periods of 5–8 h just for carbon removal. In the pure oxygen activated sludge process, which was developed around 1970, oxygen transfer is not limited. MLSS can be raised to, e.g., 5 kg m$^{-3}$, and the retention period can be shortened to, e.g., 2 h to achieve over 90% BOD removal [26]. The covered aeration tanks necessary to recycle the oxygen gas are advantageous, since the plant looks clean and no emissions are released. The disadvantage is the high carbon dioxide concentration in the gas, which may cause corrosion of the concrete and, furthermore, nitrification may be inhibited by a too-low pH. In Germany only a few pure oxygen activated sludge plants were built in the 1970s. Most of them have since been converted to nitrogen removal plants with diffused air aeration.

Today in some conventionally aerated activated sludge plants, pure oxygen is used at periods of peak oxygen demand. The tanks are not covered and oxygen is not recycled. The oxygen gas is either introduced by special hoses with fine apertures, which release very small bubbles, or by jet type aerators. At the new plant of Bremen Farge, pure oxygen in addition to diffused air is used not only during peak loads but also during periods of power failure. This was calculated to be more economical than the installation of one more turbine blower and a much larger emergency power station.

#### 3.1.5.2 Attached growth material in activated sludge aeration tanks

To increase the biomass in an aeration tank or to enhance nitrification, elements on which biomass can grow are immersed in the mixed liquor. At first, corrugated plastic sheets like that used in trickling filters, e.g., flocor, was installed (Fig. 3.6) [27]. It was believed that nitrifiers would grow on the material. Schlegel [28], however, demonstrated that almost no nitrifiers were attached, but a high number of protozoa were. He postulated that at such plants nitrification was enhanced because of lower sludge production caused by the high number of protozoa and the possibility to operate with higher MLSS because of the improved settleability (low sludge volume index) of the mixed liquor. Another material installed in aeration tanks is ring-lace, which is vertically fixed cords with loops [29]. However, massive growth of worms attracted by the attached protozoa was observed [30]. In the Linpor® process porous plastic foam cubes occupy 10%–30% of the aeration tank volume. Due to the attached biomass, the total amount of mixed liquor solids can be increased, and since the solids stay in the aeration tanks the clarifiers may not become overloaded [31]. In Japan the ANDP process was developed, by which up to 40% of the volume of the aerobic compartment is filled with pellets (short polypropylene hoses 4 mm in diameter and 5 mm long) which are assumed to contain immobilized nitrifiers [32].

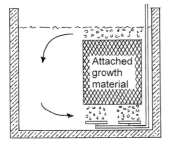

**Fig. 3.6** Cross section of an aeration tank with a module for attached growth.

### 3.1.5.3  High-rate reactors

High-rate reactors are characterized by a high volumetric removal rate, e.g., 10–60 kg COD $m^{-3}$ $d^{-1}$ and consequently by a high volumetric oxygen transfer rate. Loop reactors are preferred. The ICI Deep Shaft is such a reactor, as is the HCR reactor [33]. In the Hubstrahl reactor several plates with holes oscillate up and down with a high frequency to enhance oxygen transfer and to cause high turbulence [34].

High-rate processes are preferred for high-strength wastewater with a very high fraction of soluble and readily biodegradable organics. Because of the high loading rates used for economic reasons the overall removal is restricted to 60%–80% COD.

### 3.1.5.4  Membrane separation of mixed liquor

After Kayser and Ermel [35] successfully applied the membrane technique to separation of sample flow for continuous monitoring of, e.g., nitrate from mixed liquor, Krauth and Staab [36] installed tubular membrane units instead of a final clarifier in a pressurized aeration tank (Fig. 3.7). The process is marketed as the Biomembrat process. As a result of recent improvements in the membrane technique, numerous experiments using microfilter units are underway. The advantages are that it is possible to operate such systems with rather high MLSS (up to 15 kg $m^{-3}$) and that the permeate is almost free of suspended matter. The disadvantage is that the flux rate is only on the order of 20 L $m^{-2}$ $h^{-1}$. The process is therefore still costly and can only be used when there are special effluent requirements [37].

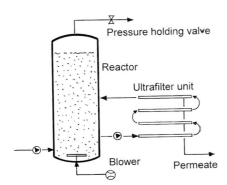

**Fig. 3.7**  Flow diagram of the Biomembrat process.

## 3.2
## Technological and microbiological aspects

### 3.2.1
### Wastewater characteristics

Wastewater to be treated can contain organic carbon predominantly as a single soluble substance (e.g., an alcohol), as a mixture of soluble substances, or as a mixture of solids and numerous soluble organic substances. Many industrial wastewaters are mixtures of soluble organic substances, but some may contain mainly one organic substance. Municipal wastewater and wastewater from most food processing industries is always a mixture of soluble and particulate organic matter.

Microbial degradation of organic carbon requires certain amounts of nitrogen, phosphorous, calcium, sodium, magnesium, iron, and other essential trace elements to grow biomass. In industrial wastewater treatment plants, missing elements have to be added. Since domestic wastewater contains all the necessary elements in excess, it is advantageous to treat special industrial wastewater together with municipal wastewater.

The concentration of organic matter in wastewater is measured as the biochemical oxygen demand ($BOD_5$, or $BOD_{20}$, incubation period of 5 or 20 d, nitrification inhibited), chemical oxygen demand (COD), or total organic carbon (TOC). Raw municipal wastewater can be characterized by the ratios shown in the first column of Table 3.1. Since BOD reflects only biodegradable matter and COD and TOC also include nonbiodegradable components, the ratios in well treated effluent ($BOD_5 = 10$–$20$ mg $L^{-1}$) differ [38].

$BOD_5$ (or in Scandinavia, $BOD_7$) has been the most common parameter for characterizing wastewater. The advantage of BOD is that it includes only biodegradable organics; the disadvantage is that $BOD_5$ measures only a fraction of the biodegradable organics because the measurement of $BOD_{20}$ is economically not feasible. After the development of simplified methods to measure COD, this parameter is mainly analyzed today. It allows mass balances and is therefore widely used for modeling biological processes [39]. Because of the mercury used in the analysis, some countries ban COD measurements and instead prefer determining TOC, which is advantageous also from the microbiological point of view but has analytical problems with solids.

**Table 3.1** Ratios of various organic parameters.

| Ratio | Influent | Effluent |
|---|---|---|
| COD/TOC | 3.2 to 3.5 | 3.0 to 3.5 |
| COD/$BOD_5$ | 1.7 to 2.0 | 3.0 to 6.0 |
| $BOD_5$/TOC | 1.7 to 2.0 | 0.5 to 1.0 |

3.2.2
**Removal of organic carbon**

Many attempts have been undertaken to describe the removal of organics according to Michaelis–Menten or Monod kinetics. For many single substances the $k_m$ or $k_s$ value is rather small; therefore, in batch experiments zero-order removal of single substances is observed. Mixed substrates like municipal wastewater, however, show first-order removal reactions in batch experiments. Wuhrmann and von Beust [40] explained this phenomena as a series of different zero-order reactions. Tischler [41] later demonstrated the removal of glucose, aniline, and phenol in batch tests. In tests with the separate substances, removal of the substances as well as decrease in COD follow different zero-order reactions (Fig. 3.8, left). It is important to notice that, although the substances are completely removed, a COD of about 30 mg L$^{-1}$ remains. In a test in which the three substances were mixed with adapted mixed liquor, a quasi first-order reaction was observed (Fig. 3.8, right). Again, a nonbiodegradable COD of about 60 mg L$^{-1}$ remained. The remaining COD can be visualized as nonbiodegradable compounds produced by bacteria; hence, at least a fraction of the nonbiodegradable COD of any wastewater does not originate from the wastewater itself.

The rate constant of the first-order removal reaction of organics from municipal and many other types of wastewater depends on the wastewater characteristics as well as on the loading rate (F/M). This is because, even at the same MLVSS, the number and types of microorganisms may be different for various types of wastewater as well as under different load conditions.

Since batch experiments indicate that the removal of organics depends on the retention period and the mixed liquor volatile suspended solids (MLVSS), the food/microorganism ratio (F/M ratio) is widely used for design (Eq. 1).

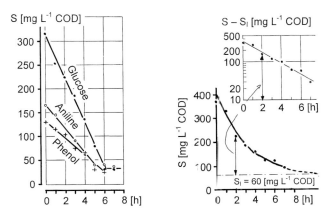

**Fig. 3.8** Removal of single substances [41]. Left: separate substances; right: mixed substances.

$$F/M = \frac{Q_d \cdot C}{V \cdot MLVSS \cdot 1000} = \frac{C}{t^* \cdot MLVSS \cdot 1000} \quad [d^{-1}] \tag{1}$$

In Germany F/M is expressed as $BOD_5$ sludge loading rate ($B_{TS}$) based on MLSS. The extent of removal of organics depends strongly on the F/M ratio. Unfortunately, similar dependencies are observed only when treating the same type of wastewater.

Today, the sludge age or the solids retention time (SRT) is more frequently used in design:

$$SRT = \frac{V \cdot MLSS}{M_{exc}} \quad [d] \tag{2}$$

Combining Eqs. 1 (with MLSS instead of MLVSS) and 2 leads to Eq. 3:

$$F/M = \frac{Q_d \cdot C}{SRT \cdot M_{exc} \cdot 1000} \quad [d^{-1}] \tag{3}$$

Since the mass of excess sludge from organic carbon removal ($M_{exc,C}$) is a function of the organic load ($Q_d \cdot C$), the loading rate F/M increases with decreasing SRT. High F/M ratios require an adequate volumetric oxygen transfer rate, which in the past was the bottleneck in high-rate processes, but new reactor developments enable this problem to be overcome (see Section 3.1.5.3). The specific energy (kW m$^{-3}$) required for oxygen and mass transfer, however, may be considerable.

To estimate the excess sludge production and the oxygen consumption for organic carbon removal and to finally calculate the reactor volume, the COD balance can be used. The following calculations are based partly on the activated sludge models ASM1 and ASM3 [39] and on considerations of Gujer and Kayser [42]. Some factors are restricted to almost complete biodegradation of organics, which may be obtained for municipal wastewater by sludge ages of >5 d.

Since the suspended solids of the final effluent are regarded as a fraction of the excess sludge for COD balancing only $S_{COD,e}$ is of interest.

The total COD of the influent can be divided into soluble COD ($S_{COD,0}$) and particulate COD ($X_{COD,0}$):

$$C_{COD,0} = S_{COD,0} + X_{COD,0} \tag{4}$$

The biodegradable COD ($C_{COD,bio}$) can be expressed by Eq. 5:

$$C_{COD,bio} = C_{COD,0} - S_{COD,I,0} - X_{COD,I,0} \tag{5}$$

The inert fraction of the particulate COD ($X_{COD,I,0}$) can be estimated as 20%–35% of the total particulate COD ($X_{COD,0}$). The soluble inert influent COD ($S_{COD,I,0}$) can be determined experimentally, assuming $S_{COD,I,e} = S_{COD,I,0}$. Generally for municipal wastewater it is in the range of 5%–10% of the total influent COD ($S_{COD,I,0} = 0.05$ to $0.1\ C_{COD,0}$).

As the result of biological treatment, the effluent COD ($S_{COD,e}$) and the excess sludge ($X_{COD,exc}$) remain; the difference represents the oxygen consumed ($OU_C$) for biological degradation of organics (Eq. 6) (Fig. 3.9).

$$C_{COD,0} = S_{COD,e} + OU_C + X_{COD,exc} \tag{6}$$

The excess sludge COD ($X_{COD,exc}$) (Eq. 7) consists of the biomass ($X_{COD,BM}$), the remaining inert particulate matter from endogenous decay of biomass ($X_{COD,BM,I}$), and the inert influent particulate COD ($X_{COD,I,0}$):

$$X_{COD,exc} = X_{COD,BM} + X_{COD,BM,I} + X_{COD,I,0} \tag{7}$$

The biomass produced with a temperature factor for decay $F = 1.072^{(T-15)}$ is obtained by Eq. 8:

$$X_{COD,BM} = C_{COD,bio} \cdot Y \cdot \frac{1}{1 + SRT \cdot b \cdot F} \tag{8}$$

The remaining inert COD of biomass decay ($X_{COD,BM,I}$) can be assumed to be the 20% of the biomass lost by decay (Eq. 9).

$$X_{COD,BM,I} = 0.2 \cdot X_{COD,BM} \cdot b \cdot SRT \cdot F \tag{9}$$

Note that $X_{COD,exc}$, $X_{COD,BM}$, and $X_{COD,BM,I}$ are in units of mg L$^{-1}$ COD, and $OU_C$ is in mg L$^{-1}$ oxygen ($O_2$) based on the daily wastewater flow $Q_d$ [m$^3$ d$^{-1}$].

Assuming that mixed liquor is 80% organic and that 1 mg organic particulate matter is equivalent to 1.45 mg COD, the excess sludge ($SS_{exc}$) as mg L$^{-1}$ suspended solids is obtained (1.45 · 0.8 = 1.16) with Eq. 10:

$$SS_{exc} = \frac{X_{COD,exc}}{1.16} + ISS_0 \tag{10}$$

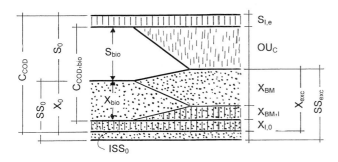

**Fig. 3.9**  Change in COD during biological treatment.

The oxygen consumption (Eq. 11) is derived from Eq. 6:

$$OU_C = C_{COD,0} - S_{COD,e} - X_{COD,exc} \tag{11}$$

The daily mass of excess sludge solids $M_{exc,C}$ [kg d$^{-1}$] is calculated with Eq. 12, and the daily mass of oxygen $M_{O,C}$ [kg d$^{-1}$] required with Eq. 13.

$$M_{exc,C} = Q_d \cdot SS_{exc}/1000 \tag{12}$$

$$M_{O,C} = Q_d \cdot OU_C/1000 \tag{13}$$

If phosphate is removed, the additional excess sludge ($M_{exc,P}$) has to be taken into account (see Section 3.2.5). In plants with a low COD/N ratio and nitrification the biomass of autotrophs produced ($M_{exc,N}$) can also be considered as additional excess sludge. The mass of oxygen required for nitrification (Eq. 16) and the oxygen 'gained' from denitrification (Eq. 25) always must be determined separately.

If SRT is selected according to the desired effluent quality, the reactor volume $V$ can be calculated with Eq. 14,

$$V = \frac{SRT \cdot M_{exc}}{MLSS} \tag{14}$$

where $M_{exc} = M_{exc,C} + M_{exc,P} + M_{exc,N}$.

## 3.2.3
## Nitrification

The main step in nitrogen removal is nitrification, which is assumed to be performed in two steps. Frst ammonia is converted to nitrite by *Nitrosomonas* and then nitrite is converted to nitrate by *Nitrobacter*. The overall reactions are written as:

$$NH_4^+ + 1.5\ O_2 \rightarrow NO_2^- + H_2O + 2\ H^+$$

$$NO_2^- + 0.5\ O_2 \rightarrow NO_3^-$$

$$NH_4^+ + 2.0\ O_2 \rightarrow NO_3^- + H_2O + 2\ H^+$$

The stoichiometric oxygen demand is 3.43 mg $O_2$ per mg of ammonia nitrogen ($S_{NH4}$) converted to nitrite nitrogen ($S_{NO2}$) and 1.14 mg $O_2$ per mg $S_{NO2}$ converted to nitrate nitrogen ($S_{NO3}$). Thus, the overall demand is 4.57 mg $O_2$ per mg $S_{NO3}$ formed. Since there is some buildup of autotrophic biomass, the total actual oxygen consumption is reported to be in the range of 4.2–4.3 mg mg$^{-1}$ [43]. For design calculations, an overall oxygen demand of 4.3 mg $O_2$ per mg $S_{NO3}$ formed is widely used. The ammonia available for nitrification $S_{NH4,N}$ is obtained from Eq. 15:

$$S_{NH4,N} = C_{TKN,0} - C_{TKN,e} - X_{N,exc} \tag{15}$$

For municipal wastewater the non-oxidized nitrogen in the effluent may be assumed to be 2 mg L$^{-1}$ organic nitrogen and 1 mg L$^{-1}$ ammonia nitrogen ($C_{TKN,e}$ = 3 mg L$^{-1}$). The nitrogen contained in the excess sludge may be assumed to be $X_{N,exc}$ = 0.02 $C_{COD,0}$. The daily mass of oxygen consumed for nitrification is calculated with Eq. 16:

$$M_{O,N} = 4.3 \cdot S_{NH4,N} \cdot Q_d / 1000 \tag{16}$$

The two moles of hydrogen released per mole of ammonia converted to nitrite destroy two moles of alkalinity, which equals 0.14 mmol mg$^{-1}$ $S_{NO2}$. If all alkalinity is destroyed, the pH may drop and the flocs of the mixed liquor may disintegrate (see Section 3.2.6.2).

The growth rate of nitrifiers is much lower than that of heterotrophic bacteria. The net maximum growth rates $\mu^*_{max}$ determined by Knowles et al. [44] are widely adopted (Eqs. 17 and 18).

| | | |
|---|---|---|
| *Nitrosomonas:* | $\mu^*_{max} = 0.47 \cdot 1.10^{(T-15)}$ | (17) |
| *Nitrobacter:* | $\mu^*_{max} = 0.78 \cdot 1.05^{(T-15)}$ | (18) |

Since at reactor temperatures below T = 30 °C the growth rate $\mu^*_{max}$ of *Nitrosomonas* is lower than $\mu^*_{max}$ of *Nitrobacter*, Eq. 17 is generally used for design calculations. To avoid a washout of nitrifiers, the inverse net maximum growth rate must be larger than the aerobic sludge age (SRT$_{aer}$). To obtain a low ammonia concentration in the effluent, a series of Monod terms may be used to estimate the desired growth rate and necessary sludge age:

$$\mu = \mu_{max} \cdot \left( \frac{X_{NH4}}{k_{NH4} + S_{NH4}} \cdot \frac{S_{O2}}{k_{O2} + S_{O2}} \cdot \frac{S_{alk}}{k_{alk} + S_{alk}} \right) - k_{D,A} \tag{19}$$

Typical values of kinetic parameters are taken from ASM2 and ASM3 [39] (Table 3.2). If one assumes that each of the three terms in Eq. 19 will not be lower than 0.8 (80% of maximum rates), the operating parameters shown in the second column of Table 3.2 should be maintained.

At 10 °C the growth rate under these conditions would become $\mu$ = 0.13 d$^{-1}$, and the aerobic sludge age required should be SRT$_{aer}$ > 1/0.13 = 7.7 d. The values for

**Table 3.2** Typical values of kinetic and operating parameters.

| Kinetic Parameter Values | Operating Parameters |
|---|---|
| $\mu_{max} = 0.52 \cdot 1.1^{(T-15)}$ d$^{-1}$ | |
| $k_{D,A} = 0.05 \cdot 1.072^{(T-15)}$ d$^{-1}$ | |
| $k_{NH4} = 1$ mg L$^{-1}$ $S_{NH4}$ | $S_{NH4} = 4$ mg L$^{-1}$ |
| $k_{O2} = 0.5$ mg L$^{-1}$ DO | $S_{O2} = 2$ mg L$^{-1}$ |
| $k_{alk} = 0.5$ mmol L$^{-1}$ | $S_{alk} = 2$ mmol L$^{-1}$ |

DO and alkalinity reflect the usual optimum process conditions. The average concentration of 4 mg L$^{-1}$ ammonia nitrogen is rather high, because practical experience indicates that, for an aerobic sludge age (SRT$_{aer}$) of 8–10 d, almost complete nitrification at a reactor temperature of 10 °C is achieved. For design purposes a more simplified approach, such as Eq. 20, may be applied [45].

$$SRT_{aer} = SF_0 \cdot SF_1 \cdot SF_2 \cdot (1/0.47) \cdot 1.10^{(T-15)} \tag{20}$$

A safety factor SF$_0$ of 1.5 enables the growth of nitrifiers; when SF$_1$ = SF$_2$ = 1.0. The safety factor SF$_1$ takes into account any inhibition of the growth rate; a value of SF$_1$ = 1.25 may be chosen. The safety factor SF$_2$ reflects the ammonia load fluctuations. Depending on the size of the plant and the load fluctuations, SF$_2$ = 1.3–1.6 seems to be appropriate. For a temperature of 10 °C this again leads to SRT$_{aer}$ = 8–10 d.

Nitrification is sensitive to pH, as shown by Anthonisen et al. [46] and Nyhuis [47]. *Nitrosomonas* activity is inhibited by low concentrations of HNO$_2$, the result being a buildup of ammonia nitrogen (S$_{NH4}$). Anthonisen et al. [46] suggested that inhibition begins at 0.8 to 2.8 mg L$^{-1}$ HNO$_2$; Nyhuis [47] found values of 0.02 to 0.1 mg L$^{-1}$ HNO$_2$. Free ammonia (NH$_3$) inhibits *Nitrobacter* starting at 0.1 to 1 mg L$^{-1}$ NH$_3$ [46] or 1.0 to 10 mg L$^{-1}$ NH$_3$ [47], leading to a buildup of nitrite nitrogen (S$_{NO2}$). Higher concentrations of free ammonia also inhibit *Nitrosomonas* (10 to 150 mg L$^{-1}$ NH$_3$ [46]; 40 to 200 mg L$^{-1}$ NH$_3$ [47]). Since the concentrations of HNO$_2$ and NH$_3$ depend on pH and the S$_{NO2}$ or S$_{NH4}$ of the mixed liquor, Anthonisen et al. developed a graph indicating the range of inhibitory conditions (Fig. 3.10). The arrows show which parameter may accumulate. It is evident that at low values of S$_{NH4}$ and S$_{NO2}$ and a pH of 6 to 7, as it is usual when treating municipal wastewater, inhibition does not occur (shaded area). When treating high strength wastewater such as that from rendering plants, disturbances are frequently observed, due to a buildup of nitrite (S$_{NO2}$).

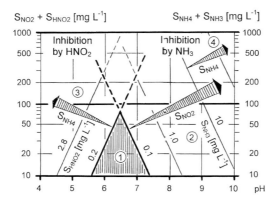

**Fig. 3.10** Inhibition of nitrification by free ammonia and nitrous acid.

## 3.2.4
**Denitrification**

From an engineering point of view denitrification can be visualized as heterotrophic respiration using nitrate instead of dissolved oxygen as the electron acceptor. Nitrate is finally converted to gaseous nitrogen ($N_2$) by numerous reduction steps (Eq. 21).

$$NO_3^- \rightarrow NO_2^- \rightarrow NO \rightarrow N_2O \rightarrow N_2 \tag{21}$$

Since nitrous oxide ($N_2O$) is a greenhouse gas, some concern about biological nitrogen removal has arisen. Investigations in Germany indicate that the contribution of $N_2O$ from nitrogen removal of all German wastewater would be in the range of 2% of the total German $N_2O$ emissions [48, 49].

A simplified overall reaction of the conversion of nitrate to molecular nitrogen may be written as:

$$2\ NO_3^- + 2\ H^+ \rightarrow N_2 + H_2O + 2.5\ [O_2] \tag{22}$$

If a substrate (acetate) is added on the right side of Eq. 22, Eq. 23 is obtained:

$$5\ CH_3COOH + 8\ NO_3^- \rightarrow 4\ N_2 + 8\ HCO_3^- + 2\ CO_2 + 6\ H_2O \tag{23}$$

Both equations indicate that, for one mole of nitrate denitrified, one mole of alkalinity is gained and 1.25 moles of $O_2$ equivalent becomes available for heterotrophic respiration, which is equal to 2.9 mg of oxygen equivalent gained per mg of $S_{NO3}$ to be denitrified. If only nitrite has to be denitrified, 1.7 mg of oxygen equivalent is gained per mg of $S_{NO2}$ denitrified.

Since bacterial synthesis is not considered in Eq. 23, the actual substrate requirement is higher. Experience indicates that 4–6 kg COD of external substrate is necessary to remove 1 kg of nitrate nitrogen.

Nitrate to be denitrified may be calculated with Eq. 24:

$$S_{NO3,D} = S_{NH4,N} + S_{NO3,0} - S_{NO3,e} \tag{24}$$

$S_{NH4,N}$ is calculated with Eq. 15. The effluent nitrate nitrogen $S_{NO3,e}$ should be about 2/3 of the value to be maintained. The daily mass of oxygen equivalents 'gained' from denitrification is obtained with Eq. 25:

$$M_{O,D} = 2.9 \cdot S_{NO3,D} \cdot Q_d / 1000 \tag{25}$$

The reactor volume for denitrification ($V_D$) may be estimated
- from empirical denitrification rates
- by applying anoxic heterotrophic growth kinetics
- by using the aerobic heterotrophic oxygen uptake rate in combination with empirical slow-down factors

Empirical denitrification rates are listed in the EPA Manual [50], among others. The estimation based on heterotrophic growth kinetics is outlined by Stensel and Barnard [43].

Since the driving force for nitrate uptake is similar to the driving force for dissolved oxygen uptake, according to Gujer and Kayser [42], for pre-anoxic zone denitrification the concentration of nitrate to be denitrified can be calculated with Eq. 26:

$$S_{\text{NO3,D}} = \frac{0.9\,(1-Y)\,S_{\text{COD,read}}}{2.9} + \frac{V_D}{V} \frac{0.6\,[OU_C - (1-Y)\,S_{\text{COD,read}}]}{2.9} \tag{26}$$

The first term indicates that readily biodegradable COD ($S_{\text{COD,read}}$) is almost immediately (factor 0.9) removed by nitrate and is therefore almost independent of the size of the anoxic tank volume. The remaining slowly biodegradable COD is removed at a slower rate (factor 0.6) and depends on the anoxic fraction of the reactor ($V_D/V$).

In the intermittent denitrification process the fraction of readily biodegradable COD that enters the reactor during aeration periods cannot be considered for denitrification:

$$S_{\text{NO3,D}} = \frac{V_D}{V} \frac{0.9\,(1-Y)\,S_{\text{COD,read}} + 0.6\,[OU_C - (1-Y)\,S_{\text{CD,read}}]}{2.9} \tag{27}$$

The readily biodegradable COD ($S_{\text{COD,read}}$) can be estimated, e.g., by the procedure of Kappeler and Gujer [51]. For municipal wastewater a value of 0.05 to 0.2 for $S_{\text{COD,read}}/S_{\text{COD,0}}$ can be assumed.

The sludge age (SRT) for the calculation of $OU_C$ in Eqs. 26 and 27 is derived from the required aerobic sludge age for nitrification:

$$SRT = \frac{1}{1 - V_D/V}\,SRT_{\text{aer}} \tag{28}$$

For a design temperature of 10–12 °C the values of $V_D/V$ may be taken from Figure 3.11.

Denitrification becomes limited depending on the ratio $V_D/V$ and $OU_C$ or on the COD/TKN ratio. Since enlarging anoxic reactor ($V_D$) increases the sludge age and decreases the volumetric carbonaceous oxygen uptake rate, it is not economical to raise the anoxic reactor ratio above $V_D/V = 0.5$. The addition of external organic carbon such as methanol or acetate under such conditions is one possible way of achieving the desired effluent nitrate concentration. Another way to gain additional organic carbon is prefermentation of sludge from primary sedimentation [52]. If external carbon is the only source of organics for denitrification, it is advisable to perform experiments to determine the size of the anoxic reactor and the substrate requirement. As a very rough estimate, one can assume, for methanol or acetate as carbon source, denitrification rates of 0.1–0.2 mg $S_{\text{NO3,D}}$ per mg VSS per day at tem-

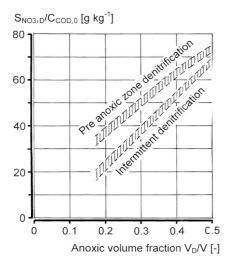

$S_{NO3,D}/C_{COD,0}$ [g kg$^{-1}$]

**Fig. 3.11** Nitrate to be denitrified as a function of the influent COD and the type of denitrification process.

Anoxic volume fraction $V_D/V$ [-]

peratures of 15–20 °C. Since using methanol requires the development of special bacteria, it is not recommended for temporary dosage, e.g., during peak hours.

Dissolved oxygen inhibits denitrification processes, since it removes readily biodegradable organic carbon at a higher rate than nitrate can. If, in the pre-anoxic zone denitrification process, the internal recycling ratio is, e.g., 4 $Q$, with 3 mg L$^{-1}$ DO denitrification of 4 · 3/2.9, about 4 mg L$^{-1}$ nitrate nitrogen is prevented. On the other hand it is possible, in an aeration tank operated with, e.g., 1 mg L$^{-1}$ DO, for some nitrate to disappear. This is attributed to zero DO within the flocs of mixed liquor where some nitrate is then denitrified.

## 3.2.5
## Phosphorus Removal

Phosphorus removal in most cases is required today; usually a concentration of $C_{P,e} < 1$–2 mg L$^{-1}$ has to be maintained. Excess phosphorous is removed by simultaneous precipitation, frequently in combination with enhanced biological phosphorous removal. With a mixing tank upstream of the biological reactor for nitrification–denitrification with a retention time of 15 to a maximum of 30 min, an enhanced biological phosphorous uptake of $X_{P,enh} \sim 0.005$–$0.007 \cdot C_{COD,0}$ can be achieved. For the buildup of heterotrophic biomass, $X_{P,BM} = 0.005 \cdot C_{COD,0}$ can be assumed. Under these conditions, the concentration of phosphorous to be precipitated is calculated with Eq. 29:

$$X_{P,prec} = C_{P,0} - C_{P,e} - X_{P,BM} - X_{P,enh} \tag{29}$$

The effluent concentration should be assumed to be $C_{P,e} = 0.6$–0.7 of the effluent standard to be maintained. The average requirement for precipitant is calculated as 1.5 mol $Me^{3+}$ $mol^{-1}$ $X_{P,prec}$. With these requirements, Eqs. 30 and 31 are obtained:

Precipitation with iron: $\qquad$ 2.7 mg $S_{Fe}$ $mg^{-1}$ $X_{P,prec,Fe}$ $\qquad\qquad$ (30)

Precipitation with aluminum: $\qquad$ 1.3 mg $S_{Al}$ $mg^{-1}$ $X_{P,prec,Al}$ $\qquad\qquad$ (31)

The additional mass of excess sludge resulting from phosphorous removal can be calculated with Eq. 32:

$$M_{exc,P} = Q_d \cdot (3 \cdot X_{P,enh} + 6.8 \cdot X_{P,prec,Fe} + 5.3 \cdot X_{P,prec,Al})/1000 \qquad (32)$$

For details on phosphorous removal, see Schön and Jardin [53].

## 3.2.6
## Environmental factors

### 3.2.6.1 Dissolved oxygen

The dissolved oxygen concentration (DO), the pH, and toxic substances in the wastewater are considered as environmental factors that may inhibit the biological reactions.

In plants with properly designed aeration systems, process disturbances can occur only because of inadequate control of DO. The adverse effects are higher with respect to nitrification than with respect to organic carbon removal, because some organic carbon (even at low DO) may be removed by adsorption, but ammonia cannot be adsorbed and furthermore the growth rate of nitrifiers is reduced at low DO (Eq. 19).

### 3.2.6.2 Alkalinity and pH

The pH in biological reactors depends partly on the pH of the incoming wastewater, if the wastewater contains inorganic acids. The pH is influenced much more by the remaining alkalinity and the concentration of carbon dioxide in the mixed liquor. The remaining alkalinity, which should not drop below $S_{alk,e} = 2$ mmol $L^{-1}$, can be calculated with Eq. 33:

$$S_{alk,e} = S_{alk,0} - 0.07 \cdot (S_{NH4,0} - S_{NH4,e} - S_{NO3,e}) - 0.06 \cdot S_{Fe3+}$$
$$- 0.04 \cdot S_{Fe2+} - 0.11 \cdot S_{Al3+} + 0.03 \cdot (S_{P,0} - S_{P,e}) \qquad (33)$$

Alkalinity is expressed in mmol $L^{-1}$, the nitrogen and phosphorous concentrations are in mg $L^{-1}$, and the possible precipitants for phosphate removal (iron and aluminum) are in mg $L^{-1}$ of incoming wastewater flow. If the precipitants contain free acids or bases, these have to be taken into account separately.

The pH in the reactor is a function of the remaining alkalinity (Eq. 33), the production of carbon dioxide, and the stripping effect of aeration (Table 3.3) [54, 55].

**Table 3.3**  pH in the biological reactor, calculated according to Nowak [54].

| $S_{alk,e}$ mmol $L^{-1}$ | Oxygen Transfer Efficiency (OTE) | | | | |
|---|---|---|---|---|---|
| | 6% | 9% | 12% | 18% | 24% |
| 1.0 | 6.6 | 6.4 | 6.3 | 6.1 | 6.0 |
| 1.5 | 6.8 | 6.6 | 6.5 | 6.3 | 6.2 |
| 2.0 | 6.9 | 6.7 | 6.6 | 6.4 | 6.3 |
| 2.5 | 7.0 | 6.8 | 6.7 | 6.5 | 6.4 |
| 3.0 | 7.1 | 6.9 | 6.8 | 6.6 | 6.5 |

The daily average oxygen transfer efficiency (OTE) is shown as percentages in Table 3.1. Assuming 300 g $O_2$ per m³ air, 1.2 kg $O_2$ to be transferred per kg oxygen uptake (to maintain a DO of >1.5 mg $L^{-1}$), and 10 g of $O_2$ transferred per m³ of air and per meter of diffuser below the water surface ($h_{air}$), the OTE is obtained from Eq. 34:

$$OTE \ [as \ \%] = 100 \cdot \frac{10 \cdot h_{air}}{300 \cdot 1.2} \tag{34}$$

### 3.2.6.3  Toxic substances

Biodegradable and nonbiodegradable toxic substances must be distinguished. Nonbiodegradable toxic substances should be retained by on-site treatment or can be removed by pretreatment (which is usually not possible).

Biodegradable organic toxic substances such as phenols and cyanides can be almost completely removed if

• They are continuously present in the wastewater so that the microorganisms can synthesize the appropriate enzymes.

and

• Their concentration in the mixed liquor is kept as low as possible at any time; shock loads should therefore be avoided.

A balancing tank for the wastewater stream containing the toxics is the most appropriate measure to overcome shock-load problems. The biological reactor should be completely mixed so as to avoid higher concentrations in any section. Intensive chemical analysis of the reactor, preferably by online monitoring, must be performed to detect anomalies such as increasing concentrations of the toxic substance as early as possible. If this happens, the wastewater feed must be decreased or completely shut down until conditions in the reactor are stabilized.

Nitrifiers are far more sensitive to toxic substances than are heterotrophs. If organic carbon removal is not inhibited, denitrification also is usually not inhibited. But since some biodegradable organic toxic substances cannot react with nitrate instead of dissolved oxygen, they may accumulate in denitrification tanks and inhibit denitrification.

### 3.2.7
**Properties of mixed liquor**

The separation of mixed liquor from treated wastewater in final clarifiers is important for the whole process. The concentration of mixed liquor suspended solids (MLSS) and the sludge volume after 0.5 h of settling ($SV_{30}$) are two parameters that describe sludge properties. Since, due to wall effects and bridging in determining SV in 1-L measuring cylinders, erroneous results are obtained at $SV_{30} > 300$–400 ml L$^{-1}$, either the diluted $SV_{30}$ ($DSV_{30}$) or the stirred $SV_{30}$ ($SSV_{30}$) should be measured. For $DSV_{30}$, the mixed liquor should be diluted with final effluent to obtain 150 mL L$^{-1}$ < SV > 250 mL L$^{-1}$. Considering the dilution factor, $DSV_{30}$ is then calculated from the measured SV. Note that mixed liquor with poor settling characteristics may have a $DSV_{30}$ value >1000 mL L$^{-1}$. The dilution method is a common practice in Germany. In other countries the stirring method is preferred; several bars (diameter ~2 mm) are rotated in the measuring cylinder at a speed of 1–2 rpm.

Combining MLSS and SV leads to the sludge volume index SVI, which was proposed by Mohlmann [56]:

$$SVI = \frac{SV_{30}}{MLSS} \quad SVI = \frac{SV_{30}}{MLSS} \tag{35}$$

If $DSV_{30}$ or $SSV_{30}$ are used in Eq. 35, SVI is called DSVI or SSVI.

The sludge volume index is an overall parameter used to characterize the sludge thickening behavior. An $SVI < 100$ mL g$^{-1}$ is regarded as 'good', but if the SVI exceeds 150 mL g$^{-1}$ the sludge is called 'bulking'. Bulking is mainly caused by the growth of filamentous organisms. It is frequently observed at low-loaded plants, e.g., with nitrogen removal and in plants treating wastewater with a high fraction of readily biodegradable organics. Bulking depends not only on loading and wastewater characteristics but also on the mixing conditions of the biological reactor. In low-loaded completely mixed tanks, sludge tends more to bulking than in plug-flow tanks.

From an engineering point of view, for prevention of sludge bulking it is important to create reactor configurations with a high concentration gradient, as in plug-flow tanks or cascaded tanks, and to combine completely mixed tanks with a selector. Even with such precautions the growth of some filamentous organisms, such as *Microthrix parvicella*, may not be prevented. The literature on causes and control of sludge and bulking foaming is broad [e.g., 57]. Recent observations indicate that the use of aluminum as a phosphate precipitant may inhibit the excessive growth of *Microthrix parvicella*.

**3.3**
**Plant Configurations**

3.3.1
**Typical Tanks for mixing and aeration**

Mixing tanks for denitrification can be square, rectangular, or circular and have
mixers in the center or propellers (Fig. 3.12). Rectangular tanks can be visualized as
a series of square tanks. In closed-loop, tanks either propellers or vertical shaft
impellers maintain the circulating flow.

Fine-bubble diffused-air aeration systems can be installed in almost any type of
tank. Tanks for vertical shaft surface aerators are either square or rectangular in
which the length is a multiple of the width (Fig. 3.13). Tanks for cyclic aeration are
preferably circular or of closed loop type. Aeration equipment as well as appropriate
mixers must be installed. Mixing can also be performed by rotating bridges on
which diffusers are mounted. (Fig. 3.14). Since the aeration is switched off cyclical-
ly, only non-clogging aeration systems, e.g., membrane diffusers for fine bubble
aeration are appropriate. Simultaneous nitrification and denitrification in practice
is performed mainly in closed-loop tanks equipped with horizontal-axis surface aerators, e.g., mammoth rotors or vertical shaft surface aerators, as in the carousel process. Simultaneous denitrification can also be performed with air diffusers arranged
in 'fields' in closed-loop tanks (Fig. 3.15).

Vertical shaft mixers (cross section)

Propeller mixers (view from top)

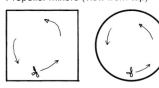

Closed loop tank with propellers

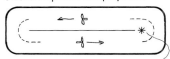

Alternative: Vertical shaft impeller        **Fig. 3.12**  Typical mixing tanks for denitrification.

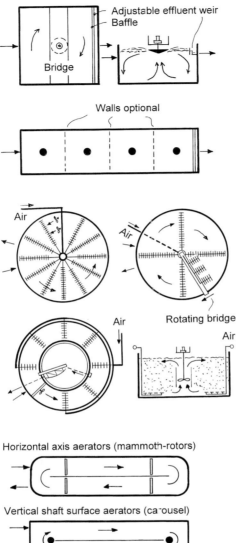

Adjustable effluent weir
Baffle
Bridge

Walls optional

Air

Air

Air

Rotating bridge

Air

**Fig. 3.13** Aeration tanks for vertical shaft surface aerators.

**Fig. 3.14** Typical tanks for cyclic aeration.

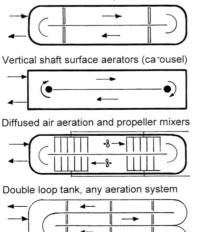

Horizontal axis aerators (mammoth-rotors)

Vertical shaft surface aerators (carousel)

Diffused air aeration and propeller mixers

Double loop tank, any aeration system

**Fig. 3.15** Typical tanks for simultaneous nitrification and denitrification.

## 3.3.2
## Carbon removal processes

In industrialized countries today, removal of both carbon and nitrogen is increasingly required; however, in developing countries and for pretreatment of industrial and trade wastewater, removal of organic carbon only is important.

As a rough design value, a sludge age SRT ~ 4 d may be chosen to achieve a low effluent $BOD_5$ of 10–30 mg $L^{-1}$. Assuming a specific sludge production of 0.4 kg SS per kg COD, the loading rate must be on the order of $F/M_{COD}$ 0.6 kg COD (kg $MLSS)^{-1}$ $d^{-1}$. For the treatment of specific industrial wastewater a higher sludge age may be appropriate. The mixed liquor's suspended solids concentration may be as high as MLSS 3–4 kg $m^{-3}$.

In designing, provisions should be made for
- a sludge that settles well
- a sufficiently high volumetric oxygen transfer rate
- a robust aeration system, especially in developing countries and in industrial plants

Sludge settling is improved by plug-flow type aeration tanks, aeration tanks constructed as a cascade, and/or use of a selector. Two-stage plants may be a choice if the wastewater is highly concentrated. In Germany, numerous plants that receive wastewater with a high fraction of readily biodegradable organics have been operated with a plastic media trickling filter as the first stage followed by an activated sludge plant. Often, intermediate settling to remove trickling filter sludge was not implemented.

Although the aeration efficiency (expressed as kg oxygen transferred per kWh at zero DO) of surface aeration systems measured in clean water may be lower than of fine bubble diffused air systems, surface aeration is in some respects preferable. First, no problems with diffuser clogging or destruction have to be considered; and second, under process conditions the aeration efficiency of fine bubble systems is lower than in clean water, which is not true of surface aeration systems. The weak points of surface aerators are the bearings and the gears. If these are properly designed, maintenance is limited to lubrication and changing gear oil.

## 3.3.3
## Nitrogen removal processes

### 3.3.3.1 Introduction
The single-stage activated sludge process for nitrogen removal, when the organic matter of the wastewater is used for denitrification, incorporates the dilution of ammonia nitrogen to a concentration equivalent to the desired effluent concentration of nitrate nitrogen. Consequently, the organics are diluted by the same ratio. Whether conditions favorable to the development of filamentous bacterial growth are created depends on the wastewater characteristics, the dilution ratio, and the process configuration. Often anaerobic contact tanks for enhanced biological phosphate removal are considered for the purpose of suppressing filamentous growth.

Unfortunately, such tanks are not successful with respect to some detrimental filamentous organisms like *Microthrix parvicella*.

The processes for nitrogen removal can be divided into three groups:

- Subdivided tanks with distinct compartments for denitrification and nitrification, e.g., pre-anoxic zone denitrification, step-feed process, or post-denitrification process.
- Completely mixed or closed-loop tanks in which conditions for nitrification or denitrification are established periodically, e.g., intermittent nitrification–denitrification, alternating nitrification–denitrification (Bio-Denitro process), or intermittent nitrification–denitrification with intermittent wastewater feeding (JARV process).
- Closed-loop tanks in which anoxic zones for denitrification and aerobic zones for nitrification are established at the same time (simultaneous nitrification–denitrification).

Activated sludge plants for nitrification only can suffer from a too-low residual alkalinity. Especially if the aeration tank is completely mixed due to denitrification in the final clarifier, sludge can float and harm the final effluent quality. To overcome such problems, implementing provisions for some denitrification even when nitrification only is required is strongly recommended.

### 3.3.3.2 Pre-anoxic zone denitrification

The activated sludge tank is divided into two main parts, the anoxic zone and the aerobic zone. Since biodegradation of organic carbon follows first-order kinetics, the anoxic zone may be subdivided (Fig. 3.16). Furthermore. the last one or two anoxic compartments may also be equipped with aeration installations, for flexibility.

The aeration tank may have any configuration, as shown in Section 3.3.1. Since it is often difficult to keep the dissolved oxygen concentration in the outlet zone as low as desired, a final nonaerated zone, from which the internal recirculation flow is

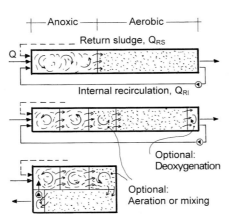

**Fig. 3.16** Pre-anoxic zone denitrification tank configurations.

withdrawn, may be appropriate. If the tank is arranged in a U-shape, the recycling pump merely has to push the flow through the dividing wall.

The hydraulic head loss along such activated sludge tanks is in the range of centimeters if the top of the dividing walls end 0.10–0.30 m below the water level.

The internal recycling flow ($Q_{IR}$) at municipal treatment plants is in the range of 3–5 $Q$, depending on the wastewater strength and the effluent requirements. Since the head loss is small, it is hard to size the recycling pump, therefore, a variable-speed drive to adjust the flow is appropriate. The total recycle flow ($Q_R$) is defined with Eq. 36:

$$Q_R = Q_{RS} + Q_{IR} \tag{36}$$

If one assumes no nitrification in the anoxic zone, no denitrification in the aerobic zone, and no nitrate in the effluent from the anoxic zone, the required flow to be recirculated can be calculated by mass balance (Eq. 37), and the denitrification efficiency is obtained with Eq. 38 which are derived as follows:

$$Q \cdot S_{NH4,N} = (Q + Q_R) \cdot S_{NO3,e}$$

$$\frac{S_{NO3,e}}{S_{NH4,N}} = \frac{1}{1 + Q_R/Q} \tag{37}$$

$$\eta_N = \frac{S_{NH4,N} - S_{NO3,e}}{S_{NH4,N}} = 1 - \frac{1}{1 + Q_R/Q} \tag{38}$$

Eq. 37 indicates that, if a low effluent nitrate concentration is required at a high influent concentration and consequently a high concentration of nitrate must be denitrified, the recycling ratio ($Q_R/Q$) must be high. Since not only nitrate but also dissolved oxygen is transferred to the anoxic zone via the recycled flow, some organic carbon is removed aerobically and thus is lost for denitrification.

Process control may be limited to automatic control of the aeration intensity, to maintain a preset dissolved oxygen concentration. If monitors for ammonia and nitrate are installed at the aeration tank outlet or – even better – at the outlet of a group of aeration tanks, the signals can be used for additional control measures, the purposes of which may be:

- Saving energy: If the concentration of ammonia is zero and that of nitrate is in the desired range, aeration of the denitrification cells can be switched off or, if they are already off, the set-point for aeration control can be lowered. If ammonia increases the opposite measures have to be taken.
- Keeping ammonia low in the winter: If at very low temperatures of the mixed liquor the concentration of ammonia increases, one might try to improve nitrification by raising the set-point for aeration control above the usual 1.5–2.0 mg L$^{-1}$ of dissolved oxygen. This was successful at some plants.
- Improving nitrate removal: If the concentration of ammonia is about zero but the concentration of nitrate is at its upper limit, aeration in the denitrification cells can be stopped and/or the internal recycling flow can be increased.

Numerous large municipalities in Germany, including Berlin, Stuttgart, and Bremen, and also smaller cities are using single-stage activated sludge plants that implement the pre-anoxic zone denitrification process.

### 3.3.3.3 **Step-feed denitrification process**

The activated sludge tank in the step-feed denitrification process consists of two to three pre-anoxic zone units in series: the return sludge is diverted to the first denitrification zone and the wastewater is distributed to each denitrification zone (Fig. 3.17). Although the first experiments on this process were performed in the UK [58] and later in Japan [59], the first results for a full-scale plant were reported by Schlegel [60] in Germany.

Again, considering complete nitrification in the aeration zones and complete denitrification in the anoxic zones, the effluent nitrate concentration for the last denitrification zone can be calculated by mass balance (Eq. 39):

$$(x \cdot Q) \cdot S_{NH4,N} = (Q + Q_{RS} + Q_{IR,n}) \cdot S_{NO3,e} \tag{39}$$

The flow $(x \cdot Q)$ is the fraction of the total wastewater flow entering the last denitrification zone, and $x$ must not be equal to the inverse number of units.

$$\frac{S_{NO3,e}}{S_{NH4,N}} = \frac{x}{1 + (Q_{RS}/Q_{IR,n})/Q} \tag{40}$$

$$\eta_N = 1 - \frac{x}{1 + (Q_{RS} + Q_{IR,n})/Q} \tag{41}$$

To increase nitrate removal in the step-feed process, it is necessary to decrease the fraction of wastewater diverted to the last denitrification zone (decrease $x$) or to increase the return sludge flow and/or the internal recirculation of the last unit $(Q_{IR,n})$ (Eqs. 40 and 41). In existing plants, however, internal recirculation is generally not used with the step-feed process. Since increasing the return sludge flow $(Q_{RS})$ can hinder the final clarification, it seems more appropriate to decrease $x$. This is possible only if the organic carbon contained in the wastewater flow $(x \cdot Q)$ is sufficient to denitrify the incoming nitrate load.

Due to the stepwise 'dilution' of the return sludge with wastewater, the MLSS drops from unit to unit. The result is a higher average concentration of MLSS than

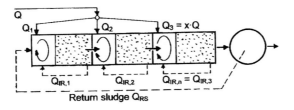

**Fig. 3.17** Step-feed denitrification tank configuration.

in the effluent of the last aeration tank. This is considered an advantage, since the MLSS of the effluent determines the size of the final clarifiers. To maintain the same loading rates in the three compartments it is possible to chose appropriate tank volumes or, which is easier, to distribute the wastewater flow appropriately. This would occur when $Q_1 = 0.4\ Q$, $Q_2 = 0.33\ Q$, and $Q_3 = 0.27\ Q$. The MLSS would then be, respectively, 4.3 kg m$^{-3}$, 3.5 kg m$^{-3}$, and 3.0 kg m$^{-3}$. Without internal recirculation, the ratio $C_{COD}/S_{NO3}$, which determines the degree of nitrate removal, in the three denitrification zones differs considerably. With $Q_{IR,1} = 1.0$–$1.5\ Q$, $Q_{IR,2} = 0.4$–$0.5\ Q$, and $Q_{IR,3} = 0$, $C_{COD}/S_{NO3}$ takes the same value. In the three-step process with $Q_{RS} = Q$, $Q_{IR,3} = 0$, and $x = 0.27$, the denitrification efficiency is 86% (Eq. 41). If with more-concentrated wastewater a low nitrate concentration must be maintained, $Q_{IR,3}$ (Eq. 41) has to be selected. $Q_{IR,1}$ and $Q_{IR,2}$ then have to be increased appropriately.

The similarity of the pre-anoxic zone denitrification process and the step-feed process is obvious when Eqs. 38 and 41 are compared. To achieve 86% denitrification efficiency in the pre-anoxic zone process, the recycling ratio must be $Q_R/Q = 6.4$. The differences between the pre-anoxic zone process and the step-feed process are illustrated in Figure 3.18.

The advantages of the pre-anoxic zone process are that each of the three tanks is operated independently and that the tanks all have the same water level and depth. In the step-feed process some head loss occurs as the water flows from one tank to the next; therefore, either the water depth differs from tank to tank (same bottom

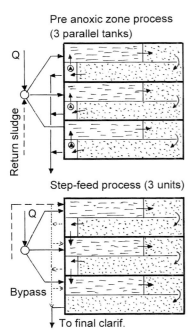

**Fig. 3.18** Comparison of step-feed and pre-anoxic zone processes.

level) or the water depth is kept constant (different bottom levels). For maintenance, it is necessary to have a bypass for each tank in the step-feed process.

In both processes, dissolved oxygen that enters the denitrification zone removes organic carbon and hence decreases denitrification. Therefore, non-aerated outlet zones are shown in the pre-anoxic zone tanks. In the step-feed process only the first and second aeration tanks may be equipped with a non-aerated zone.

Several plants with the step-feed process are operated by the Emschergenossenschaft and, e.g., the cities of Wolfsburg and Bremerhaven.

### 3.3.3.4 Simultaneous nitrification and denitrification

The key to nitrogen removal by the simultaneous nitrification–denitrification process is to appropriately set the aerators so as to establish sufficiently large aerobic and anoxic zones simultaneously (Fig. 3.19). Since the load of any wastewater treatment plant fluctuates diurnally, the concentrations of nitrate and ammonia vary inversely when the aerator setting is constant. To achieve the desired nitrogen removal, a process control is therefore required.

Pasveer in 1964 [16] was the first to report on simultaneous denitrification in an oxidation ditch. He achieved this by setting the optimal immersion depth of the surface aerator so as to create a sufficiently large anoxic zone. At the Vienna Blumenta plant (Section 3.1.3), the oxygen uptake rate was continuously measured with a special (homemade) respirometer. The respirometer output was used to switch the appropriate number of aerators to achieve the desired nitrogen removal [61].

The first real process control for simultaneous denitrification was developed by Ermel [62]. A continuous sample flow was separated from the mixed liquor by ultrafiltration and diverted to a nitrate monitor. At the Salzgitter Bad plant the two closed-loop aeration tanks are operated in parallel. Each tank is equipped with three mammoth rotor surface aerators. One rotor in each tank is operated continuously to create sufficient circulating flow. The two other rotors of each tank are automatically switched on if the nitrate concentration in the sample flow drops to, e.g., $S_{NO3}$ = 3 mg L$^{-1}$. The aerators were stopped if, e.g., $S_{NO3}$ reached 6 mg L$^{-1}$ and were switched on again after the set point of $S_{NO3}$ = 3 mg L$^{-1}$ was reached. Figure 3.20 shows a daily time course of ammonia and nitrate levels measured in one of the aer-

High load (small aerobic volume)

Low load (small anoxic volume)

Low load, two aerators off

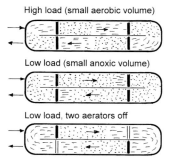

**Fig. 3.19** Simultaneous nitrification and denitrification under different loading conditions.

ation tanks. Due to the still-low denitrification rate during the period of high ammo-nia load (8 to 12 AM) the concentration of ammonia rises to about 4 mg L$^{-1}$ N. As the denitrification rate increases after noon, the off-periods of the two aerators become shorter and the ammonia concentration decreases.

A more sophisticated control system based on monitoring both nitrate and ammonia was used in the wastewater treatment plant in Hildesheim [63].

In Salzgitter Bad an ORP controller was tested as a very simple, low-cost control method. Since the oxidation–reduction potential (ORP) drops sharply, if at zero DO nitrate reaches zero, the controller switches the additional aerators on at a certain slope of ORP. The additional aerators are then operated for a preset period of time. Figure 3.21 shows the time course of ORP. The effluent during the time when the ORP controller was used was as good as when the nitrate controller was used [64]. At about the same time, similar experiments with ORP were conducted in Canada [65].

Some newer plants for simultaneous nitrification and denitrification are equipped with propellers (in addition to surface aerators) to maintain a sufficient flow velocity independent of the number of aerators in operation. Such plants can also be operated in intermittent nitrification–denitrification mode.

In addition to the cities already mentioned, plants with simultaneous nitrifica-tion–denitrification are also operated in, e.g., Osnabrück, Münster, Gera, Pader-born, Tel-Aviv, and in numerous smaller communities in which oxidation ditches were converted to the simultaneous mode by means of automatic control.

Whether and how much of the readily biodegradable organics are oxidized by dis-solved oxygen (and therefore lost for denitrification) depend on the placement of the wastewater inlet and its distance to the next aerator. It is advisable to introduce the flow near the tank bottom, since anoxic conditions are predominant there.

The closed-loop circulating flow tanks for simultaneous nitrification–denitrifica-tion can be regarded as completely mixed. In addition, due to the high dilution of

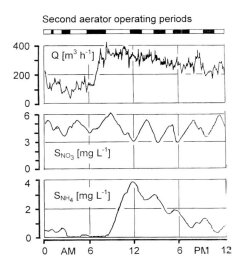

**Fig. 3.20** Time courses of ammonia and nitrate levels in one aeration tank at the Salzgitter Bad plant.

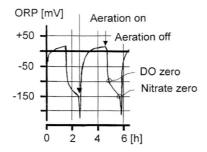

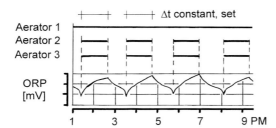

Fig. 3.21 ORP to control simultaneous denitrification. Top: with intermittent aeration, showing the important points; bottom, denitrification control at the Salzgitter Bad plant.

the incoming wastewater by the circulating flow, conditions favorable for filamentous bacterial growth are established. If the wastewater contains a higher fraction of readily biodegradable organics, a selector or an anaerobic mixing tank for enhanced biological phosphate removal is recommended.

### 3.3.3.5 Intermittent nitrification–Denitrification process

Aeration tanks for intermittent nitrification–denitrification must be equipped with an aeration system and mixing devices. The design of the aeration system has to take into consideration the aeration-off periods. To improve the aeration efficiency of plants with diffused air aeration, it may be advisable to also operate the propellers during the aeration periods.

In intermittent nitrification–denitrification plants the fraction of readily biodegradable organics, which enters the tank while aeration is operating, may be oxidized by dissolved oxygen and therefore lost for denitrification. This must be considered when calculating the nitrate to be denitrified. Precautions against the growth of filamentous organisms should be taken, because the tanks are completely mixed.

The duration of the aeration periods and the aeration-off periods can be set with a timer. Any other control system described in Section 3.3.3.4 can be used to achieve a more stable effluent quality. In Germany, especially in the large number of extended aeration plants, an ORP controller is frequently used. NitraReg is another low-cost control system, which requires only continuous measurement of the dissolved oxygen concentration [66]. By knowing the oxygen transfer capacity, the oxygen uptake rate is automatically calculated and used as the control parameter.

The time course of nitrate concentration during the anoxic period ($t_D$) depends on the activity of the denitrifiers, and the time courses of ammonia and nitrate levels during the aerobic period ($t_N$) depend on the activity of the nitrifiers. Since the reactor is completely mixed, the time course of ammonia concentration during the anoxic period can be constructed with Eq. 42.

$$S_{NH4,t} = S_{NH4,N} \cdot [1 - \exp(-t/t^*)] \tag{42}$$

Neglecting the effluent loss of ammonia during the anoxic period, a straight line shows how ammonia would increase (Eq. 43).

$$S_{NH4,t} = S_{NH4,N} \cdot (t/t^*) \tag{43}$$

The deviation between Eqs. 42 and 43 is <5% if $t_D/t^*$ is below 0.25, which is true in practice in activated sludge plants. Figure 3.22 therefore shows only the straight line calculated according to Eq. 43. Since ammonia is converted to nitrate during the aerobic period, the nitrate concentration at the end of a cycle is the starting concentration for the anoxic period of the next cycle.

The average effluent concentration of nitrogen ($S_{NO3,e} + S_{NH4,e}$) can be derived from Figure 3.22, to obtain Eq. 44:

$$S_{NO3,e} + S_{NH4,e} = 0.5 \cdot S_{NH4,N} \cdot \frac{2 \cdot t_D + t_N}{t^*} \tag{44}$$

The efficiency of nitrogen removal can be calculated with Eq. 45:

$$\eta_N = \frac{S_{NH4,N} - (S_{NH4,e} + S_{NO3,e})}{S_{NH4,N}} = 1 - \frac{t_D + 0.5 \cdot t_N}{t^*} \tag{45}$$

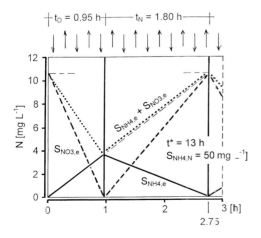

**Fig. 3.22** Time courses of ammonia and nitrate levels during intermittent nitrification and denitrification.

To achieve a nitrogen removal efficiency of 86% ($\eta_N = 0.86$) at, e.g., $t^* = 13$ h and $t_D/(t_N + t_D) = 0.35$, the cycle period should be $t_N + t_D = 2.75$ h.

Eq. 45 indicates that, by variation of the cycle period, technically any removal efficiency can be tuned, but as a prerequisite the nitrate removal capacity of the process must be observed. Cycle periods of less than $t_N + t_D = 1$ h, however, are not recommended. From the ratio $V_D/V$ (Eq. 27) the ratio $t_D/(t_N + t_D)$ can be calculated:

$$t_D/(t_N + t_D) = V_D/V \tag{46}$$

### 3.3.3.6 Intermittent nitrification–denitrification processes with intermittent wastewater feeding

Intermittent wastewater feeding is used in three processes:
- alternating nitrification–denitrification process (Bio-Denitro)
- Tri-Cycle process
- Jülich wastewater treatment process (JARV)

The alternating nitrification–denitrification process consists of pairs of activated tanks, each equipped with aeration equipment and mixing devices. The two parallel interconnected tanks are alternately fed with wastewater and return sludge. While one tank is fed, the mixed liquor is discharged from the other to the final clarifier (Fig. 3.23). This process was developed in Denmark at a plant with two parallel oxidation ditches. It was named the Bio-Denitro process [67].

The time courses of ammonia and nitrate levels during a cycle theoretically looks similar to those obtained with intermittent nitrification–denitrification (Fig. 3.24). The effluent concentration of nitrogen can be calculated in a similar way as for intermittent nitrification–denitrification.

The alternating nitrification–denitrification process is usually controlled with a timer. However, Sørensen [68] demonstrated that inline monitoring of nitrate and ammonia and use of appropriate control measures improved the effluent quality.

In Denmark numerous plants are designed for the alternating nitrification–denitrification process. There are also some plants in Germany, e.g., a newer one in Stuttgart Feuerbach.

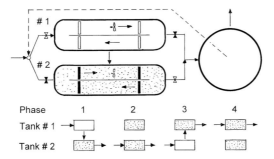

**Fig. 3.23** Flow diagram of the alternating nitrification–denitrification process.

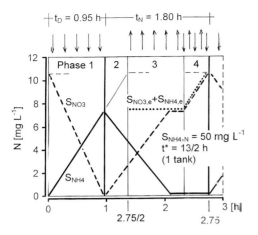

**Fig. 3.24** Time courses of ammonia and nitrate levels in one of the tanks in the alternating nitrification–denitrification process.

In the Tri-Cycle process three interconnected aeration tanks are consecutively (controlled by a timer) fed with wastewater and return sludge. A cycle begins with a mixing period for denitrification which normally lasts as long as the feeding period (1/3 of the cycle period). Valves in the connecting pipes and at the inlet and outlet of the tanks guide the mixed liquor through the three tanks. Mixed liquor is discharged to the final clarifier from the aeration tank that is aerated for the longest period. This process is used in at least six plants in Germany. Unfortunately only brochures of the company (GVA, Wülfrath, Germany) that developed this process are available.

The Jülich Wastewater Treatment Process (JARV) was developed to overcome problems with bulking sludge in completely mixed aeration tanks. The key of the process is a rather short feeding period compared to the cycle period [69]. This requires a balancing tank upstream of the aeration tank and an effluent gate to equalize the flow to the final clarifiers. Balancing is avoided if, e.g., four parallel aeration tanks are consecutively fed with wastewater. The effluent from the aeration tanks may contain considerable ammonia peaks. It is, therefore, advisable to build a small post-aeration basin between the aeration tanks and the final clarifiers. Only two plants in Germany are operated by this process.

The differences of the three processes relative to the intermittent nitrification–denitrification process are that almost all readily biodegradable organics are available for denitrification, that there is a higher concentration gradient by which the sludge volume index can be improved and that during the discharge period the mixed liquor is aerobic (except in the JARV process if a post aeration tank is omitted). The anoxic tank fraction ($V_D/V$) can be estimated as for the pre-anoxic zone denitrification process (Eq. 26).

### 3.3.3.7 Special processes for low COD/TKN ratio

Wastewater from the food industry may be characterized by high concentrations of organics as well as considerable nitrogen concentrations. Anaerobic pretreatment of such sewages is favorable because of the negligible sludge production and energy requirement. If nitrogen removal is required, the COD/TKN ratio can however decrease too much for sufficient denitrification.

Since denitrification of nitrite requires about 35% less organics than denitrification of nitrate, a process in which *Nitrosomonas* were inhibited was developed by Abeling [70]. The pre-anoxic zone process was operated with a control system to maintain the pH at about pH = 8 by dosing the aerobic tank with NaOH and to keep ammonia at $S_{NH4} = 10$ mg L$^{-1}$ by appropriate aeration control. According to the findings of Anthonisen et al. [45] and Nyhuis [46] (Fig. 3.10), these are conditions under which *Nitrobacter* are inhibited. Because there is always some ammonia and some nitrite in the effluent, a second stage using a fixed-film reactor was implemented.

A different method of removing high nitrogen concentrations without any organic substrate was published by Jetten et al. [71]. In a continuous-flow stirred reactor as a first stage, about 50% of ammonia was converted to nitrite; and in a second stage a fixed-film reactor converted nitrite to nitrogen gas (N$_2$) by autotrophic bacteria using ammonia as electron donor. However, this is not regarded as an activated sludge process.

### 3.3.3.8 Post-denitrification with external organic carbon

Few plants around the world use external carbon as the sole carbon source for denitrification in single-stage activated sludge plants. One such plant was constructed to treat the whole wastewater of the Salzgitter Steel Works in Germany.

The wastewater originates from treated blast furnace gas (high ammonia loads), treated coke oven gas (phenols, cyanides, ammonia, etc.), and several other discharges. The total flow of about 50 000 m$^3$ d$^{-1}$ contains about 35 mg L$^{-1}$ ammonia nitrogen. Because soft water is used for cooling, the alkalinity is low. Therefore, experiments were first carried out using the pre-anoxic zone process to gain as much alkalinity as possible. Since the organic carbon content of the wastewater was too low, methanol had to be added to ensure sufficient nitrate removal, but the process was very unstable due to the toxic components of the wastewater.

The process was therefore changed to post-denitrification, with satisfying experimental results [72]. The full-scale plant is shown in Figure 3.25. It is possible to operate the first two aeration tanks (#1 and #2) in parallel, to prevent a strong concentration gradient of the toxic substances, but this has not been necessary. The center zone of tank #4 (1500 m$^3$), to which methanol is added in response to the measured nitrate concentration in the tank, serves as a denitrification zone. In the final aerated zone (1500 m$^3$) accidentally overdosed methanol is oxidized. Tank #3, for maintenance purposes, is constructed like tank #4. During normal operation both parts of this tank are aerated.

To remove 1250 kg of nitrate nitrogen per day, 3500 kg of methanol per day are consumed, which is in the practical range of 2.5 to 3.0 kg methanol per kg nitrate nitrogen to be denitrified [72].

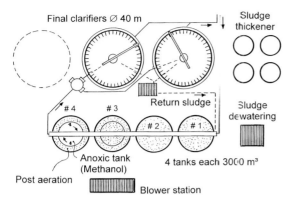

### 3.3.4
### Interactions between the biological reactors and the final clarifiers

The biological reactor and the final clarifier, linked by the return sludge flow, form a unit process. The conditions shown in Figure 3.26 for rectangular tanks with scrapers are similar to those in circular tanks with scrapers. For steady-state conditions, suspended solids concentration ($SS_{RS}$) of the return sludge is a function of MLSS and the return sludge flow ratio ($Q_{RS}/Q$):

$$SS_{RS} = MLSS \cdot \frac{Q + Q_{RS}}{Q_{RS}} = MLSS \cdot \left(1 + \frac{Q}{Q_{RS}}\right) \qquad (47)$$

Since MLSS is kept constant by withdrawing excess sludge, the suspended solids concentration ($SS_{RS}$) of the return sludge increases with decreasing return sludge flow ($Q_{RS}$). The return sludge flow is the sum of the flow of thickened sludge ($Q_{sl}$ with $SS_{sl}$) and the short circuit flow ($Q_{short}$ with MLSS). Since the return sludge flow is set by the return sludge pumping rate and the flow of thickened sludge to the hopper depends on the speed and other factors of the scraper as well as the height of the bottom sludge layer ($h_S$), the short circuiting flow is obtained with Eq. 48:

$$Q_{short} = Q_{RS} - Q_{sl} \qquad (48)$$

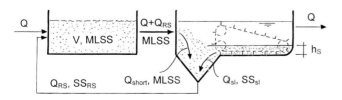

**Fig. 3.26** Interaction between biological reactor and final clarifier.

As long as $Q_{RS} > Q_{sl}$ there will be some short circuit flow, but if

$$Q_{sl} \cdot SS_{sl} \geq Q_{RS} \cdot SS_{RS} \tag{49}$$

the sludge in the clarifier may accumulate (increasing $h_S$) if the wastewater flow and hence the solids load in the final clarifier increases. In practice, therefore, $Q_{RS}$ should be higher than $Q_{sl}$.

For return sludge pumping two strategies are common:
- constant flow of $Q_{RS}$ at least during dry-weather periods
- constant ratio $Q_{RS}/Q$

Because of the diurnal fluctuation in wastewater flow, at constant $Q_{RS}$ the bottom sludge layer height $h_S$ and $SS_{RS}$ fluctuate approximately in parallel with $Q$, and MLSS fluctuates inversely to the wastewater flow variation. If the return sludge ratio $Q_{RS}/Q$ is kept constant $SS_{RS}$ fluctuates only a little. But if, e.g., during sudden storm water flows, the return sludge flow is immediately increased so as to keep $Q_{RS}/Q$ constant, the short circuit flow increases and $SS_{RS}$ decreases accordingly. It therefore makes sense to keep the ratio of $Q_{RS}$ to the 1–2 h sliding average of $Q$ constant.

The main design parameter for horizontal-flow clarifiers is the sludge volume surface load $q_{sv}$ expressed as L of sludge per m2 of clarifier surface ($A_{clar}$ in m²) and hour ($Q_h$ is the hourly peak flow in m³ h⁻¹):

$$q_{sv} = \frac{Q_h \cdot MLSS \cdot ISV}{A_{clar}} \; [L\,m^{-2}\,h^{-1}] \tag{50}$$

These and all further design calculations for clarifiers have been compiled by Ekama et al. [73]. The construction of the mixed liquor inlet, as well as of the clear water outlet and the sludge collection system, are also important for the performance of clarifiers.

If aeration tanks having a depth of more than 6 m precede the final clarifier, mixed liquor may float due to over-saturation with nitrogen. Measures to prevent floating include [74]:
- Stripping of nitrogen gas by an overflow weir cascade at the aeration tank outlet or by intensive coarse-bubble aeration in a shallower outlet zone of the aeration tank.
- Applying deep final clarifiers in which mixed liquor is introduced near the bottom.

## 3.4
### Design procedure

Detailed design information may be taken from handbooks [e.g., 75, 76]. In addition, for final clarifier design, Ref. 73 is recommended. The main design steps for single-stage activated sludge plants comprise:

1. Determination of the design loads (e.g., $BOD_5$, COD, suspended solids, nitrogen, phosphorous), the average alkalinity, the wastewater flow (daily and peak), and the peak storm water flow ($m^2\ h^{-}$). Existing data should be checked for annual fluctuations in the wastewater temperature, the loads, and the flows. If nitrification is required, the design load should be selected in combination with the wastewater temperature. The time period with the highest two-week average COD load at the lowest wastewater temperature can be selected to determine the critical sludge age and the sludge production. The highest oxygen demand may however have to be calculated for another period (e.g., summer).

2. Required effluent quality as percent removal or as concentrations ($C_{BOD,e}$, $C_{COD,e}$, $S_{NH4,e}$, $S_{NO3,e}$, $S_{P,e}$, $SS_e$). Mode of inspection, e.g., grab sample or 24-h composite sample. Value to be maintained, e.g., the annual average, 80% of all days, or 80% of grab samples.

3. Selection of the process configuration. Should it be with primary sedimentation? Aerobic selector? Anaerobic mixing tank for enhanced phosphate removal? Phosphate precipitation (and what type of precipitant)? Process for nitrogen removal? Addition of external organic carbon for nitrogen removal – continuously or temporarily? Necessity of adding N, P, and trace elements to specific industrial wastewater? Sludge disposal, separate stabilization, or costabilization of sludge?

4. An aerobic selector can be designed for a volumetric load of 20 kg COD ($m^3\ d^{-1}$). For enhanced biological phosphate removal, an anaerobic contact period of 0.25–0.5 h (based on $Q + Q_{RS}$) can be assumed.

5. Selection of the aerobic sludge age ($SRT_{aer}$) considering the degree of treatment, e.g., organic carbon removal only; nitrification at any time, at certain time periods, or at periods beyond a certain reactor temperature (e.g., $T$ 10 °C); costabilization of sludge. The following values may be used: Organic carbon removal only: SRT = 4 d; nitrification at $T \geq 10°$ °C: $SRT_{aer}$ = 8–10 d.

6. For denitrification: estimation of the nitrate to be denitrified (Eq. 24). Assumption of anoxic tank volume fraction $V_D/V$ (Eqs. 26 or 27 or values from Figure 3.11). Calculation of sludge age with Eq. 28.

7. For costabilization of sludge (extended aeration) and nitrogen removal, a sludge age of SRT = 25 d can be assumed.

8. Calculation of the remaining alkalinity with Eq. 33 and, if aeration tanks deeper than 6 m are planned, checking of the resulting pH (Section 3.2.6.2).

9. Determination of the MLSS considering the sludge volume index ISV to be expected. It is recommended to first check the final clarifier design to decide upon the appropriate MLSS. Generally MLSS = 3.0–5.0 kg $m^{-3}$ is used in design calculations. Higher values should be used only if a sludge volume index of ISV = 100 mL $L^{-1}$ or less can be expected.

10. Calculation of the daily mass of excess sludge with Eqs. 12 ($M_{exc,C}$) and 32 ($M_{exc,P}$). For settled municipal wastewater, as a rough estimate one can assume 0.4 to 0.5 kg excess sludge solids per kg COD influent. Finally, the reactor volume is calculated with Eq. 14.

11. For nitrogen removal only: estimation of the required internal recycle flow (Eq. 37) or the wastewater distribution in the step-feed process (Eq. 40) or the cycle period for intermittent nitrification–denitrification processes (Eq. 44).
12. Design of the aeration installation, considering the minimum and maximum oxygen uptake rates (Eqs. 13, 16, and 25).
13. Design of the final clarifier. As a rough estimate for the surface area of horizontal flow clarifiers, a volumetric sludge surface loading rate of $q_{SV} \leq 450$ L m$^{-2}$ h$^{-1}$ (Eq. 50) can be assumed. The side wall water depth should not be less than 3.00 m, and the average water depth should be in the range of 3.50–4.50 m (except, e.g., downstream of deep aeration tanks).

For larger plants it is recommended to check the results of static dimensioning with dynamic simulation. This should be considered with the collection of load data.

### References

1 von der Emde, W., Die Geschichte des Belebungsverfahrens, *gwf Wasser Abwasser* **1964**, *105*, 755–780.
2 Ardern, E., Lockett, T., Experiments on the oxidation of sewage without the aid of filters, *J. Soc. Chem. Ind.* **1914**, *33*, 523–529.
3 Pasveer, A., Abwasserreinigung im Oxidationsgraben, *Bauamt und Gemeindebau* **1958**, *31*, 78–85.
4 Zeper, J., De Man, A., New developments in the design of activated sludge tanks with low BOD loadings, in: *Advances in Water Pollution Research*, Vol. 1 (Jenkins, S.H., Ed.), pp. II-8/1–10, Oxford **1970**: Pergamon.
5 Pasveer, A., Sweeris, S., A new development in diffused air aeration. *T. N. O. Working Report A* **1962**, *27*, Delft, NL: TNO.
6 Scherb, K., Vergleichende Untersuchungen über das Sauerstoffeintragsvermögen verschiedener Belüftungssysteme auf dem Münchener Abwasserversuchsfeld, *Münchener Beiträge zur Abwasser-, Fischerei und Flußbiologie* **1965**, *12*, 330–350.
7 Wagner, M., Darstellung von Schadensfällen durch belegte EPDM-Membranen und Lösungsmöglichkeiten, *Schriftenreihe WAR TH Darmstadt* **2001**, *134*, 71–84.
8 Mueller, J. A., Boyle, W. C., Pöpel, H. J., *Aeration: Principles and Practice*. Boca Raton FL **2002**: CRC Press.
9 Jübermann, O., Krause, G., Die Zentralkläranlage der Erdölchemie GmbH und der Farbenfabriken Bayer AG in Dormagen; Ejektorbelüftung in der biologischen Abwasserreinigung, *Chem. Ing. Techn.* **1968**, *40*, 288–291.
10 Ullrich, A. J., Smith, M. W., The biosorption process of sewage and waste treatment, *Sewage. Ind. Wastes* **1951**, *23*, 1248–1252.
11 Imhoff, K., Das zweistufige Belebungsverfahren für Abwasser, *gwf Wasser Abwasser* **1955**, *96*, 43–45.
12 Böhnke, B., Möglichkeiten der Abwasserreinigung durch das Adsorptions-Belebungsverfahren. *Gewässerschutz–Wasser–Abwasser, RWTH Aachen* **1978**, *25*, 437–466.
13 Bringmann, G., Vollständige biologische Stickstoffeliminierung aus Klärwässern im Anschluß an ein Hochleistungsnitrifikationsverfahren, *Ges. Ing.* **1961**, *82*, 233–235.
14 Wuhrmann, K., Stickstoff- und Phosphorelimination, Ergebnisse von Versuchen im technischen Maßstab, *Schweiz. Z. Hydrol.* **1964**, *26*, 520–558.
15 Ludzack, F. J., Ettinger, M.B., Controlling operation to minimize activated sludge effluent nitrogen, *J. WPCF* **1962**, *35*, 920–931.
16 Pasveer, A., Über den Oxidationsgraben, *Schweizerische Z. Hydrol.* **1964**, *26(2)*, 466–484.

17  Kienzle, K. H., Untersuchungen über BSB₅ und Stickstoffelimination in schwachbelasteten Belebungsanlagen mit Schlammstabilisation. *Stuttgarter Berichte zur Siedlungswasserwirtschaft*, 47. München **1971**: R. Oldenbourg.

18  Barnard, J. L., Biological denitrification, *Water Pollut. Contr.* **1973**, *72*, 705–720.

19  Barnard, J. L., Cut P and N without chemicals, *Water Wastes Eng.* 11, 33–36.

20  Krauth, K. H., Staab, K.F., **1982**, Stickstoffelimination durch vorgeschaltete Denitrifikation, *Wasserwirtschaft* **1974**, *72*, 251–252.

21  von der Emde, W., Die Kläranlage Wien Blumental, *Oesterr. Wasserwirtschaft* **1971**, *23*, 11–18.

22  Matché, N. F., The Elimination of nitrogen in the treatment plant of Vienna–Blumental. *Water Res.* **1972**, *6*, 485–486.

23  Thomas, E. A., Verfahren zur Entfernung von Phosphaten aus Abwässern, *Swiss Patent* No. 361 543 **1962**.

24  Morgenroth, E., Wilderer, P. A., Continuous flow and sequential processes in municipal wastewater treatment, in: *Biotechnology: A Multi-volume Comprehensive Treatise, Vol. 11a, Environmental Processes I*, (Winter, J., Ed.), pp. 312–334. Weinheim **1999**: Wiley-VCH.

25  Kayser, R., Bemessung von Belebungs- und SBR-Anlagen, *ATV-DVGW-Kommentar zum ATV-DVGW-Regelwerk*, Hennef, Germany **2001**: ATV-DVWK.

26  Kulpberger, R. J., Matsch, L. C., Comparison of treatment of problem wastewater with air and high purity oxygen activated sludge systems, *Prog. Water Technol.* **1977**, *8(6)*, 141–151.

27  Lang, H., Nitrifikation in biologischen Klärstufen mit Hilfe des Bio-2-Schlamm-Verfahrens, *Wasserwirtschaft* **1981**, *71*, 166–169.

28  Schlegel, S., Der Einsatz von getauchten Festbettkörpern beim Belebungsverfahren, *gwf Wasser Abwasser* **1986**, *127*, 421–428.

29  Lessel, T. H., First practical experiences with submerged rope-type biofilm reactors, *Water Sci. Technol.* **1991**, *23*, 825–834.

30  Heinz, A., Röthlich, H., Leesel, T., Kopmann, T., Erfahrungen mit ring-lace Festbettreaktoren in der kommunalen Abwasserreinigung, *Wasser Luft Boden* **1996**, *40(4)*, 18–20.

31  Morper, M. R., Upgrading of activated sludge systems for nitrogen removal by application of the LINPOR® process, *Water Sci. Technol.* **1994**, *29(12)*, 107–112.

32  Takahashi, M., Suzuki, Y., Biological enhanced phosphorus removal process with immobilized microorganisms, *Veröffentlichungen des Instituts für Siedlungswasserwirtschaft und Abfallwirtschaft Universität Hannover* **1995**, *92*, 22/1–22/12.

33  Vogelpohl, A., Hochleistungsverfahren und Bioreaktoren für die biologische Behandlung hochbelasteter industrieller Abwässer, in: *Behandlung von Abwässern, Handbuch des Umweltschutzes und der Umweltschutztechnik*, Vol. 4 (Brauer, H. Ed.), pp. 391–413. Berlin **1996**: Springer-Verlag.

34  Brauer, H., Aerobe und anarobe biologische Behandlung von Abwässern im Hubstrahl-Bioreaktor, in: *Behandlung von Abwässern, Handbuch des Umweltschutzes und der Umweltschutztechnik*, Vol. 4 (Brauer, H., Ed.), pp. 414–504. Berlin **1996**: Springer-Verlag.

36  Krauth, Kh., Staab, K. F., Substitution of the final clarifier by membrane filtration within the activated sludge process: initial findings, *Desalination* **1988**, *68*, 179–189.

37  van der Roest, H. F., Lawrence, D. P., van Betem A. G. N., Membrane bioreactors for municipal wastewater treatment, *Water and Wastewater Practitioner Series: STOWA Report*, London **2002**: IWA.

38  Begert, A., Summen- und Gruppenparameter für organische Stoffe von Wasser und Abwasser, *Wiener Mitteilungen Wasser, Abwasser, Gewässer* **1985**, *57*, G1–G29.

39  Henze, M., Gujer, W., Takahshi, M., van Loosdrecht, M., Activated sludge models ASM1, ASM2, ASM3d and ASM 3. *IAWQ Scientific and Technical Report No. 9*. London **2000**: IAWQ.

40  Wuhrmann, K., von Beust, F., Zur Theorie des Belebtschlammverfahrens, II. Über den Mechanismus der Elimination gelöster organischer Stoffe aus Abwasser, *Schweiz. Z. Hydrol.* **1958**, *20*, 311–330.

41  Tischler. L. F., Linear Removal of Simple Organic Compounds in the Activated Sludge Process, Thesis, The University of Texas, Austin, TX **1968**.

42 Gujer, W., Kayser, R., Bemessung von Belebungsanlagen auf der Grundlage einer CSB-Bilanz, *Korrespondenz Abwasser* **1998**, *45*, 944–948.

43 Stensel, H. D., Barnard, J. L., Principles of biological nutrient removal, in: *Design and Retrofit of Wastewater Treatment Plants for Nutrient Removal* (Randal, C. W., Barnard, J. L., Stensel, H. D., Eds), pp. 25–84. Lancaster, PA **1992**: Technomic Publishing.

44 Knowles, G., Downing, A. L., Barrett, M. J., Determination of kinetic constants for nitrifying bacteria in mixed culture, *J. Gen. Microbiol.* **1965**, *38*, 263–278.

45 Kayser, R., A 131: Quo vadis, *Wiener Mitt. Wasser, Abwasser, Gewässer* **1997**, *114*, 149–166.

46 Anthonisen, A. C., Loehr, R. C., Prakasam, T. B. S., Srinath, E. G., Inhibition of nitrification by ammonia and nitrous acid, *J. WPCF* **1976**, *24*, 835–852.

47 Nyhuis, G., Beitrag zu den Möglichkeiten der Abwasserbehandlung bei Abwässern mit erhöhten Stickstoffkonzentrationen. *Veröffentlichungen des Instituts für Siedlungswasserwirtschaft und Abfalltechnik der Universität Hannover 61* **1985**.

48 Wicht, H., Beier, M., $N_2O$-Emissionen aus nitrifizierenden und denitrifizierenden Kläranlagen, *Korrespondenz Abwasser* **1995**, *42*, 404–413.

49 Wicht, H., $N_2O$-Emissionen durch den Betrieb biologischer Kläranlagen, *Veröffentlichungen des Instituts für Siedlungswasserwirtschaft TU Braunschweig, 58* **(1995)**.

50 EPA *Nitrogen Removal, Process Design Manual.* Environmental Protection Agency, Cincinnati, OH **1975**.

51 Kappeler, J. und Gujer, W., Estimation of kinetic parameters of heterotrophic biomass under aerobic conditions and characterization of wastewater for activated sludge modelling, *Water Sci. Technol.* **1992**, *25(6)*, 125–139.

52 Barnard, J. L., Design of prefermentation process, in: *Design and Retrofit of Wastewater Treatment Plants for Nutrient Removal* (Randal, C. W., Barnard, J. L., Stensel, H. D., Eds), pp. 85–96. Lancaster PA **1992**: Technomic Publishing.

53 Schön, G., Jardin, N., Biological and chemical phosphorus elimination, in:

*Biotechnology: A Multi-volume Comprehensive Treatise, Vol. 11a, Environmental Processes I* (Winter, J., Ed.), pp. 285–319. Weinheim **1999**: Willey-VCH.

54 Nowak, O., Nitrifikation im Belebungsverfahren bei maßgebendem Mischwasserzufluss. *Wiener Mitt. Wasser, Abwasser, Gewässer 135* **(1996)**.

55 Kayser, R., Activated sludge process, in; *Biotechnology: A Multi-volume Comprehensive Treatise, Vol. 11a, Environmental Processes I* (Winter, J., Ed.), pp. 253–283. Weinheim **1999**: Willey-VCH.

56 Mohlman, F. W., The sludge index, *Sewage Works J.* **1934**, *6*, 119–122.

57 Jenkins, D., Daigger, G. T., *Manual on the Causes and Control of Activated Sludge Bulking, Foaming and Other Solids Separation Problems*, 3rd Edn. Boca Raton FL **2003**: CRC Press.

58 Cooper, P. F., Collinson, B., Green, M. K., Recent advances in sewage effluent denitrification: Part II, *Water Pollut. Control* **1977**, *76*, 389–398.

59 Miyaji, Y., Iwasaki, M., Serviga, Y., Biological nitrogen removal by step-feed process, *Prog. Water Technol.* **1980**, *12(6)*, 193–202.

60 Schlegel, S., Nitrifikation und Denitrifikation in einstufigen Belebungsanlagen; Betriebsergebnisse der Kläranlage Lüdinghausen, *gwf Wasser Abwasser* **1983**, *124*, 428–434.

61 Matsché, N. F., Removal of nitrogen by simultaneous nitrification–denitrification in an activated sludge plant with mammoth rotor aeration, *Prog. Water Technol.* **1977**, *8(4–5)*, 625–637.

62 Ermel, G., Stickstoffentfernung in einstufigen Belebungsanlagen: Steuerung der Denitrifikation. *Veröffentlichungen des Instituts für Stadtbauwesen TU Braunschweig 35* **1983**.

63 Seyfried, C. F., Hartwig, P., Großtechnische Betriebserfahrungen mit der biologischen Phosphatelimination in den Klärwerken Hildesheim und Husum, *Korrespondenz Abwasser* **1991**, *38*, 184–190.

64 Kayser, R., Process control and expert systems for advanced wastewater treatment plants. in: *Instrumentation and Control of Water and Wastewater Treatment and Transport Systems* (Briggs, R., Ed.), pp. 203–210. Oxford **1990**: Pergamon.

65 Wareham, D. G., Hall, K. J., Mavinic, D. S., Real-time control of aerobic anoxic sludge digestion using ORP, *ASCE J. Env. Eng.* **1993**, *119*, 120–136.

66 Boes, M., Stickstoffentfernung mit intermittierender Denitrifikation: Theorie und Betriebsergebnisse, *Korrespondenz Abwasser* **1991**, *38*, 228–234.

67 Tholander, B., An example of design of activated sludge plants with denitrification, *Prog. Water Technol.* **1977**, *9(4–5)*, 661–672.

68 Sørensen, J., Optimization of a nutrient removing wastewater treatment plant using on-line monitoring, *Water Sci. Technol.* **1996**, 33(1), 265–273.

69 Zanders, E., Groeneweg, J., Soeder, C. J., Blähschlammverminderung und simultane Nitrifikation/Denitrifikation, *Wasser-Luft-Betrieb* **1987**, *31(9)*, 20–24.

70 Abeling, U., Stickstoffelimination aus Industrieabwässern; Denitrifikation über Nitrit. *Veröffentlichungen des Instituts für Siedlungswasserwirtschaft und Abfalltechnik der Universität Hannover 86* (**1994**).

71 Jetten, M. S. M., Horn, S. J., van Loosdrecht, M. C. M., Towards a more sustainable municipal wastewater treatment system, *Water Sci. Technol.* **1997**, *35(6)*, 171–180.

72 Zacharias, B., Biologische Stickstoffelimination hemmstoffbelasteter Abwässer am Beispiel eines Eisenhüttenwerkes, *Veröffentlichungen des Instituts für Siedlungswasserwirtschaft TU Braunschweig, 60* (**1996**).

73 Ekama, G. A., Barnard, J. L., Günthert, F. W., Krebs, P., McCorquodale, J. A., Parker, D. S., Wahlberg, E. J., Secondary settling tanks: theory, modelling, design and operation. *IAWQ Scientific and Technical Report No. 6*, London **1997**: IAWQ.

74 ATV, Hinweise zu tiefen Belebungsbecken, *Korrespondenz Abwasser* **1996**, *43*, 1083–1086.

75 Randal, C. W., Barnard, J. L., Stensel, H. D., *Design and Retrofit of Wastewater Treatment Plants for Nutrient Removal.* Lancaster, PA **1992**: Technomic Publishing.

76 ATV, *ATV Handbuch: Biologische und weitergehende Abwasserreinigung*, 4th Edn. Berlin **1997**: Ernst & Sohn.

# 4
# Modeling of Aerobic Wastewater Treatment Processes

Mogens Henze

## 4.1
## Introduction

The trend of improving the effluent quality from wastewater treatment plants leads to an increasing complexity in the design and operation of the plants. To design the plants and optimize and control the operation, it is necessary to use dynamic models. Most models used for wastewater treatment plants today are deterministic, the exception being some models for control, which can be of the black-box type. The deterministic models aim to give a realistic description of the main processes of the plant. However, models are never true in the process sense and always simplify the complicated processes occurring in a biological treatment plant. The models we have at hand today include the phenomena for which we think we have a reasonable explanation. A major gap in modeling is lack of knowledge about the development of the microbiology in treatment plants. Thus, phenomena like bulking and foaming are still awaiting more basic understanding before we can develop reliable models for them. Below, various elements of modeling are discussed. For further information on modeling in general and on modeling of biofilters, please see Chapter 13, Volume 4, of the Biotechnology series (Rehm et al., 1991).

## 4.2
## Purpose of Modeling

Models are used for several purposes with regard to wastewater treatment plants:
- design
- control
- operational optimization
- teaching
- organizing tool

The intended use of a model has implications for its structure. Models used for design and for control need not be identical.

*Environmental Biotechnology. Concepts and Applications* Edited by H.-J. Jördening and J. Winter
Copyright © 2005 WILEY-VCH Verlag GmbH & Co. KGaA, Weinheim
ISBN: 3-527-30585-8

Models are no better than the assumptions upon which they are built. Several factors influence their behavior. The way they handle detailed wastewater composition is important. Wastewater characterization has a strong impact on real plant behavior as well as on its modeling. Errors in characterization of the wastewater or changes in the composition of the wastewater can result in erroneous modeling results.

## 4.3
## Elements of Activated Sludge Models

An activated sludge model consists of various elements:
- transport processes
- components
- biological and chemical processes
- hydraulics

To this can be added a framework of component conversion and data presentation tools. The amount of detail needed depends on the intended use of the model and on the amount of information available on the wastewater and the treatment plant.

### 4.3.1
### Transport Processes and Treatment Plant Layout

Transport processes deal only with movement of water and do not include any chemical or biological processes. The flow scheme of the treatment plant, which describes the movement of water and sludge, must be known, and a model of the flow scheme must be part of the overall model. Often it is necessary to simplify the flow scheme, due to a high complexity of the plant or to limitations to the number of tanks that can be applied in the model. Simplification of a flow scheme can be very beneficial for improving the overview of the situation in the plant. Parallel tanks can be modeled as one tank, and tanks in series might also be modeled as one. However, one must be aware that the more simplified the flow scheme model is, the greater the model calculations may deviate from the real world.

The flow pattern of wastewater and sludge must be known. Not all treatment plant operators can give this information directly, at least not correctly.

#### 4.3.1.1  **Aeration**
Aeration in each of the tanks must be described, either with a fixed oxygen concentration or as an aeration coefficient, $K_La$, operating uncontrolled or controlled together with a control strategy. Aeration of tanks without mechanical aeration should also be taken into account. The oxygen penetration through the surface of, e.g., denitrification tanks can be significant.

The use of a fixed oxygen concentration is a simplification that does not allow for simultaneous nitrification–denitrification processes to be modeled, nor the impact of oxygen in recycle flows. Thus, the $K_La$ model is recommended for professional modeling.

### 4.3.1.2 **Components**

Components are the soluble and suspended substances that one wants to model, including substances present in the raw wastewater as well as substances found within the treatment plant. The components to model depend on the purpose of the model. If the purpose is to model a nitrifying plant, one phosphorous component will suffice, but if modeling of biological phosphorous removal is the objective, then at least two phosphorous components are needed: polyphosphate in the biomass ($X_{PP}$) and a soluble component, orthophosphate ($S_{PO4}$).

### 4.3.1.3 **Processes**

Processes include modeling of chemical and biological processes. The model to be used is always a strong simplification of reality, but these simplified models often give reasonable modeling results, but sometimes not. For example, a simple phosphorous precipitation model, like the one applied in ASM2 (Henze et al., 1995), can model the part of phosphorous removal that is not related to assimilation and biological excess uptake. On the other hand, a model that does not account for the denitrification that occurs in nitrification tanks, the so-called simultaneous denitrification, gives effluent nitrate concentrations that are too high. Thus, the degree of simplification must be considered. If one wants to model denitrification, there must be at least one process for this (anoxic growth of heterotrophic biomass). But it is also possible to apply many processes for the description of denitrification. This occurs for a model that is aimed at predicting intermediates like nitrite, dinitrogen oxide, nitrous oxide, etc., in the denitrification process.

### 4.3.1.4 **Hydraulic Patterns**

The hydraulics of the various tanks have to be modeled. Any tanks can be modeled as ideal mixing, but some may need to be modeled as a series of ideally mixed tanks. More complex models of the hydraulics are seldom needed.

## 4.4
## Presentation of Models

The complex interrelationship between the various components and processes are best presented in a matrix. A very simple model for an activated sludge process for removal of organic matter is shown in the matrix in Table 4.1. With this table it is possible to obtain a quick overview of the conversions in the described process.

**Table 4.1.** Simple model matrix for activated sludge organic removal[a].

| Process | Component | | | Process Rate |
|---|---|---|---|---|
| | $S_{O2}$ | $S_S$ | $X_H$ | |
| Aerobic growth | ±1 | – | +1 | $m_{m,H} X_H$ |
| Decay | | +1 | –1 | $b_H X_H$ |

[a] $S_{O2}$: soluble oxygen, g $O_2$ m$^{-3}$; $S_S$: soluble organic substrate, g COD m$^{-3}$; $X_H$: biomass, g COD m$^{-3}$; $Y_H$: yield coefficient, g COD g$^{-1}$ COD; $m_{m,H}$: maximum growth rate, d$^{-1}$; $K_S$: half saturation constant, g COD m$^{-3}$; $b_H$: decay constant, d$^{-1}$.

### 4.4.1
### Mass Balances

The rows in the matrix give the mass balance of the process. The second row in Table 4.1 shows, e.g., that decay removes biomass, $X_H$ (the stoichiometric coefficient is negative, –1), and produces soluble substrate, $S_S$ (the stoichiometric coefficient is +1).

### 4.4.2
### Rates

The rate equations of the processes are found in the right column (Table 4.1). To find the rate for the change in biomass in relation to decay, $r_{XH}$, the rate equation must be multiplied by the stoichiometric coefficient in the matrix; that is:

$$r_{XH} = -1 \cdot b_H X_H \tag{1}$$

### 4.4.3
### Component Participation

The columns in the matrix (Table 4.1) give an overview of which processes the various components participate in (according to the model). The second column shows that the substrate, $S_S$, is removed by aerobic growth and is produced by decay.

### 4.5
### The Activated Sludge Models Nos. 1, 2 and 3 (ASM1, ASM2, ASM3)

Most models for activated sludge processes are based on the IAWQ Activated Sludge Model No. 1, called ASM1 (Henze et al., 1987). This model is used as a platform for further model development and is today the international standard for advanced activated sludge modeling. Because the ASM contains only a very simplified model, most modelers expand the ASM1, depending on the degree of complexity needed to model the actual problem. In some situations more complex kinetic equations are needed, in others more processes or more components have to be included. Nutrient limitation of the growth processes is usually included.

## 4.5.1
## Activated Sludge Model No. 1 (ASM1)

The ASM1 model describes activated sludge processes with nitrification and denitrification. It is the experience of more than 10 years of use that the model gives a good description of the processes, as long as the wastewater has been characterized in detail and is of domestic or municipal origin. The model was not developed to fit industrial wastewater treatment. When used for industrial wastewater, great care should be taken in calibration and in interpretation of the results.

The ASM1 can calculate several details in the plant:
- oxygen consumption in the tanks
- concentration of ammonia and nitrate in the tanks and in the effluent
- concentration of COD in the tanks and in the effluent
- MLSS in the tanks
- solids retention time
- sludge production

As with all models, ASM1 also gives erroneous results if it is fed erroneous or overly simplified information. Table 4.2 shows the process matrix for ASM1. All processes, reaction kinetics, mass balances, and stoichiometry in the model are included in the matrix. If understood, the matrix notation is a useful tool to obtain an easy overview of the model.

In the ASM1 the processes for organic matter removal and nitrification are slightly coupled, as seen in Figure 4.1.

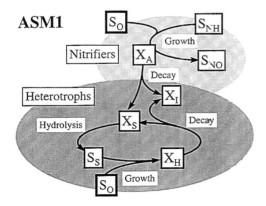

**Fig. 4.1** Processes for heterotrophic and nitrifying bacteria in the Activated Sludge Model No. 1 (ASM1) (Gujer et al., 1999).

**Table 4.2.** The process matrix for Activated Sludge Model No. 1 (ASM1), based on Henze et al. (1997).[a]

| | 1 $S_I$ | 2 $S_S$ | 3 $X_I$ | 4 $X_S$ | 5 $X_{B,H}$ | 6 $X_{B,A}$ | 7 $X_P$ | 8 $S_{O2}$ | 9 $S_{NO3}$ | 10 $S_{NH4}$ | 11 $S_{ND}$ | 12 $X_{ND}$ | 13 $S_{ALK}$ | Process Rate[b] |
|---|---|---|---|---|---|---|---|---|---|---|---|---|---|---|
| 1. Aerobic growth of heterotrophs | | $-1/Y_H$ | | | 1 | | | $-(1-Y_H)/Y_H$ | | $-i_{XB}$ | | | $-i_{XB}/14$ | $m_{max,H} \cdot M_2 \cdot M_8 \cdot X_{B,H}$ |
| 2. Anoxic growth of heterotrophs | | $-1/Y_H$ | | | 1 | | | | $-(1-Y_H)/2.86\,Y_H$ | $-i_{XB}$ | | | $\dfrac{(1-Y_H)/14 -}{2.86\,Y_H} - i_{XB}/14$ | $m_{max,H} \cdot M_2 \cdot I_8 \cdot M_9 \cdot h_g \cdot X_{B,H}$ |
| 3. Aerobic growth of autotrophs | | | | | | 1 | | $-(4.57 - Y_A)/Y_A$ | $1/Y_A$ | $i_{XB} - 1/Y_A$ | | | $-i_{XB}/14 - 1/(7\,Y_A)$ | $m_{max,A} \cdot M_{10} \cdot M_8 \cdot X_{B,A}$ |
| 4. Decay of heterotrophs | | | | $1 - f_P$ | $-1$ | | $f_P$ | | | | | $i_{XB} - f_P \cdot i_{XP}$ | | $b_H \cdot X_{B,H}$ |
| 5. Decay of autotrophs | | | | $1 - f_P$ | | $-1$ | $f_P$ | | | | | $i_{XB} - f_P \cdot i_{XP}$ | | $b_A \cdot X_{B,A}$ |
| 6. Ammonification of soluble organic nitrogen | | | | | | | | | | 1 | $-1$ | | $1/14$ | $k_s \cdot S_{ND} \cdot X_{B,H}$ |
| 7. Hydrolysis of entrapped organics | 1 | | | $-1$ | | | | | | | | | | $k_h \cdot sat\,(M_8 + h_h \cdot I_8 \cdot M_9)$ |
| 8. Hydrolysis of entrapped organic nitrogen | | | | | | | | | | | 1 | $-1$ | | rate 7 $(X_{ND}/X_S)$ |
| Unit | COD | COD | COD | COD | COD | COD | COD | negative COD | N | N | N | N | Mole | |

[a] For detailed explanation, see Henze et al. (1987)

[b] $M_x$ = Monod kinetics for component $x$ $(x/(K+x))$; $I_x$ = inhibition kinetics for component $x$ $(K/(K+x))$; $sat$ = saturation kinetics $k_h$

## 4.5.2
### Activated Sludge Model No. 2 (ASM2)

The ASM2 model incorporates biological phosphorous uptake (Henze et al., 1995). Its combination with denitrification results in a model that is much more complex than ASM1. To describe the life of PAOs (phosphate-accumulating organisms), internal storage compounds of phosphorous and organic matter are needed. Modeling the growth of the heterotrophic organisms requires at least three different organic substrates:
- acetic acid-like compounds, $S_A$
- fermentable substrate, $S_F$
- slowly degradable substrate, $X_S$

The ASM2 assumes that growth occurs only on substrate $S_A$ and that the other organic substrates are hydrolyzed and fermented so as to be converted to $S_A$.

## 4.5.3
### Activated Sludge Model No. 3 (ASM3)

The ASM3 model represents an alternative process description for heterotrophic bacteria. Under transient loading conditions, heterotrophic bacteria can store organic matter as polymeric compounds like polyhydroxyalkanoate (PHA) or glycogen (GLY). This storage can influence the overall process of activated sludge by supplying organic material under starvation conditions. The Activated Sludge Model No. 3 (ASM3) includes this storage in its processes and applies a simplified maintenance/decay process. Figure 4.2 shows the processes used to describe heterotrophic bacteria; as can be seen, the substrate flows through the organisms during oxygen consumption for storage, growth, and maintenance.

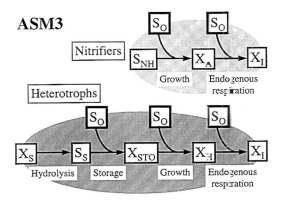

**Fig. 4.2** Processes for heterotrophic organisms in the Activated Sludge Model No. 3 (ASM3) (Gujer et al., 1999).

## 4.6
## Wastewater Characterization

Wastewater characterization is an important element in dealing with models. The detailed wastewater composition has a significant influence on the performance of the treatment processes (Henze, 1992). The degree of detail needed in the wastewater characterization depends on the objective of modeling. Figure 4.3 shows a detailed fractionation of municipal wastewater in relation to biological phosphorous removal. As can be seen, what is considered 'soluble' in the model (substances with the symbol S) is not equivalent to analytically determined soluble material, because part of the particulate organic material is easily degradable. Thus, from a reaction point of view, it cannot be distinguished from the analytically soluble material.

Table 4.3 shows a detailed fractionation of wastewater, which allows for modeling denitrification as well as biological phosphorous removal processes.

The characterization of wastewater can be simplified by combining existing data on BOD, COD, soluble fractions, etc. An example of this is from the Dutch river boards, which use the procedure shown below (Roeleveld and Kruit, 1998).

Influent:

$$COD_{total} = S_S + S_I + X_S + X_I \tag{2}$$
$$COD_{soluble} = S_S + S_I \ (0.1 \ \mu m \ filter) \tag{3}$$
$$COD_{suspended} = X_S + X_I \tag{4}$$

where $S_I$ = soluble COD (0.1 μm filter) in the effluent from a low loaded biological treatment plant.

**Table 4.3.** Wastewater components in ASM2. Note that modeled levels of soluble organic material are higher than analytically determined levels.

| Symbol | Component | Typical Level | Unit |
|---|---|---|---|
| **'Soluble' compounds:** | | | |
| $S_F$ | fermentable organic matter | 20–250 | g COD m$^{-3}$ |
| $S_A$ | acetate and other fermentation products | 10–60 | g COD m$^{-3}$ |
| $S_{NH4}$ | ammonium | 10–100 | g N m$^{-3}$ |
| $S_{NO3}$ | nitrate + nitrite | 0–1 | g N m$^{-3}$ |
| $S_{PO4}$ | orthophosphate | 2–20 | g P m$^{-3}$ |
| $S_I$ | inert soluble organic matter | 20–100 | g COD m$^{-3}$ |
| **'Suspended' compounds:** | | | |
| $X_I$ | inert suspended organic matter | 30–150 | g COD m$^{-3}$ |
| $X_S$ | slowly degradable organic matter | 80–600 | g COD m$^{-3}$ |
| $X_H$ | heterotrophic biomass | 20–120 | g COD m$^{-3}$ |
| $X_{PAO}$ | phosphorous-accumulating biomass (PAO) | 0–1 | g COD m$^{-3}$ |
| $X_{PP}$ | stored polyphosphate in PAO | 0–0.5 | g P m$^{-3}$ |
| $X_{PHA}$ | stored organic polymer in PAO | 0–1 | g COD m$^{-3}$ |
| $X_{AUT}$ | nitrifying biomass | 0–1 | g COD m$^{-3}$ |

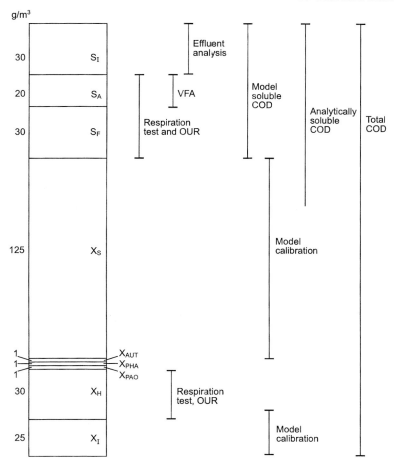

**Fig. 4.3** COD fractionation in the Activated Sludge Model No. 2 (ASM2). The figure shows a typical composition of primary settled municipal wastewater. Various analytical methods that can be used for the determination of single components are shown. Organic inert suspended material, $X_I$, and organic, slowly degradable suspended material, $X_S$, can be determined only by modeling and calculation (Henze et al., 1995). At present, no methods exist to directly determine these two fractions independently. For inert suspended organic matter, $X_I$, the reason is that its concentration in the sludge originates in two sources: it is a component of the raw wastewater and is also produced during the decay of microorganisms. These two contributions cannot be separated analytically at present. Slowly degradable suspended organic matter, $X_S$, cannot be determined analytically except from calculation based on determinations of the other fractions of organic matter.

$S_S + X_S$ is determined from a $BOD_5$ measurement or from a $BOD_{20}$ measurement (or calculation):

$$S_S + X_S = 1.5 \ BOD_5 \qquad (5)$$
$$S_S + X_S = BOD_{20}/0.85 \qquad (6)$$

A simplified evaluation of the nitrogen components is done based on the content of nitrogen in the organic components:

$$TN = f_{NX} \cdot X_S + f_{NS} \cdot S_S + S_{NH4} \qquad (7)$$

where $f_{NX}$ and $f_{NS}$ are the fractions of nitrogen (g N g$^{-1}$ COD) and are typically around 0.03 g N g$^{-1}$ COD (Henze et al., 1997).

## 4.7
## Model Calibration

Calibration of models is difficult. It is mandatory for a successful calibration that the content and the functioning of the model be understood. When calibrating by hand, it is important to follow a strategy and be aware of which parameters are to be used for which part of the calibration. It is equally important to know the realistic intervals within which the various parameters should vary, in order to still have a realistic model. Table 4.4 shows the steps to follow for calibration of the Activated Sludge Model No. 1. Constants not shown should, under normal circumstances, not be changed.

For processes without changes in performance or loading, calibration is easy but not very reliable. Often, many sets of calibration constants give a nice model fit. The

**Table 4.4.** Calibration steps for the activated sludge model no. 1 (ASM1).

| Parameter | Calibration Based on: | Comments |
|---|---|---|
| Heterotrophic growth rate, $m_{H,max}$ | OUR | often kept unchanged |
| Half saturation constant, $K_S$ | COD in effluent | used to model the hydraulics of the tanks |
| Growth rate for nitrifying biomass, $m_{A,max}$ | $NH_4$ in effluent | if $NH_4$ for all conditions is low, $m$ should not be changed |
| Half saturation constant, $K_{NH4}$ | $NH_4$ in effluent | only if $NH_4$ is low |
| Half saturation constant, $K_{O,A}$ | $NH_4$ in effluent | only if $S_{O2}$ varies |
| Denitrification factor, $h_{NO3}$ | $NO_3$ in effluent | |
| Half saturation constant, $K_{NO3}$ | $NO_3$ in effluent | only if $NO_3$ is low |

more dynamic the data for calibration, the better the calibration will be, and the greater the chance will be for the model to respond correctly to situations outside the range of those used in the calibration. Many treatment plants have only small dynamic changes. For such plants, single dynamic events (overloading, mechanic breakdown, snow melt, etc.) are excellent for calibration purposes.

## 4.8
## Computer Programs

Several computer programs that implement various models exist, and many include one or more of the models in the ASM family. Table 4.5 gives a short overview.

## 4.9
## Use of Models

Models can be used for many purposes. An important field is control of treatment plants, for which it is often possible to use simplified models. Complex processes, like nitrification– denitrification combined with biological phosphorous removal and external carbon source addition, can make it necessary to use even more complex models than those in the ASM family (Meinholt et al., 1998).

Optimizing the operation of a treatment plant is another use for models. Figure 4.4 shows a model calculation for a treatment plant that receives varying loads of ammonia from the sludge treatment. This often creates a problem with overload of the nitrification process, resulting in breakthrough of ammonia peaks into the effluent from the plant. A model calculation can help find the optimal operational strategy for handling ammonia-rich sidestreams.

**Table 4.5.** Selected simulation programs for activated sludge plants.

| Program | Models Included | Characteristics |
| --- | --- | --- |
| ASIM | ASM1, ASM2, ASM3 | simple program, allows for quick modeling |
| EFOR | ASM1, ASM2 | program with a detailed wastewater calibration model, in which available data (including BOD) is used for calculation of both raw and treated wastewater composition; choice between various settler models |
| GPX | ASM1, ASM2, etc. | program in which different models can be selected for the simulation of both biology and settling |
| SIMBA based on Matlab-Simulink | ASM1, ASM2 | program widely used in the Netherlands and Germany |

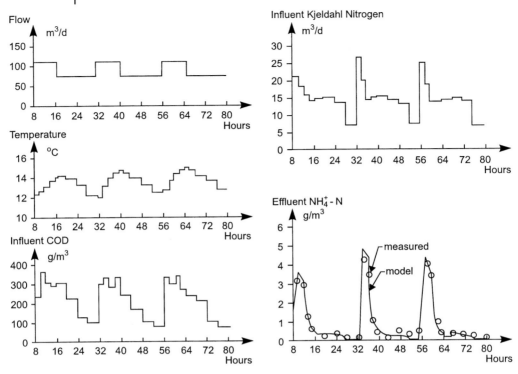

**Fig. 4.4** Use of ASM1 for modeling effluent variations for ammonia dynamically loaded with ammonia-rich supernatant (Gujer, 1985).

Figure 4.5 shows a model calculation with ASM1 in relation to startup of a nitrifying activated sludge treatment process. The calculations are made to evaluate the possible effect of inoculation with nitrifying biomass. The results show that the inoculation should have no significant impact on the startup.

Figure 4.6 shows a model calculation of a process for denitrification with methanol. The process has an aerobic and an anoxic tank. By using two heterotrophic populations with different growth characteristics during modeling, a result is obtained that explains the differences in methanol removal rate. If the process is run solely aerobically for a period, the group of organisms that, under aerobic conditions, metabolizes methanol quickly will increase. This group has a slow methanol conversion rate under anoxic conditions. In the second part of the experiment, from day 40 onward, the process is operated with 30% anoxic reaction time and 70% aerobic reaction time. Under these operational conditions, the other group of organisms will dominate, resulting in higher methanol conversion rates under anoxic conditions.

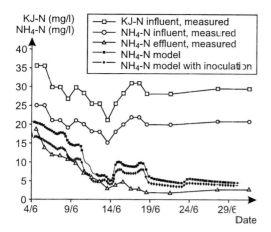

**Fig. 4.5** Model calculation of startup of an activated sludge process with and without inoculation with nitrifying organisms (Finnson, 1994).

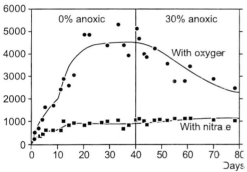

**Fig. 4.6** Modeling denitrification rate in the presence of methanol under varying operational conditions in an activated sludge process (Purtschert and Gujer, 1999).

## References

Finnson, A., Computer simulations of full-scale activated sludge processes. Thesis, Kungl. Tekniska Högskolen, Department of Water Resources Engineering (**1994**).

Gujer, W., *Ein dynamisches Modell für die Simulation von komplexen Belebtschlamm-verfahren*. Dübendorf, Switzerland **1985**: ERWAG.

Gujer, W., Henze, M., Activated sludge modelling and simulation, *Water Sci. Technol.* **1991**, *23*, 1011–1023.

Gujer, W., Henze, M., Loosdrecht, M., Mino, T., Activated sludge model no. 3, *Water Sci. Technol.* **1999**, *3(1)*, 183–193.

Henze, M., Characterization of wastewater for modelling of activated sludge processes, *Water Sci. Technol.* **1992**, *25(6)*, 1–15.

Henze, M., Grady, C. P. L., Gujer, W., Marais, G. v. R., Matsuo, T., *Activated-Sludge Model No. 1*, IAWPRC Scientific and Technical Reports, No. 1. London **1987**: IAWPRC.

Henze, M., Gujer, W., Mino, T., Matsuo, T., Wentzel, M. C., Marais, G. v. R., *Activated Sludge Model No. 2*, IAWQ Scientific and Technical Reports, No. 3. London **1995**: IAWQ.

Henze, M., Harremoës, P., la Cour Jansen, J., Arvin, E., *Wastewater Treatment: Biological and Chemical Processes*, 2nd Edn. Berlin **1997**: Springer-Verlag.

Meinholt, J., Arnold, E., Isaacs, S., Pedersen, H. R., Henze, M., Effect of continuous addition of an organic substrate to the anox phase on biological phosphorus removal, *Water Sci. Technol.* **1988**, *38(1)*, 97–105.

Purtschert, I., Gujer, W., Population dynamics by methanol addition in denitrifying waste water treatment plants, *Water Sci. Technol.* **1999**, *39(1)*, 43–50.

Rehm, H.-J., Reed, G., Pühler, A., Stadler, P., *Biotechnology* 2nd Edn., Vol. 4, pp. 407–439. Weinheim **1991**: VCH.

Roeleveld, P. J., Kruit, J., Richtlinien für die Charakterisierung von Abwasser in den Niederlanden, *Korrespondenz Abwasser* **1998**, *45*, 465–468.

# 5
# High-rate Anaerobic Wastewater Treatment

Hans-Joachim Jördening and Klaus Buchholz

## 5.1
## Introduction

Anaerobic wastewater treatment is a very old method of water purification. The Sumerians were already familiar with this method of wastewater purification. The first full-scale application of the anaerobic treatment of domestic wastewater is described for the 1860s (McCarty, 2001).

Anaerobic wastewater treatment has some advantages in comparison to aerobic treatment. The specific productivity of an anaerobic system is much higher and the engineering is simple. Whereas much energy is needed for aeration in aerobic wastewater treatment plants, anaerobic treatment produces energy in the form of usable biogas. Anaerobic treatment produces only low amounts of excess biomass (5% on COD reduction in comparison to nearly 30%–50% in aerobic wastewater treatment) and the nutrient requirements are therefore small compared with aerobic treatment. On the other hand, this also means that the growth of anaerobic bacteria is slow, and for this reason in particular the start-up of an anaerobic plant is slow. Anaerobically treated water usually cannot be released directly into the environment. The COD concentration in the effluent is usually not as low as is achievable with aerobic plants and is often odorous. For this reason high-loaded wastewaters are often treated by a combination of anaerobic and aerobic treatment. The second (aerobic) step is then, if necessary, designed also for N-elimination.

Anaerobic treatment of wastewaters includes acid formation and methane formation, which differ significantly in terms of nutrient needs, growth kinetics, and sensitivity to environmental conditions (Demirel and Yenigün, 2002). Therefore, these two steps are preferentially separated physically. For many industrial wastewaters and hydraulic retention times (HRT) that are not too short, acid formation can easily be done without specific installations for bacteria retention or recycling, whereas these procedures are absolutely necessary for methane formation.

Figure 5.1 shows some basic types of reactor for methane formation.

*Environmental Biotechnology. Concepts and Applications.* Edited by H.-J. Jördening and J. Winter
Copyright © 2005 WILEY-VCH Verlag GmbH & Co. KGaA, Weinheim
ISBN: 3-527-30585-8

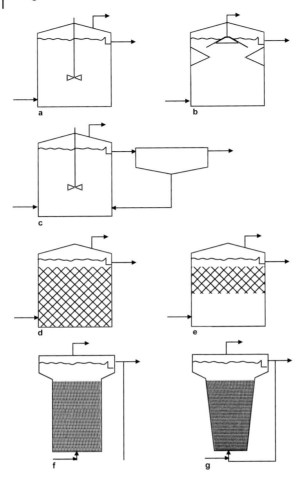

**Fig. 5.1** Reactor systems for anaerobic wastewater treatment. a, Completely stirred tank reactor; b, upflow anaerobic sludge blanket; c, contact process; d, fixed film stationary bed; e, combination of a and d; f, fluidized bed; g, tapered bed.

The classical digester or completely stirred tank reactor (CSTR) (Fig. 5.1a) without biomass retention is hardly used except in the treatment of sewage sludge. Industrial wastewater treatment always aims to be highly productive, and for that purpose, high concentrations of the biocatalyst, as is possible with the reactor systems shown in Figure 5.1b–g, are required.

The most widely used system worldwide is the upflow anaerobic sludge blanket (UASB) (Fig. 5.1b) (McCarty, 2002). The incoming wastewater is equally distributed over the cross section by a system of tubes. At low superficial upflow velocities (1–2 m h$^{-1}$) the wastewater flows vertically through a sludge bed. A three-phase separation unit, which prevents disintegration of the sludge pellets, is integrated in the upper part of the reactor. A disadvantage of this system is the limited knowledge concerning the formation of sludge pellets. These pellets are essential for the success of the system, but are not always formed with every wastewater. To avoid long times for start up, it is desirable to 'seed' new plants with sludge pellets from existing plants treating similar wastewaters. The productivity lies in the range of $10$–$15\,kg\,m^{-3}\,d^{-1}$ COD-degradation.

The so-called contact process (Fig. 5.1c) includes a CSTR and an external device for separation and recycling of anaerobic bacteria. For the separation, settling tanks or a lamella clarifier are used. Bacterial aggregates with adhering biogas bubbles cannot be held back by these methods. Therefore, a degasser is installed between the anaerobic reactor and the settling unit. Furthermore, a reaction in the settling unit has to be avoided, which can be achieved by low COD concentrations or a decrease in temperature. COD degradation in the whole plant (including the settling unit) ranges from 3 to $6\,kg\,m^{-3}\,d^{-1}$. An interesting report concerning this process and its application is given by Kroiss and Svardal (1999).

Fixed- and fluidized-bed reactors offer the advantage of high-load systems, requiring much less volume and space, and hence less investment than conventional systems. Furthermore, these systems tend to operate more stably under transient conditions such as fluctuations in substrates and pH. For these reasons, this chapter deals especially with these modern systems for high-rate anaerobic wastewater treatment.

The advantages mentioned are of interest to those industries that produce large amounts and/or highly concentrated wastewaters, notably the food, paper and pulp industries. Several fixed-film system reactor configurations are shown schematically in Figure 5.1d–g. Although most fixed-film systems can be associated with this scheme, the number of variations is high as regards the flow mode (up- or downflow), the fluid distribution system at the reactor inlet, the support material and expansion (for fluidized beds), and the configuration of the reactor outlet (gas–liquid–support separation). Recycling is in general provided for dilution of substrate and fluidization.

Several problems, however, have hindered the application of these systems. These are longer startup times if no specific inoculum is available, requirement for a more sophisticated process control, and the cost of the support material (Weiland and Rozzi, 1991). Appropriate solutions are, in principle, available to overcome these problems.

Nevertheless, there has been considerable progress in application as a result of research and pilot plant investigations, as well as data published on industrial systems. A large amount of literature has accumulated, mainly over the last two decades, so that only part of it can be mentioned here This chapter concentrates on basic principles of anaerobic fixed- and fluidized-bed systems and on recent experience accumulated on the laboratory, pilot, and industrial scales. One should, however, keep in mind that industrial applications must be based on pilot plant experience at the factory site for every new substrate or new source of wastewater or any specific problem.

A very broad and most successful application of high-performance anaerobic treatment in fixed- and fluidized-bed reactors relates to wastewater from industries based on agricultural and forestry products, which typically have high concentrations of organic substrates readily degraded by anaerobic bacteria (Austermann-Haun et al., 1993). They may result from raw material washing procedures, blanching, extraction, fermentation, or enzyme processing. Original substrates are usually carbohydrates, such as sugar, starch, cellulose and hemicellulose proteins, and fats, which readily undergo bacterial degradation to fatty acids, mainly acetic, propionic, butyric, and lactic acids. The majority of installations are in the potato, starch and sugar industries, in fruit, vegetable, and meat processing, in cheese, yeast, alcohol, citric acid, and pectin manufacturing and in the paper and pulp industries. The concentrations of substrates are typically in the range 5–50 kg (COD) m$^{-3}$, which are diluted by recirculation (loop reactor) to less than 2 kg (COD) m$^{-3}$.

In some countries, such as China, systems for the treatment of domestic sewage also play a major role (Yi-Zhang and Li-Bin, 1988). In the chemical and petrochemical industries, the implementation of anaerobic treatment plants for complex wastewaters started in the 1990s (Macarie, 2001). Furthermore, considerable efforts were made also to treat inhibitory or toxic substances. Thus, cyanide-, formaldehyde-, ammonium-, nickel- and sulfide-containing wastewaters were investigated, and it was shown that methanogens can accommodate to rather high concentrations of such toxins, depending on the retention time (Parkin and Speece, 1983). Furthermore, organochlorine compounds in kraft bleaching effluents and in pesticide-containing water, including chloroform, chlorophenols, chlorocatechols and similar compounds, as well as chlorinated resin acids, could be treated to the stage of mineralization by adapted biofilms (Salkinoja-Salonen et al., 1983). These positive results were obtained on the laboratory scale. The degradation of furfural in sulfite evaporator condensate can proceed to a conversion of about 90% (Ney et al., 1989).

## 5.2
## Basic Principles

Reactors are tubes with fixed bed internals or fluidized suspended particles, which serve as a support for biomass immobilization. The dimensions range from 10 to 500 m$^3$ with a ratio of height to diameter of 1–5. In general, an external loop recycles part of the effluent to the inlet, where mixing with the wastewater provides for its dilution to noninhibitory substrate concentrations and pH. Rarely, tapered beds have been used; most of the fluidized beds are provided with a settling zone with a larger diameter at the top of the reactor.

### 5.2.1
### Biofilm Formation

The basis for the use of packed-bed and fluidized-bed systems is the immobilization of bacteria on solid surfaces. Many species of bacteria (and other microorganisms) have the ability to adhere to supporting matrices.

Although immobilized bacteria have been used in aerobic wastewater treatment since the beginning of the 20. century, the application of these systems to anaerobic wastewater treatment is relatively new.

The fundamentals of bacterial adhesion to and growth on solid surfaces are discussed by Wingender and Flemming (1999). Here, only some aspects concerning anaerobic fixed films are considered. The preconditioning of solid surfaces is influenced by both environmental conditions (e.g. pH, temperature) and the surface itself (e.g. hydrophobicity, surface charge). The initial anaerobic biofilm attachment can be improved by the addition of cationic polymers (Stronach et al., 1987; Diaz-Baez, 1988) or slime-producing bacteria (Diaz-Baez, 1988), but the biofilm development is worse than in systems lacking these components. Jördening (1987) reported a positive effect resulting from supplementation of calcium.

The primary adhesion of cells to the surface is due to hydrogen bonds, van der Waals forces, and/or electrostatic interactions. This reversible form of adhesion can become irreversible due to the production of exopolymeric substances (EPS), which act as a glue (Wingender and Flemming, 1999). Experiments on the first steps of the formation of anaerobic biofilms have given different results: although Sanchez et al. (1996) found facultative anaerobic bacteria to be primary colonizers, Sreekrishnan et al. (1991) observed that biofilm formation was initiated by methanogenic bacteria.

After a lag phase, which seems to be necessary for adaptation of the microorganisms to the new environment, exponential growth of bacteria begins. The growth rate is mainly determined by substrate transport and temperature (Heijnen et al., 1986).

## 5.2.2
### Biofilm Characteristics

In view of the wide range of possible biofilm compositions, it is obvious that biofilm thickness does not correspond to the activity of the biocatalyst. Hoehn (1970) reported that the highest biofilm density occurs when the total biofilm thickness corresponds to the active biofilm thickness, i.e., the substrate-penetrated part of the biofilm.

## 5.2.3
### Kinetics and Mass Transfer

The reaction kinetics for any process changes with immobilization of the catalyst. In general, the following mass transfer processes have to be considered:
1.  transport of substrate from the fluid to the surface of the support through the boundary layer (external mass transfer)
2.  transport of substrate from the surface into the pores of the biocatalyst
3.  reaction
4, 5. transport of products in the opposite direction of steps 1 and 2

The transport of substrates and products through the reactor is related to the hydrodynamic characteristics of the system and is generally much faster than steps 2–4. Mass transfer is mostly reduced by diffusion limitation.

Depending on the mode of limitation, one can distinguish between film diffusion and pore diffusion and combined limited systems. Figure 5.2 schematically illustrates the resulting substrate profiles.

For the description of the activity changes resulting from immobilization, an effectiveness factor is used, defined as

$$\eta = \frac{\text{observed reaction rate}}{\text{reaction rate in bulk liquid conditions}} \tag{1}$$

### 5.2.3.1 External Mass Transfer

In passing a solid surface, the fluid characteristics change from turbulent to laminar flow and produce a boundary layer around the surface. The flux $j_1$ through the boundary layer is equal to the mass transfer coefficient $k_e$ and the concentration gradient from the outer shell $c_{is}$ to the particle surface $c_{bi}$:

$$j_1 = k_e \left( c_{is} - c_{bi} \right) \tag{2}$$

An analytical solution for $k_e$ can be given only for the ideal case of a single particle at infinite dilution. The mass transfer coefficient is given by

$$k_e = 2 \frac{D}{d_p} \tag{3}$$

with $D$-diffusion coefficient and $d_p$-particle diameter.

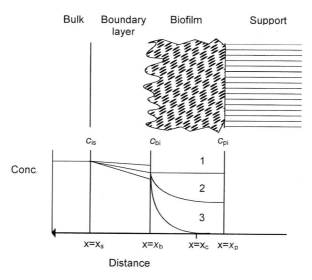

**Fig. 5.2** Substrate concentration profiles at an immobilized biocatalyst surface. 1, Reaction rate-controlled system; 2, combination of reaction rate- and diffusion-controlled system; 3, diffusion-controlled system.

For more complex systems, such as packed- or fluidized-bed reactors, the external mass transfer is usually described by dimensionless analysis, which leads to the correlation characterized by the Sherwood number $Sh$:

$$Sh = f(Re, Sc) = \frac{ke\, d_p}{D} \tag{4}$$

$$Re = \frac{d_p\, u \rho}{\eta} \tag{5}$$

$$Sc = \frac{\eta}{D\rho} \tag{6}$$

with the Reynolds number ($Re$) and the Schmidt-number ($Sc$).

Mulcahy and La Motta (1978) calculated the external mass transfer for a fluidized biofilm with a correlation given by Snowden and Turner (1967):

$$Sh = \frac{0.81}{\varepsilon} = Re^{0.5}\, Sc^{0.33} \tag{7}$$

Gonzalez-Gil et al. (2001) showed that the external mass transfer does not play an important role for superficial liquid velocities higher than 1 m h$^{-1}$.

### 5.2.3.2 Internal Mass Transfer

The diffusion transport of substrate in the biofilm may reduce the reaction rate. For steady-state conditions the net diffusion rate is equal to the reaction rate. Writing the resulting mass balance in a general form gives a second-order differential equation:

$$D_{eff}\left(\frac{d^2 c_i}{dx^2} + \frac{a}{x}\frac{dC_i}{dx}\right) - r = 0 \tag{8}$$

where $D_{eff}$ is the effective pore diffusion coefficient, $c_i$ the concentration of substrate $i$, $x$ the position in the porous particle, $r$ the reaction rate, and $a$ a geometrical factor: for a plate $a = 0$, for a cylinder $a = 1$, and for a sphere $a = 2$.

The diffusion coefficient in the biofilm ($D_{eff}$) may be smaller than in water. By comparing acetate diffusion with lithium diffusion, Kitsos et al. (1992) found a value of 7%, related to the diffusivity in water ($6.6 \times 10^{-10}$ m$^2$ s$^{-1}$). Ozturk et al. (1989) calculated an effective diffusion coefficient of $1.7 \times 10^{-10}$ m$^2$ s$^{-1}$ from measurements with inactive anaerobic biofilms. These values are lower than those determined for glucose or oxygen in aerobic biofilms, where the values are nearly the same as for water (Horn and Hempel, 1996). Kitsos et al. (1992) attribute this disagreement to differences between aerobic and anaerobic biofilms with regard to the symbiosis between the bacterial groups in anaerobic biofilms.

Two boundary conditions can be defined, at the support interface and at the interface between the biofilm and the boundary layer:

$$\frac{dc_i}{dx} = 0 \quad x = x_p \tag{9}$$

$$c_i = c_{is} \quad x = x_b \tag{10}$$

If a zero-order reaction is assumed, integration of Eq. (5) with the boundary conditions yields the expression given by Shieh and Keenan (1986):

$$\left(\frac{x_c}{x_b}\right)^3 - 1.5 \left(\frac{x_c}{x_b}\right)^2 + \left(\frac{1}{2} - \frac{3}{\phi^2}\right) = 0 \tag{11}$$

where the Thiele modulus $\phi$ is defined as

$$\phi = x_b \sqrt{\frac{\rho \, k_0}{D_{eff} \, c_i}} \tag{12}$$

Hence, the effectiveness factor can be calculated for $c_S > 0$.

For first-order reactions Shieh et al. (1982) showed good agreement between $\eta$ and the Thiele modulus given by

$$\eta = \frac{\coth(3\,\phi_{tm})}{\phi_{tm}} - \frac{1}{3\,\phi_{1m}^2} \tag{13}$$

with a modified first-order Thiele modulus $\phi_{1m}$:

$$\phi_{1m} = (\rho k_1 c_i)^{0.5} \, \frac{x_b^3 - x_p^3}{x_b^3} \tag{14}$$

Since anaerobic reactors are generally run at substrate concentrations that are high in relation to the substrate affinity constant, zero-order kinetics are a useful way for calculating diffusion limitation.

## 5.2.4
## Support Characteristics

Many different support materials have been tested for use in fixed-film stationary and fluidized-bed reactors. The major factors for bacterial attachment and growth to both systems are their roughness and porosity. The process configurations are discussed separately, because of differences in the relative importance of these factors and because additional factors may apply to only one of the systems.

### 5.2.4.1  Stationary Fixed-film Reactors

Supports in fixed-bed reactors must meet the following specific requirements for scale-up: Biogas must be separated from the fluid phase, and its transport must be possible through the full-scale reactor, for distances of up to several meters and under conditions of gas holdup of up to 3%, without major pressure drop. Supports with small dimensions (a few mm or less) are therefore not suited for this reactor type. Most supports tested with success up to the pilot or industrial scale offer sufficient hydrodynamic radii (in general >20 mm) and a very high void volume (over 70%, mostly more than 90%). As a consequence the support surface per volume is rather low, the concentration of biomass in the surface fixed film is low as well and the suspended biomass in the void volume contributes considerably to the activity (Weiland and Wulfert, 1986). The maximal volumetric load is distinctly lower in general than in expanded or fluidized-bed reactors.

Supports utilized in most applications are typically internals, such as Raschig or Pall rings, Berl or Intalox saddles, plastic cylinders, clay blocks or potter's clay of dimensions typically in the range of 20–60 mm (Henze and Harremoes, 1983; Young and Dahab, 1983; Weiland et al., 1988). Trends in application favor supports with a void volume of >90% and a surface area in the range 100–300 m$^2$ m$^{-3}$ (Austermann-Haun et al., 1993).

### 5.2.4.2 Fluidized-bed Reactors

The choice of support material determines the process engineering much more than for packed bed reactors, because the fluidization characteristics depend on the density and the diameter of the support.

To calculate the fluidization behavior, considering only ideally spherical particles without a biolayer, one starts with the terminal settling velocity for a single particle at infinite dilution:

$$u_{\mathrm{T}} = \sqrt{\frac{4\,g\,d_{\mathrm{p}}\,\rho_{\mathrm{p}} - \rho_{\mathrm{L}}}{3\,C_{\mathrm{D}}\,\rho_{\mathrm{L}}}} \tag{15}$$

where $g$ is the gravitational acceleration. The terminal settling velocity $u_{\mathrm{T}}$ depends on the particle diameter $d_{\mathrm{p}}$, the drag coefficient $C_{\mathrm{D}}$, and the difference in densities between the particle $\rho_{\mathrm{p}}$ and the liquid $\rho_{\mathrm{L}}$. $C_{\mathrm{D}}$ correlates with the particles' Reynolds number $Re_{\mathrm{p}}$:

$$Re_{\mathrm{p}} = \frac{u_{\mathrm{T}}\,d_{\mathrm{p}}}{v_{\mathrm{L}}} \tag{16}$$

Equations describing the relation between $C_{\mathrm{D}}$ and $Re_{\mathrm{p}}$ can be found in the literature. Bird et al. (1960) gave a generally accepted formula for the intermediate range of Reynolds numbers ($1 < Re_{\mathrm{P}} < 50$):

$$C_{\mathrm{D}} = \frac{18.5}{Re_{\mathrm{p}}^{0.6}} \tag{17}$$

From these equations, the single-particle settling velocity can be calculated by iteration.

However, to calculate the fluidized-bed expansion additional effects of the reactor wall, as well as the characteristics of flow and of adjacent particles, have to be considered. Usually the fluidization is then described with empirical equations. The correlation used most frequently for fluidized beds is given by Richardson and Zaki (1954):

$$u = u_T * \varepsilon^n \tag{18}$$

where $u$ is the superficial liquid velocity, $u_T^*$ is the settling velocity of the particle swarm (equal to $u_T \times 10^{-d_P/d_R}$), $\varepsilon$ is the bed voidage, and $n$ is an expansion index. $n$ is given as follows, provided that the particle diameter is much smaller than that of the bed:

$$
\begin{aligned}
n &= 4.65 & Re_T &\leq 0.2 \\
n &= 4.35 \times Re_T^{-0.03} & 0.2 &\leq Re_T \leq 1 \\
n &= 4.45 \times Re_T^{-0.1} & 1 &\leq Re_T \leq 500 \\
n &= 2.39 & 500 &\leq Re_T \leq 7000
\end{aligned}
$$

For anaerobic fluidized beds, $n$ can normally be calculated for particle terminal Reynolds number in the range 1–500.

Most of the materials (e.g., granular activated carbon, pumice, sepiolite) used as supports for anaerobic fluidized beds are not ideal spheres and show a distribution of particle diameters. Thus the Sauter diameter $d_{V/S}$ should be used in the above calculations; this is defined as

$$d_{V/S} = \frac{\int x\, q_r(x)\, dx}{\int q_r(x)\, dx} \tag{19}$$

where $q_r$ is the amount of particles (of volume or surface fraction) with diameter $x$. A sphericity factor $\Phi$ can be determined by microscopic comparison of particle shape with model geometrical figures given by Rittenhouse (1943). From these values the voidage can be described with a correlation of Wen and Xu (1966):

$$\frac{1-\varepsilon}{\varepsilon^3} = 11\,\Phi \tag{20}$$

Volume contraction has to be considered for the calculation of $\varepsilon$, if a mixture of different particle sizes is used. A detailed calculation procedure is given by Ouchijama and Tanaka (1981).

The easiest way to determine the fluidization behavior is to conduct experiments in laboratory-scale reactors. To obtain representative data it is necessary that the ratio of the reactor diameter to the particle diameter be at least 100. However, the growing biofilm and possible precipitation of inert solids may cause significant changes in the fluidization behavior. Therefore, it is very important to use real wastewater and to control the fluidization until a dynamic steady state is reached.

## 5.3
## Reactor Design Parameters

The development of any anaerobic system requires evaluation of optimal conditions affecting several factors. For fixed-film systems this includes especially the choice of support, the reactor geometry, the startup procedure, and the handling of excess sludge or inert support.

### 5.3.1
### Scale-up

Concepts for scale-up have been summarized by Kossen and Oosterhuis (1985); dimensional analysis and rules of thumb may be mentioned, since they provide guidance and recourse to practical experience. Fluid flow and fluidization can be treated with the aid of Reynolds, Peclet, and Froude numbers so as to estimate regimes appropriate for technical-scale operation (Mösche, 1998). However, a rational design seems very difficult because of the high complexity of the systems. Therefore, empirical rules are mostly used in practice to design technical reactors (Henze and Harremoes, 1983).

The most important parameter is the load of biodegradable organics in terms of COD. This must be correlated with the active biomass in the reactor. So a load of $1$–$1.5 \text{ kg (COD) kg}^{-1} \text{ (VSS) d}^{-1}$ is considered the upper limit for stable operation, and the following correlation can be used for guidance (Henze and Harremoes, 1983):

$$B_{V,COD} = \frac{X/\tau_X - B_{V,inert}}{Y} \tag{21}$$

where $B_{V,COD}$ denotes the volumetric loading rate, $X$ the biomass concentration, $\tau_X$ the biomass retention time, $B_{V,inert}$ the volumetric loading rate of non-biodegradable solids and $Y$ the yield coefficient.

A range of additional parameters of high significance should be taken into account: geometrical dimensions and $HD^{-1}$; recirculation rate, determining the substrate dilution and pH (and their gradient); residence time and distribution; mixing behavior; flow rate, pressure drop, and energy requirements; fluidization and bed expansion.

The recirculation rate, inlet substrate concentration, and pH and its gradient are correlated with one another, and they are highly important aspects since the stability of the stationary operation greatly depends on them, as discussed below (Burkhardt and Jördening, 1994; Mösche, 1998).

An example for modeling an industrial fluidized-bed reactor as a guide to scale-up and optimization of its operation was presented by Schwarz et al. (1998). The model comprises those aspects that were most sensitive in affecting the results: material balance equations for substrates and products in the gas and liquid phases, kinetics of biological degradation, mass transfer between gas and liquid phases, chemical equilibria, as well as convection and dispersion (Fig. 5.3).

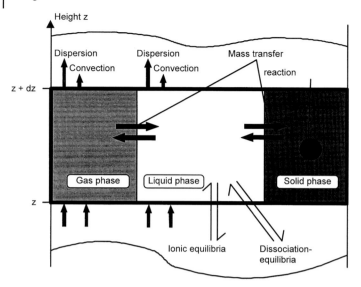

**Fig. 5.3** Finite volume in a fluidized-bed reactor and fluxes considered (Schwarz et al., 1996).

Maximum individual reaction rates for the different substrate components turned out to be the most sensitive to substrate and product concentrations as well as to pH. With scale-up, the ratio of feed and recirculation increases, since the flow rate is limited by the carrier settling velocity. Axial gradients of pH were calculated and turned out to be most sensitive to the load (Fig. 5.3) and the buffer capacity. Thus, at high loading rates, dynamic changes in the feed composition could lead to a decrease in pH in the lower part of the reactor and to breakdown of the system. Backmixing turned out to be of minor importance in this system (500 m$^3$ volume and 18 m height of the fluidized bed). A distribution of substrate feeding at two positions, the bottom and at medium height, overcame problems with greater concentration gradients. The corresponding increase in the superficial upflow velocity restricts this possibility only for highly concentrated wastewaters. Otherwise, problems caused by a higher segregation of the bed would occur. For pumice as support material, an increase in the superficial upflow velocity of 10% in the upper half of the fluidized bed did not cause any problems (Mösche, 1998).

The buffer capacity of the water is significantly influenced by $CO_2$ and its equilibrium concentration as a function of pressure and of ions such as calcium present in the wastewater.

### 5.3.2
### Support

Many support materials have been investigated in laboratory-scale reactors for use in packed- or fluidized-bed systems (Henze and Harremoes, 1983). Despite this, the number of supports that are used even in technical-scale systems is rather low. Cer-

tainly, one big problem for the implementation of new processes is due to the difficulty of cooperation with manufacturers. But the problems often result from shortcomings in the studies with respect to requirements of large-scale application. For successful use of any material as a support in fixed-film stationary or fluidized-bed reactors on the technical scale, the following general requirements should be met:
- availability of the material in large quantities (>1000 m$^3$)
- low cost of the material (related to the achievable performance; it should in general be less than 150 \$US m$^{-3}$)
- inert behavior (mechanically and microbially stable) without toxic effects and easy disposal
- low pressure drop (low energy demand for mixing or fluidization)

### 5.3.2.1 Stationary-bed Reactors
A selection of supports, which seem to have been applied successfully on the pilot and industrial scale, is summarized in Table 5.1; examples are shown in Figure 5.4.

The biofilm thickness is of limited significance in fixed-bed reactors since it makes up a minor part of the biomass. Published data are mostly in the range 1–4 mm, the upper limit relating to the inflow zone (top of a reactor operated in downflow) (Switzenbaum, 1983; Andrews, 1988).

More important is the biomass concentration in the reactor, which is distributed into two fractions, one immobilized in the biofilm, the other suspended in the void volume of the reactor (Hall, 1982; Weiland and Wulfert, 1986). The organic dry biomass in pilot reactors is mostly in the range 5–15 kg m$^{-3}$ (Henze and Harremoes, 1983). Gradients in the concentration of biomass are observed, with the maximum near the reactor inlet [e.g., about 15 kg m$^{-3}$ in the lower and 4 kg m$^{-3}$ in the upper part for upstream operation (Weiland and Wulfert, 1986)]. Gradients were also found by Hall (1982), which depended on the flow direction: 6 kg m$^{-3}$ of fixed biomass and 9 suspended for upstream, and 9 fixed and 4 suspended for downstream operation (all in kg m$^{-3}$ organic dry matter). For an industrial starch-processing

**Table 5.1** Typical supports for fixed beds.

| Support | Diameter (mm) | Surface Area (m$^2$ m$^{-3}$) | Bed Porosity | Equivalent Pore Diameter | Reference |
|---|---|---|---|---|---|
| Raschig rings | 10–16 | 45–49 | 0.76–0.78 | | Carrondo et al. (1983) |
| Pall rings corrugated | 90 | 102 | 0.95 | 20 | Young and Dahab (1983) |
| modular blocks | | 98 | 0.95 | 46 | |
| Pall rings | 25 | 215 | | | Schultes (1998) |
| Hiflow 90® | 90 | 65 | 0.965 | >30 mm | Weiland et al. (1988) |
| Plasdek C. 10® | | 148 | 0.96 | >30 mm | Weiland and Wulfert (1986) |
| Flocor R® | | 320 | 97 | >30 mm | |
| Ceramic Raschig rings | 25 | 190 | 0.74 | | Andrews (1988) |

**Fig. 5.4** Typical support materials for anaerobic stationary fixed-film reactors: a, Hiflow 90® (uncovered and biofilm-covered); b, Flocor R®; c, Plasdek C®. 10 (Weiland and Wulfert, 1986).

plant a biomass concentration of 20 kg m$^{-3}$ was reported (Schraewer, 1988). A high biomass concentration was reported for a special support, sintered glass fillings (expensive): up to 65 kg m$^{-3}$ in laboratory-scale experiments, of which 31 kg m$^{-3}$ was found to be immobilized within the inner pores of the support (Ney et al., 1989).

### 5.3.2.2 Fluidized-bed Reactors

The energy demand for fluidization of the support is often said to be very high, in contrast to the energy requirements of other anaerobic techniques. In general, much higher volumetric flow rates have to be achieved in comparison to those in a CSTR. Nevertheless, the energy demand is relatively low, because only the additional pressure drop of the support has to be overcome. Hence, with respect to higher loading rates, the overall energy demand is in the same range or lower than in CSTRs, depending on the support density. Table 5.2 shows some materials that have been tested for use as supports in anaerobic fluidized-bed systems. All materials have particle diameters significantly less than 1 mm, which results in a greater surface area for colonization, decreased superficial upflow velocity and no diffusion limitation, even for porous materials.

Porous materials have the advantage of lower superficial upflow velocities than nonporous materials have. In addition, biomass gradients mainly occur in reactors filled with nonporous materials (Anderson et al., 1990; Jördening, 1987). Franklin et al. (1992) reported that a sand fluidized-bed reactor contained only uncovered bare sand in the bottom part. They attributed this to the extreme shear forces, which have a large influence on bacteria on the surface. Bacteria in pores are protected against shear forces, and therefore, such pronounced gradients are not known when porous materials are used. Figure 5.5 shows some porous and nonporous supports. Nakhla and Suidan (2002) showed close agreement for the detachment of bacteria in differ-

**Table 5.2** Support materials for anaerobic fluidized-bed systems.

| Support | Diameter (10$^{-3}$ m) | Density (kg m$^{-3}$) | Surface Area (m$^2$ m$^{-3}$) | Porosity | Upflow Velocity (m h$^{-1}$) | Biomass (kg m$^{-3}$) | Reference |
|---|---|---|---|---|---|---|---|
| Sand | 0.5 | 2540 | 7100[a] | 0.41 | 30 | 4–20 | Anderson et al. (1990) |
| Sepiolite | 0.53 | 1980 | 20 300[b] | | | 32 | Balaguer et al. (1992) |
| GAC | 0.6 | | | | | 34 | Chen et al. (1995) |
| Biomass granules | | | | | 2–6.5 | | Franklin et al. (1992) |
| Sand | 0.1–0.3 | 2600 | | | 16 | 40 | Heijnen (1985) |
| Biolite | 0.3–0.5 | 2000 | | | 5–10 | 30–90 | Ehlinger (1994), Holst et al. (1997) |
| Pumice | 0.25–0.5 | 1950 | $2.2 \times 10^6$ | 0.85 | 10 | | Jördening (1996), Jördening and Küster (1997) |

[a] Calculated from the given data with the assumption of ideal sphericity.
[b] Calculated with data given in Sanchez et al. (1994).

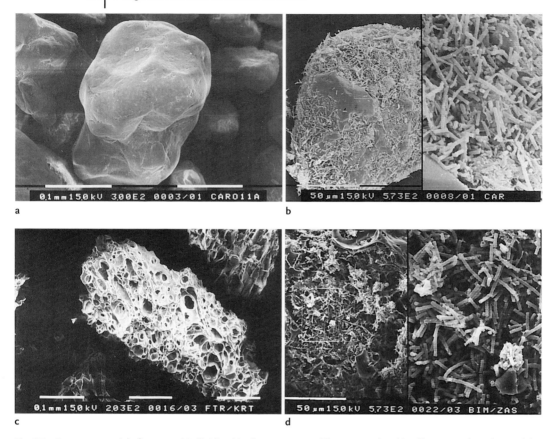

**Fig. 5.5** Support materials for anaerobic fluidized-bed reactors (a and b uncovered and biofilm-covered sand, c and d uncovered and biofilm-covered pumice particles).

ent reactors by normalizing the coefficients of the shear force and the daily methane production per unit biofilm area.

A special type of fluidized system uses granular pellets without a solid support. These pellets, used in so-called expanded granular sludge beds (EGSB) have a diameter of 1–3 mm and can therefore also be expanded by higher superficial upflow velocities of up to 20 m h$^{-1}$ (Lettinga et al. (1999)).

### 5.3.3
**Wastewater**

Wastewater should be acidified to a high degree (>80%, related to COD). Otherwise, acidifying bacteria could lower the pH, overgrow the methanogenic biofilm and thus reduce the methanogenic activity. Thus two-stage systems are considered superior, since their performance in terms of stability and space–time yield is superior to that of one-stage systems. The load of reactors fed with volatile fatty acids (in a system with a separate acidification reactor) can be higher by a factor of 4–5 than that of reactors fed with complex substrates (Henze and Harremoes, 1983).

Inhibitory substances such as sulfur compounds may play a major role, as in yeast processing (Friedmann and Märkl, 1994). It is also essential that results concerning load refer to an average of a stable continuous process rather than to a singular maximum. Even results obtained in laboratory reactors differ in general from pilot-plant and full-scale reactors operating at the factory site with variations in substrate quality and concentrations and additional fluctuations.

### 5.3.3.1 Solids in Stationary Fixed-film Reactors
Suspended solids and even suspended biomass can cause reactor clogging, which can be reinforced by extracellular polysaccharides secreted by acidogenic bacteria (Ehlinger et al., 1987). Therefore, backwash and excess sludge removal must be provided for in the reactor design. Gas-phase desorption and transport through the reactor must be possible. Fixed-bed systems are not feasible for wastewater having high solids content or components that tend to precipitate, such as calcium ions.

### 5.3.3.2 Solids in Fluidized-bed Reactors
Solids from the wastewater can cause clogging, especially at the entrance region of the reactor. Holst et al. (1997) recommend solids concentrations $<0.5$ kg m$^{-3}$ to prevent problems of clogging in the distribution system, but Mösche (1998) reported that even solid concentrations up to 1.7 kg m$^{-3}$ did not cause problems. This difference can be explained by differences in the composition of the solids as well as in the construction of the reactor inlet.

Some inorganic compounds, such as calcium carbonate and ammonium magnesium phosphate, precipitate mainly onto the support in the reactor, when the actual concentrations are beyond the equilibrium. At higher concentrations of precipitated solids on the support, diffusion limitation occurs. In such situations it is necessary to enable removing this material and replacing it with uncovered new support. Sand and other materials with a high settling velocity in relation to the support must either be removed before entering the reactor or be removed by a device at the bottom of the reactor (as described under the first subheading in section 5.3.4.2).

### 5.3.4
### Reactor Geometry and Technological Aspects

Some general aspects of reactor design are dealt with in Chapter 6 of this book.

### 5.3.4.1 Fixed-bed Reactors
Upflow reactors tend to be favored, because they allow clumps of biomass to be retained in the filter by gravity, and the start-up period may be shorter (e.g., 3–4 months compared to 4–6 months for downflow reactors) (Andrews, 1988; Weiland et al., 1988). The height is limited by the gradients of the biomass and reaction rate.

Fixed-bed reactors do not require major specific design considerations. The ratio of height to diameter is usually in the range 1–2. The inlet must provide equal distribution of the wastewater by means of distribution devices, generally a system of tubes with nozzles, about one for each 5–10 m$^2$ (Lettinga et al., 1983). The fluid flow should in general be about 1 m h$^{-1}$, up to a maximum of 2 m h$^{-1}$ (Austermann-Haun et al., 1993).

### 5.3.4.2 Fluidized-bed Reactors

Fluidized-bed reactors are taller than agitated tanks or stationary-bed reactors. The height–diameter ratio for technical plants varies from 2 to 5. Figure 5.6 shows a technical plant with 500 m$^3$ volume. The ratio of height to diameter should not be too high with respect to axial concentration gradients, which increase with the height of the reactor. But difficulties concerning uniform fluidization of the support increase with increasing reactor diameter (Couderc, 1985). Therefore, a compromise has to be found.

Whereas fluidized bed reactors are mostly cylindrical, the use of tapered fluidized be reactors has also been investigated and showed advantages concerning the performance (Huang et al., 2000)

**Fig. 5.6** Technical-scale anaerobic fluidized-bed reactor (sugar factory in Clauen, Germany).

## Fluidization of the Support

One of the key factors for the development of a fluidized-bed system is the fluidization zone. This zone has to provide a homogenous distribution of support and substrate by the incoming feed to prevent any dead-zone formation and to avoid high shear forces. Most laboratory-scale reactors work with inlet tubes in the downward direction or sieves, sometimes also with glass beads for providing a uniform upflow distribution. These are not applicable to full-scale reactors. The use of a multitube system could be a solution for technical reactors, but such a system is expensive and may be subject to problems with solids (blocking) or precipitation (lime or other inorganic compounds) (Iza, 1991).

Problems in distribution systems are sometimes discussed in the literature (Franklin et al., 1992; Oliva et al., 1990), but new developments and improvements of existing distribution systems are rarely described in detail.

A fluidized-bed bottom used in a 500 m³ BMA (Braunschweiger Maschinenbauanstalt) reactor with sugar factory wastewater is shown in Figure 5.7 (Jördening et al., 1996). It has a conical shape with 12 concentric pipes for the incoming water. Due to an inner double cone, the superficial liquid velocity is twice that of the reaction zone. So only material with a high settling velocity can settle in this part. A valve at the lower end of the cone allows these particles to be removed.

## Bed Height and Loss of Support

The fluidized bed's height is determined by the flow rate and depends on the support. It varies over time with growth of the biofilm, the gas production rate and possible precipitates on the support. Changes occur only slightly with time.

Strategies for preventing support loss are different in several systems on the technical scale:

- For the Anitron process (sand as support) the bed height can be controlled by a stationary support/biomass-separating device at the maximum bed height, which contains the removal of support from the reactor by a centrifugal pump which dislodges the biomass from the support. While the support is fed back to the reactor,

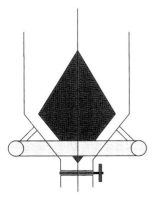

**Fig. 5.7** Flow distribution system for fluidized-bed reactors (Jördening et al., 1996).

excess biomass is discarded. The maximum bed height is at least 1.5 m below the overflow section.

- For the Gist-Brocades (sand as support) and the Anaflux processes (Biolite® as support) the bed height is controlled only by the fluidization velocity. Biocatalyst loss is prevented by means of three-phase separation constructions. Although the Gist-Brocades system does not contain a unit for removing support, the Anaflux and the Anitron systems are provided with a centrifugal pump for removal of excess biomass from the separator. Therefore, automatic control is not necessary, but daily control via a sample port is useful to prevent support loss.

## 5.4
## Reactor Operation

### 5.4.1
### Start-up Procedure

The start-up of fixed-bed reactors is governed by several parameters (Weiland et al., 1988; Burkhardt and Jördening, 1994):
- size and quality of the inoculum, notably the activity of slow-growing methanogens
- degree of adaptation, mainly the content of bacteria adapted to adhesion
- degree of biomass retention

The limited activity, and thus slow growth of propionic or lactic acid converting bacteria are of crucial importance, and the load must be controlled so as to maintain the substrates below growth-limiting concentrations. The initial load should be low, in the range of 0.1 kg(COD) kg$^{-1}$(VSS) d$^{-1}$ (Henze and Harremoes, 1983). Start-up is facilitated by Ca$^{2+}$ ions in the concentration range 100–200 g m$^{-3}$. The load may be increased at a rate of 5%–10% d$^{-1}$. The overall time required from start-up to full load may be 1–3 months. It can be much less when an adapted inoculum and a sufficient amount of biomass are used. In industries working in campaign periods, restart after storage of the biomass (in the reactor at ambient temperature) is rather straightforward reaches full load within several days.

The time required for the development of a well-attached film on the support in fluidized beds is sometimes very long in contrast to that in suspended systems (Heijnen et al., 1986).

Because continuous inoculation seems to be impractical for technical systems, batch inoculation is usually used. This means that digester sludge is added and the continuous flow of wastewater to the reactor is started after several days or weeks of adaptation. If suspended bacteria are used as an inoculum, the start of the continuous work will cause a significant loss of activity. Hence, it is advantageous to use – whenever possible – immobilized inoculum from comparable plants to reduce the time for adaptation and the loss of inoculum. Jördening et al. (1991) reported that the use of immobilized inoculum could reduce the start-up period by about 30%. Heijnen et al. (1986) showed, by comparing data in the literature, that the loading rate profile is an additional key factor affecting rapid biofilm development: the use of a so-called 'maximum load profile' reduces the start-up time up to a half that achieved with the so-called 'maximum efficiency profile'. The 'maximum load profile' means that the load is increased if the concentration of volatile fatty acids in the reactor is high (>2000 ppm); the 'maximum efficiency profile' means that the load is increased only if the concentration of volatile fatty acids in the reactor is decreased to a minimum (<100 ppm) and therefore conversion to the extent possible has been achieved.

## 5.4.2
### Operation Results: Stationary Bed

Several reviews have been published: e.g., by Henze and Harremoes (1983), Switzenbaum (1983) and Austermann-Haun et al. (1993). Table 5.3 shows selected data for typical substrates and carriers on the laboratory and pilot scale, which also seem appropriate for scale-up to the industrial level. One example is included representing very high load, however, with an expensive carrier (Siran®). Most typical loading rates (kg m$^{-3}$ d$^{-1}$ COD) are in the range 3–10 for common wastewater from agricultural and food processes with high conversion (over 80% of degradable COD), but

**Table 5.3** Data from laboratory- and pilot-scale anaerobic stationary-bed systems.

| Substrate | COD$_{in}$ (kg m$^{-3}$) | Carrier | Load (kg m$^{-3}$ d$^{-1}$) | Conversion (%) | Reference |
|---|---|---|---|---|---|
| Sucrose 30%, ethanol 65% | 3–6 | modular blocks pall rings | 2–4 | 83–85 | Young and Dahab (1983) |
| Distillery effluents[a] | <10 | Plasdek® Flocor® Hiflow 90® | 8–10 | 90–95 | Weiland and Wulfert (1986) |
| Volatile acids | 6 | | 1.7–3.4 | 87–98 | Young and McCarty (1969)[b] |
| Protein–carbohydrate waste | | | 3.1–3.3 | 90–50 | Mueller and Mancini (1975)[c] |
| Sludge heat-treatment liquor | 10 | | 6.5 | 55–65 | Donovan (1981)[c] |
| Chemical plant waste | 16 | | | 65 | Ragan (1981)[c] |
| Sulfite evaporator condensate[d] | 37 | Siran® Raschig rings | 45 | 84 | Ney et al. (1989) |

[a] Acidification prior to methanogenesis.
[b] Test in different reactors.
[c] Taken from Switzenbaum (1983).
[d] Sulfite evaporator condensate: acetic acid 425, methanol 75, furfural 28 m.

there are also examples of loading rates of about 20 and even 40 kg m$^{-3}$ d$^{-1}$ COD. However, these rather exceptional results may not be suitable for scale-up.

A considerable number of technical-scale systems with volumes of several 10 m$^3$ to several 1000 m$^3$ are in operation. The sites are small- and medium-sized agricultural installations, typically treating wastewater from poultry, pig, and cattle farms and distilleries; medium- and large-scale reactors in food industries such as sugar, potato, starch, and dairy processing, paper and pulp manufacturers, and the chemical industry. Some typical data are summarized in Table 5.4. Acidification often occurs in the wastewater stream prior to feeding it to the methane reactor even when this is not mentioned in detail; notably, for substrates that easily undergo microbial fermentation. For example, this occurs in sugar industry wastewater, containing sucrose as the main substrate, which is stored in lagoons before being fed to a methane reactor. More examples and data have been published by Austermann-Haun et al. (1993).

Overloading may be possible without inactivation of the biomass (if an appropriate pH – least 6 – is maintained), but conversion decreases according to the maximal

**Table 5.4** Data from industrial-scale anaerobic stationary bed systems.

| Wastewater | Reactor Size (m$^3$) | Support Material | Load (kg m$^{-3}$ d$^{-1}$) | Removal (%) | Reference |
|---|---|---|---|---|---|
| Meat processing | 22[a] | porous glass (Siran®) | 10–50 | up to 80 | Breitenbücher (1994) |
| Dairy | 260 | plastic rings (Biofar®) | 10 | | Weiland et al. (1988) |
| Dairy | 362 | plastics (Flocor® and cloisonyle) | 10–12 | 70–80 | Austermann-Haun et al. (1993) |
| Sugar | [b] | plastics | [b] | 90 | Camilleri (1988), Henry and Varaldo (1988) |
| Sugar | 1400 | plastic rings (Flocor) | 13 | | Weiland et al. (1988) |
| Potato processing | 660 | pallrings (100 mm) | 3–6 | | Weiland et al. (1988) |
| Starch | 4300[c] | lava rock | 25 | >70 | Schraewer (1988) |
| Soft drink production | 85 | modular plastic blocks (Plasdek®) | 4–6 | 90 | Austermann-Haun et al. (1993) |
| Distillery | 13 000 | modular plastic blocks | 8–12 | | Weiland et al. (1988) |
| Distillery | 92 | modular plastic blocks (BIO-NET®) | 13 | 78 | Austermann-Haun et al. (1993) |
| Chemical industry | 1900 | plastics | 16–20 | 90 | Henry and Varaldo (1988) |

[a] Two-stage system with acidification (40 m$^3$) as the first stage.
[b] Up to 2000 m$^3$ d$^{-1}$ wastewater with 16 t d$^{-1}$ COD.
[c] Two-stage system with acidification reactor of 1000 m$^3$ as the first stage.

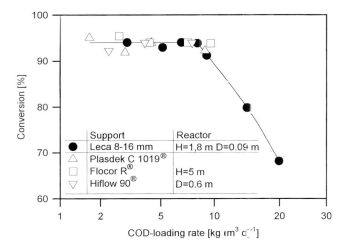

**Fig. 5.8**  Influence of volumetric loading rate and support material on reactor conversion (Weiland and Wulfert, 1986).

activity of biomass (in general 1–2 kg COD m$^{-3}$ kg$^{-1}$ dry biomass; Figure 5.8) (Weiland and Wulfert, 1986).

### 5.4.3
### Operational Results: Fluidized-bed Reactors

The performance of fluidized-bed reactors on the laboratory and pilot scale is sometimes excellent: Keim et al. (1989) used Siran® spheres in a fluidized-bed reactor for the treatment of evaporation condensate and reported a loading rate up to 315 kg m$^{-3}$ d$^{-1}$ with 79% removal. Jördening et al. (1991) used another sintered glass (Poraver®) to treat a sugar wastewater with loading rates up to 183 kg m$^{-3}$ d$^{-1}$ and achieved 89% removal. Data from some laboratory- and pilot-scale plants are given in Table 5.5.

Despite these results on the laboratory scale, the data reported for performance on the technical scale (Table 5.6) are in a significantly lower range: 15–50 kg m$^{-3}$ d$^{-1}$ (Fig. 5.9) (see e.g. Ehlinger, 1994; Franklin et al., 1992; Jördening, 1996; Jördening and Küster, 1997; Oliva et al., 1990).

### 5.5
### Conclusions

As a general conclusion, we can state that both fixed- and fluidized-bed reactors are well established on the industrial scale, notably in the food and related industries (e.g., breweries and distilleries) and in the paper and pulp industry, where the substrates are of natural origin, but also in other areas, e.g., the chemical industry.

**Table 5.5** Data of laboratory and pilot-scale anaerobic fluidized bed systems.

| Waste | Support | Reactor Volume (m³) | Ratio[a] | Loading (kg m⁻³ d⁻¹) | Concen- tration (kg m⁻³) | HRT (h) | Removal (%) | Reference |
|---|---|---|---|---|---|---|---|---|
| Spoiled beer | sand | $64\times10^{-3}$ | 18 | 1–14.8 | 1–12 | 0.2–7 | 75–87 | Anderson et al. (1990) |
| Synthetics | GAC[b] | $1.2\times10^{-3}$ | 5.6 | 5.3–108 | 0.5–9 | 0.5–8 | 75–98 | Chen et al. (1995) |
| Synthetics | sand | $70\times10^{-3}$ | 20 | 3–20 | 13 | 1.5 | 98 | Jördening (1987) |
| Fruits and vegetables | biolite | | 25 | 38 | | | >80 | Menon et al. (1997) |
| Brewery | sand | $63\times10^{-3}$ | 18.7 | 14.6 | 1–3.5 | 5–34 | 70[b] | Ozturk et al. (1989) |

[a] Height/diameter.
[b] Granular activated carbon.

**Table 5.6** Data from technical-scale anaerobic fluidized-bed systems.

| Waste | Company | Support | Reactor Volume (m³) | Ratio[a] | Loading (kg m⁻³ d⁻¹) | Concen- tration (kg m⁻³) | HRT (h) | Removal (%) | Reference |
|---|---|---|---|---|---|---|---|---|---|
| Soybeans | Dorr–Oliver | sand | 304 | 2.0 | 14–21 | 0.8–10 | | 75–80 | Sutton et al. (1982) |
| Several | Degremont | biolite | 210–480 | | 16–21 | 3.8–5 | 0.25 | 50–60 | Ehlinger (1994) |
| Yeast and pharma- ceuticals | Gist- Brocades | sand | 400 | 4.4 | 8–30 | 1.9–4 | 0.14– 0.41 | 95–98 | Franklin et al. (1992) |
| Evapora- tion con- densates and osmotic permeates | Gist- Brocades | granules | 125 | | 21–35 | 5–22.5 | | | Franklin et al. (1992) |
| Sugarbeets | BMA | pumice | 500 | 5.0 | 20–60 | 1–5 | 1.5 | 65–78 | Jördening (1996) |
| Brewery | Degrement | biolite | 165 | 3.2 | | | | | Oliva et al. (1990) |

[a] Height/diameter.

Specific limitations, however, must be taken into consideration:
- Acidification inside the methane reactor requires a higher residence time than in two-stage systems with acidification as a first stage; this also provides safer and more stable processing, e.g., when a shock load occurs – a quite common situation in practice.

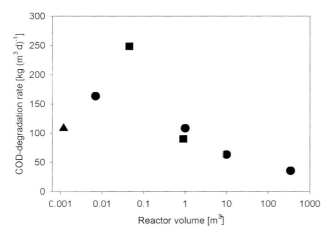

**Fig. 5.9** Changes in reactor performance as a function of reactor volume (■ Keim et al. 1989, ● Burkhardt and Jördening 1994, ▲ Chen et al. 1995).

- Wastewater free of suspended solids is required for fixed-bed reactors. No precipitation of solids should occur inside the reactor.
- Advantages of both types of systems include higher density of biomass and, therefore, higher reaction rates and a considerably smaller reactor volume required than in conventional systems such as stirred tanks. This is especially true for fluidized-bed reactors, since they can provide a much larger support surface and, hence, a high biomass concentration per volume. Much less space is required due to the relatively small reactor volume and the high ratio of height to diameter of the reactors.
- For these reasons, investment costs are lower than for systems without biomass immobilization.

## References

Anderson, G. K., Ozturk, I., Saw, C. B., Pilot-scale experiences on anaerobic fluidized bed treatment of brewery wastes, *Water Sci. Technol.* **1990**, *22*, 157–166.

Andrews, G. F., Design of fixed film and fluidized bed bioreactors, in: *Handbook on Anaerobic Fermentations* (Erickson, L. E., Fung, Y.-C., eds.), pp. 765–802. New York **1988**: Marcel Dekker.

Austermann-Haun, U., Kunst, S., Saake, M., Seyfried, C. F., Behandlung von Abwässern, in: *Anaerobtechnik* (Böhnke, B., Bischofsberger, W., Seyfried, C. F., eds.), pp. 467–696. Berlin **1993**: Springer-Verlag.

Balaguer, M. D., Vicent, M. T., Paris, J. M., Anaerobic fluidized bed with sepiolite as support for anaerobic treatment of vinasse, *Biotechnol. Lett.* **1992**, *5*, 333–338.

Bird, R. B., Stewart, W. E., Lightfood, E. N., *Transport Phenomena*. New York **1960**: Wiley.

Breitenbücher, K., Hochleistung durch mehr Biomasse, *UTA Umwelttechnik Aktuell* **1994**, *5*, 372–374.

Burkhardt, C., Jördening, H.-J., Maßstabsvergrößerung und Betriebsdaten von anaeroben Hochleistungs-Fließbettreaktoren, in: *Anaerobe Behandlung von festen und flüssigen Rückständen* (Märkl, H., Stegmann, R., eds.), pp. 145–162. Weinheim **1994**: VCH.

Camilleri, C., Start-up of fixed film stationary bed anaerobic reactors, in: *Anaerobic Digestion 1988* (Hall, E. R., Hobson, P. N., eds.), pp. 407–412. Oxford **1988**: Pergamon.

Carrondo, M. J. T., Silva, J. M. C., Figuera, M. I. I., Ghano, R. M. B., Oliveira, J. F. S., Anaerobic filter treatment of molasses fermentation wastewater, *Water Sci. Technol.* **1983**, *15*, 117–126.

Chen, S. J., Li, C. T., Shieh, W. K., Performance evaluation of the anaerobic fluidized bed system. I. Substrate utilization and gas production, *J. Chem. Technol. Biotechnol.* **1995**, *35B*, 101–109.

Couderc, J.-P., Incipient fluidization and particulate systems, in: *Fluidization*, 2nd edn. (Davidson, J. F., Clift, R., eds.), pp. 1–46. London **1985**: Academic.

Demirel, B., Yenigün, O., Two-phase anaerobic digestion processes: a review, *J. Chem Technol. Biotechnol* **2002**, *77*, 743–755

Diaz-Baez, M. C., A study of factors affecting attachment in the start up and operation of anaerobic fluidized bed reactors, in: *Poster-Papers 5th Int. Conf. Anaerobic Digestions* (Tilche, A., Rozzi, A., eds.), pp. 105–108. Bologna **1988**: Monduzzi Editore.

Donovan, E. J., Treatment of high strength wastes by an anaerobic filter, *Proc. Semin.: Anaerobic Filters: An Energy Plus for Wastewater Treatment*, pp. 179–187, Howey In The Hills, FL **1981**. ANL/CSNV-TM-50, 9–10 January **1980**.

Ehlinger, F., Anaerobic biological fluidized beds: operating experiences in France, pp. 315– 323, *Proc. 7th Int. Symp. Anaerobic Digestion*, Cape Town, South Africa **1994**, 23–27 January 1994.

Ehlinger, F., Audic, J. M., Verrier, D., Faup, G. M., The influence of the carbon source on microbial clogging in an anaerobic filter, *Water Sci. Technol.* **1987**, *19*, 261–273.

Franklin, R. J., Koevoets, W. A. A., van Gils, W. M. A., van der Pas, A., Application of the biobed upflow fluidized bed process for anaerobic waste water treatment, *Water Sci. Technol.* **1992**, *25*, 373–382.

Friedmann, H., Märkl, H., Der Einfluß der Produktgase auf die mikrobiologische Methanbildung, *gwf Wasser Abwasser* **1994**, *6*, 302–311.

Gonzalez-Gil, G., Seghezzo, L., Lettinga, G., Kleerebezem, R., Kinetics and mass-transfer phenomena in anaerobic granular sludge, *Biotechnol. Bioeng.* **2000**, *73*, 125–134

Hall, E. R., Biomass retention and mixing characteristics in fixed-film and suspended growth anaerobic reactors, pp. 371–396, *Proc. JAWPR Specialized Semin. Anaerobic Treatment Waste Water in Fixed Film Reactors*, 16–18 June 1982, Copenhagen, Denmark **1982**.

Heijnen, J. J., *US Patent* 4560479 (**1985**).

Heijnen, J. J., Mulder, A., Enger, W., Hoeks, F. (**1986**), Review on the application of anaerobic fluidized bed reactors in waste water treatment, *Proc. EWPCA Conf. Anaerobic Wastewater Treat.* pp. 159–173.

Henry, M., Varaldo, C., Anaerobic digestion treatment of chemical industry wastewaters at the Cuise–Lamotte (Oise) plant of Société Française Hoechst 479, in: *Anaerobic Digestion 1988* (Hall, E. R., Hobson, P. N., eds.), pp. 479–486. Oxford **1988**: Pergamon.

Henze, M., Harremoes, P., Anaerobic treatment of wastewater in fixed film reactors: a literature review, *Water Sci. Technol.* **1983**, *15*, 1–101.

Hoehn, D. C., *The Effects of Thickness on the Structure and Metabolism of Bacterial Films*, Thesis, University of Missouri **1970**.

Holst, T. C., Truc, A., Pujol, R., Anaerobic fluidized beds: ten years of industrial experience, *Water Sci. Technol.* **1997**, *36*, 415–422.

Horn, H., Hempel, D. C., Modellierung von Substratumsatz und Stofftransport in Biofilmsystemen, *gwf Wasser/Abwasser* **1996**, *137*, 293–299.

Huang, J.-S., Yan, J.-L., Wu, C.-S., Comparative bioparticle and hydrodynamic characteristics of conventional and tapered anaerobic fluidized bioreactors, *J Chem. Technol. Biotechnol.* **2000**, *75*, 269–278

Iza, J., Fluidized bed reactors for anaerobic wastewater treatment, *Water Sci. Technol.* **1991**, *24*, 109–132.

Jördening, H.-J., *Untersuchungen an Hochleistungsreaktoren zum anaeroben Abbau von calciumhaltigen Abwässern*, Thesis, Technical University of Braunschweig, Germany (**1987**).

Jördening, H.-J., Scaling-up and operation of anaerobic fluidized bed reactors, *Zuckerindustrie* **1996**, *121*, 847–854.

Jördening, H.-J., Küster, W., Betriebserfahrungen mit einem anaeroben Fließbettreaktor zur Behandlung von Zuckerfabriksabwasser, *Zuckerindustrie* **1997**, *122*, 934–936.

Jördening, H.-J., Jansen, W., Brey, S., Pellegrini, A., Optimierung des Fließbettsystems zur anaeroben Abwasserreinigung, *Zuckerindustrie* **1991**, *116*, 1047–1052.

Jördening, H.-J., Mansky, H., Pellegrini, A., *German Patent* 19502615 C2 (**1996**).

Keim, P., Luerweg, M., Aivasidis, A., Wandrey, C., Entwicklung der Wirbelschichttechnik mit dreidimensional kolonisierbaren Trägermaterialien aus makroporösem Gas am Beispiel der anaeroben Abwasserreinigung, *Korrespondenz Abwasser* **1989**, *36*, 675–687.

Kitsos, H. M., Roberts, R. S., Jones, W. J., Tornabene, T. G., An experimental study of mass diffusion and reaction rate in an anaerobic biofilm, *Biotechnol. Bioeng.* **1992**, *39*, 1141–1146.

Kossen, N. W. F., Oosterhuis, N. M. G., Modelling and scaling-up of bioreactors, in: *Biotechnology*, 1st edn., Vol. 2 (Rehm, H.-J., Reed, G., eds.), pp. 571–605. Weinheim **1985**: VCH.

Kroiss, H., Svardal, K., CSTR reactors and contact processes in industrial wastewater treatment, in: *Biotechnology* 2nd edn. Vol. 11a (Rehm, H.-J., Reed, G., Pühler, A. Stadler, P., eds.) pp. 65–84, Weinheim **1999**, Wiley VCH

Lettinga, G., Hobma, L. W., Hulshoff-Pol, L. W., De Zeeuw, W., De Jong, P., Design operation and economy of anaerobic treatment, *Water Sci. Technol.* **1983**, *15*, 177–196.

Lettinga. G., Hulshoff Pol, L.W., Lier, J.B.B.V., Zeeman, G., Possibilities and potential of anaerobic waste water treatment using anaerobic sludge bed (ASB) reactors, in: *Biotechnology* 2nd edn. Vol. 11a (Rehm, H.-J., Reed, G., Pühler, A. Stadler, P., eds.) pp. 65–84, Weinheim **1999**, Wiley VCH

Macarie, H., Overview of the application of anaerobic treatment to chemical and petrochemical wastewaters, *Water Sci. Technol.* **2001**, *44*, 201–214

McCarthy, P.L., The development of anaerobic treatment and its future, *Water Sci. Technol.* **2001**, *44*, 149–156

Menon, R., Delporte, C., Johnstone, D. S., Pilot treatability testing of food processing wastewaters using the Anaflux™ anaerobic fluidized-bed reactor, *52nd Ind. Waste Conf.*, Purdue University, West Lafayette, IN, USA **1997**.

Mösche, M., *Anaerobe Reinigung von Zuckerfabriksabwasser in Fließbettreaktoren*, Thesis, Technical University of Braunschweig, Germany **1998**.

Mueller, J. A., Mancini, J. L., Anaerobic filter kinetics and application, *Proc. 30th Ind. Waste Conf.*, pp. 423–447, Purdue University, West Lafayette, IN, USA **1975**.

Mulcahy, L. T., La Motta, E.J., Mathematical model of the fluidized bed biofilm reactor, *Report No. 59-78-2*, Dept. Civil Eng., University of Massachusetts Amherst, USA **1978**.

Nakhla, G., Suidan, M.-T., Determination of biomass detachment rate coefficients in anaerobic fluidized bed GAC reactors, *Biotechnol. Bioeng* **2002**, *80*, 660–669

Ney, U., Schobert, S. M., Sahm, H., Anaerobic degradation of sulfite evaporator condensate by defined bacterial mixed cultures, *Dechema Biotechnol. Conf.*, Vol. 3, Part B, pp. 889–892. Weinheim **1989**: VCH.

Oliva, E., Jacquart, J. C., Privot, C., Treatment of waste water at the El Aguila brewery (Madrid, Spain): methanization in fluidized bed reactors, *Water Sci. Technol.* **1990**, *22*, 483–490.

Ouchijama, N., Tanaka, T., Porosity of a mass of solid particles having a range of sizes, *Ind. Eng. Chem. Fundam.* **1981**, *20*, 66–71.

Ozturk, I., Anderson, G. K., Saw, C. B., Anaerobic fluidized bed treatment of brewery wastes and bioenergy recovery, *Water Sci. Technol.* **1989**, *21*, 1681–1684.

Parkin, G. F., Speece, R. E., Attached versus suspended growth anaerobic reactors: Response to toxic substances, *Water Sci. Technol.* **1983**, *15*, 261–289.

Ragan, J. L., Celanese experience with anaerobic filters, *Proc. Semin.: Anaerobic Filters: An Energy Plus for Wastewater Treatment*, pp. 129–135, 9–10 January 1980, Howey In The Hills, FL **1981**, ANL/CSNV-TM-50.

Richardson, J. F., Zaki, W. N., Sedimentation and fluidization: Part 1, *Trans. Inst. Chem. Eng.* **1954**, *32*, 35–53.

Rittenhouse, G., A visual method of estimating two-dimensional sphericity, *J. Sediment. Petrol.* **1943**, *13*, 79–81.

Salkinoja-Salonen, M. S., Nyns, E.-J., Sutton, P. M., Van Den Berg, L., Wheatley, A. D., Starting up of an anaerobic fixed film system, *Water Sci. Technol.* **1983**, *15*, 305–308.

Sanchez, J. M., Arijo, S., Munoz, M. A., Morinigo, M. A., Borrego, J. J., Microbial colonization of different support materials used to enhance the methanogenic process, *Appl. Microbiol. Biotechnol.* **1994**, *41*, 480–486.

Sanchez, J. M., Rodriguez, F., Valle, L., Munoz, M. A., Morinigo, M. A., Borrego, J. J., Development of methanogenic consortia in fluidized bed batches using sepiolite of different particle size, *Microbiologia* **1996**, *12*, 423–434.

Schraewer, R., DasAnfahrverhalten von anaeroben Bioreaktoren zur Reinigung hochbelasteter Stärkeabwässer, *Starch/Stärke* **1988**, *40*, 347–352.

Schultes, M., Füllkörper oder Packungen? *Chem. Ing. Tech.* **1998**, *70*, 25–261.

Schwarz, A., Yahyavi, B. M., Mosche, M., Burkhardt, C., Jördening, H.-J. et al., Mathematical modeling for supporting scale-up of an anaerobic wastewater treatment in a fluidized bed reactor, *Wat. Sci. Tech.* **1996**, *34*, 501–508.

Schwarz, A., Mösche, M., Jördening, H.-J., Buchholz, K. Reuss, M., Modellgestützte Betrachtungen zur Reaktionstechnik und Fluiddynamik im industriellen Anaerob-Fließbettreaktor, in: *Technik anaerober Prozesse* (Märkl, H., Stegmann, R., eds.). Frankfurt **1998**, Dechema.

Shieh, W. K., Keenan, J. D., Fluidized bed biofilm reactor for wastewater treatment, *Adv. Biochem. Eng. Biotechnol.* **1986**, *33*, 132–169.

Shieh, W. K., Mulcahy, L. T., LaMotta, E. J., Mathematical model for the fluidized bed biofilm reactor, *Enzyme Microb. Technol.* **1982**, *4*, 269–275.

Snowdon, C. B., Turner, J. C. R., Mass transfer in liquid-fluidized beds of ion-exchange resin beads, in: *Proc. Int. Symp. Fluidization* (Drinkenburg, A. A. H., ed.), pp. 599–608. Eindhoven **1967**.

Sreekrishnan, T. R., Ramachandran, K. B., Ghosh, P., Effect of operating variables on biofilm formation and performance of an anaerobic fluidized-bed bioreactor, *Biotechnol. Bioeng.* **1991**, *37*, 557–566.

Stronach, S. M., Diaz-Baez, M. C., Rudd, T., Lester, J. H., Factors affecting biomass attachment during start-up and operation of anaerobic fluidized beds, *Biotechnol. Bioeng.* **1987**, *30*, 611–620.

Sutton, P. M., Li, A., Evans, R. R., Korchin, S., Dorr-Oliver's fixed film and suspended growth anaerobic systems for industrial wastewater treatment and energy recovery, *37th Ind. Conf.*, Purdue University, West Lafayette, IN, USA **1982**.

Switzenbaum, M. S., A comparison of the anaerobic filter and the anaerobic expanded/ fluidized bed process, *Water Sci. Technol.* **1983**, *15*, 345–358.

Weiland, P., Rozzi, A., The start up, operation and monitoring of high-rate anaerobic treatment systems: discussers' report, *Water Sci. Technol.* **1991**, *24*, 257–277.

Weiland, P., Wulfert, K., Festbettreaktoren zur anaeroben Reinigung hochbelasteter Abwässer: Entwicklung und Anwendung, *BTF* **1986**, *3*, 152–158.

Weiland, P., Thomsen, H., Wulfert, K., Entwicklung eines Verfahrens zur anaeroben Vorreinigung von Brennereischlempen unter Einsatz eines Festbettreaktors, in: *Verfahrenstechnik der mechanischen, thermischen, chemischen und biologischen Abwasserreinigung, Part 2: Biologische Verfahren* (GVC–VDI, ed.), pp. 169–186. Düsseldorf **1988**: VDI-Gesellschaft für Verfahrenstechnik und Chemieingenieurswesen.

Wen, C. Y., Yu, Y. H., A generalized method for predicting the minimum fluidization velocity, *AIChE J.* **1966**, *12*, 610–612.

Wingender, J. Flemming, H.-C., Autoaggregation of Microorganisms: Flocs and Biofilms, in: *Biotechnology* 2nd edn. Vol. 11a (Rehm, H.-J., Reed, G., Pühler, A. Stadler, P., eds.) pp. 65–84, Weinheim **1999**, Wiley VCH

Yi-Zhang, Z., Li-Bin, W., Anaerobic treatment of domestic sewage in China, in: *Anaerobic Digestion 1988* (Hall, E. R., Hobson, P. N., eds), pp. 173–184. Oxford **1988**: Pergamon.

Young, J. C., Dahab, M.F., Effect of media design on the performance of fixed-bed anaerobic reactors, *Water Sci. Technol.* **1983**, *15*, 369–384.

Young, J. C., McCarty, P. L., The anaerobic filter for waste treatment, *J. Water Pollut. Control Fed.* **1969**, *41*, R161.

# 6
# Modeling of Biogas Reactors

Herbert Märkl

## 6.1
## Introduction

As early as 1973 Graef and Andrews (1973) published a mathematical model of the anaerobic digestion of organic substrates. The authors assumed that the conversion of volatile organic acids by methanogenic bacteria to methane was the rate-limiting step in the sequence of biological reactions and that all volatile acids can be represented as acetic acid. Acetic acid is dissociated to a large extent at the relevant pH range between 6.6 and 7.4. The authors assumed that only the undissociated form of acetic acid is the limiting substrate for the microbial production of methane.

The pH value in this mathematical model is calculated from the ion balance, assuming electroneutrality in the fermentation broth. In this context the concentration of dissolved $CO_2$, which itself is partly dissociated to $HCO_3^-$ and $CO_3^{2-}$, must also be known. In the mathematical model of Graef and Andrews, the concentration of dissolved $CO_2$ is calculated on the basis of the mass transport from the liquid phase to the gas phase of the system. This transport phenomenon proved to be very important for the mathematical modeling of technical biogas reactors, as is shown later in this chapter.

The mathematical model of Graef and Andrews describes a homogenous, completely mixed system as far as the liquid phase of the reactor is concerned.

## 6.1.1
## Elements of the Mathematical Model

This chapter shows that the mentioned elements of the model of Graef and Andrews are an excellent basis for the quantitative analysis of modern biogas reactor systems. In the research work of different groups following this first publication, it was proved that more knowledge is necessary, especially about the microbial kinetics of biogas production (Section 6.3). Most of the real biogas reactors are far from being completely mixed systems, as demonstrated by the example of a biogas tower reactor (BTR) (Fig. 6.1). This type of reactor is described in more detail by Märkl and Re-

*Environmental Biotechnology. Concepts and Applications.* Edited by H.-J. Jördening and J. Winter
Copyright © 2005 WILEY-VCH Verlag GmbH & Co. KGaA, Weinheim
ISBN: 3-527-30585-8

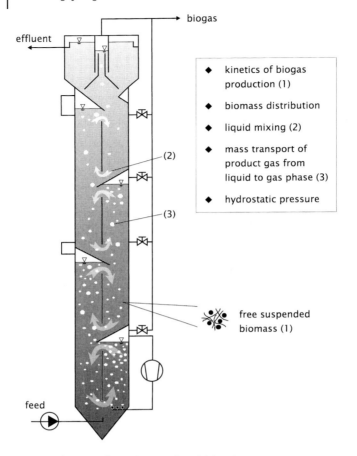

effluent

biogas

- ◆ kinetics of biogas production (1)
- ◆ biomass distribution
- ◆ liquid mixing (2)
- ◆ mass transport of product gas from liquid to gas phase (3)
- ◆ hydrostatic pressure

(2)

(3)

free suspended biomass (1)

feed

**Fig. 6.1** Elements of a mathematical model for a biogas tower reactor (BTR).

inhold (1994). The typical features of the BTR are its tower shape, the modular structure, and the internal installations. Gas collecting devices are used to withdraw the fermentation gas from different levels of the reactor. These gas collectors separate the reactor into modules along the height. By means of these devices, which are equipped with valves, the gas loading and the mixing intensity can be controlled separately in each module. To avoid flotation of active biomass due to excessive gas load, an effective biomass accumulation is generated within the reactor.

The biomass is represented in this reactor by free suspended microorganisms which are associated in the form of sedimentable more-or-less loose pellets (flocs). One very important point, when modeling such a reactor, is to describe the local distribution of active biomass within the reactor (Section 6.4.2).

In a mathematical model the mixing behavior of the liquid phase is of similar importance (Section 6.4.1). The mixing of the BTR is caused by internal airlift loop units. Since the reactor is designed in the shape of a tower and is only fed at the bottom (Fig. 6.1), the question of mixing and supplying the microorganisms in each part of the reactor with nutrients is very important during the scale-up of such a system. Besides, hydraulic mixing is important to prevent toxic concentrations of substrate near the inlet of the reactor.

As pointed out by Graef and Andrews (1973), the mass transport of product gas from the liquid phase to the gas phase is an essential element of the mathematical model. Local transport data as a function of the hydrodynamics for the example of the biogas tower reactor is given for $CO_2$ and $H_2S$ in Section 6.5 of this chapter. The last element of a mathematical model which is of high importance, especially when discussing tall reactors, is represented by the influence of the local hydrostatic pressure on the kinetics of biogas generation (Section 6.6).

## 6.1.2
### Scale-Up Strategy

Parameters of the mathematical model elements must be identified by experiments, and the models have to be evaluated with respect to their suitability for the design of technical scale biogas reactors. The scale-up strategy is demonstrated in Figure 6.2 with explanations in Table 6.1. Basic experiments are done in the laboratory. The hydrodynamic behavior and the sedimentation of non-active biomass were studied first in a small reactor unit with a height of 6.5 m. This laboratory reactor has a diameter of 0.4 m and consists of only two modules. To understand the influence of

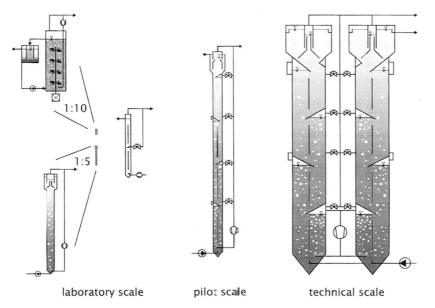

laboratory scale          pilot scale          technical scale

**Fig. 6.2** Strategy for scale-up of the biogas tower reactor, specification of reactors according to Table 6.1.

**Table 6.1** Specification of biogas reactors used for the experiments.

| | Experiments / Reactors | Kinetics | Hydro-dynamic and Liquid Mixing | Local Distribution of Biomass | Mass Transport of Product Gas | Influence of Hydrostatic Pressure |
|---|---|---|---|---|---|---|
| Labor-atory | Stirred tank reactor, active volume 16 L | ✔ | | | ✔ | ✔ |
| | Two UASB reactors with biogas recirculation, active volume 36.5 L and 39.5 L, height 1.7 m and 1.95 m, respectively | ✔ | | | ✔ | |
| | Laboratory tower reactor, height 6.5 m, diameter 0.4 m, two modules | | ✔ | ✔ | | |
| Pilot | Pilot scale tower reactor, height 20 m, diameter 1 m, volume 15 m³, four modules, photo of reactor shown in Figure 6.3 | ✔ | ✔ | ✔ | ✔ | ✔ |

**Fig. 6.3** Biogas tower reactor, pilot scale, at the site of Deutsche Hefe-werke in Hamburg, built by Preus-sag Wassertechnik GmbH, Bremen.

**Table 6.2** Characteristic data of the wastewater used for the experiments.

|  | Average | Range | Unit |
|---|---|---|---|
| Total organic carbon (TOC) | 13.8 | 11–19 | g L$^{-1}$ |
| Chemical oxygen demand (COD) | 31 | 25–36 | g L$^{-1}$ |
| COD (incl. betaine) | 44 |  | g L$^{-1}$ |
| Betaine | 5.7 | 4–10 | g L$^{-1}$ |
| Sulfate | 3.9 | 2.5–6.5 | g L$^{-1}$ |
| Acetic acid | 58 | 0.6–120 | mmol L$^{-1}$ |
| Total suspended solids (TSS) | 2.5 | 0.8–3 | g L$^{-1}$ |
| pH | 4.5 | 4.3–5 | – |

the height of a reactor, a pilot scale biogas reactor was built at the site of a company producing bakers' yeast (Fig. 6.3). The reactor was designed on the basis of the laboratory scale experiments and investigations on the kinetics of wastewater digestion from bakers' yeast production. The characteristic data of the wastewater are given in Table 6.2. The active volume of the reactor is 15 m$^3$. It is divided into four modules and at the top of the reactor a settling zone is integrated to maintain a high sludge concentration. The pilot reactor is well equipped with instruments for measuring concentrations and velocities along the height of the reactor. Because the pilot scale reactor reaches almost the height of a technical reactor (in Fig. 6.2 two reactor units with a height of 25 m and a diameter of 3.5 m are shown as examples) it is assumed that the mathematical model, which is evaluated on the pilot scale with respect to hydraulics and takes into account the influence of hydrostatic pressure, will permit prediction of the behavior of a technical scale reactor.

## 6.2
## Measuring Techniques

To understand the microbial degradation process and the system behavior of a biogas reactor and to establish reliable mathematical models, measuring techniques must be available – if possible – to analyze online special substances in the fermentation broth. Therefore, some new measuring techniques have been developed during the last years.

### 6.2.1
### Online Measurement Using a Mass Spectrometer

The anaerobic treatment of wastewater loaded with high concentrations of sulfate is often accompanied by problems due to inhibition by hydrogen sulfide (H$_2$S). It is well established that the inhibition is caused by undissociated hydrogen sulfide present in the liquid phase and not by the total sulfide content. This concentration of undissociated hydrogen sulfide depends on sulfate concentration in the wastewater, pH, and ionic strength. The concentration of undissociated hydrogen sulfide in the liquid phase undergoes great changes in the pH range used for anaerobic digestion, which is between 6.6 and 7.4. Since the actual concentration of dissolved $H_2S$ cannot be calculated on the basis of the $H_2S$ partial pressure in the biogas, because gas phase and liquid phase are not in equilibrium as is demonstrated in Section 6.5, this concentration has to be measured.

Measurement on the basis of ion-sensitive electrodes proved to be very problematic, especially with respect to their stability for a long period of time. Meyer-Jens et al. (1995) developed a new type of probe connected to a mass spectrometer. The probe consists of a small tube (outer diameter 0.9 mm, inner diameter 0.7 mm) of polydimethylsilicon. The tube can be inserted directly into the fermentation liquid of the biogas reactor. The $H_2S$ dissolved in the liquid phase penetrates the membrane and is transported by a carrier gas (technical nitrogen) to the mass spectrometer.

A very simple silicon membrane probe is shown in Figure 6.4. The silicon membrane tube is arranged in a short bypass loop outside of the reactor. The fermentation liquid flows outside of the tube while the carrier gas passes inside. In Figure 6.5 calibration of the measurement system is demonstrated. The calibration curve is perfectly linear within the relevant $H_2S$ concentrations. The different gradient for

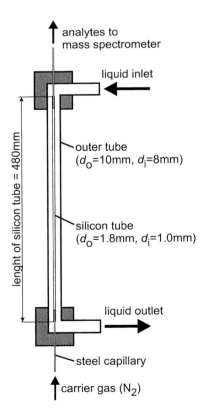

**Fig. 6.4** Configuration of a silicon membrane probe operated in a short bypass at the reactor (after Polomski, 1998).

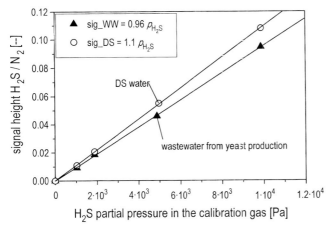

**Fig. 6.5** Calibration of the silicon membrane probe coupled with a mass spectrometer measuring the concentration of dissolved $H_2S$ in desalinated water (DS water) and in the wastewater of bakers' yeast production. The ordinate gives the value of the signal with reference to the $N_2$ signal of the carrier gas. The abscissa is the partial pressure of $H_2S$, with which the liquid is in equilibrium (after Polomski 1998).

desalinated water and wastewater indicates the difference in the solubility (Henry coefficient) of $H_2S$ in desalinated water (78.2 mmol $L^{-1}$ $bar^{-1}$) and wastewater (68.5 mmol $L^{-1}$ $bar^{-1}$) at 37 °C (data given by Polomski, 1998).

The measurement system can also serve well for the detection of dissolved $CO_2$ in the biogas reactor. Figure 6.6 shows the calibration curve for $CO_2$ in a sample of desalinated water. Also with $CO_2$ the solubility (Henry coefficient) for wastewater ($2.21 \cdot 10^{-4}$ mmol $L^{-1}$ $Pa^{-1}$) is 12% smaller than the literature data given for desalinated water ($2.52 \cdot 10^{-4}$ mmol $L^{-1}$ $Pa^{-1}$). Matz and Lennemann (1996) used a similar silicon membrane probe for online monitoring of biotechnological processes by gas chromatographic mass spectrometric analysis. The sensitivity of the probe system could be improved considerably by an instantaneous heating of the membrane to a temperature of 200 °C. This quasi online measuring procedure had a sampling period of 15 min between two successive measuring cycles. Substances like dimethylsulfide, butanethiol, cresol, phenol, ethylphenol, and indole at concentrations between 10 and 200 ppm were detected.

### 6.2.2
### Online Monitoring of Organic Substances
### with High-Pressure Liquid Chromatography (HPLC)

Zumbusch et al. (1994) reported the simultaneous online determination of betaine (trimethylglycine), *N,N*-dimethylglycine (*N,N*-DMG), acetic and propionic acid during the anaerobic fermentation of wastewater from bakers' yeast production. Ammonia and chloride could also be measured. The determinations were carried out by

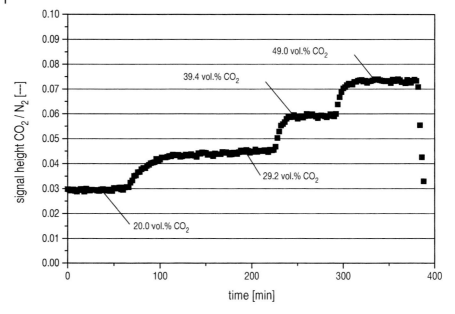

**Fig. 6.6** Calibration of the silicon membrane probe coupled to mass spectrometer measuring the dissolved $CO_2$ in desalinatec water (after Polomski, 1998).

means of isocratic cation exchange chromatography. For the detection of betaine and $N,N$-DMG an ultraviolet detector was used. All other substances were detected by conductivity.

Besides other organic components, such as acetic and propionic acid, betaine is the dominant organic substance in the wastewater. Up to 33% of the TOC is represented by betaine. The use of HPLC analysis combined with an automated ultrafiltration setup for online process monitoring gives interesting information about the dynamic behavior of the anaerobic process, as demonstrated in Figure 6.7. The graphs show the effect of hydrogen sulfide elimination from the fermentation broth. The concentration of undissociated $H_2S$ was decreased by a factor of 2 (from 300 to 150 mg L$^{-1}$). During the same time biogas production increased. The concentration of acetic acid decreased while the other acids measured remained more or less stable. Betaine was completely degraded only at lower $H_2S$ concentrations. Interestingly, the concentration of $N,N$-DMG, which is a metabolic product of the anaerobic degradation of betaine, is also decreased.

## 6.3
## Kinetics

The anaerobic degradation of organic substances is performed in a sequence of biological reactions in a synthrophic cooperation of different microorganisms

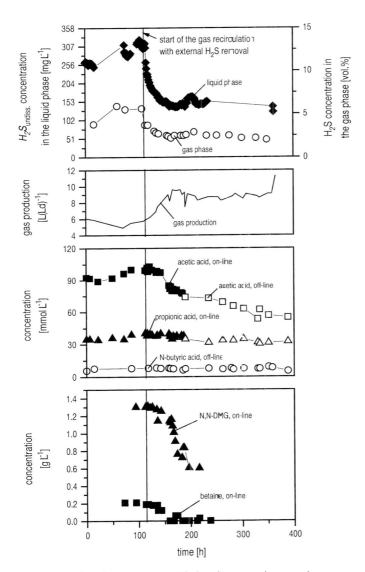

**Fig. 6.7** Combined measurement with the silicon membrane probe mass spectrometer and online high-pressure liquid chromatography in a laboratory upflow anaerobic sludge blanket (UASB) reactor with and without elimination of H₂S in an external biogas recirculation loop over a period of some hundred hours (after Polomski, 1998). Online measurements were carried out in cooperation with Zumbusch et al. (1994). H₂S concentrations were detected by the silicon membrane probe and mass spectrometer.

(Fig. 6.8). In a first step, complex biopolymers (lipids, proteins, carbohydrates) are split into the monomers which can be assimilated by the microbial cell. This first breakdown is catalyzed by extracellular enzymes. In a second step, fermentative bac-

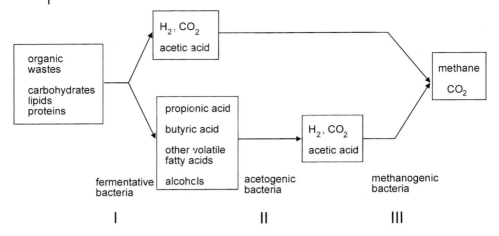

**Fig. 6.8** Pathway of anaerobic biodegradation.

teria form hydrogen, carbon dioxide, acetic acid, propionic acid, and further intermediary products, predominantly volatile fatty acids.

Acetic acid, carbon dioxide, and hydrogen can be directly converted to methane
and carbon dioxide (biogas) by methanogenic bacteria. Propionic acid as well as the
other intermediates cannot be directly transformed to biogas. In an intermediate
step, the so-called acetogenic bacteria have to generate acetic acid, carbon dioxide,
and hydrogen which then can be converted to biogas again. Thauer et al. (1977)
showed that for thermodynamic reasons the degradation of propionic acid is only
possible at very low concentrations of hydrogen (1 Pa). From this argument it is
clear that the acetogenic reaction which generates hydrogen can be performed only
by a direct and close synthrophic cooperation with methanogenic bacteria which
consume hydrogen. Sulfate, a frequent component in almost all food industries'
wastewater, is almost totally converted to hydrogen sulfide.

The dynamic behavior of the system is mainly controlled by the conversion of acetic acid to methane and carbon dioxide. This reaction is rather slow and, therefore, it
is the bottleneck of the anaerobic digestion, as demonstrated by experiments like
those shown in Figure 6.9. Whey which contains about 5% lactose, is continuously
digested to biogas. At the beginning of the experiment the concentration of the different volatile acids denoted in Figure 6.9 represent a steady state. In this situation
the reactor was heavily overloaded during the next 2.5 h. As a result, the concentration of acetic acid increased by a factor of 5. The concentrations of the other acids remained more or less constant. This finding supports the idea of Graef and Andrews
(1973) that the conversion of acetic acid to methane and $CO_2$ is the rate-limiting step
and that a mathematical model of the anaerobic degradation process can be reduced
to this dominating reaction.

This simple idea is expressed in Figure 6.10 and Eqs. 1 and 2, which describe the
concentration of microorganisms $x$ and the concentration of acetic acid $Ac$ as a function of time.

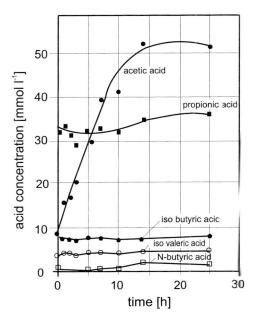

**Fig. 6.9** Concentration of the different volatile acids as a function of the time during the continuous anaerobic degradation of whey, a byproduct of cheese production (after Märkl et al., 1983). The experiments were performed in a 20 L stirred tank reactor. The graph shows the dynamic behavior of the system after being heavily overloaded with whey during the first 2.5 h of the experiment.

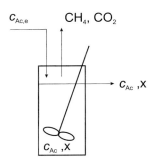

**Fig. 6.10** Mathematical model of anaerobic digestion in a stirred tank reactor. The reactor is directly fed by acetic acid (concentration in the inlet: $c_{Ac,e}$). The concentration of acetic acid in the reactor is $c_{Ac}$, $x$ denotes the concentration of biomass.

$$\frac{dx}{dt} = \mu\left(c_{Ac}, c_i\right) x - \frac{\dot{V}_{feed}}{V_{reactor}} x \qquad (1)$$

$$\frac{dc_{Ac}}{dt} = \frac{\dot{V}_{feed}}{V_{reactor}}\left(c_{Ac,e} - c_{Ac}\right) - \frac{1}{Y}\ \mu\left(c_{Ac}, c_i\right) x \qquad (2)$$

The reactor is directly fed by acetic acid $c_{Ac,e}$. The growth rate of the microorganisms is denoted as $\mu\left(c_{Ac}, c_i\right)$, and the biogas production can be assumed to be directly proportional to $\mu$. At a first glance, this model looks very simple. The real problem when solving the equation is that we need information on the kinetics of the reaction. The growth rate $\mu$ is a function of the acetic acid concentration $c_{Ac}$ but may also depend on the concentrations of other substances $c_i$. Before going into detail of formulating this kinetic equation one should understand more about the physicochemical state of the fermentation broth.

Acetic acid is dissociated to a large extent in the relevant pH range between 6.6 and 7.4. The pH value is calculated on the basis of the activity of $H^+$, which equals $\gamma_1 \cdot c_{H^+}$.

The concentration of $H^+$ ions $c_{H^+}$ can be calculated according to Eq. 3 which represents the electroneutrality of the fermentation broth:

$$c_{OH^-} + c_{Ac^-} + c_{Pro^-} + c_{HCO_3^-} + c_{HS^-} = c_{NH_4^+} + c_{H^+} + Z \tag{3}$$

The amounts of negative and positive ions must be balanced.

The most important ion concentrations in this calculation are the concentrations of dissociated acetic acid $c_{Ac^-}$ and propionic acid $c_{Pro^-}$, $OH^-$ ion $c_{OH^-}$, hydrocarbon ion $c_{HCO_3^-}$, and dissociated hydrogen sulfide $c_{HS^-}$. Besides the $H^+$ ion, the ammonium ion (concentration $c_{NH_4^+}$) has a positive charge. $Z$ represents the pool of not extra specified ions which are necessary to balance the equation and are assumed to be constant during the different stages of fermentation.

The concentrations of these ions can be calculated according to Eqs. 4–14.

$$HAc \leftrightarrows Ac^- + H^+ \qquad\qquad c_{Ac} = c_{HAc} + c_{Ac^-} \tag{4}$$

$$H\,Pro \leftrightarrows Pro^- + H^+ \qquad\qquad c_{Pro} = c_{H\,Pro} + c_{Pro^-} \tag{5}$$

$$\begin{aligned} H_2O + CO_2 &\leftrightarrows HCO_3^- + H^+ \\ HCO_3^- &\leftrightarrows CO_3^{2-} + H^+ \end{aligned} \qquad c_{C,tot} = c_{CO_2} + c_{HCO_3^-} + c_{CO_3^{2-}} \tag{6}$$

$$\begin{aligned} H_2S &\leftrightarrows HS^- + H^+ \\ HS^- &\leftrightarrows S^{2-} + H^+ \end{aligned} \qquad c_{S,tot} = c_{H_2S} + c_{HS^-} + c_{S^{2-}} \tag{7}$$

$$H_2O + NH_3 \leftrightarrows NH_4^+ + OH^- \qquad c_{N,tot} = c_{NH_3} + c_{NH_4^+} \tag{8}$$

$$c_{OH^-} = \frac{1}{c_{H^+}} \cdot 10^{-14}\ \text{mol}^2\ \text{L}^{-2} \tag{9}$$

$$c_{Ac^-} = \frac{K_{D,Ac}}{K_{D,Ac} + c_{H^+}\,\gamma_1^2}\, c_{Ac} \tag{10}$$

$$c_{Pro^-} = \frac{K_{D,Pro}}{K_{D,Pro} + c_{H^+}\,\gamma_1^2}\, c_{Pro} \tag{11}$$

$$c_{HCO_3^-} = \frac{K_{D,CO_2}^I}{K_{D,CO}^I + c_{H^+}\,\gamma_1^2}\, c_{C,tot} \tag{12}$$

$$c_{HS^-} = \frac{K_{D,H_2S}^I}{K_{D,H_2S}^I + c_{H^+}\,\gamma_1^2}\, c_{S,tot} \tag{13}$$

$$c_{NH_4^+} = \frac{K_{D,NH_3}^I}{K_{D,NH_3}^I + c_{OH^-}\,\gamma_1^2}\, c_{N,tot} \tag{14}$$

For example, the acetic acid HAc is dissociated to $Ac^-$ and $H^+$ according to Eq. 4. The amount of total acetic acid (concentration: $c_{Ac}$) is calculated by the sum of dissociated ($c_{Ac^-}$) and undissociated species ($c_{HAc}$). From these two equations and the knowledge of the dissociation constant $K_{D,Ac}$ and the activity coefficient $\gamma_1$ the concentration of the dissociated species can be calculated according to Eq. 10. The activity coefficient is calculated by an approximation after Davis which can be found in Loewenthal and Marais (1976). The activity coefficients depend on the ionic strength, which was determined to be 0.235 mol $L^{-1}$ in wastewater from bakers' yeast production. The calculation of the dissociation constants was performed after Beutier and Renon (1978). Data for the discussed wastewater are given in Table 6.3.

The calculation for the dissociation of $CO_2$ and $H_2S$ is in principal more complex because it is performed in two steps according to Eqs. 6 and 7, which are qualitatively shown in Figure 6.11. But from these graphs it is clear that both the concentra-

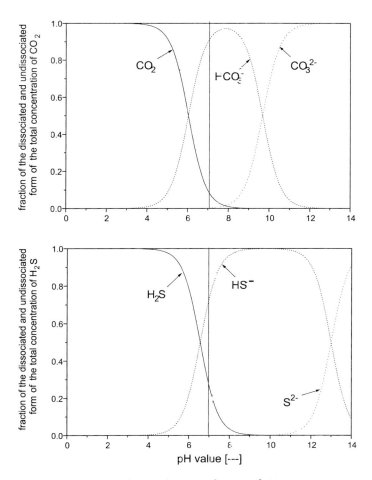

**Fig. 6.11** Dissociation of $CO_2$ and $H_2S$ as a function of pH.

**Table 6.3** Equilibrium constants for wastewater of bakers' yeast production, temperature 37.4 °C.

| | | | Activity Coefficients | | | |
|---|---|---|---|---|---|---|
| $\gamma_1$ | | 0.73 | | Monovalent ion | | |
| $\gamma_2$ | | 0.28 | | Bivalent ion | | |
| | | | Dissociation Constant [mol L$^{-1}$] | | | |
| $K_{D,Ac}$ | $K_{D,Pro}$ | $K^I_{D,Ac}$ | $K^{II}_{D,CO_2}$ | $K^I_{D,H_2S}$ | $K^{II}_{D,H_2S}$ | $K_{D,NH_3}$ |
| $1.17 \cdot 10^{-5}$ | $1.3 \cdot 10^{-5}$ | $4.94 \cdot 10^{-7}$ | $5.79 \cdot 10^{-11}$ | $1.44 \cdot 10^{-7}$ | $2.7 \cdot 10^{-14}$ | $6.83 \cdot 10^{-6}$ |

tions of $CO_3^{2-}$ and $S^{2-}$ can be neglected in the relevant range of pH between 6.6 and 7.4. Therefore, calculation can be handled like a 1-step dissociation. Eqs. 3, 9–14 give the pH value in implicit form. The pH can be evaluated by a stepwise iterative procedure.

### 6.3.1
### Acetic Acid, Propionic Acid

Many experiments have been reported on the growth rate of acetic acid converting organisms with respect to the concentration of acetic acid in the fermentation broth. A reliable collection of those data is presented in Figure 6.12. Because it is very difficult to measure the growth rate of methane producing organisms directly (in a mixed culture it may be impossible) and because it is assumed that this growth rate is directly proportional to the methane production rate, usually the latter value is taken as a measure of bacterial activity. In Figure 6.12 the methane production rate, related to the maximum methane production rate of a biogas reactor, is denoted on the ordinate of the graph.

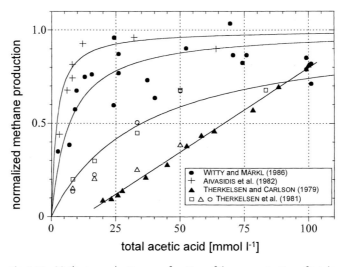

**Fig. 6.12** Methane production as a function of the concentration of total acetic acid.

On the abscissa the concentration of the total acetic acid measured in the fermentation broth is denoted. Data of different systems are collected. Witty and Märkl (1986) reported the digestion of waste mycelium from antibiotics production, Therkelsen and Carlson (1979) and Therkelsen et al. (1981) digested a complex artificial substrate, and Aivasidis et al. (1982) studied a pure culture of *Methanosarcina barkeri* which was directly fed with acetic acid. Due to the complexity of the synthrophic reaction system and also due to the unstable nature of the biogas reaction, the data points reflecting the experimental results of the different authors are somewhat scattered. But apart from this phenomenon, it can be stated that the general behavior of the different systems differs to a large extent from one to the other.

In Figure 6.13 exactly the same experimental data are presented. Here, the abscissa denotes only that part of the acetic acid that is not dissociated.

The undissociated acetic acid can be calculated with the help of Eq. 15 if the total acetic acid and the pH value are known.

$$c_{HAc} = \frac{c_{H^+} \gamma_1^2}{K_{D,Ac} + c_{H^+} \gamma_1^2} \, c_{Ac} \tag{15}$$

It can be easily seen that this parameter integrates all experimental data into a clear picture. From this finding it was concluded that the microorganisms use and only 'see' the undissociated form of acetic acid. The graph drawn in Figure 6.13 has the form of Michaelis–Menten kinetics, the $K_s$ value is 0.07 mmol L$^{-1}$.

The $K_s$ value represents the concentration of substrate that corresponds to 50% of the maximal reaction rate. Because the undissociated acetic acid represents the relevant substrate, we can also write $K_s = K_{HAc}$.

To find out if this $K_{HAc}$ value is of a general nature or if it depends on the type of microorganism that is active in the conversion of acetic acid to methane, some more

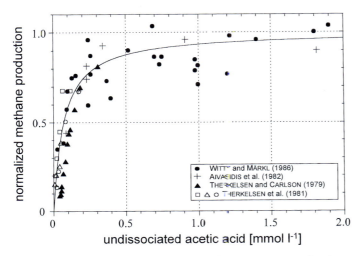

**Fig. 6.13** Methane production as a function of the concentration of undissociated acetic acid.

kinetic data of authors who worked with more defined microbial systems are studied. These data are presented in Figure 6.14. It can be seen that experiments performed with *Methanothrix* sp. result in a smaller $K_{HAc}$ value ($K_{HAc}$ = 0.02 mmol L$^{-1}$) than those performed with *Methanosarcina barkeri*, which correspond to a somewhat higher value ($K_{HAc}$ = 0.12 mmol L$^{-1}$). This result may be due to the fact that the relative surface of the filamentous *Methanothrix* sp. is larger than that of the rod-shaped *Methanosarcina* sp.

Altogether, the $K_{HAc}$ value based on undissociated acetic acid is roughly in the range between 0.02 mmol L$^{-1}$ and 0.12 mmol L$^{-1}$. In a mixed culture an average value of 0.07 mmol L$^{-1}$ may be adequate. If this mixed culture contains mainly *Methanothrix* sp. to convert acetic acid, the value will be smaller. If *Methanosarcina* sp. dominate, the value of $K_{HAc}$ should be higher. The author of this paper used $K_{HAc}$ = 0.07 mmol L$^{-1}$.

Witty and Märkl (1986) reported that propionic acid inhibits the conversion of acetic acid to methane and $CO_2$. Experiments with different amounts of propionic acid resulted in an inhibition coefficient of $K_{H\,Pro}$ = 0.97 mmol L$^{-1}$. Therefore, a concentration of 0.97 mmol L$^{-1}$ of undissociated propionic acid causes an inhibition of 50% of the rate without propionic acid (Eq. 18).

$$c_{H\,Pro} = \frac{c_{H^+}\,\gamma_1^2}{K_{D,Pro}+c_{H^+}\,\gamma_1^2}\,c_{Pro} \qquad (16)$$

$$c_{H_2S} = \frac{c_{H^+}\,\gamma_1^2}{K_{D,H_2S}+c_{H^+}\,\gamma_1^2}\,c_{S,tot} \qquad (17)$$

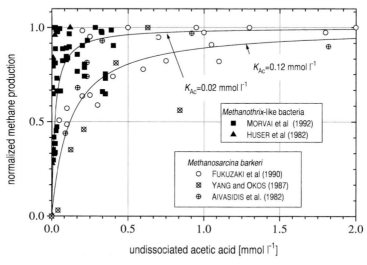

**Fig. 16.14** Methane production as a function of the concentration of the undissociated acetic acid. Experimental data of cultures of *Methanothrix*-like sp. and *Methanosarcina barkeri*.

$$\mu = \mu_{max} \frac{c_{HAc}}{c_{HAc}+K_{HAc}} \cdot \frac{K_{H\,Pro}}{K_{H\,Pro}+c_{H\,Pro}} \cdot \frac{K_{H_2S}}{K_{H_2S}+c_{H_2S}}$$ (18)

$K_{HAc} = 0.07$ mmol $L^{-1}$
$K_{H\,Pro} = 0.97$ mmol $L^{-1}$
$K_{H_2S} = 2.5$ mmol $L^{-1}$

## 6.3.2
## Hydrogen Sulfide

The inhibition of methane generation from acetic acid by $H_2S$ is well known. Because the medium sulfate concentration in the wastewater from bakers' yeast production is as high as 3.9 g $L^{-1}$, which is almost totally converted to hydrogen sulfide, the problem of $H_2S$ inhibition is very substantial. Friedmann and Märkl (1994) reported an inhibition constant of 2.8 mmol $L^{-1}$ (95.2 mg $L^{-1}$). Recent experiments of Polomski (1998) on both the laboratory and pilot scale (for experimental apparatus see Figure 6.2) resulted in an inhibition constant of 2.5 mmol $L^{-1}$ (85 mg $L^{-1}$). The results of these experiments are shown in Figure 6.15.

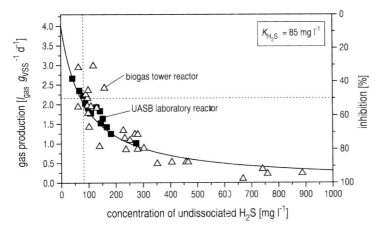

**Fig. 6.15** Specific gas production of a UASE laboratory reactor and a biogas tower reactor on pilot scale (for both reactors, see Figure 6.2 and Table 6.1) as a function of the concentration of undissociated $H_2S$. The data of gas production are corrected for the influence of acetic acid and propionic acid concentrations according to Eq. 18. VSS: volatile suspended solids (biomass).

## 6.3.3
## Conclusions

As far as the kinetics of the conversion of acetic acid to methane and $CO_2$ are concerned, the results are summarized in Eq. 18. The data given are valid only near the steady state of the system; this means in a pH range between 6.6 and 7.4 and in the usual range of substrate concentrations. If the system was heavily overloaded for some time and was out of the described range, the microbial population might need some recovering time, which cannot be described by Eq. 18. Figure 6.16 shows a sketch of different configurations in which active microorganisms exist in biogas reactors. They can exist in a freely dispersed form, i.e., as single organisms, or as in the biogas tower reactor in the form of loose pellets. The microbial system can also be immobilized in the form of a film on the wall or as a dense pellet. In the second case the conversion efficiency is also influenced by transport phenomena. The data given in Eq. 18 hold only for freely dispersed organisms. When immobilized microorganisms dominate in a biogas reactor the constants (both saturation and inhibition constants) are larger.

Eq. 18 can also be used to calculate the combined influence of two parameters as demonstrated in Figure 6.17. This graph shows the gas production as a function of

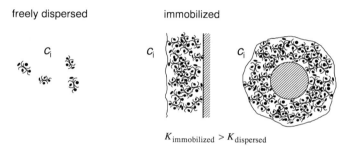

**Fig. 6.16** Configuration of the active microbial system.

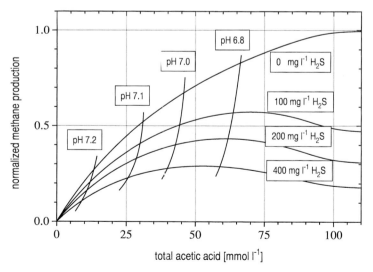

**Fig. 6.17** Methane production as a function of the concentration of total acetic acid at different concentrations of total hydrogen sulfide ($c_{HPro} = 0$).

the concentration of total acetic acid. It can be seen that the behavior changes completely if it is compared to the graph in Figure 6.12. Adding $H_2S$ to the fermentation broth changes the kinetics from a Michaelis–Menten type to a substrate inhibition type. The effect can be explained by the continuous decrease of the pH value if the concentration of acetic acid is increased. At smaller values of pH more of the total hydrogen sulfide exists in the form of undissociated $H_2S$ and the inhibitory effect increases along the abscissa.

## 6.4
## Hydrodynamic and Liquid Mixing Behavior of the Biogas Tower Reactor

Mixing of the liquid phase with respect to the scale-up of a reactor has to be known to understand the nutrient supply of all of the active reactor regions and to avoid local overloading of the reactor. The biogas tower reactor of Figure 6.1 consists of four modules. The active volume of one reactor unit, as shown in Figure 6.2, is 280 m$^3$ (height 25 m, diameter 3.5 m). A settler at the top of the reactor helps to maintain high biomass concentrations within the reactor.

The principal function of one module is described in Figure 6.18. Biogas produced below this module rises and is collected under the collecting device (1), where it can be withdrawn. The quantity of gas withdrawn is controlled by a valve (2). If less gas is taken off than collected, the remaining gas rises through the open cross-sectional area (3) into the next module and into the next gas collecting device (5). The baffle (4) separates the space between two gas collecting devices into two channels joining at the top and the bottom of each module. Because of the rising gas on one side of the baffle, there is a difference in the gas holdup in two channels, which corresponds to a fluid circulation along the baffle (Blenke, 1985). The upflow channel is

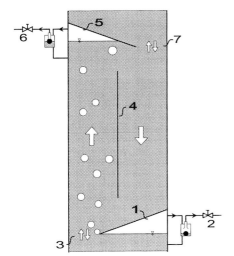

**Fig. 6.18** One module of the BTR (1,5: gas collecting devices; 2,6: gas control units; 3,7: connecting cross-sectional areas; 4: baffle).

called a "riser" and the downflowing part is called a "downcomer". The circulation velocity is strongly linked to the amount of gas passing through the open cross-sectional area (3). Hence, by controlling the gas outlets (2) and (6) it is possible to control the liquid circulation velocity and the mixing time in a module. If no gas is withdrawn, both the circulation velocity and the mixing intensity are high. In contrast, if all collected gas is withdrawn, the circulation slows and good settling conditions for biomass set in. In the lower zones of the reactor it might be advantageous to support the mixing characteristics, in the higher zones good settling conditions should be established. As in mixing within one module, the mixing between two neighboring modules depends, as experiments show, on the gas flow rate through the open cross-sectional area ( 3) between the two modules. If gas rises through this area, turbulence is generated, causing a convective transport between the two modules. If the gas flow rate through the connecting area (3) is low, the mass transport between the modules decreases.

## 6.4.1
### Mixing of the Liquid Phase

Reinhold et al. (1996) provided a theoretical and analytical analysis of the mixing behavior of a BTR. Experiments were performed on a laboratory (Fig. 6.19) and a pilot

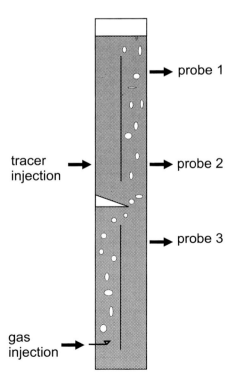

**Fig. 6.19** Laboratory reactor representing a model of two modules of the BTR. It was filled with 0.1% ethanol in tap water, and air was injected through a sparger at the bottom to simulate biogas production (height: 6.5 m; diameter 0.4 m; volume 0.7 m³; height of a module 3.5 m; length of the baffle 2.89 m).

scale (Fig. 6.20). To study mixing, a saline solution was injected in the form of a Dirac impulse as a tracer at positions indicated in Figures 6.19 and 6.20. The time course of the salt concentration was continuously monitored by conductivity probes at different positions. For the experiments the reactors were filled with tap water. To reduce the coalescence of bubbles and to obtain small bubbles as observed in biogas reactors, 0.1% ethanol was added. Instead of biological gas production, air was injected at the bottom of the reactors. It was assumed that the dominating effect on the circulating flow was caused by gas coming from the lower module. The results of the experiments can be summarized as follows. The mixing times for a 95% homogeneity after a tracer impulse are

- mixing within one module: $(1–2) \cdot 10^2$ s
- mixing from one module to a neighboring module: $(0.5–2) \cdot 10^3$ s

Because the heights of a module in the laboratory and in the pilot scale reactor do not differ much, the situation on the pilot scale is more or less the same as in the laboratory scale reactor.

Mixing within one module due to the axial dispersion in the circulating flow is shown in Figure 6.21, presenting laboratory scale experiments. The concentration profiles over time for the four modules of the pilot scale reactor after applying a Dirac impulse in the third module are recorded in Figure 6.22. The experimental behavior is simulated very well by a mathematical model.

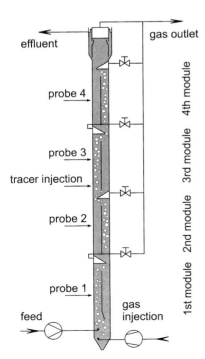

**Fig. 6.20** Pilot scale BTR (height 20 m; diameter 1 m; volume 15 m³; height of a module 4 m; length of the baffle 3 m).

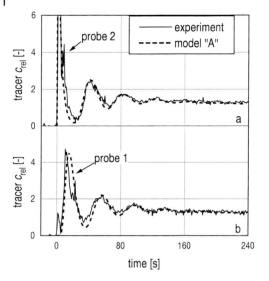

**Fig. 6.21** Mixing of a Dirac impulse in the upper module of the laboratory scale reactor (Fig. 6.19).

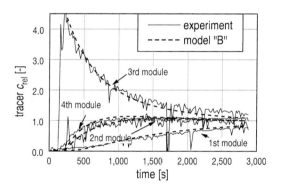

**Fig. 6.22** Mixing of a Dirac impulse (3rd module) in the pilot scale BTR (Fig. 6.20).

The structure of the mathematical model developed is shown in Figure 6.23. The structures of the two models A and B follow the modular structure of the reactor concept. Both models couple two neighboring modules by the flow of the feed ($\dot{V}_{feed}$), which is constant within the tower reactor and is directed from the bottom to the top of the reactor. The central idea of both models is to predict the coupling between two modules by the so-called "exchange flow rate". This virtual flow rate ($\dot{V}_{exchange}$) accounts for the situation of a turbulent exchange of volume elements in the area between two modules ($A_{exchange}$). This turbulent flow is driven by the bubbles passing this area. $\dot{V}_{feed}$ is kept at zero during the experiment. Also, in reality the contribution of $\dot{V}_{feed}$ is usually small compared to the turbulent exchange flow $\dot{V}_{exchange}$.

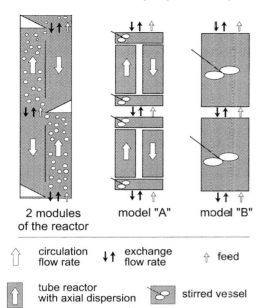

**Fig. 6.23** Structure of models A and B as compared to the BTR for describing the mixing behavior of the liquid phase.

2 modules of the reactor    model "A"    model "B"

⇑ circulation flow rate    ↓↑ exchange flow rate    ⇑ feed

▯ tube reactor with axial dispersion    ⌀ stirred vessel

### 6.4.1.1 Model A

Model A, which is a more detailed one, is able to describe mixing within a module as well as intermixing between modules. Each module is composed of four compartments. Riser and downcomer are mathematically replaced by tube reactors with axial dispersion. At the bottom and the top of each module, riser and downcomer join in mixing zones which are regarded as stirred vessels. The four zones are linked by the circulation flow rate $\dot{V}_{circ}$. Since the cross-sectional area of the riser $A_{riser}$ and the downcomer $A_{downc.}$ as well as the flow rate through the riser $\dot{V}_{riser}$ and the downcomer $\dot{V}_{downc.}$ are equal, the mean circulation velocity $w_m$ is given by Eq. 19.

$$w_m = \frac{\dot{V}_{riser}}{A_{riser}} = \frac{\dot{V}_{downc.}}{A_{downc.}} \tag{19}$$

Each real module of the reactor is thus replaced by a "ring structure" of four reactors. The transport and mixing mechanisms in this ring structure are convection and axial dispersion. The mathematical equations of model A can be derived from the mass balances of each of the four compartments. Model A needs three parameters for calculating the mixing behavior: the liquid circulation velocity $w_m$, the axial dispersion coefficient $D_{ax}$, and the exchange rate $\dot{V}_{exchange}$. All these parameters depend on the gas flow rate entering the module from the lower module. For more details of the model, the mathematical structure, and the procedure of solving the partial differential equations, the publication of Reinhold et al. (1996) should be consulted.

### 6.4.1.2 Model B

Model B, which is simpler, consists of only one stirred vessel per module (Fig. 6.23). This model is able to describe the mixing behavior of the whole reactor if the mixing intensity within one module is high compared to the tracer transport from one module to another. This assumption holds true in the real situation, as shown earlier. Model B links the stirred vessels in the same way as model A. The intermixing between two neighboring modules is again modeled by an exchange flow rate $\dot{V}_{\text{exchange}}$ going up and down. Eq. 20 shows the material balance for one module $i$.

$$V_i \frac{dc_i}{dt} = \dot{V}_{\text{exchange}, i-1} (c_{i-1} - c_i) + \dot{V}_{\text{exchange}, i} (c_{i+1} - c_i) + \dot{V}_{\text{feed}} (c_{i-1} - c_i) \tag{20}$$

For describing the mixing of one compound in a BTR with model B, the number of ordinary differential equations is equal to the number of modules. It is an initial-value problem which can be solved by the Runge–Kutta method. The only necessary parameter for calculating the mixing behavior is the exchange flow rate $\dot{V}_{\text{exchange}}$.

Experiments in both laboratory and pilot scale reactors are carried out at different gas flow rates $\dot{V}_{\text{gas}}$ to investigate the dependence of the hydrodynamic parameters on the gas loading. Although there are many investigations on airlift loop reactors and bubble columns in general, the results cannot be applied to this type of reactor because the gas and liquid loading of the BTR are far below those of the airlift loop and bubble column reactors found in the literature. Since the present BTR has unique characteristics, it was necessary to determine the exchange flow rates $\dot{V}_{\text{exchange}}$. Consequently, studies on the circulation velocity $w_m$ and the axial dispersion $D_{\text{ax}}$ were carried out.

The characteristic parameters $D_{\text{ax}}$, $w_m$, $\dot{V}_{\text{exchange}}$, are a function of the gas loading of the system. The parameters can be obtained by using the least-squares method to fit the simulation to the experimental data. The results are shown in Figures 6.24–6.26.

In Figure 6.24 the axial dispersion coefficient is plotted against the superficial gas velocity of the riser $u_{\text{riser}}$ which is given by Eq. 21.

$$u_{\text{riser}} = \frac{\dot{V}_{\text{gas}}}{A_{\text{riser}}} \tag{21}$$

This coefficient was only determined on the laboratory scale. Figure 6.25 shows the mean circulation velocity $w_m$ of the laboratory scale and the pilot scale reactors as a function of the superficial gas velocity of the riser $u_{\text{riser}}$, From a momentum balance the mean circulation velocity $w_m$ can be calculated on the basis of the difference in gas holdup between riser and downcomer $\Delta\varepsilon$ and the knowledge of the total pressure loss (friction) coefficient $\xi$ according to Eq. 22.

$$w_m = \sqrt{\frac{2\,g\,L}{\xi}\,\Delta\varepsilon} \tag{22}$$

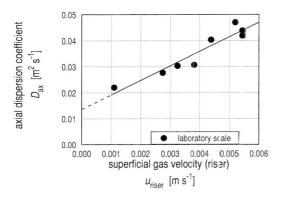

**Fig. 6.24** Axial dispersion coefficient as a function of the superficial gas velocity in the riser.

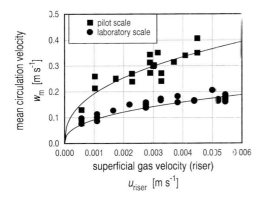

**Fig. 6.25** Mean circulation velocity of the liquid as a function of the superficial gas velocity of the riser.

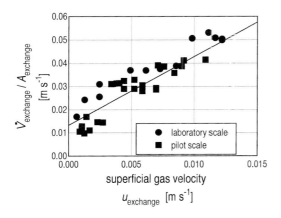

**Fig. 6.26** Effect of the superficial gas velocity on the exchange flow rate between two modules.

$L$ is the length of the baffle and $g$ is the gravitational acceleration. The gas holdup can be correlated from the superficial gas velocity $u_{riser}$ according to Weiland (1978). Due to experiments of Reinhold et al. (1996), it can be calculated according to Eqs. 23 and 24).

$$\Delta\varepsilon = 0.27 \, u_{riser}^{0.8} \quad \text{(laboratory scale)} \tag{23}$$

$$\Delta\varepsilon = 0.76 \, u_{riser}^{0.8} \quad \text{(pilot scale)} \tag{24}$$

The pressure loss coefficients were determined to be $\xi = 6.9$ on the laboratory scale and to be $\xi = 4.8$ on the pilot scale. As expected, the total friction coefficient $\xi$ slightly decreases during scale-up.

Figure 6.26 shows the exchange flow rate $\dot{V}_{exchange}$ as a function of the superficial gas velocity $u_{exchange}$, which can be calculated according to Eq. 25.

$$u_{exchange} = \frac{\dot{V}_{gas}}{A_{exchange}} \tag{25}$$

It is quite remarkable that the findings of Figure 6.26 are invariant with the scale of the reactor. Because $\dot{V}_{exchange}$ dominates the mixing behavior of the system (indeed, it is the only experimental parameter necessary for model B) this result is very important for the scale-up of biogas tower reactors.

### 6.4.2
### Distribution of Biomass within the Reactor

The wastewater is passing the BTR from the bottom to the top of the reactor. With this feeding procedure the upflow velocity due to the feed increases – at a constant mean residence time of wastewater in the reactor – linearly with the height of the reactor. Therefore, in high reactors it is essential to have reliable mechanisms to retain the active biomass. These mechanisms and the resulting distribution of biomass within the reactor were studied by Reinhold and Märkl (1997).

#### 6.4.2.1 Experiments
Experiments were performed with a two-modular laboratory tower reactor (height 6.5 m) and in a pilot scale tower reactor (height 20 m) as shown in Figure 6.2 and described in Table 6.1. The laboratory tower reactor was filled with tap water and anaerobic pelletized sludge. The sludge was biologically, inactive, as no substrate was present in the tap water, i.e., it did not produce any fermentation gas. Air was injected at the bottom of the column instead of real biogas. As the reactor was built out of transparent material, visual observations were possible in addition to measurements from samples. The following observations were made during these experiments:

- The suspended solids concentration in the lower module is always higher than in the upper module.

- The macroscopic liquid turbulence backmixing flow ($\dot{V}_{\text{exchange}}$), which causes the interaction between adjoining modules, was visualized through the movement of the sludge pellets as indicated by the arrows (7) in Figure 6.27.
- The most remarkable observation is also indicated in Figure 6.27: At the lower end of the baffle (4) the suspension turns 180° (6) and from this flow a continuously downward-directed solid mass flow (5) is observed, falling at the downward-sloping flat blade (1) and from there sliding to the module below. This may prove to be a very effective mechanism of retaining active sludge in the reactor.

Measurements of sludge distribution were also performed in the pilot scale tower reactor digesting wastewater of the bakers' yeast company. The BTR was inoculated with pelletized sludge, and the space loading was increased in 75 d to 8.8 kg TOC m$^{-3}$ d$^{-1}$ with a TOC removal efficiency of 60%. The TOC conversion rate by the sludge was 1.2 kg TOC (kg SS)$^{-1}$ d$^{-1}$. The size of the pellets in the reactor ranged from 0.08 to 1 mm. Volatile suspended solids were about 50%–60% of the suspended solids. Elementary analysis showed that 50% of the dry matter was carbon. Two main observations concerning the sludge distribution were made during these experiments on the pilot scale:

- There is no difference in the concentration of suspended solids between riser and downcomer. Each module can be regarded as a completely mixed reactor.
- The concentration of suspended solids in the BTR decreases gradually in upward direction from module to module.

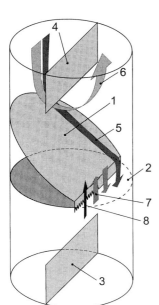

**Fig. 6.27** Flows in the coupling zone of two adjacent modules: (1) gas collecting device; (2) connecting area between the modules; (3) upper end of the lower baffle; (4) lower end of the upper baffle; (5) sedimenting mass flow ($\dot{M}_{\text{sedi}}$); (6) circulating flow ($\dot{V}_{\text{circ}}$); (7) backmixing flow ($\dot{V}_{\text{exchange}}$); (8) feed flow ($\dot{V}_{\text{feed}}$).

### 6.4.2.2  Mathematical Modeling

The structure of the mathematical models is shown in Figure 6.28. The mass balance of one module ($i$), including the interaction with an upper ($i+1$) and a lower module ($i-1$), leads to Eq. 26.

$$V_i \frac{dSS_i}{dt} = \dot{V}_{\text{exchange},i-1} \left( SS_{i-1} - SS_i \right) + \dot{V}_{\text{exchange},i} \left( SS_{i+1} - SS_i \right)$$
$$+ \dot{V}_{\text{feed}} \left( SS_{i-1} - SS_i \right) - \dot{M}_{\text{sedi},i-1} + \dot{M}_{\text{sedi},i} + \dot{M}_{\text{prod}} \frac{V_i \, SS_i}{\sum\limits_{k=1}^{n} V_k \cdot SS_k} \tag{26}$$

In this equation $V_i$ denotes the volume of a module, $SS_i$ the concentration of suspended solids, $\dot{V}_{\text{feed}}$, the volumetric feed rate of the reactor, and $\dot{M}_{\text{sedi},i}$ the sedimenting mass flow between the modules. The produced biomass $\dot{M}_{\text{prod}}$ according to the growth of organisms could be calculated according to the TOC consumed (Eq. 27).

$$\dot{M}_{\text{prod}} = 0.1 \cdot \dot{V}_{\text{feed}} \left( TOC_{\text{in}} - TOC_{\text{out}} \right) \tag{27}$$

The volumetric exchange flow $\dot{V}_{\text{exchange},i}$ is calculated with Eq. 28 on the basis of measurements in Figure 6.26.

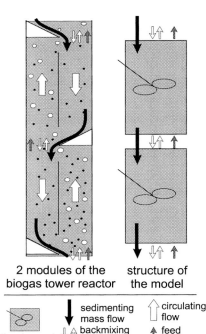

2 modules of the          structure of
biogas tower reactor      the model

| stirred vessel | ⬇ sedimenting mass flow | ⇕ backmixing flow | ⇧ circulating flow | ⬆ feed flow |

**Fig. 6.28**  Structure of the mathematical model compared to two modules of the BTR.

$$\frac{\dot{V}_{\text{exchange}}}{A_{\text{exchange}}} = 0.014 \text{ m s}^{-1} + 3.1 \cdot \frac{\dot{V}_{\text{gas}}}{A_{\text{exchange}}} \tag{28}$$

The only parameter that is not known a priori, is the sedimenting mass flow rate $\dot{M}_{\text{sedi}}$. This value was determined by regression analysis from the experiments performed as shown in Fig. 6.29. It can be clearly seen that the sedimentation capacity of active gassing sludge is much smaller compared to inactive (dead) sludge as used in the laboratory experiment. The sedimentation of active sludge follows Eqs. 29 and 30 where $A_{\text{reactor}}$ is the cross-sectional area of the reactor.

$$SS < 3 \text{ kg m}^{-3}: \frac{\dot{M}_{\text{sedi}}}{A_{\text{reactor}}} = 0.0025 \text{ m s}^{-1} \cdot SS \tag{29}$$

$$SS > 3 \text{ kg m}^{-3}: \frac{\dot{M}_{\text{sedi}}}{A_{\text{reactor}}} = 0.04 \text{ kg s}^{-1} \text{ m}^{-2} + 0.015 \text{ m s}^{-1} \cdot SS \tag{30}$$

In Figure 6.30 results of experiments and the calculation on the basis of the mathematical model are compared. The agreement between simulated data and measured suspended solid concentration is quite good. The relative deviation between measured and calculated values increases with decreasing concentration of suspended solids. The relative deviation ranges from 10%–15% of the concentrations in the BTR. To investigate the scattering of the measured concentration of suspended solids, 10 samples were taken at one sampling point during a laboratory scale experiment. The detected suspended solid concentrations deviated by up to 8% from the average value. This shows that part of the deviation might be due to sampling errors.

Reinhold and Märkl (1997) also demonstrated that the mathematical model allows for realistic predictions of the dynamic behavior of the system, as was shown by simulating the start-up of a BTR.

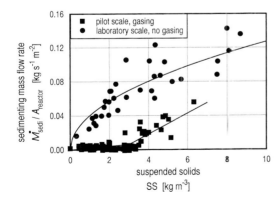

**Fig. 6.29** Sedimenting mass flow rate between adjacent modules on the laboratory scale and pilot scale as a function of the suspended solid concentration in the upper module.

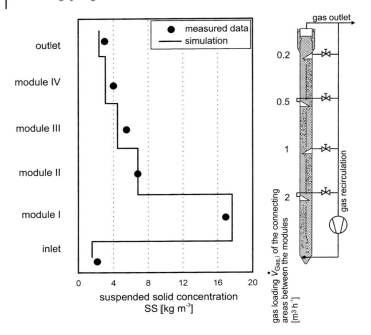

**Fig. 6.30** Concentration of suspended solids as a function of reactor height in the 15 m³ pilot scale BTR. $\dot{V}_{feed}$ = 315 L h⁻¹; total gas production of the reactor 2.3 m³ h⁻¹; gas recirculated 1.8 m³ h⁻¹; $\Delta TOC$ = 7.3 kg m⁻³; gas loading $V_{gas, i}$ of the connecting areas between the modules as indicated.

## 6.5
## Mass Transport from the Liquid Phase to the Gas Phase

As demonstrated in Section 6.3, it is necessary to have quantitative data for the concentrations of the product gases, like $CO_2$, $NH_3$, and $H_2S$ in the fermentation broth, to calculate the pH value within the fermentation broth and get information on the kinetics of methane production. To calculate these concentrations it is necessary to have additional information about the transport of these gases from the liquid phase – where they are produced by the microorganisms – to the gas phase. Basic ideas of this calculation will be discussed with the example of $H_2S$ because of its strong influence on biogas production. Polomski (1998) performed experiments on the desorption of $H_2S$ from the fermentation liquid broth in laboratory and pilot scale reactors.

The experimental setup of the laboratory experiment is shown in Figure 6.31. The UASB reactor is supplied with wastewater at the bottom. A part of the generated biogas is withdrawn at the top of the reactor and recycled to the bottom where it is injected by means of a perforated tube. The recycled biogas is either recycled directly or over a $H_2S$ adsorber. The $H_2S$ concentration in the fermentation liquid and the partial presure of $H_2S$ in the gas were measured by means of a silicon membrane probe and the mass spectrometer (see also Section 6.2.1).

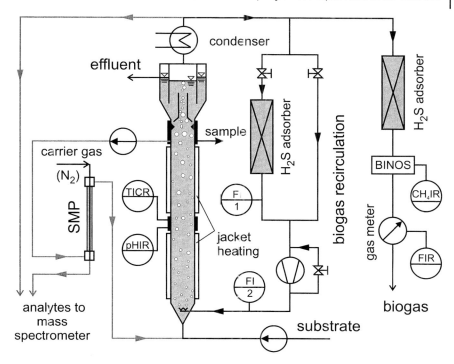

**Fig. 6.31** Experimental setup of a modified laboratory scale UASB reactor (active volume: 36.5 L) equipped with a bypass membrane probe and a mass spectrometer.

The effect of gas recirculation can be explained with the example of one result of the experiment shown in Figure 6.32. At the ordinate on the left side the $H_2S$ partial pressure is denoted; on the right side the $H_2S$ concentration in the liquid phase is denoted which is in equilibrium with the partial pressure in the gas according to Eq. 31 and the Henry coefficient $H_{H_2S} = 2.330 \cdot 10^{-2}$ mg $L^{-1}$ $Pa^{-1}$ (equal to $6.85 \cdot 10^{-4}$ mmol $L^{-1}$ $Pa^{-1}$).

$$c_i = p_i \cdot H_i \qquad (31)$$

It can be easily seen that in this type of reactor without gas recirculation (first 23 h of the experiment) $H_2S$ in the liquid and the gas phases is far from equilibrium. The concentration in the liquid phase is overconcentrated by 75% compared to the gas phase. After recirculating the gas and concurrently producing a larger interfacial area between the liquid and gas phases the overconcentration is diminished to 18%. After starting recirculation of biogas 8 h are required to come from one steady state to the next. In Figure 6.33 the result of a similar experiment is shown. Different from the first one, the recirculated biogas passes a $H_2S$ absorber and is eliminated from the system. Although here the sulfate content in the wastewater is extremely

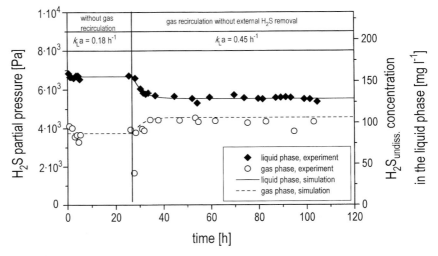

**Fig. 6.32** Effect of recycling biogas on the $H_2S$ concentration in the gas phase and the liquid phase in the laboratory UASB reactor. Experimental data are compared to results of the calculation with the mathematical model of Eqs. 32–34 (gas recirculation: 7.5 L h$^{-1}$; gas production: 7.45 L h$^{-1}$; pH value: 7.2 (constant); sulfate concentration in the wastewater: 3.85 g L$^{-1}$; residence time of wastewater: 2 d).

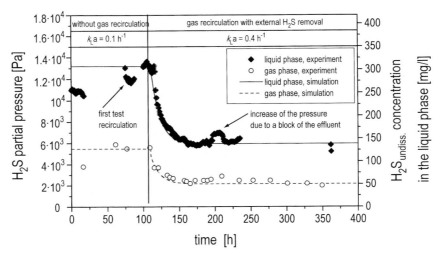

**Fig. 6.33** Different from Figure 6.32 an $H_2S$ absorber was integrated in the recycled biogas (gas recirculation 27 L$^{-1}$; gas production: 7.6–13.7 L h$^{-1}$; pH value: 7.1–7.3; sulfate concentration: 5.13 g L$^{-1}$; mean residence time of wastewater in reactor: 1 d).

high (5.13 g $L^{-1}$), the concentration of $H_2S$ in the liquid phase can be lowered to 130 mg $L^{-1}$.

Figures 6.32 and 6.33 also demonstrate that model calculations fit well with the experimental data. The $H_2S$ content in the liquid and the gas phases can be calculated by using $H_2S$ balances for the different compartments.

## 6.5.1
## Liquid Phase

In Eq. 32 $V_{reactor}$ is the volume of the liquid in the reactor, $c_{S,tot}$ the concentration of the total sulfur in the liquid according to Eq. 7, $\dot{V}_{feed}$ is the volumetric feed rate of the reactor and $c_{S,tot,e}$ the concentration of the total sulfur in the reactor feed.

$$V_{reactor} \cdot \frac{dc_{S,tot}}{dt} = \dot{V}_{feed}(c_{S,tot,e} - c_{S,tot}) - V_{reactor} \cdot k_L a \quad (c_{H_2S} - H_{H_2S} \cdot p_{H_2S,bubble}) \quad (32)$$

$k_L a$ denotes the transport capacity between the liquid phase and the gas phase (a: interfacial area per volume $m^2$ $m^{-3}$), $c_{H_2S}$ is the concentration of the undissociated portion of the $H_2S$ in the liquid, and $p_{H_2S,bubble}$ the partial pressure of $H_2S$ in the bubbles rising in the reactor liquid. $H_{H_2S}$ is the Henry coefficient for $H_2S$.

## 6.5.2
## Gas Bubbles

Eq. 33 gives the $H_2S$ balance for the gas bubbles in the reactor suspension:

$$V_{bubble} \cdot \frac{dp_{H_2S,bubble}}{dt} = \dot{V}_{recyc.} \cdot p_{H_2S,recyc.} - \dot{V}_{bubble} \cdot p_{H_2S,bubble}$$
$$+ V_{reactor} \cdot k_L a (c_{H_2S} - H_{H_2S} \cdot p_{H_2S,bubble}) \cdot R \cdot T \quad (33)$$

$V_{bubble}$ is the total volume of all the bubbles, $\dot{V}_{recyc.}$ denotes the volumetric flow rate of the biogas recycled from the top to the bottom of the reactor, $p_{H_2S,recyc.}$ is the partial pressure of $H_2S$ in this flow. $\dot{V}_{bubble}$ is the volumetric gas production rate of the reactor.

## 6.5.3
## Head Space

Eq. 34 gives the $H_2S$ ($p_{H_2S}$) partial pressure in the head space located over the reactor liquid, which is equal to the partial pressure of $H_2S$ in the biogas outlet of the reactor:

$$\frac{dp_{H_2S}}{dt} = \frac{\dot{V}_{bubble}}{V_{gas}}(p_{H_2S,bubble} - p_{H_2S}) \quad (34)$$

$V_{gas}$ is the volume of the head space. Assuming steady state, the partial pressure of $H_2S$ in the gas bubbles rising in the biogas suspension $p_{H_2S,bubble}$ and the partial pressure of the produced biogas $p_{H_2S}$ are equal.

The $k_La$ value in Eqs. 32 and 33 is not known a priori. It can be estimated by a regression analysis, fitting the model calculation to the experimental data. The resulting $k_La$ values are denoted in the Figures 6.32 and 6.33. It can be seen that recirculation of biogas in the experimental reactor of Figure 6.31 increases the $k_La$ value by a factor of 3–4 compared to the situation without gas recirculation.

Applying Eqs. 32–34 to each module of the biogas tower reactor and paying additional regard to the liquid mixing, which is described in Section 6.4.1, the $H_2S$ content in the liquid phase and the gas phase in the different sections of the reactor can be simulated as shown in Figure 6.34. It can be seen that the highest concentration of $H_2S$ is found in the liquid phase at the bottom of the reactor (module I). The cal-

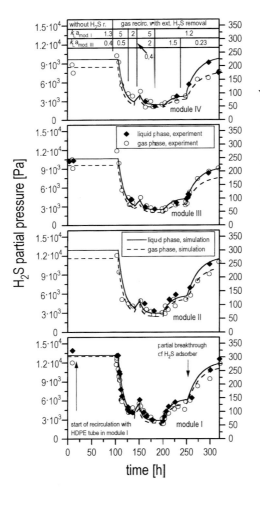

**Fig. 6.34** Effect of recycling biogas on the $H_2S$ concentration in the gas phase and the liquid phase in the pilot scale biogas tower reactor. Wastewater was simultaneously supplied to modules I, II, and III. Gas was recycled to module I: 2.3, 13, 5, 13, 1.8, 1.8 $m^3$ $h^{-1}$ and to module II: 0, 0, 2, 2, 3, 0 $m^3$ $h^{-1}$ in the sequence of the experiment according to the denoted time intervals; sulfate content of wastewater: 3.4–4.5 g $L^{-1}$; mean residence time of wastewater in the reactor: 2.1 d), r., recirc. recirculation, ext. external, mod. module.

culated $k_L a$ values in the pilot scale biogas tower reactor are larger by a factor of about 5–10 than those in the UASB reactor. Therefore, the differences between the liquid and the gas phase concentrations are much smaller. By recirculating biogas using an external $H_2S$ adsorber, values as small as 50 mg $L^{-1}$ undissociated $H_2S$ in the reactor liquid can be realized.

The $k_L a$ values identified are proportional to the gas holdup $\varepsilon_G$ in the reactor suspension according to Eq. 35 as shown in Figure 6.35.

$$k_L a = k \cdot \varepsilon_G \tag{35}$$

The proportional coefficient $k$ depends largely on the size of the bubbles produced. The smallest interfacial area is found when bubbles are generated only by an overflow of the gas collection devices $k_{II} = k_{III} = k_{IV} = 100 \ h^{-1}$. $k_L a$ is improved by using a nozzle for the gas injection $k_{III,nozzle} = 200 \ h^{-1}$. In a gas injection over a frit made of high density polyethylene, $k_{I,frit} = 400 \ h^{-1}$ is found.

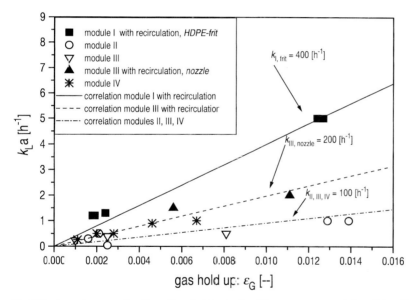

**Fig. 6.35** $k_L a$ values as a function of gas holdup in the different reactor modules of the pilot scale biogas tower reactor. Gas was supplied to module I by means of a HDPE frit and to module III by means of a simple nozzle.

## 6.6
## Influence of Hydrostatic Pressure on Biogas Production

Knowledge of the effect of higher hydrostatic pressure on the anaerobic digestion is important for the understanding of anaerobic microbiological processes in landfills, sediments of lakes, or deeper parts of high technical biogas reactors, especially of biogas tower reactors. Anaerobic reactors for the treatment of sewage sludge may also have a height up to 50 m.

In the literature very little has been published on this important effect. The reason may be found in the high complexity of performing experiments under steady-state conditions and elevated pressure. A theoretical and experimental study was published by Friedmann and Märkl (1993, 1994). It was shown that the most important effect of elevated pressure is due to the higher solubility of the product gases, like $CO_2$, $H_2S$, and $NH_3$. According to Eqs. 3–14 the pH may change because of a higher solubility of these gases and, therefore, the equilibrium of dissociation of all the relevant substances will shift, with a significant effect on the kinetics of biogas production.

If a higher percentage of the produced gases is dissolved in the liquid phase of the reactor broth at higher pressures, a smaller amount of gas $G$ will remain compared to the gas volume at a usual pressure of $10^5$ Pa $G_0$. Because this effect is different for each gas component, the composition of the produced biogas will change when increasing hydrostatic pressure.

Assuming the number of gas bubbles remains constant, increasing the pressure leads to a smaller size of bubbles (bubble diameter: $d_B$) according to Eq. 36.

$$d_b \sim \sqrt[3]{\frac{G}{G_0} \frac{1}{p_{hydrost}}} \tag{36}$$

The bubble diameter decreases because the relative content of dissolved gas is higher at higher pressures, which results in a smaller $G$; furthermore, the pressure $p_{hydrost}$ itself reduces the bubble size. Highbie's penetration theory (Eq. 37) and Stoke's theory (Eq. 39) of bubble rising velocity ($w_b$) show the influence on the mass transfer coefficient $k_L$ (Eq. 40).

$$k_L = 2\sqrt{\frac{D}{\pi \, \tau}} \tag{37}$$

$$\tau = \frac{d_b}{w_b} \tag{38}$$

$$w_b \sim d_b^2 \tag{39}$$

$$k_L \sim \sqrt{d_b} \tag{40}$$

It is assumed that the diffusion coefficient $D$ in Eq. 37 is constant at the different pressures and the boundary layer renewal time $\tau$ can be calculated according to Eq. 38. Due to the fact that the interfacial contact area $a$ between liquid and gas will also be smaller according to Eq. 41, the $k_La$ value with respect to the hydrostatic pressure $p_{hydrost}$ can be calculated by Eq. 42.

$$a \sim d_b^2 \tag{41}$$

$$k_L a \sim \left(\frac{G}{G_0} \frac{1}{p_{hydrost}}\right)^{0.83} \tag{42}$$

Altogether at a higher hydrostatic pressure the transport capacity of the product gases from the liquid phase to the gas phase (Section 6.5, Eqs. 32 and 33) decreases and, therefore, the concentration of these gases in the liquid phase will increase additionally.

Friedmann and Märkl (1993) give a solution of this complex physicochemical system (39 nonlinear equations have to be solved simultaneously). Figure 6.36 gives an example of the results, showing the relative change in the composition of produced biogas as a function of pressure.

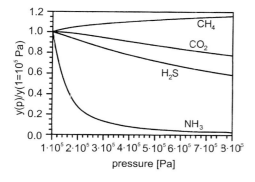

**Fig. 6.36** Relative change in the composition of the produced biogas due to an alteration in the absolute pressure (model calculation for the digestion of wastewater of bakers' yeast production).

Experiments were performed in a 16 L stirred tank reactor. The reactor was equipped for continuous experiments under elevated pressures (up to $8 \cdot 10^5$ Pa). The pH value, the concentration of undissociated $H_2S$ in the liquid, and the $H_2S$ content in the gas were measured in situ and continuously. Concentrations of the undissociated acetic acid and propionic acid were also detected (Fig. 6.37).

A comparison of the total experimental data investigated with model calculations is given in Figure 6.38 showing a reasonable degree of agreement.

## 6.7
## Outlook

The mathematical model was developed and its use was demonstrated with the example of a biogas tower reactor. The hydrodynamic and liquid mixing behavior (Section 6.4), as well as the application of the mathematical model for reactor control (Section 6.7) are directly attributed to this reactor type.

But it should also be stated that the structure of the model and the basic understanding of the system behavior are of a more general nature and can be applied to all reactor types. This holds especially true for the measuring techniques (Section 6.2) and the microbial kinetics (Section 6.3). In each biogas reactor the transport of the gas phase (Section 6.5) will plays an important role. The influence of gas oversaturation which is a function of that transport phenomenon, is of particular impor-

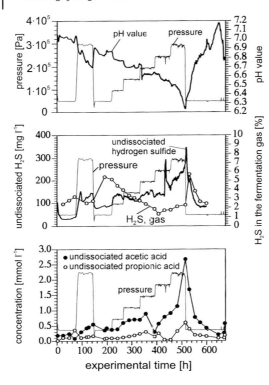

**Fig. 6.37** Experiments digesting wastewater of bakers' yeast production at different pressures (pressure difference to normal pressure) in a stirred tank reactor (active volume 16 L). Continuous feeding of wastewater (TOC = 12 g L⁻¹, concentration of sulfate 2.8 g L⁻¹). Pressure was kept constant ±2500 Pa within a certain time interval to achieve steady state conditions.

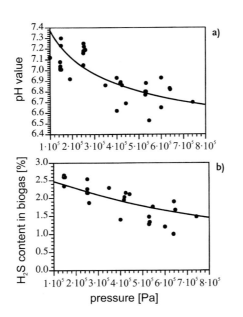

**Fig. 6.38a, b** Comparison between model calculation and experiment. **a** pH as a function of the absolute pressure, **b** $H_2S$ content in the biogas as a function of the absolute pressure.

tance in the production of gases which have an inhibitory effect on methane generation like $H_2S$ and $NH_3$.

The influence of hydrostatic pressure on biogas production (Section 6.6) should be taken into consideration for most large technical scale bioreactors with a height of more than 5 m.

## List of Symbols

| | | |
|---|---|---|
| $A$ | area | $m^2$ |
| $a$ | interfacial area per volume | $m^2\ m^{-3}$ |
| $ci$ | concentration of component $i$ | $mol\ L^{-1}$ |
| $D$ | diffusion coefficient | $m^2\ s$ |
| $d_b$ | diameter, bubble | $m$ |
| $G$ | gas volume | $m^3$ |
| $H_i$ | Henry coefficient of component $i$ | $mol\ L^{-1}\ Pa^{-1}$ |
| $K_i$ | kinetic constant of component $i$ | $mol\ L^{-1}$ |
| $K_{D,i}$ | dissociation coefficient of component $i$ | $mol\ L^{-1}$ |
| $k$ | proportional coefficient | $h^{-1}$ |
| $k_L a$ | volume specific mass transport coefficient | $h^{-1}$ |
| $\dot{M}_{prod}$ | produced biomass | $g\ h^{-1}$ |
| $\dot{M}_{sedi,i}$ | sedimenting mass flow from module $i$ | $g\ h^{-1}$ |
| $p_i$ | partial pressure of component $i$ | $Pa$ |
| $R$ | universal gas constant, $R=8.314\ kJ\ kg^{-1}\ K^{-1}$ | $kJ\ kg^{-1}\ K^{-1}$ |
| $SS_i$ | concentration of suspended solids in module $i$ | $g\ L^{-1}$ |
| $T$ | absolute temperature | $K$ |
| TOC | total organic carbon | $g\ L^{-1}$ |
| $u$ | velocity | $m\ s^{-1}$ |
| $w$ | velocity | $m\ s^{-1}$ |
| $V$ | volume | $m^3$ |
| $\dot{V}_{exchange}$ | exchange flow rate | $m^3\ h^{-1}$ |
| $\dot{V}_{bubble}$ | volumetric gas production rate | $m^3\ h^{-1}$ |
| $\dot{V}_{feed}$ | feed flow rate | $m^3\ h^{-1}$ |
| $\dot{V}_{gas}$ | gas flow rate | $m^3\ h^{-1}$ |
| $\dot{V}_{recyc.}$ | recycling gas flow rate | $m^3\ h^{-1}$ |
| $x$ | concentration of biomass | $g\ L^{-1}$ |
| $Y$ | yield coefficient | $g\ mol^{-1}$ |

## Greek

| | | |
|---|---|---|
| $\gamma_i$ | activity coefficient [–] | |
| $\varepsilon_G$ | gas hold up [–] | |
| $\Delta\varepsilon$ | difference of gas hold up [–] | |
| $\tau$ | boundary layer renewal time s | |

## Special Symbols of Chemical Components

| | |
|---|---|
| Pro | total propionic acid |
| HPro | undissociated propionic acid |
| $Pro^-$ | dissociated propionic acid |

## References

Aivasidis, A., Bastin, K., Wandrey, C., Anaerobic Digestion 1981, p. 361. Amsterdam **1982**: Elsevier.

Beutier, D., Renon, H., Representation of $NH_3—H_2S—H_2O$, $NH_3—CO_2—H_2O$ and $NH_3—SO_2—H_2O$ vapor–liquid equilibria, *Ind. Eng. Chem. Process Des.* **1978**, *17*, 220–230.

Blenke, H., Biochemical loop reactors, in: *Biotechnology* 1st Edn., Vol. 2, *Fundamentals of Biochemical Engineering*, pp. 465–517. Weinheim **1985**: VCH.

Friedmann, H., Märkl, H., Der Einfluß von erhöhtem hydrostatischen Druck auf die Biogasproduktion, *Wasser-Abwasser gwf* **1993**, *134*, 689–698.

Friedmann, H., Märkl, H., Der Einfluß der Produktgase auf die mikrobiologische Methanbildung, *Wasser-Abwasser gwf* **1994**, *6*, 302–311.

Fukuzaki, S., Nishio, N., Nagai S., Kinetics of the methanogenic fermentation of acetate, *Appl. Environ. Microbiol.* **1990**, *56*, 3158–3163.

Graef, S. P., Andrews, J. F., Mathematical modeling and control of anaerobic digestion, *AIChE Symp. Series* **1973**, *70*, 101–131.

Huser, B. A., Wuhrmann, K., Zehnder, A. J. B., *Methanothrix soehngenii* gen. nov., a new acetotrophic non-hydrogen oxidizing methane bacterium, *Arch. Microbiol.* **1982**, *132*, 1–9.

Loewenthal, R. E., Marais, G. v. R., *Carbonate Chemistry of Aquatic Systems: Theory and Application*, Vol. 1. Ann Arbor, MI **1976**: Science Publishers.

Märkl, H., Reinhold, G., Biogas-Turmreaktor, ein neues Konzept in der anaeroben Abwasserreinigung, *Chem.-Ing. Tech.* **1994**, *66*, 534–536.

Märkl, H., Mather, M., Witty, W., Meß- und Regeltechnik bei der anaeroben Abwasserreinigung sowie bei Biogasprozessen. *Münchener Beiträge zur Abwasser-, Fischerei- und Flußbiologie*, Vol. 36, pp. 49–64. München, Wien **1983**: R. Oldenbourg Verlag.

Matz, G., Lennemann, F., On-line monitoring of biotechnological processes by gas chromatographic mass-spectrometric analysis of fermentation suspensions, *J. Chromatogr. A* **1996**, *750*, 141–149.

Meyer-Jens, T., Matz, G., Märkl, H., Online measurement of dissolved and gaseous hydrogen sulphide in anaerobic biogas reactors, *Appl. Microbiol. Biotechnol.* **1995**, *43*, 341–345.

Morvai, L., Miháltz, P., Holló, J., Comparison of the kinetics of acetate biomethanation by raw and granular sludges, *Appl. Microb. Biotechnol.* **1992**, *36*, 561–567.

Polomski, A., Einfluß des Stofftransports auf die Methanbildung in Biogas-Reaktoren. Thesis, Technical University Hamburg-Harburg, Germany (**1998**).

Reinhold, G., Märkl, H., Model-based scale-up and performance of the Biogas Tower Reactor for anaerobic waste water treatment, *Water Res.* **1997**, *31*, 2057–2065.

Reinhold, G., Merrath, S., Lennemann, F., Märkl, H., Modeling the hydrodynamics and the fluid mixing behavior of a Biogas Tower Reactor, *Chem. Eng. Sci.* **1996**, *51*, 4065–4073.

Thauer, R. K., Jungermann, K., Decker, K., Energy conservation in chemotrophic anaerobic bacteria, *Bact. Rev.* **1977**, *41*, 100–180.

Therkelsen, H. H., Carlson, D. A., Thermophilic anaerobic digestion of a strong complex substrate, *J. Water Pollut. Control. Fed.* **1979**, *51*, 7, 1949–1964.

Therkelsen, H. H., Sörensen, J. E., Nielsen, A. M., *Thermophilic Anaerobic Digestion of Wastewater Sludge and Farm Manure*, Project Brief of COWI-Consult in 45 Teknikerbyen, DK-2830 Virum (**1981**).

Weiland, P., Untersuchungen eines Airliftreaktors mit äußerem Umlauf im Hinblick auf seine Anwendung als Bioreaktor. Thesis, University of Dortmund, Germany (**1978**).

Witty, W., Märkl, H., Process engineering aspects of methanogenic fermentation on the example of fermentation of *Penicillium mycelium*, *Ger. Chem. Eng.* **1986**, *9*, 238–245.

Yang, S. T., Okos, M. R., Kinetic study and mathematical modeling of methanogenesis of acetate using pure cultures of methanogens, *Biotechnol. Bioeng.* **1987**, *30*, 661–667.

Zumbusch, P. von, Meyer-Jens, T., Brunner, G., Märkl, H., On-line monitoring of organic substances with high-pressure liquid chromatography (HPLC) during the anaerobic fermentation of waste water, *Appl. Microbiol. Biotechnol.* **1994**, *42*, 140–146.

# 7

# Aerobic Degradation of Recalcitrant Organic Compounds by Microorganisms

Wolfgang Fritsche and Martin Hofrichter

## 7.1
### Introduction: Characteristics of Aerobic Microorganisms Capable of Degrading Organic Pollutants

The most important classes of organic pollutants in the environment are mineral oil constituents and halogenated products of petrochemicals. The capacities of aerobic microorganisms are of particular relevance for the biodegradation of such compounds and are described as examples with reference to the degradation of aliphatic and aromatic hydrocarbons as well as their chlorinated derivatives. The most rapid and complete degradation of the majority of pollutants is brought about under aerobic conditions.

The following are essential characteristics of aerobic microorganisms degrading organic pollutants (Fig. 7.1):
- Metabolic processes for optimizing the contact between the microbial cells and the organic pollutants. The chemicals must be accessible to the organisms having biodegrading activities. For example, hydrocarbons are water-insoluble and their degradation requires the production of biosurfactants.
- The initial intracellular attack on organic pollutants is an oxidative process; the activation and incorporation of oxygen is the enzymatic key reaction catalyzed by oxygenases and peroxidases.
- Peripheral degradation pathways convert the organic pollutants step by step into intermediates of the central intermediary metabolism, e.g., the tricarboxylic acid cycle.
- Biosynthesis of cell biomass from the central precursor metabolites, e.g., acetyl-CoA, succinate, pyruvate. Sugars required for various biosyntheses and growth must be synthesized by gluconeogenesis.

A huge number of bacterial and fungal genera possess the ability to degrade organic pollutants. Biodegradation is defined as the biologically catalyzed reduction in complexity of chemical compounds (Alexander 1994). It is based on two processes: growth and cometabolism. In growth, an organic pollutant is used as sole source of

*Environmental Biotechnology. Concepts and Applications.* Edited by H.-J. Jördening and J. Winter
Copyright © 2005 WILEY-VCH Verlag GmbH & Co. KGaA, Weinheim
ISBN: 3-527-30585-8

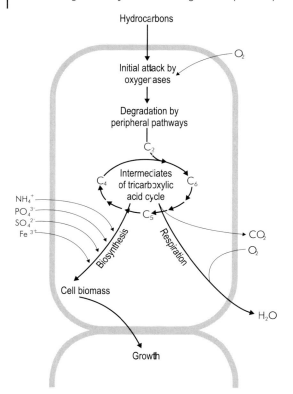

**Fig. 7.1** Main characteristics of aerobic degradation of hydrocarbons: processes associated with growth of microorganisms.

carbon and energy. This process results in a complete degradation (mineralization) of organic pollutants, as demonstrated in Section 7.2.2. Cometabolism is defined as the metabolism of an organic compound in the presence of a growth substrate that is used as the primary carbon and energy source. The principle is explained in Section 7.2.4.

Key enzymatic reactions of aerobic biodegradation are oxidations catalyzed by oxygenases and peroxidases. Oxygenases are oxidoreductases that use $O_2$ to incorporate oxygen into the substrate. Degradative organisms need oxygen at two metabolic sites – the initial attack on the substrate and the end of the respiratory chain (Fig. 7.1). Certain higher fungi have developed a unique oxidative system for the degradation of lignin based on extracellular ligninolytic peroxidases and laccases. This enzymatic system possesses increasing significance for the cometabolic degradation of persistent organopollutants. Thus, the basidiomycetous fungi require deeper insights and extensive consideration. Therefore, this chapter is divided into two sections: bacterial and fungal degradation.

## 7.2
## Principles of Bacterial Degradation

### 7.2.1
### Typical Aerobic Degrading Bacteria

The predominant degraders of organopollutants in the oxic zone of contaminated areas are chemo-organotrophic species, which are able to use a huge number of natural and xenobiotic compounds as carbon sources and as electron donors for the generation of energy. Although many bacteria are able to metabolize organic pollutants, a single bacterial species does not possess the enzymatic capability to degrade all or even most of the organic compounds in a polluted soil. Mixed microbial communities have the most powerful biodegradative potential, because the genetic information of more than one organism is necessary to degrade the complex mixtures of organic compounds present in contaminated areas. The genetic potential and certain environmental factors such as temperature, pH, and available nitrogen and phosphorous sources, therefore, seem to determine the rate and the extend of degradation.

The predominant bacteria in polluted soils belong to a spectrum of genera and species (Table 7.1). The table lists only bacteria that can be cultured on nutrient-rich media. We have to consider that the majority of bacteria present in soils cannot yet be cultivated in the laboratory.

The pseudomonads, aerobic gram-negative rods that never show fermentative activity, seem to have the highest degradative potential, e.g., *Pseudomonas putida* and *P. fluorescens*. Further important degraders of organic pollutants are found within the genera *Comamonas*, *Burkholderia*, and *Xanthomonas*. Some species utilize >100 different organic compounds as carbon sources. The immense potential of the pseudomonads does not solely depend on the catabolic enzymes, but also on their capability for metabolic regulation (Houghton and Shanley 1994). A second important group of degrading bacteria are the gram-positive rhodococci and coryneform bacteria. Many species, now classified as *Rhodococcus* spp., were originally described as *Nocardia* spp., *Mycobacterium* spp., and *Corynebacterium* spp.. Rhodococci are aero-

**Table 7.1** Predominant bacteria in soil samples polluted with aliphatic and aromatic hydrocarbons, polycyclic aromatic hydrocarbons and chlorinated compounds.[a]

| *Gram-negative bacteria* | *Gram-positive bacteria* |
| --- | --- |
| *Pseudomonas* spp. | *Nocardia* spp. |
| *Acinetobacter* spp. | *Mycobacterium* spp. |
| *Alcaligenes* sp. | *Corynebacterium* spp. |
| *Flavobacterium/Cytophaga* group | *Arthrobacter* spp. |
| *Xanthomonas* spp. | *Bacillus* spp. |

[a] Bacteria have been reclassified based on phylogenetic markers, resulting in changes in some genera and species; thus, the names of species are not mentioned.

bic actinomycetes that show considerable morphological diversity. A certain group of these bacteria possesses mycolic acids on the external surface of the cell. These compounds are unusual long-chain alcohols and fatty acids, esterified to the pepti-doglycan of the cell wall. These lipophilic cell structures probably are important for the affinity of rhodococci to lipophilic pollutants. In general, rhodococci have high and diverse metabolic activities and can synthesize biosurfactants.

## 7.2.2
**Growth-associated Degradation of Aliphatics**

The aerobic initial attack on aliphatic and cycloaliphatic hydrocarbons requires molecular oxygen. Two types of enzymatic reactions are involved in these processes (Fig. 7.2); whether a monooxygenase or a dioxygenase reaction occurs depends on the nature of the substrate and the enzymes possessed by the microorganisms. The n-alkanes are the main constituents of mineral oil contaminations. Long-chain n-alkanes ($C_{10}$–$C_{24}$) are degraded most rapidly by the pathways shown in Figure 7.3. Short-chain alkanes (less than $C_9$) are toxic to many microorganisms, but they evap-

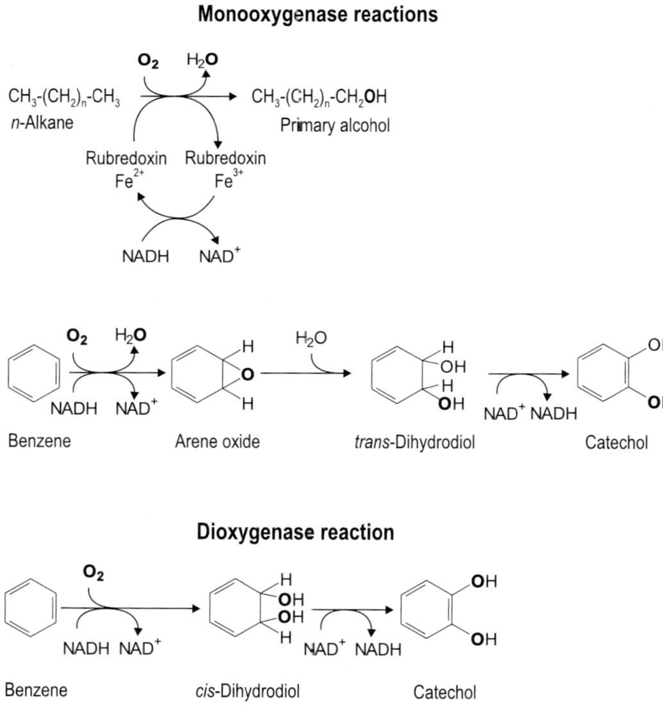

**Fig. 7.2** Initial attack on xenobiotics by oxygenases. Monooxygenases incorporate one atom of dioxygen ($O_2$) into the substrate, the second oxygen atom is reduced to $H_2O$ by means of reduction equivalents. Dioxygenases incorporate both atoms of $O_2$ into the substrate.

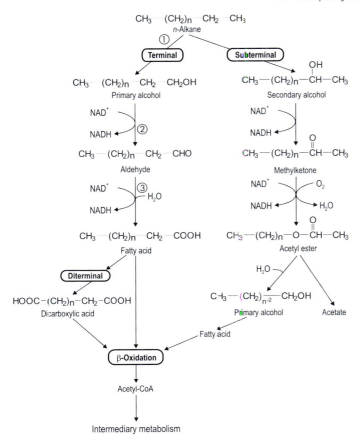

**Fig. 7.3** Pathways of alkane degradation. The main pathway is their terminal oxidation to fatty acids catalyzed by ① n-alkane monooxygenase, ② alcohol dehydrogenase, and ③) aldehyde dehydrogenase.

orate rapidly from petroleum-contaminated sites. Oxidation of alkanes is classified as terminal or diterminal. Monoterminal oxidation is the main pathway and proceeds by formation of the corresponding alcohol, aldehyde, and fatty acid. Beta oxidation of fatty acids results in the formation of acetyl-CoA. n-Alkanes having an odd number of carbon atoms are degraded to propionyl-CoA, which is in turn carboxylated to methylmalonyl-CoA and further converted to succinyl-CoA. Fatty acids having a physiological chain length may be directly incorporated into membrane lipids, but most degradation products are fed into the tricarboxylic acid cycle. Subterminal oxidation occurs with lower ($C_3$–$C_6$) and longer alkanes, with formation of a secondary alcohol and subsequently of a ketone. Unsaturated 1-alkenes are oxidized at the saturated end of the chains. A minor pathway proceeds via an epoxide, which is converted to a fatty acid. Branching generally reduces the rate of biodegradation. Methyl side groups do not noticeably decrease the biodegradability, whereas complex branched chains, e.g., the tertiary butyl group, hinder the action of the degradative enzymes.

Cyclic alkanes represent minor components of mineral oil and are relatively resistant to microbial attack. The absence of an exposed terminal methyl group complicates the primary attack. A few species can use cyclohexane as a sole carbon source, but it is more commonly cometabolized by mixed cultures. The pathway of cyclohexane degradation is shown in Figure 7.4. In general, the presence of alkyl sidechains on cycloalkanes facilitates their degradation.

Aliphatic hydrocarbons become less water soluble with increasing chain length; those with a chain length of $C_{12}$ or more are virtually water-insoluble. Two mechanisms are involved in microbial uptake of these lipophilic substrates: attachment of microbial cells to oil droplets and production of biosurfactants. The uptake mechanism linked to attachment of the cells is still unknown, but the effect of biosurfac-

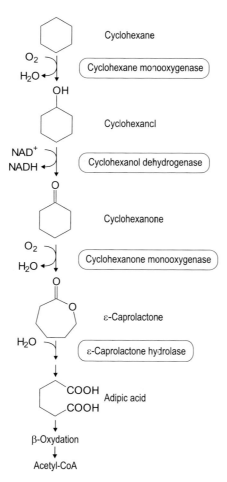

**Fig. 7.4** Metabolic pathways for degradation of cycloaliphatic compounds (cycloparaffins).

tants has been studied well (Fig. 7.5). Biosurfactants are molecules consisting of a hydrophilic and a lipophilic moiety. They act as emulsifying agents by decreasing the surface tension and by forming micelles. The latter can be encapsulated by the hydrophobic microbial cell surface.

The products of hydrocarbon degradation that are fed into the central tricarboxylic acid cycle have a dual function: as substrates of energy metabolism and building blocks for the biosynthesis of cell biomass (Fig. 7.1). Synthesis of amino acids and proteins requires nitrogen and sulfur sources, that of nucleotides and nucleic acids a phosphorous source. Biosynthesis of the bacterial cell wall requires activated sugars synthesized by gluconeogenesis. The cells act as complex biocatalysts of degradation.

Products of growth-associated degradation are $CO_2$, $H_2O$, and cell biomass. The cell biomass can be mineralized after exhaustion of degradable pollutants in a contaminated site.

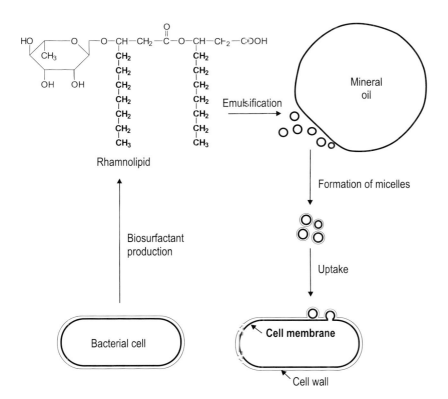

**Fig. 7.5** Involvement of biosurfactants in the uptake of hydrocarbons. The emulsifying effect of a rhamnolipid produced by *Pseudomonas* spp. within the oil–water interphase and the formation of micelles are shown. Lipid phases are printed in bold.

7.2.3
## Diversity of Aromatic Compounds: Unity of Catabolic Processes

Aromatic hydrocarbons, e g., benzene, toluene, ethylbenzene and xylenes (BTEX compounds), and naphthalene, belong to the large-volume petrochemicals, which are widely used as fuels and industrial solvents. Phenols and chlorophenols are released into the environment as products and waste materials from industry. Aromatic compounds are formed in large amounts by all organisms, e.g., as aromatic amino acids, phenols, or hydroquinones/quinones. Thus, it is not surprising that many microorganisms have evolved catabolic pathways to degrade aromatic compounds. In general, man-made organic chemicals (xenobiotics) can be degraded by microorganisms when they are similar to natural compounds. The range of man-made aromatics shown in Figure 7.6 can be converted enzymatically to the natural intermediates of degradation: catechol and protecatechuate. In general, benzene and related compounds are characterized by greater thermodynamic stability than aliphatics are. The first step in benzene oxidation is a hydroxylation that is catalyzed by a dioxygenase (Fig. 7.2). The product, a diol, is then converted to catechol by a dehydrogenase. These initial reactions, hydroxylation and dehydrogenation, are also common

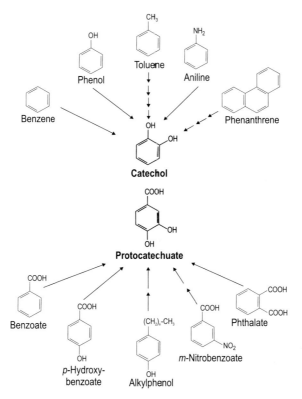

**Fig. 7.6** Degradation of a broad spectrum of natural and xenobiotic aromatic compounds into two central intermediates: catechol and protocatechuate.

to pathways of degradation of other aromatic hydrocarbons, e.g., of 3- and 4-ring polycyclic aromatic hydrocarbons (PAHs; Kästner 2000).

The introduction of a substituent onto the benzene ring makes it possible to attack sidechains or to oxidize the aromatic ring by alternative mechanisms. The versatility and adaptability of bacteria are based on the existence of catabolic plasmids. Catabolic plasmids have been found that encode enzymes that degrade such naturally occurring aromatics as camphor, naphthalene, and salicylate. Most of the catabolic plasmids are self-transmissible and have a broad host range. The majority of gram-negative soil bacteria isolated from polluted areas possess degradative plasmids, mainly the so-called TOL plasmids. These pseudomonads can grow on toluene, *m*- and *p*-xylene, and *m*-ethyltoluene. The main reaction involved in the oxidation of toluene and related arenes is methyl-group hydroxylation. The methyl group of toluene is oxidized stepwise to the corresponding alcohol, aldehyde, and carboxylic group. Benzoate formed or its alkylated derivatives are then oxidized by toluate dioxygenase and decarboxylated to catechol.

The oxygenolytic cleavage of the aromatic ring occurs via *ortho* or *meta* cleavage. The significance of the diversity of degradative pathways and of the few key intermediates is still under discussion. Both pathways may be present in one bacterial species. "Whenever an alternative mechanism for the dissimilation of any compound becomes available (*ortho* versus *meta* cleavage of ring structures, for example), control of each outcome must be imposed' (Houghton and Shanley 1994). The metabolism of a broad spectrum of aromatic compounds by one species requires the metabolic isolation of intermediates into distinct pathways. This kind of metabolic compartmentation seems to be implemented by metabolic regulation. The key enzymes in the degradation of aromatic substrates are induced and synthesized in appreciable amounts only when the substrate or structurally related compounds are present. The enzyme induction depends on the concentration of the inducing molecules. The substrate-specific concentrations represent the threshold of utilization and growth and are in the micromolar range. A report on the regulation of TOL catabolic pathways was published by Ramos et al. (1997).

Figure 7.7 shows the pathways of oxygenolytic ring cleavage to intermediates of the central metabolism. At the branchpoint, catechol is oxidized by either intradiol *ortho* cleavage or extradiol *meta* cleavage. Both ring-cleavage reactions are catalyzed by specific dioxygenases. The product of the *ortho* cleavage – *cis,cis*-muconate – is transformed into an unstable enol-lactone, which is in turn hydrolyzed to oxoadipate. This dicarboxylic acid is activated by transfer to CoA, followed by thiolytic cleavage to acetyl-CoA and succinate. Protocatechuate is metabolized by a homologous set of enzymes. The additional carboxylic group is decarboxylated, and the double bond is simultaneously shifted to form oxiadipate enol-lactone. The oxygenolytic *meta* cleavage yields 2-hydroxymuconic semialdehyde, which is metabolized by hydrolytic enzymes to formate, acetaldehyde, and pyruvate. These are then utilized in the central metabolism. In general, a wealth of aromatic substrates are degraded by a limited number of reactions: hydroxylation, oxygenolytic ring cleavage, isomerization, and hydrolysis. The inducible nature of the enzymes and their substrate specificity enable bacteria having high degradation potential, e.g., pseudomonads and

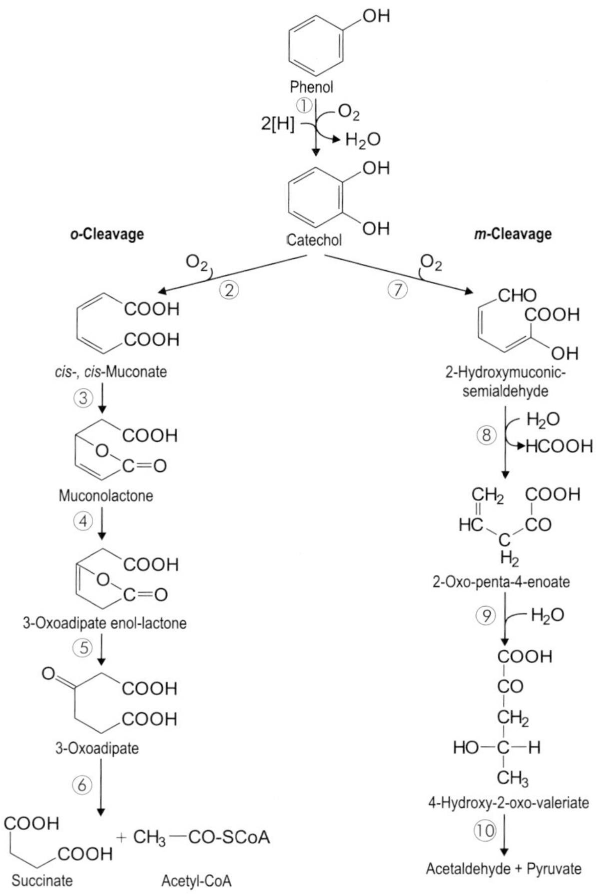

**Fig. 7.7** The two alternative pathways for aerobic degradation of aromatic compounds: *ortho* and *meta* cleavage. ① phenol monooxygenase, ② catechol 1,2-dioxygenase, ③ muconate-lactonizing enzyme, ④ muconolactone isomerase, ⑤ oxoadipate enol-lactone hydrolase, ⑥ oxoadipate suc-cinyl-CoA transferase, ⑦ catechol 2,3-dioxygenase, ⑧ hydroxymuconic semialdehyde hydrolase, ⑨ 2-oxopent-4-enoic acid hydrolase, ⑩ 4-hydroxy-2-oxovalerate aldolase.

rhodococci, to adapt their metabolism to the effective utilization of substrate mixtures in polluted soils and to grow at a high rate.

## 7.2.4
## Extension of Degradative Capacities

### 7.2.4.1 Cometabolic Degradation of Organopollutants

Cometabolism, the transformation of a substance without nutritional benefit in the presence of a growth substrate, is a common phenomenon in microorganisms. It is the basis of biotransformations (bioconversions) used in biotechnology to convert a substance to a chemically modified form. Microorganisms growing on a particular substrate can gratuitously oxidize a second substrate (cosubstrate). The cosubstrate is not assimilated, but the product can be available as a substrate for other organisms in a mixed culture.

The prerequisites for cometabolic transformations are the enzymes of the growing cells and the synthesis of cofactors necessary for enzymatic reactions, e.g., of hydrogen donors (reducing equivalents, NAD(P)H) for oxygenases. For example, methanotrophic bacteria can use methane or other $C_1$ compounds as a sole source of carbon and energy. They oxidize methane to $CO_2$ via methanol, formaldehyde, and formate. The assimilation requires special pathways, and formaldehyde is the intermediate that is assimilated. The first step in methane oxidation is catalyzed by methane monooxygenase, which attacks the inert $CH_4$ (Fig. 7.8). Methane monooxygenase is unspecific and also oxidizes various other compounds, e.g., alkanes, aromatic compounds, and trichloroethylene (TCE). The proposed mechanism of TCE transformation according to Henry and Grbic-Gallic (1994) is shown in Figure 7.8. TCE is oxidized to an epoxide that is excreted from the cell. The unstable oxidation product breaks down to compounds that can be used by other microorganisms.

Methanotrophic bacteria are indigenous aerobic bacteria in soils and aquifers, but methane has to be added as growth substrate and inducer for the development of methanotrophic biomass, which limits their usefulness in bioremediation.

Cometabolism of chloroaromatics is a widespread activity of bacteria in mixtures of industrial pollutants. Knackmuss (1997) demonstrated that the cometabolic transformation of 2-chlorophenol gives rise to dead-end metabolites, e.g., 3-chlorocatechol. This reaction product can be autoxidized or polymerized in soils to humic-like

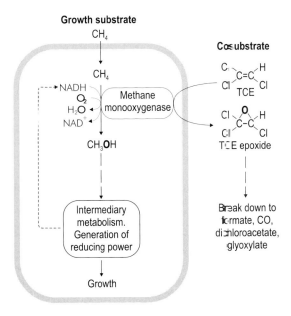

**Fig. 7.8** Cometabolic degradation of trichloroethylene (TCE) by the methane monooxygenase system of methanotrophic bacteria.

structures. Irreversible binding of dead-end metabolites may fulfill the function of detoxification. (The accumulation of dead-end products within microbial communities under selective pressure is the basis for the evolution of new catabolic traits (Reinecke 1994).)

### 7.2.4.2 Overcoming Persistent Pollutants by Cooperation Between Anaerobic and Aerobic Bacteria

As a rule, the recalcitrance of organic pollutants increases with the degree of halogenation. Substitution of halogen as well as nitro and sulfo groups on the aromatic ring is accompanied by increasing electrophilicity. These compounds resist electrophilic attack by oxygenases of aerobic bacteria. Compounds that persist under oxic conditions include PCBs (polychlorinated biphenyls), chlorinated dioxins, and some pesticides, e.g., DDT and lindane.

To overcome the relatively high persistence of halogenated xenobiotics, reductive attack by anaerobic bacteria is important. Degradation of environmental pollutants by anaerobic bacteria is the subject of Chapter 8 of this volume. Reductive dehalogenation as the first step in the degradation of higher halogenated compounds by anaerobic bacteria is a significant finding, because it is either a gratuitous reaction or a new type of anaerobic respiration. The process reduces the degree of chlorination and thus makes the products more accessible to mineralization by aerobic bacteria.

A sequence of anaerobic and aerobic bacterial activities for the mineralization of chlorinated xenobiotics is shown in Figure 7.9. PCBs, which are selected as an example for the degradation of halogenated compounds, are well-studied substances. The implementation of the reactions depends on the structure of the chemical compounds as well as on the microorganisms and conditions in a polluted ecosystem. We have to distinguish between the general degradation potential and the actual conditions necessary for it to occur. Reductive dehalogenation, the first step in PCBs degradation, requires anaerobic conditions and organic substrates acting as electron donors. The PCBs function as electron acceptors to allow the anaerobic bacteria to transfer electrons to these compounds. Anaerobic bacteria capable of catalyzing reductive dehalogenation seem to be relatively ubiquitous in nature. Most dechlorinating cultures are mixed cultures (consortia). Anaerobic dechlorination is always incomplete; products are di- and monochlorinated biphenyls. These products can be metabolized further by aerobic microorganisms. The substantial decrease in concentration of PCBs by sequential anaerobic and aerobic treatment has been demonstrated in the laboratory (Abramowicz 1990).

The principles of aerobic dehalogenation reactions of chloroaromatics are illustrated in Figure 7.9. Hydrolytic dechlorination has been elucidated by using 4-chlorobenzoate as the substrate for *Pseudomonas* and *Nocardia* spp.. A halidohydrolase is capable of replacing the halogen substituent by a hydroxy group originating from water. This type of reaction seems to be restricted to halobenzoates substituted in the *para* position. Dechlorination after ring cleavage is a common reaction in the *ortho* pathway of chlorocatechols, catalyzed by catechol 1,2-dioxygenases to produce chloromuconates. Oxygenolytic dechlorination is a rare fortuitous reaction catalyzed

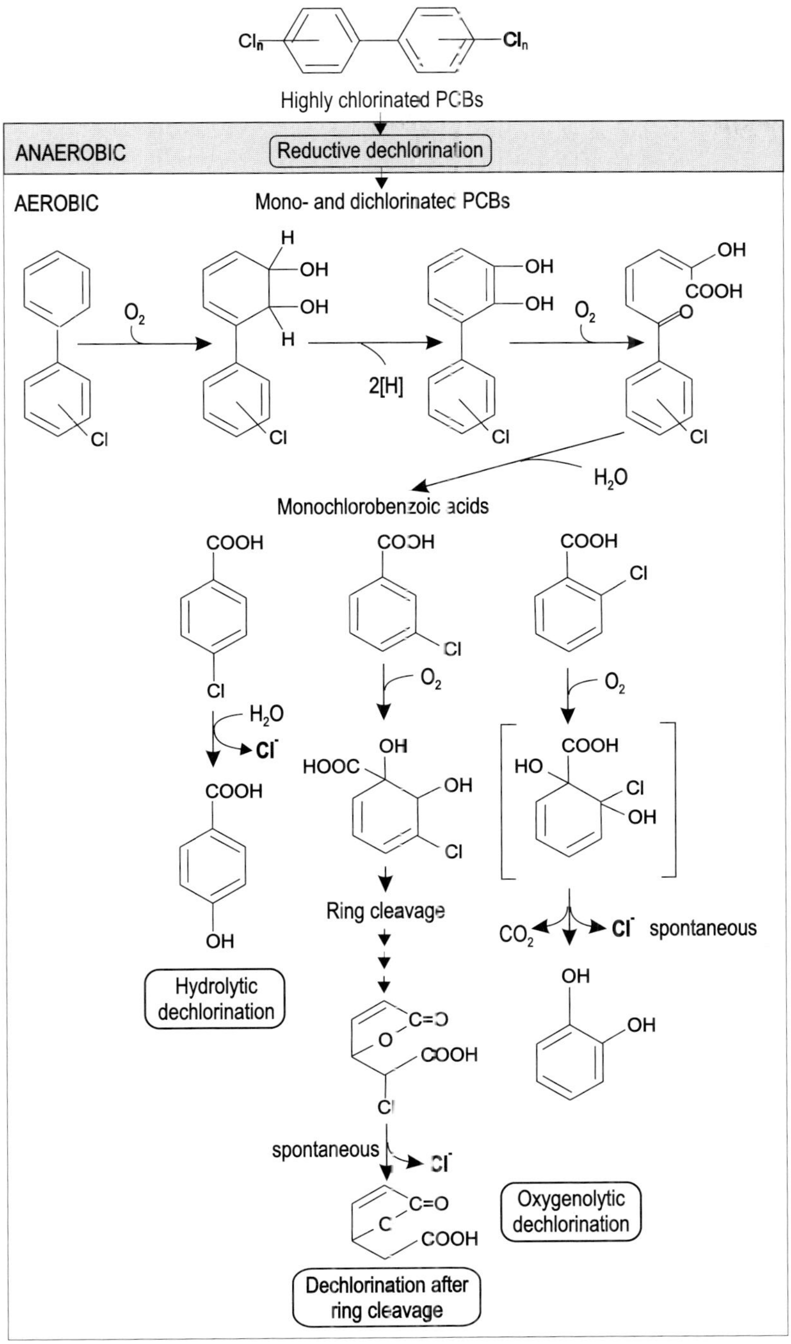

**Fig. 7.9** Principles of dehalogenation: degradation of PCBs by a sequence of anaerobic and aerobic bacterial processes.

by mono- and dioxygenases. During this reaction, the halogen substituent is replaced by an oxygen of $O_2$.

Higher-chlorinated phenols, e.g., pentachlorophenol, have been widely used as biocides. Several aerobic bacteria that degrade chlorophenols have been isolated (*Flavobacterium, Rhodococcus*). The degradation mechanism has been elucidated in some instances (McAllister et al. 1996). For example, *Rhodococcus chlorophenolicus* degrades pentachlorophenol through a hydrolytic dechlorination and three reductive dechlorinations, producing trihydroxybenzene. The potential of these bacteria is limited to some specialists and to specific conditions. Therefore, the use of polychlorinated phenols has been banned in many countries.

## 7.3
## Degradative Capacities of Fungi

As saprophytic decomposers, members of the fungal kingdom permeate the living scene. An enormous number of fungi exist, perhaps as many as 1.5 million species, a magnitude that is comparable only with the huge biodiversity of insects. Fungi exist in a wide range of habitats: in freshwater and the sea, in soil, litter, decaying remains of plants and animals, in dung, and in living organisms. Fungi have been estimated to account for more than 60% of the living microbial biomass in certain habitats (e.g., forest and other soils; Dix and Webster 1995). Last but not least, fungi possess important degradative capabilities that have implications for the recycling of recalcitrant polymers (e.g., lignin) and for the elimination of hazardous wastes from the environment. Below, some aspects of the fungal degradation of organopollutants are discussed.

### 7.3.1
### Metabolism of Organopollutants by Microfungi

For practical reasons, yeasts and molds can be grouped together into the microfungi (Gravesen et al. 1994). Taxonomically, they belong to the ascomycetous, deuteromycetous, and zygomycetous fungi. Yeasts preferentially grow as single cells or form pseudomycelia, whereas molds typically grow as mycelia-forming real hyphae. Microfungi colonize various habitats, and some genera are typical soil microorganisms (*Aspergillus, Penicillium, Trichoderma, Candida, Trichosporon*).

#### 7.3.1.1   Aliphatic Hydrocarbons
Biodegradation of aliphatic hydrocarbons occurring in crude oil and petroleum products has been investigated well, especially for yeasts. The *n*-alkanes are the most widely and readily utilized hydrocarbons, with those between $C_{10}$ and $C_{20}$ being most suitable as substrates for microfungi (Bartha 1986). However, the biodegradation of *n*-alkanes having chain lengths up to *n*-$C_{24}$ has also been demonstrated (Fritsche and Hofrichter 2000). Typical representatives of alkane-utilizing yeasts include

*Candida lipolytica, C. tropicalis, Rhodoturula rubra*, and *Aureobasidion* (*Trichosporon*) pullulans; examples of molds using *n*-alkanes as growth substrates are *Cunninghamella blakesleeana, Aspergillus niger*, and *Penicillium frequentans* (Watkinson and Morgan 1990). In this context, one should keep in mind that in the 1960s and 1970s British Petrol (BP) and other companies ran large plants for the production of single-cell protein (SCP) from the wax fraction of crude oil. BP used two yeasts, *Candida lipolytica* and *C. tropicalis* to produce a product called Toprina that was marketed as a replacement for fish meal in high-protein feeds.

Yeast and molds preferentially oxidize long-chain *n*-alkanes, because short-chain liquid alkanes ($n$-$C_5$–$C_9$) are toxic. The toxic effects of short-chain alkanes can be reversed by adding nontoxic long-chain hydrocarbons. Thus, *Candida* spp. grew on *n*-octane if 10% pristane was present (Britton 1984). The soil fungus *Penicillium frequentans* is capable of utilizing monohalogenated *n*-alkanes (e.g., 1-fluorotetradecane) and dehalogenates them completely. Since all aliphatic hydrocarbons are nearly insoluble in water, fungi produce biosurfactants, which disperse the substrates into oil-in-water emulsions, increasing the interfacial area and thereby enhancing the bioavailability of hydrocarbons (Hommel 1990).

In microfungi, the alkanes are mostly terminally oxidized to their corresponding primary alcohols (*n*-alkan-1-ols) by a monooxygenase enzyme complex containing cytochrome P450 and NAD(P)H–cytochrome P450 reductase (Britton 1984):

$$R\text{–}CH_2\text{–}CH_3 + O_2 + NAD(P)H_2 \rightarrow R\text{–}CH_2\text{–}CH_2OH + NAD(P) + H_2O$$

Subterminal oxidation yielding various secondary alcohols is a rarer type of primary attack on *n*-alkanes and has been documented in certain molds (*Aspergillus* spp., *Fusarium* spp.; Rehm and Reiff 1982). Peroxisomal enzymes carry out the remaining degradation of alkanols to intermediates that are transferable to the mitochondria (Britton 1984). After terminal oxidation, the alcohol produced is normally oxidized to the corresponding aldehyde and fatty acid by means of pyridine nucleotide-linked dehydrogenases. In some *Candida* spp. and several molds, alcohol oxidases have been shown to be present instead of dehydrogenases. Secondary alcohols resulting from subterminal oxidation systems are oxidized to the corresponding ester and hydrolytically cleaved to acetic acid and an alcohol, which is subsequently also converted to a fatty acid. The fatty acids produced are always further metabolized by β-oxidation and finally to $CO_2$ via the tricarboxylic acid cycle.

In contrast to *n*-alkanes, fungi cannot utilize branched alkanes or cycloaliphatic compounds as sole sources of carbon and energy (Britton 1984; Morgan and Watkinson 1994).

### 7.3.1.2 Aromatic Compounds

Several yeasts and molds can utilize aromatic compounds as growth substrates (Table 7.2), but more important is their ability to convert aromatic substances cometabolically. Figure 7.10 illustrates this principle with the example of 3,4-dichlorophenol oxidation in the presence of phenol as the actual growth substrate in the soil-

**Table 7.2** Some species of yeasts and molds that utilize aromatic compounds as growth substrates (according to Fritsche and Hofrichter 2000).

| Species | Growth substrates |
| --- | --- |
| **Yeasts** | |
| *Aureobasidium pullulans* | phenol, o-cresol, p-cresol, benzoic acid |
| *Candida maltosa* | phenol, catechol, benzoic acid |
| *Exophiala jeanselmei* | phenol, styrene, benzoic acid, acetophenone |
| *Rhodotorula glutinis* | phenol, m-cresol, benzoic acid |
| *Trichosporon cutaneum* | phenol, p-cresol, benzoic acid, salicylic acid |
| **Molds** | |
| *Aspergillus niger* | 2,4-dichloro-phenoxy acetic acid, benzoic acid, salicylic acid, monochlorobenzoic acids |
| *Aspergillus fumigatus* | phenol, p-cresol, 4-ethylphenol, phenylacetic acid |
| *Fusarium flocciferum* | phenol, resorcinol |
| *Penicillium frequentans* | phenol, p-cresol, resorcinol, phloroglucinol, anisole, benzyl alcohol, benzoic acid, salicylic acid, gallic acid, phenylacetic acid, 1-phenylethanol acetophenone |
| *Penicillium simplicissimum* | phenol, phloroglucinol, monofluorophenols |

dwelling mold *Penicillium frequentans* (Hofrichter et al. 1994). Whereas phenol is completely converted into biomass, carbon dioxide, and water, the chlorinated phenol is only transformed to a catechol that cannot be further degraded and remains as a dead-end product.

Table 7.2 is a partial list of microfungal species that have been shown to utilize aromatic compounds as sole sources of carbon and energy. Some species – such as the soil yeast *Trichosporon cutaneum* – possess specific energy-dependent uptake systems for aromatic substrates (e.g., for phenol; Mörtberg and Neujahr 1985). They cleave aromatic rings exclusively via the *ortho* pathway analogously to many bacteria (Wright 1993; compare Figure 7.7). To this end, they first insert activating hydroxyl groups into the aromatic ring. Phenol hydroxylase, benzoate-4-hydroxylase, and 4-hydroxybenzoate-3-hydroxylase are examples of enzymes hydroxylating the aromatic ring in the *ortho* or *para* position. They are $NADPH_2$-dependent monooxygenases and have been described in both yeasts and molds (e.g., *Trichosporon cutaneum*, *Penicillium frequentans*, *Aspergillus fumigatus*; Fritsche and Hofrichter 2000). After activation, dioxygenases cleave the aromatic ring to form *cis,cis*-muconic acids. The latter are lactonized, isomerized, and hydrolyzed, resulting in the formation of β-ketoadipate, which is degraded to $CO_2$ via the tricarboxylic acid cycle.

Hydroxylating and ring-cleaving enzymes of yeasts and molds are relatively unspecific and usually convert also related compounds, including halogenated and nitro aromatics. Thus, depending on the substitution pattern, fluoro- and chlorophenols are converted to the corresponding catechols, cleaved to halogenated muconic acids, or dehalogenated. The alkane- and phenol-utilizing mold *Penicillium frequentans*

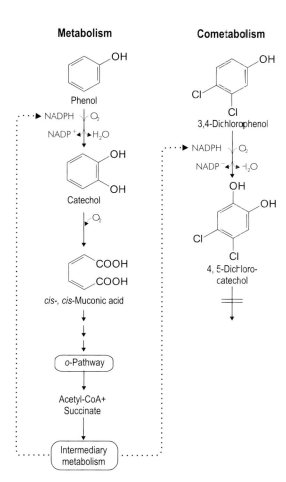

**Metabolism**

OH

Phenol

NADPH O₂

NADP⁺ H₂O

OH
OH

Catechol

O₂

COOH
COOH

*cis-, cis-*Muconic acid

*o*-Pathway

Acetyl-CoA+
Succinate

Intermediary
metabolism

**Cometabolism**

OH

Cl

Cl

3,4-Dichlorophenol

NADPH O₂

NADP H₂O

OH
OH

Cl

Cl

4, 5-Dichloro-
catechol

**Fig. 7.10** Principle of comet-abolic transformation of organo-pollutants by microfungi: hydrox-ylation of 3,4-dichlorophenol by the mold *Penicillium frequentans* growing on phenol.

has been shown to degrade a mixture of phenol, *p*-cresol, and two chlorophenols without additional growth substrate in soil. In addition to halogenated aromatic compounds, microfungi transform numerous other aromatic organopollutants cometabolically, including polycyclic aromatic hydrocarbons (PAHs) and biphenyls, dibenzofurans, nitro aromatics, various pesticides, and plasticizers (Fritsche and Hofrichter 2000). Typical fungal transformations are glycosylations, hydroxylations and ring cleavage, methoxylations, and the reduction of nitro groups to amino groups (Table 7.3). Hydroxylation reactions are particular important for elimination of organopollutants in soils, because they increase the reactivity of the molecules and make their subsequent covalent coupling to humus possible (Kästner 2000). The introduction of hydroxyl groups is often catalyzed by relatively unspecific cyto-chrome P450-containing monooxygenases which, for example, oxidize benzo(a)py-rene in *Aspergillus ochraceus* (Dutta et al. 1983).

**Table 7.3** Selected organopollutants and their metabolites that are cometabolically formed by yeasts and molds (according to Fritsche and Hofrichter 2000).

| Organopollutants | Metabolites |
| --- | --- |
| Mono- and dichlorophenols | chlorocatechols[a–c], chloroguaiacols[c], chlorinated muconic acids[c], chloride[c] |
| Pentachlorophenol | pentachloroanisole[f] |
| Fluorophenols | fluorocatechols[c], fluorinated muconic acids[c], fluoride[c] |
| 2,4-Dinitrophenol | monoaminonitrophenols[e] |
| 2,4,6-Trinitrotoluene | aminodinitrotoluenes[c–e], hydroxylamino dinitrotoluenes[c–e] |
| Pyrene | 1-pyrenol[c–e], 1-methoxypyrene[d], dihydroxypyrenes[c] |
| Benzo(a)pyrene | hydroxybenzo(a)pyrenes[d] |
| Dibenzofurane | 2,3-dihydroxybenzofuran[b], ring-cleavage products[b] |
| Biphenyl | 4,4′-dihydroxybiphenyl[d] |
| Atrazine | N-dealkylated products[c–f] |

[a] *Rhodotorula* spp.
[b] *Trichosporon* spp
[c] *Penicillium* spp.
[d] *Aspergillus* spp.
[e] *Fusarium* spp.
[f] *Trichoderma* spp.

## 7.3.2
## Degradative Capabilities of Basidiomycetous Fungi

Basidiomycetous fungi form characteristic macroscopic fruiting bodies and belong, like certain ascomycetes, to the macrofungi (Dix and Webster 1995). Colloquially, many of these fungi (especially the edible ones) are called mushrooms or toadstools. They preferably colonize the litter layer in woodlands and pastures as well as ligno-cellulosic materials (wood, straw). From the ecophysiological point of view, they can be broadly classified into litter-decomposing, wood-decaying, and mycorrhizal fungi, although there is, inevitably, some overlap of roles. Basidiomycetes can form huge mycelial clones, and it has been estimated that they can extend over an area of several hectares while reaching an age of 1500 y and a mycelial biomass of 100 t (Dix and Webster 1995).

### 7.3.2.1 The Ligninolytic Enzyme System

Lignin, like cellulose, is a major component of plant materials and the most abundant form of aromatic carbon in the biosphere. It provides strength and rigidity to the cell walls of all vascular plants (club mosses, ferns, horsetails, seed plants) by acting as a glue between the cellulose and hemicellulose fibers. In addition, lignin forms a barrier against microbial destruction and protects the readily degradable carbohydrates. From the chemical point of view, lignin is a heterogeneous, optically inactive polymer consisting of phenylpropanoid subunits that are linked by various covalent bonds (e.g., aryl-ether, aryl-aryl, aliphatic carbon-aryl bonds; Fengel and Wegener 1989). The polymer arises from the enzyme-initiated polymerization of phenolic precursors (coniferyl, sinapyl, *p*-cumaryl alcohol) via the radical coupling of their corresponding phenoxy radicals. Because of the types of bonds and their heterogeneity, lignin cannot be degraded by hydrolytic mechanisms as most other natural polymers can.

During the course of evolution, only certain basidiomycetous fungi have developed an efficient enzyme system to mineralize lignin substantially, so these microorganisms play an important role in maintaining the global carbon cycle (Griffin 1994). Because of their ability to remove lignin selectively while leaving behind white cellulose fibers, these fungi are also called white-rot fungi. Typical representatives are the wood degraders *Trametes versicolor*, *Phanerochaete chrysosporium*, *Pleurotus ostreatus*, and *Nematoloma frowardii*, as well as the litter-decomposers *Agaricus bisporus*, *Agrocybe praecox*, and *Stropharia coronilla*. Lignin degradation does not provide a primary source of carbon and energy for fungal growth and is therefore a cometabolic process in principle (mostly sugars released by hydrolases from hemicelluloses are used as growth substrates).

Lignin degradation is thought to be brought about by the synergistic action of several oxidoreductases (lignininolytic enzymes). Two types of enzymes, namely peroxidases (manganese peroxidase, MnP, EC 1.11.1.13; lignin peroxidase, LiP, EC 1.11.1.14) and laccase (Lacc, EC 1.10.3.2), are the key biocatalysts and are responsible for the unspecific attack on lignin (Hatakka 2001). Both peroxidases are ferric-iron–containing heme proteins requiring peroxides (e.g., $H_2O_2$) for function, and laccase belongs to the copper-containing blue oxidases that use molecular oxygen ($O_2$). The enzymes have the common ability to catalyze one-electron oxidations, resulting in the formation of free-radical species inside the lignin polymer (Fernando and Aust 1990). Afterwards, the radicals undergo spontaneous reactions leading to the incorporation of oxygen ($O_2$), bond cleavages, and finally, to the breakdown of the lignin molecule (Kirk and Farrell 1987).

LiP and laccase react directly with aromatic lignin structures, whereas MnP works via chelated $Mn^{3+}$ ions acting as low-molecular-weight redox mediators. Thus, the function of MnP is the generation of $Mn^{3+}$ from $Mn^{2+}$, which is the actual substrate of the enzyme. Similar to the catalytic cycle of other peroxidases, including LiP, that of MnP involves the formation of peroxidase compounds I and II. The latter are highly reactive and abstract one electron from $Mn^{2+}$ to form $Mn^{3+}$ (Fig. 7.11). MnP catalysis has an absolute requirement for Mn-chelating organic acids (i.e., oxalate, malate, or malonate), which both increase the affinity of $Mn^{2+}$ to the enzyme and stabilize the reactive $Mn^{3+}$ at high redox potentials (Wariishi et al. 1992). In contrast to the relatively large enzyme molecules, $Mn^{3+}$ chelates are small enough to diffuse into the compact lignocellulosic complex, where they preferably react with phenolic lignin structures. Therefore, it is supposed that – in most white-rot fungi – the primary attack on lignin is brought about by the MnP system (Wariishi et al. 1992, Hofrichter 2002).

As the result of MnP-catalyzed primary attack, water-soluble lignin fragments are formed, which are accessible for the further conversion by ligninolytic enzymes.

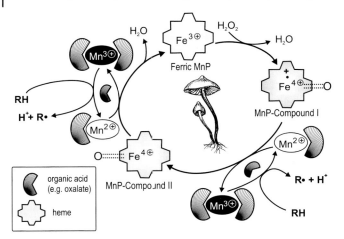

**Fig. 7.11** Catalytic cycle of manganese peroxidase (according to Wariishi et al. 1992; Hofrichter 2002).

Laccase oxidizes phenolic structures, whereas LiP preferentially cleaves recalcitrant nonphenolic lignin moieties (e.g., β-O-4 ethers). MnP is also involved in further degradation of the lignin fragments. Thus, there are indications that the MnP system even mineralizes aromatic lignin structures directly. It is the only enzyme – as far as known – that converts aromatic compounds, including lignin, partly to carbon dioxide ($CO_2$). This means that lignin can be mineralized – at least in part – outside the fungal hyphae (Hofrichter 2002). The reactions underlying the MnP-catalyzed mineralization are not yet completely understood, but it is certain that decarboxylation reactions involving $Mn^{3+}$ are the final step in this process. Furthermore, certain radicals (carbon-centered radicals, peroxyl radicals, superoxide) that derive from the autocatalytic decomposition of organic acids in the presence of $Mn^{3+}$ may also be involved in the mineralization process.

The oxidative strength of the MnP system can be enhanced in the presence of co-oxidants such as unsaturated fatty acids or thiols. These are first oxidized by $Mn^{3+}$ and form – in the presence of $O_2$ – highly reactive radical species (peroxyl and alkoxy radicals of fatty acids, thioxyl radicals), which enhance lignin mineralization and make possible the cleavage of structures that are normally not attacked by the MnP system (e.g., nonphenolic aromatic aryl ethers; Hofrichter 2002).

According to a concept of Kirk and Frarrell (1987), the process of MnP-catalyzed degradation of lignin and other substances has been described as 'enzymatic combustion', in the course of which MnP acts as a 'radical pump' (Fig. 7.12, Hofrichter 2002).

Finally, we should mention that recent studies have indicated that alternative ligninolytic systems may exist that use chloroperoxidases (CPOs) as key enzymes. Examples are the ascomycete Leptoxyphium (*Caldariomyces*) *fumago* and the basidiomycete *Agrocybe aegerita*. Both fungi colonize plant debris and produce heme-thiolate CPOs that oxidize various aromatic substrates, including aryl alcohols, nonphenolic

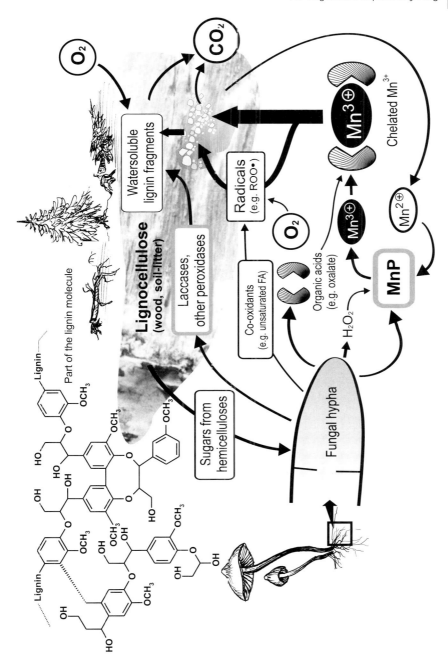

**Fig. 7.12** 'Enzymatic combustion' of lignin by manganese-peroxidase–producing basidiomycetes. The degradation of lignin is a generally aerobic, cometabolic process. Sugars and acetate derived from hemicelluloses and glucose from cellulose serve as the actual growth substrates.

lignin model compounds, and synthetic lignin preparations (Ortiz-Bermúdez et al. 2003; Ullrich et al. 2004).

### 7.3.2.2 Degradation of Organopollutants

The unique ligninolytic enzyme system of basidiomycetous fungi, being based on a highly reactive free-radical depolymerization mechanism, should be ideal for the biodegradation of organopollutants in the environment (Fernando and Aust 1990). An excellent review of the feasibility of bioremediation with white-rot fungi was published by Pointing in 2001.

Compared with other potential bioremediation systems, the extracellular, nonspecific, nonstereoselective lignin-degrading system of basidiomycetous fungi has the advantage of being applicable to a variety of recalcitrant and toxic chemicals. Examples of such chemicals and selected fungi responsible for their biodegradation are given in Table 7.4. These substances include such hazardous xenobiotic compounds as polychlorinated dibenzodioxines and dibenzofurans, other chlorinated aromatics, nitroaromatic compounds (explosives), and carcinogenic organopollutants belonging to the polycyclic aromatic hydrocarbons (Hofrichter and Fritsche 2000, Manji and Ishihara 2003). MnP, LiP, and laccase have all been shown to be involved in the oxidation of organopollutants (Bollag 1988, Barr and Aust 1994). As a result of the attack on organopollutants by ligninolytic enzymes, various metabolites are formed that either can be further degraded intracellularly, coupled to the humus, or mineralized by the MnP system. The latter has been shown to mineralize several organopollutants partly in cell-free systems, indicating that an extracellular 'enzymatic combustion' of hazardous chemicals is possible in principle (Hofrichter 2002).

Over the last decade, efforts have been made to use ligninolytic basidiomycetes in bioremediation technologies. Laboratory experiments have shown that degradation of certain organopollutants (e.g., PAHs, PCP, TNT) is stimulated by wood-inhabiting white-rot fungi in contaminated soils (*Phanerochaete chrysosporium, Lentinus*

**Table 7.4** Selected organopollutants that are transformed or mineralized by ligninolytic basidiomycetes and/or their ligninolytic enzymes (according to Barr and Aust 1994, Fritsche and Hofrichter 2000; Baciocchi et al. 2000, Haas et al. 2003).

| *Fungus* | *Organopollutants* |
|---|---|
| *Bjerkandera adusta* | Benzo(a)pyrene, other PAHs, TNT, dyes |
| *Nematoloma frowardii* | Benzo(a)pyrene, other PAHs, TNT, DCP, PCP, AsO |
| *Phanerochaete chrysosporium* | Benzo(a)pyrene, other PAHs, BTX, DNT, TNT, DDT, DCP, PCP, PCBs, DCA, dyes, polystyrenes, KCN, aromatic sulfides |
| *Phanerochaete sordida* | PAHs, polychlorinated DBDs and DBFs |
| *Phlebia radiata* | TNT, dyes |
| *Pleurotus ostreatus* | Benzo(a)pyrene, other PAHs, dibenzothiophene, TNT |
| *Stropharia rugosoannulata* | DCP, PCP, TNT |
| *Trametes versicolor* | Benzo(a)pyrene, other PAHs, DCA, DCP, PCP, dyes |

*edodes, Kuehneromyces mutabilis*; Rajarathnam et al. 1998). However, a disadvantage of these fungi is their small competitive potential in soil. Therefore, recent research has focused on litter-decomposing basidiomycetes, which naturally colonize the upper-most portion of the soil and the humus layers of forests and grasslands. *Stropharia rugosoannulata, S. coronilla, Agrocybe praecox*, and *Collybia dryophila* are examples of such fungi. *Stropharia rugosoannulata* has already been successfully tested for decontamination of soils containing TNT as well as for PAH removal (Fritsche et al. 2000, Steffen et al. 2003).

## 7.4
## Conclusions

It has been said that without microorganisms and their degradative capabilities animal life, including human life, on the earth would cease to exist within about five years. Whether or not this is an exaggeration as to the time scale, it is true in principle that we absolutely depend on microbial activities for the renewal of our environment and maintenance of the global carbon cycle (Ratledge 1991). Among the substances that can be degraded or transformed by microorganisms are a huge number of synthetic compounds (xenobiotics) and other chemicals having environmental relevance (e.g., mineral oil components). However, it has to be considered that this statement concerns potential degradabilities which – in most instances – were estimated in the laboratory by using pure cultures and ideal growth conditions. Under natural conditions in soil, the actual degradability of organopollutants is lower, due to a whole range of factors: competition with other microorganisms, insufficient supply with essential substrates (C, N, P, S sources), unfavorable external conditions ($O_2$, $H_2O$, pH, temperature), and low bioavailability of the pollutant that is to be degraded. Thus, environmental biotechnology has the important assignment of tackling and solving these problems so as to permit the use of microorganism in bioremediation technologies. For this purpose, it is necessary to support the activities of the indigenous microorganisms in polluted soils and to enhance their degradative potential by bioaugmentation. The former measure particularly applies to bacteria, whereas bioaugmentation mainly concerns basidiomycetous fungi. In this context, we should point out that neither bacteria nor fungi are 'better' degraders. In nature, both groups of microorganisms work together of course, and complement one another in their degradative capabilities.

## References

Abramowicz, D. A., Aerobic and anaerobic biodegradation of PCBs: a review, *Crit. Rev. Biotech.* **1990**, *10*, 241–251.

Alexander, M., *Biodegradation and Bioremediation*, San Diego CA **1994**: Academic Press.

Apajalahti, J. H. A., Salkinoja-Salonen, M. S., Complete dechlorination of terachlorohydroquinone by cell extracts of pentachlorophenol-induced *Rhodococcus chlorophenolicus*, *J. Bacteriol.* **1987**, *169*, 5125–5130.

Baciocchi, E., Gerini, F., Harvey, P. J., Lanzalunga, O., Mancinelli, S., Oxidation of aromatic sulfides by lignin peroxidase from *Phanerochaete chrysosporium. Eur. J. Biochem.* **2000**, *267*, 2705–2710.

Barr, D. P., Aust, S. D., Mechanisms of white rot fungi to degrade pollutants, *Environ. Sci. Technol.* **1994**, *2*, 78A–87A.

Bartha, R., Biotechnology of petroleum pollutant degradation, *Microbiol. Ecol.* **1986**, *12*, 155–172.

Bollag, J.-M., Shuttleworth, K. L., Anderson, D.H., Laccase-mediated detoxification of phenolic compounds, *Appl. Environ. Microbiol.* **1988**, *54*, 3086–3091.

Britton, L. N., Microbial degradation of aliphatic hydrocarbons, in: *Microbial Degradation of Organic Compounds* (Gibson, D. T., ed.), pp. 89–129. New York **1984**: Marcel Dekker.

Dix, N. J., Webster, J., *Fungal Ecology.* London **1995**: Chapman & Hall.

Dutta, D., Ghosh, D. K., Mishra, A. K., Samanta, T.B., Induction of benzo(a)pyrene hydroxylase in *Aspergillus ochraceus* TS: evidences of multiple forms of cytochrome P-450, *Biochem. Biophys. Res. Commun.* **1983**, *115*, 692–699.

Fengel, D., Wegener, G., *Wood: Chemistry, Ultrastructure, Reactions*, Berlin **1989**: Walter de Gruyter.

Fernando, T. Aust, S. D., Biodegradation of toxic chemicals by white rot fungi, in: *Biological Degradation and Bioremediation of Toxic Chemicals* (Chaudhry, G. R., Ed.), pp. 386–402. London **1990**: Chapman & Hall.

Fritsche, W., Hofrichter, M., Aerobic degradation by microorganisms, in: *Biotechnology.* Vol. 11b, *Environmental Processes*, (Rehm, H.-J., Reed, G., Eds.), Wiley-VCH, Weinheim **2000**, pp. 145–167.

Fritsche, W., Scheibner, K., Herre, A., Hofrichter, M., Fungal degradation of explosives: TNT and related nitroaromatic compounds, in: *Biodegradation of Nitroorganic Compounds and Explosives.* (Spain, J., Hughes, J. B., Knackmuss, H.-J., Eds.), Boca Raton FL **2000**: Lewis Publishers, pp. 213–237.

Griffin, D. H., *Fungal Physiology.* New York **1994**: Wiley-Liss.

Haas, R., Tsivunchyk, O., Steinbach, K., v. Löw, E., Scheibner, K., Hofrichter, M., Conversion of adamsite (phenarsazine chloride) by fungal manganese peroxidase. *Appl. Microbiol. Biotechnol., online first:* DOI: 10.1007/s00253-003-1453-x (**2003**).

Hatakka, A., Biodegradation of lignin, in *Biopolymers.* Vol. 1: *Lignin, Humic Substances and Coal.* (Hofrichter, M., Steinbüchel, A., Eds), Weinheim **2001**: Wiley-VCH, pp. 129–180.

Henry, S. M., Grbic-Galic, D., Biodegradation of trichloroethylene in methanotrophic systems and implications for process applications, in: *Biological Degradation and Bioremediation of Toxic Chemicals* (Chaudhry, G. R., ed.), pp. 314–344. London **1994**: Chapman & Hall.

Hofrichter, M., Bublitz, F., Fritsche, W., Unspecific degradation of halogenated phenols by the soil fungus *Penicillium frequentans* Bi 7/2, *J. Basic Microbiol.* **1994**, *34*, 163–172.

Hofrichter, M., Scheibner, K., Bublitz, F., Schneegaß, I., Ziegenhagen, D., Martens, R., Fritsche, W., Depolymerization of straw lignin by manganese peroxidase is accompanied by release of carbon dioxide. *Holzforschung* **1999**, *53*, 161–166.

Hofrichter, M., Review: lignin conversion by manganese peroxidase (MnP). *Enzyme Microbiol. Technol.* **2002**, *30*, 454–466.

Hommel, R. K., Formation and physiological role of biosurfactants produced by hydrocarbon-utilizing microorganisms, *Biodegradation* **1990**, *1*, 107–109.

Houghton, J. E., Shanley, M. S., Catabolic potential of pseudomonads: a regulatory perspective, in: *Biological Degradation and Bioremediation of Toxic Chemicals* (Chaudhry, G. R., ed.), pp. 11–32. London **1994**: Chapman & Hall.

Kästner, M., Humification process or formation of refractory soil organic matter, in: *Biotechnology* Vol. 11b, 2nd edit., (Rehm, H.-J, Reed, G., eds), pp. 89–125. Weinheim **2000**: Wiley-VCH.

Kästner, M., Degradation of aromatic and polyaromatic compounds, in: *Biotechnology* Vol. 11b, 2nd edit., (Rehm, H.-J, Reed, G., Eds), pp. 211–239. Weinheim **2000**: Wiley-VCH.

Kirk, T. K., Farrell, R. L., Enzymatic 'combustion': the microbial degradation of lignin, *Ann. Rev. Microbiol.* **1987**, *41*, 465–505.

Knackmuss, H. J., Abbau von Natur- und Fremdstoffen, in: *Umweltbiotechnologie* (Ottow, C. G., Billingmaier, W, Eds.), pp. 39–80. Stuttgart **1997**: Gustav Fischer.

Manji, S., Ishihara, A., Screening of tetrachlorodibenzo-*p*-dioxin–degrading fungi capable of producing extracellular peroxidases under various conditions. *Appl. Microbiol. Biotechnol.* **2003**, *63*, 438–444.

McAllister, K. A., Lee, H., Trevors, J. T., Microbial degradation of pentachlorophenol. *Biodegradation* **1996**, *7*, 1–40.

Morgan, P., Watkinson, R. J., Biodegradation of components of petroleum, in: *Biochemistry of Microbial Degradation* (Ratledge, C., Ed.), pp. 1–31, Dordrecht **1994**: Kluwer.

Mörtberg, M., Neujahr, H. Y., Uptake of phenol in Trichosporon cutaneum, *J. Bacteriol.* **1985**, *161*, 615–619.

Ortiz-Bermúdez, P., Srebotnik, E., Hammel, K. E., Chlorination and cleavage of lignin structures by fungal chloroperoxidase. *Appl. Environ. Microbiol.* **2003**, *69*, 5015–5018.

Pointing, S., Feasibility of bioremediation by white-rot fungi. *Appl. Microbiol. Biotechnol.* **2001**, *57*, 20–33.

Rajarathnam, S., Shashirekha, N. U., Bano, Z., Biodegradative and biosynthetic capacities of mushrooms: present and future strategies, *Crit. Rev. Biotechnol.* **1998**, *18*, 91–236.

Ramos, J. L., Marques, S., Timmis, K. N., Transcriptional control of the Pseudomonas TOL plasmid catabolic operons is achieved through an interplay of host factors and plasmid-encoded regulators, *Annu. Rev. Microbiol.* **1997**, *51*, 341–73.

Ratledge, C., Editorial, in: *Physiology of Biodegradative Microorganisms* (Ratledge C., Ed.), pp. vii–viii. Dordrecht **1991**: Kluwer.

Rehm, H. J., Reiff, I., Regulation of microbial alkane oxidation with respect to the formation of products, *Acta Biotechnol.* **1982**, *3*, 279–288 (in German).

Reinecke, W., Degradation of chlorinated aromatic compounds by bacteria: strains development, in: *Biological Degradation and Bioremediation of Toxic Chemicals* (Chaudhry, G. R., Ed.), pp. 416–454. London **1994**: Chapman & Hall.

Steffen, K., Hatakka, Hofrichter, M., Degradation of benzo(a)pyrene by the litter-decomposing basidiomycete *Stropharia coronilla*: role of manganese peroxidase. *Appl. Environ. Microbiol.* **2003**, *69*, 3957–3964

Ullrich, R., Nüske, J., Scheibner, K., Spantzel, J., Hofrichter, M., A novel peroxidase from the agaric basidiomycete *Agrocybe aegerita* oxidizing aryl alcohols and aldehydes and showing chloroperoxidase activity, **2004**, *70*, 4575–4581.

Wariishi, H., Valli, K., Gold, M. H., Manganese(II) oxidation by manganese peroxidase from the basidiomycete *Phanerochaete chrysosporium*: kinetic mechanism and role of chelators, *J. Biol. Chem.* **1992**, *267*, 23688–23695.

Watkinson, R. J., Morgan, P., Physiology of aliphatic hydrocarbon-degrading microorganisms, *Biodegradation* **1990**, *1*, 79–92.

Wright, J. D., Fungal degradation of benzoic acids and related compounds. *World J. Microbiol. Biotechnol.* **1993**, *9*, 9–16.

# 8
# Principles of Anaerobic Degradation of Organic Compounds

Bernhard Schink

## 8.1
### General Aspects of Anaerobic Degradation Processes

The vast majority of organic compounds produced in nature or through human manufacture is degraded aerobically, with molecular oxygen as terminal electron acceptor. As long as oxygen is available, it is the preferred electron acceptor for microbial degradation processes in nature.

Anaerobic degradation processes have always been considered inferior to aerobic degradation in their kinetics and capacities. They are thought to be slow and inefficient, especially with certain comparably stable types of substrates. Nonetheless, in certain anoxic environments, such as the cow's rumen, the turnover of, e.g., cellulose is much faster than in the presence of oxygen with average half-life times in the range of one day. Fermentative degradation of fibers in the rumen reaches its limit with plant tissues rich in lignin which largely withstands degradation in the absence of oxygen.

Also in waste treatment, especially with high loads of easy-to-degrade organic material, anaerobic processes have proved to be efficient and far less expensive than aerobic treatment: they require only small amounts of energy input, in contrast to treatment in aeration basins, and can produce a mixture of methane and $CO_2$ ('biogas'), which can be used efficiently for energy generation. This holds true for most waste materials that are easily accessible to degradation without the participation of oxygen, such as polysaccharides, proteins, fats, nucleic acids, etc. These polymers are hydrolyzed through specific extracellular enzymes, and the oligo- and monomers can be degraded inside the cell through enzyme reactions similar to those known in aerobic metabolism. The specific activities of such enzymes in anaerobic cultures are in the same range (0.1–1 µmol substrate per min and mg cell protein) as those of aerobic bacteria, and thus the transformation rates per unit biomass should be equivalent.

Nonetheless, anaerobic bacteria obtain far less energy from substrate turnover than their aerobic counterparts. Whereas aerobic oxidation of hexose to six $CO_2$ yields 2870 kJ per mol, dismutation of hexose to three $CH_4$ and three $CO_2$ yields on-

*Environmental Biotechnology. Concepts and Applications.* Edited by H.-J. Jördening and J. Winter
Copyright © 2005 WILEY-VCH Verlag GmbH & Co. KGaA, Weinheim
ISBN: 3-527-30585-8

ly 390 kJ per mol, about 15% of the aerobic process, and this small amount of energy has to be shared by at least three different metabolic groups of bacteria (see Schink, 1997). As a consequence, they can produce far less biomass per substrate molecule than aerobes can. Their growth yields are low, and most often growth is slower than that of aerobes. Maintaining the biomass inside specifically designed reactors (fixed bed, fluidized bed reactors, Upflow Anaerobic Sludge Blanket reactors) helps to overcome the problem of low and slow biomass production in anaerobic degradation and largely uncouples substrate turnover from biomass growth (for an overview, see, e.g., Schink, 1988). These systems allow anaerobic wastewater treatment to be nearly as efficient as and less expensive than the aerobic process, with methane as a useful product; but the microbial communities in these advanced anaerobic reactors still are comparably sluggish in reacting to changes in substrate composition or in their reestablishment after accidental population losses due to toxic ingredients in the feeding waste.

Degradation of organic matter in the absence of oxygen can be coupled to the reduction of alternative electron acceptors following a certain sequence that appears to be determined by the respective redox potentials. Molecular oxygen ($O_2/H_2O$ $E_h = +810$ mV; $E_h$ values calculated for pH 7.0) is followed by nitrate ($NO_3^-/NO_2^-$ $E_h = +430$ mV), manganese(IV) oxide ($MnO_2/Mn^{2+}$ $E_h = +400$ mV), iron(III) hydroxides ($FeOOH/Fe^{2+}$ $E_h = +150$ mV), sulfate ($SO_4^{2-}/HS^-$ $E_h = -218$ mV), and finally $CO_2$ ($CO_2/CH_4$ $E_h = -244$ mV), with the release of nitrite, ammonia, dinitrogen, manganese(II) and iron(II) carbonates, sulfides, and finally methane as products (Zehnder and Stumm, 1988). Reduction of these acceptors with electrons from organic matter (average redox potential for glucose $\rightarrow$ 6 $CO_2$ is $-0.434$ V; calculated after data of Thauer et al., 1977) provides metabolic energy in the mentioned sequence. Thus, the energy yields of the various anaerobes mentioned differ, and the availability of either high- or low-potential electron acceptors may also influence the biochemistry of anaerobic degradation processes.

Limits of anaerobic degradation become obvious with those organic compounds that accumulate in anoxic sediments or that persist in anoxic soil compartments contaminated with mineral oil or other rather recalcitrant compounds. Mineral oil consists mainly of aliphatic and aromatic hydrocarbons which, in the presence of molecular oxygen, are attacked biochemically through oxygenase reactions, which introduce molecular oxygen into the respective molecule (Lengeler et al., 1999; see also Chapter 7). Oxygenase reactions cannot be employed in the absence of oxygen, and, in particular, compounds that require oxygenases for aerobic breakdown might resist degradation under anoxic conditions. Alternatives usually exist in the anoxic world that also allow oxygen-independent degradation of such compounds.

Oxygen is not always advantageous in degradation processes. Oxygenases introduce hydroxyl groups into aromatics, and further oxygen may cause formation of phenol radicals that initiate uncontrolled polymerization and condensation to polymeric derivatives, similar to humic compounds in soil, which are very difficult to degrade further, whether anaerobically or aerobically. Therefore, anaerobic degradation processes may be used for treatment of specific wastewaters rich in phenolic compounds, e.g., from the chemical industry, to avoid formation of unwanted side products such as condensed polyphenols. In other situations, aerobic treatment may cause technical problems, e.g., by extensive foam formation during aerobic treatment of surface-active compounds such as tensides. Thus, knowledge of the limits and principles of anaerobic degradation processes under the various conditions prevailing in natural habitats might help to design suitable alternative techniques for cleanup of contaminated soils or for treatment of specific wastewaters that have so far been applied only insufficiently.

The following survey gives an overview of our present knowledge of the limits and principles of anaerobic degradation of organic compounds. The focus is on those compounds that were for long times considered to be stable in the absence of oxygen.

## 8.2
## Key Reactions in Anaerobic Degradation of Certain Organic Compounds

### 8.2.1
### Degradation of Hydrocarbons

Saturated aliphatic hydrocarbons are attacked only slowly in the absence of oxygen, and the first reliable proof of such a process was provided only about eight years ago for a culture of sulfate-reducing bacteria (Aeckersberg et al., 1991). Growth of this culture with hexadecane was very slow, with doubling times of more than one week under optimal conditions. In the meantime, several strains of alkane-oxidizing anaerobes were isolated (Aeckersberg et al., 1998; Rueter et al., 1994), which are specialized for either long-chain ($C_{12}$–$C_{20}$) or medium-chain ($C_{6-16}$) alkanes and use either sulfate or nitrate as electron acceptor.

Insight into the biochemistry of alkane activation in the absence of oxygen has been obtained only recently. The initial activation is basically similar to the corresponding reaction involved in anaerobic oxidation of toluene (see Section 8.2.6.8): the hydrocarbon is added with its subterminal carbon atom to fumarate through a radical reaction, to form an alkyl succinate derivative (Rabus et al., 2001). This strategy is used in a basically similar manner by nitrate-reducing and sulfate-reducing bacteria (Wilkes et al., 2002). In either case, anaerobic hydrocarbon degradation is very slow, but may play a role in, e.g., natural attenuation of soil sites polluted with petroleum or diesel fuel.

A special example, although not of technical interest, is the anaerobic degradation of methane, e.g., with sulfate as electron acceptor, a process that is of major importance in global carbon transformations. No bacterium that catalyzes this reaction has been isolated so far, although it is thermodynamically feasible (see Schink, 1997). Recent evidence has shown that this reaction is most probably carried out by archaea similar to methanogens, which operate methane formation in the backwards reaction, most often in syntrophic association with sulfate-reducing partner organisms (Boetius et al., 2000). Although this concept was suggested many years ago (Zehnder and Brock, 1980), experimental evidence has been obtained only recently with samples from deep-sea sources, and efficient methane oxidation requires enhanced methane pressure (Nauhaus et al., 2002).

Unsaturated long-chain hydrocarbons with terminal double bonds can be hydrated to the corresponding primary alcohols (although against the Markownikoff rule) and completely degraded (Schink, 1985a). A branched-chain unsaturated hydrocarbon such as squalene was degraded in methanogenic enrichment cultures (Schink, 1985a); however, degradation was incomplete, probably due to the formation of saturated branched derivatives. Other unsaturated isoprene derivatives such as terpenes have been shown recently to be completely degraded, with nitrate as electron acceptor (Harder and Probian, 1995; Foß and Harder, 1998; Foß et al., 1998). Although the structures of terpenes differ substantially with respect to the way of possible attack, an amazingly broad variety of terpenes was completely degraded. Some concepts of the biochemistry of degradation of these compounds have been developed (Hylemon and Harder, 1999),but experimental evidence is still lacking. Acetylene, a highly unsaturated hydrocarbon, is fermented comparably rapidly to acetate and ethanol through acetaldehyde (Schink, 1985b), which is formed by a hydratase enzyme (Rosner and Schink, 1995). Acetylene hydratase is an iron-sulfur protein containing a tungsten cofactor (Meckenstock et al., 1999); it is active only in the reduced state, but the reaction mechanism is still unknown. No anaerobic degradation has been documented so far for ethylene, propylene, propine, and higher homologs having up to six carbon atoms.

### 8.2.2
### Degradation of Ether Compounds and Nonionic Surfactants

Ether linkages are rather stable, and their chemical cleavage requires severe conditions, e.g., boiling at strongly alkaline or acidic pH. Biological ether cleavage in the presence of oxygen employs oxygen as cosubstrate in an oxygenase reaction, which transforms the ether into an unstable hemiacetal (Bernhardt et al., 1970). Thus, methyl groups of lignin monomers are released as formaldehyde, not as methanol.

Anaerobic demethylation of lignin monomers by the homoacetogen *Acetobacterium woodii* was first described by Bache and Pfennig (1981) and later was repeatedly observed with several other homoacetogens. The mechanism of this phenyl methyl ether cleavage were only recently elucidated. Studies with the homoacetogen *Holophaga foetida* showed that the methyl group is first transferred as a methyl cation to a fully reduced cob(I)alamin carrier, which later methylates the coenzyme tetrahydrofolate (Fig. 8.1; Kreft and Schink, 1993, 1994). Similar studies with *Acetobacterium woodii*, *Sporomusa ovata*, or *Acetobacterium dehalogenans* revealed that also in these species, the methyl group is transferred as a methyl cation, but that the details of further methyl transfer to coenzymes may differ with the strain studied (Berman and Frazer, 1992; Stupperich and Konle, 1993; Kaufmann et al., 1997; Kaufmann et al., 1998; Engelmann et al., 2001).

A different type of anaerobic ether cleavage was observed with the synthetic polyether polyethylene glycol (PEG). Formation of acetaldehyde as the first cleavage product, extreme oxygen sensitivity of the ether-cleaving enzyme in cell-free ex-

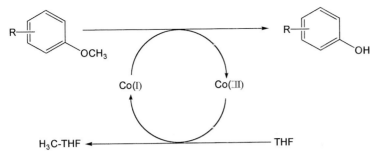

**Fig. 8.1** Anaerobic demethylation of phenyl methyl ethers. Co(I), Co(III), cobalamin in different redox states; THF, tetrahydrofolate.

tracts, and interference with cobalamins strongly suggest that the first step in this degradation is a cobalamin-dependent shift of the terminal hydroxyl group to the subterminal C atom, analogous to a diol dehydratase reaction (Fig. 8.2; Schramm and Schink, 1991; Frings et al., 1992). This reaction again transforms the ether into a hemiacetal derivative that decomposes spontaneously. Since the ether-cleaving enzyme is located in the cytoplasmic space, the polymeric PEG (which has molecular mass up to 40 kDa) has to cross the cell membrane(s) before it is cleaved inside, and the same is true for PEG-containing nonionic surfactants. Since the bacterial strains studied so far are specialized only for the degradation of the PEG chain, the lipophilic residues of the surfactants have to cross the membrane(s) again on their way back

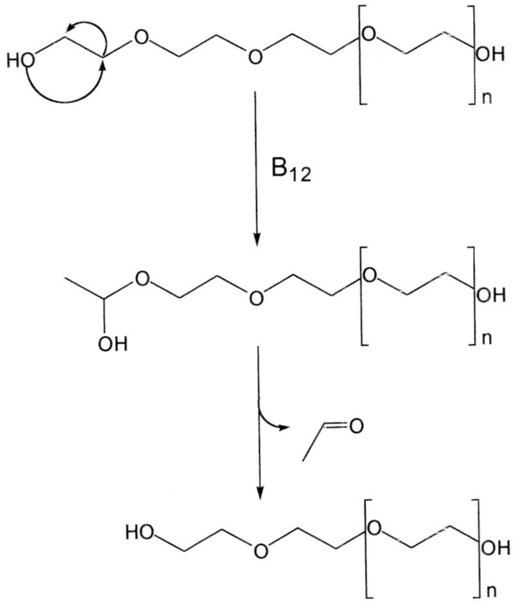

**Fig. 8.2** Anaerobic degradation of polyethylene glycol by fermenting bacteria. $B_{12}$, coenzyme $B_{12}$.

out. It is obvious that these transport steps considerably limit the applicability of such degradation capacities to treatment of, e.g., wastewaters rich in such surfactants. Nonetheless, the applicability of anaerobic fixed-bed reactors for treatment of nonionic surfactants of various types to methane, $CO_2$, and mixtures of fatty acids has been demonstrated (Wagner and Schink, 1987, 1988).

The anaerobic ether cleavage reactions described here proceed only inside the bacterial cell. In this, they differ from the corresponding reactions reported for certain basidiomycetes (Kerem et al., 1998), which use these cleavage capacities, e.g., in lignin degradation. It is not surprising, therefore, that highly polymeric, condensed-ether compounds such as lignin are not degraded to any significant extent in the absence of molecular oxygen.

## 8.2.3
### Degradation of N-Alkyl Compounds and Nitrilotriacetate

Among the natural N-alkyl compounds are, in addition to t he amino acids, several methylated amines such as trimethylamine, which is formed during the initial decay of fish tissue through reduction of trimethylamine-N-oxide by several enterobacteria and others. Under strictly anoxic conditions, methanogenic archaea have been found to efficiently demethylate trimethylamine via dimethylamine to monomethyl-amine and to ferment the methyl moieties by dismutation to methane and $CO_2$ (Hippe et al., 1979). The cleavage between the nitrogen and the neighboring carbon atom is accomplished by a nucleophilic attack by a cob(I)alamin, analogous to the demethylation of phenyl methyl ethers by homoacetogenic bacteria (see above). So far, only methanogens have been found to demethylate methylamines, whereas homoacetogens appear to be specialists for demethylation of phenyl methyl ethers which, in turn, are not attacked by methanogens. It should be emphasized at this point that the strategy of methyl cation removal by cob(I)alamin derivatives is applicable only to these one-carbon compounds. Ethyl or higher homologs cannot be cleaved this way.

N-alkyl compounds of technical and environmental concern include ethylenediaminetetraacetate (EDTA) and nitrilotriacetate (NTA); the latter has largely replaced polyphosphates as a calcium chelator in most commercial washing detergents. The main problem in degrading EDTA is the formation of strong complexes of EDTA with metal ions, which make this substrate very difficult to attack. Nonetheless, microbial EDTA degradation in the presence of oxygen has been documented (-Nörtemann, 1992), but no reports exist on possible anaerobic degradation of EDTA. NTA is degraded aerobically through oxygenase-dependent hydroxylation of one methylene carbon. The resulting hydroxy compound is unstable and releases glyoxylic acid. Removal of an additional carboxymethylene residue produces glycine as co-product. Anaerobic degradation of NTA is possible with nitrate as electron acceptor. The first degradation step is dehydrogenation to an unsaturated iminium derivative which, upon hydration, can again release glyoxylic acid to form the iminodiacetate derivative (Egli et al., 1990). Since oxidation of NTA to the unsaturated derivative has a rather high redox potential (> +100 mV), it is not surprising that so far, only nitrate reducers have been found to be able to degrade NTA. We have to assume that, under strictly reducing conditions, i.e., in deeper sediment layers, NTA is stable to microbial attack, because neither sulfate-reducing nor fermenting bacteria can release electrons arising at this high redox potential.

### 8.2.4
### Degradation of S-Alkyl Compounds

Dimethylsulfoniopropionate is an osmoprotectant found in several green algae and seaweeds. Its cleavage by anaerobic bacteria leads to acrylate and dimethylsulfide, which can escape into the atmosphere. The thioether dimethylsulfide can also be degraded anaerobically by methanogenic archaea (Kiene et al., 1986; Oremland et al., 1989). The carbon-sulfur linkage is cleaved by methyl cation removal, analogous to the demethylation reactions described above (Keltjens and Vogels, 1993).

### 8.2.5
### Degradation of Ketones

Aerobic degradation of ketones appeared to be well established. Indications of an oxygenase-catalyzed hydroxylation of acetone to acetol by aerobic bacteria were provided early (Lukins and Foster, 1963; Taylor et al., 1980), but this type of reaction was never confirmed unequivocally. Anaerobic degradation of acetone by nitrate-reducing, sulfate-reducing, or fermenting bacteria cooperating with methanogenic partners uses a carboxylation reaction as primary step of activation, leading to an acetoacetyl derivative that undergoes subsequent cleavage to two acetate moieties (Fig. 8.3; Platen and Schink, 1987, 1989; Bonnet-Smits et al., 1988; Platen et al., 1990). Unfortunately, the primary carboxylation reaction was never convincingly proved with these cultures. Very little acetone carboxylation activity was observed in enzyme assays (Platen and Schink, 1991) or in radiotracer experiments using suspensions of intact cells (Janssen and Schink, 1995). Acetone carboxylation activity in the phototrophic anaerobe *Rhodobacter capsulatus* was also very weak in vitro (Birks and Kelly, 1997). An acetone-carboxylating enzyme complex of high activity has been found in an aerobic *Xanthobacter* strain (Sluis et al., 1996) and was purified and characterized (Sluis and Ensign, 1997; Sluis et al., 2002). The reaction requires as energy source for the acetone carboxylation reaction one ATP, which is hydrolyzed to AMP plus

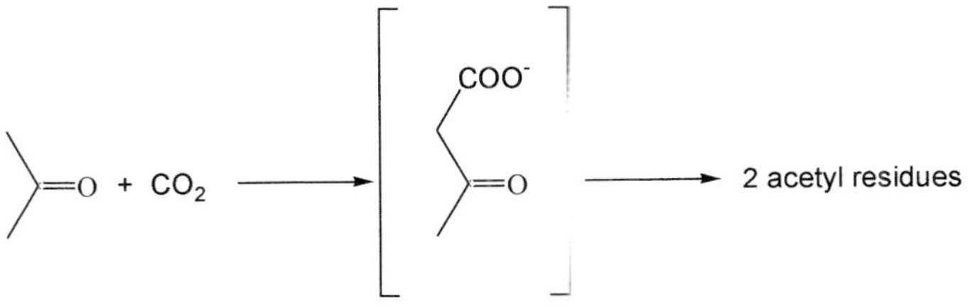

**Fig. 8.3** Anaerobic degradation of acetone through carboxylation.

two $P_i$. Whether sulfate-reducing or fermenting bacteria provide the necessary energy for carboxylation in different ways still has to be examined. The fermenting bacteria especially cannot afford to spend the equivalent of two ATP molecules on this carboxylation reaction. To our surprise, all aerobic bacteria enriched with acetone also used a carboxylation step rather than an oxygenase reaction in acetone activation, even if $CO_2$ was trapped during the enrichment process (Schink, unpublished data). Thus, the original reports of oxygenase-dependent acetone activation may describe an exceptional situation with one single strain, which is not representative of the majority of aerobic acetone degraders.

Higher homologs of acetone appear to be attacked in a similar manner by carboxylations. This applies also to acetophenone, the phenyl-substituted analog of acetone (Schink, unpublished data).

### 8.2.6
### Degradation of Aromatic Compounds

In aerobic degradation of aromatic compounds, oxygenases activate the comparably stable oxygen molecule in such a way as to produce highly electrophilic species, which add to a comparably inert aromatic compound to form hydroxylated products such as catechol (Dagley, 1971; Schlegel, 1992). A further oxygenase-dependent step opens the aromatic ring, either between or vicinal to the two hydroxyl groups of catechol, thus forming an unsaturated, open-chain carboxylic acid which undergoes further degradation, typically to an acetyl and a succinyl derivative.

That aromatic compounds can also be degraded anaerobically was documented as early as 1934: a broad variety of mononuclear aromatic compounds, such as benzoate, phenols, and several lignin monomers, was converted stoichiometrically to methane and $CO_2$ (Tarvin and Buswell, 1934). Later these observations were forgotten, and some textbooks maintained even into the 1980s the dogma that aromatics can be attacked only with oxygen as cosubstrate. During the 1970s, Evans (1977) developed the concept that destabilization of the aromatic nucleus in the absence of oxygen could proceed through a reductive rather than an oxidative reaction. Today we know at least three different pathways of anaerobic degradation of mononuclear compounds, i.e., the benzoyl-CoA pathway, the resorcinol pathway, and the phloroglucinol pathway (Evans and Fuchs, 1988; Schink et al., 1992; Fuchs et al., 1994; Heider and Fuchs, 1997; Schink et al., 2000). In all these pathways, a 1,3-dioxo structure is formed through a reduction step, either inside the ring itself or in combination with a carboxyl coenzyme A moiety. This structure allows a nucleophilic attack on one of the ring ketone carbon atoms and subsequent ring fission. Depending on the aromatic substrate, either a pimelic ($C_7$-dicarboxylic) residue bound to coenzyme A or a partly oxidized caproic ($C_6$-monocarboxylic) acid is formed, which undergoes subsequent beta oxidation to three acetyl moieties. Formation of other products in fermentative benzoate degradation, such as succinate or propionate as claimed in earlier papers, could never be reproduced with defined cultures and may have been due to uncontrolled side reactions or misinterpretations of insufficient chemical analyses.

### 8.2.6.1 Benzoate and the Benzoyl-CoA Pathway

The benzoyl-CoA pathway appears to be the most important one in anaerobic degradation of aromatics, because a broad variety of compounds enter this path, including phenol, various hydroxybenzoates, phenylacetate, aniline, certain cresols, and even the hydrocarbon toluene (Fig. 8.4; Schink et al., 1992; Heider and Fuchs, 1997; Harwood et al., 1999). Once benzoyl-CoA is formed, the stability of the aromatic ring structure is overcome by a reductive step that introduces two single electrons and protons, through radical intermediates to form cyclohexadiene carboxyl-CoA as the first identifiable product (Fig. 8.5; Koch et al., 1993; Boll and Fuchs, 1995; Boll et al. 2000a). Because reduction of the benzene ring to a cyclohexadiene derivative, even with electrons at the ferredoxin level (Boll et al., 2000b) is endergonic, it requires the investment of energy in the form of two ATP molecules (Boll and Fuchs, 1995; Harwood et al., 1999). Nitrate-reducing bacteria can recover this energy investment through the further breakdown of the $C_7$-dicarboxylic acid derivative produced upon ring cleavage, via beta oxidation to three acetyl-CoA residues that are finally oxidized in the citric acid cycle. Fermenting bacteria and sulfate-reducing bacteria recover on-

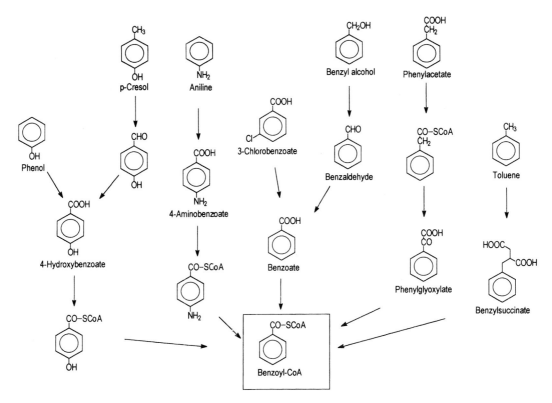

**Fig. 8.4** Overview of mononuclear compounds entering the benzoyl-CoA pathway of anaerobic degradation (courtesy of Prof. G. Fuchs, Freiburg).

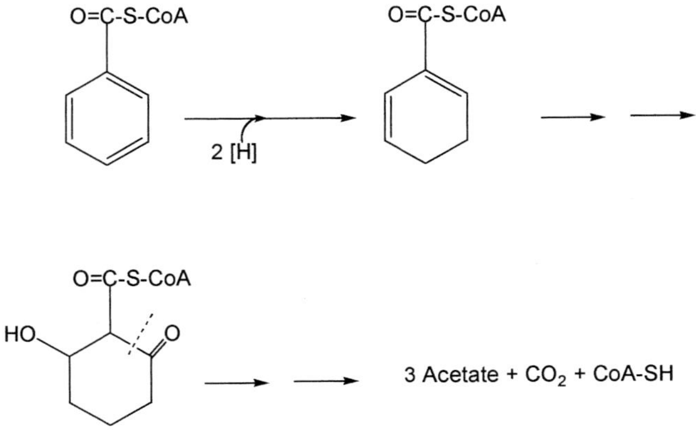

**Fig. 8.5** Initial steps in anaerobic degradation of benzoate by the nitrate-reducing bacterium *Thauera aromatica.*

ly a little energy in the further breakdown of the open-chain intermediate. They may use a different type of reaction for benzoyl-CoA dearomatization that leads to a cyclohexene carboxyl derivative (Schöcke and Schink, 1999; Harwood et al., 1999) or to a hydroxylated derivative through hydroxylations catalyzed by selenium- or molybdenum-containing enzymes (Peters et al. 2004). However, the biochemistry of these new reactions still needs to be elucidated.

### 8.2.6.2 Phenol, Hydroxybenzoates, and Aniline

Aromatic compounds that do not carry a carboxyl group, such as phenol or aniline, are first carboxylated to a *p*-hydroxy or a *p*-amino benzoic acid residue, which is subsequently activated with coenzyme A (Fig. 8.6; Tschech and Fuchs, 1989; Lack and Fuchs, 1992; Schnell and Schink, 1991). The carboxylation of phenol by a nitrate-reducing *Thauera aromatica* strain can be followed in vitro with phenyl phosphate as substrate, which is carboxylated to 4-hydroxybenzoate and further degraded as such (see below). The phosphate donor for phenol phosphorylation is still unknown (Lack and Fuchs, 1992, 1994; Breinig et al., 2000). Whether sulfate-reducing bacteria or fermenting bacteria cooperating with methanogenic partners use the same pathway for phenol degradation also remains to be examined. An H/D exchange at carbon atom 4 of phenol by cell suspensions of a methanogenic phenol-degrading enrichment culture indicates that these cultures also activate phenol through carboxylation at this position (Gallert et al., 1991). Whether the carboxylation reaction in these fermentative bacteria is also initiated by phenol phosphorylation is still an open question. The overall energy budget of fermentative phenol degradation is very tight and hardly allows spending a full ATP equivalent or even more on this carboxylation reaction. The biochemistry of phenol degradation by sulfate-reducing bacteria has not been studied so far, but it is likely to proceed basically through the same pathway.

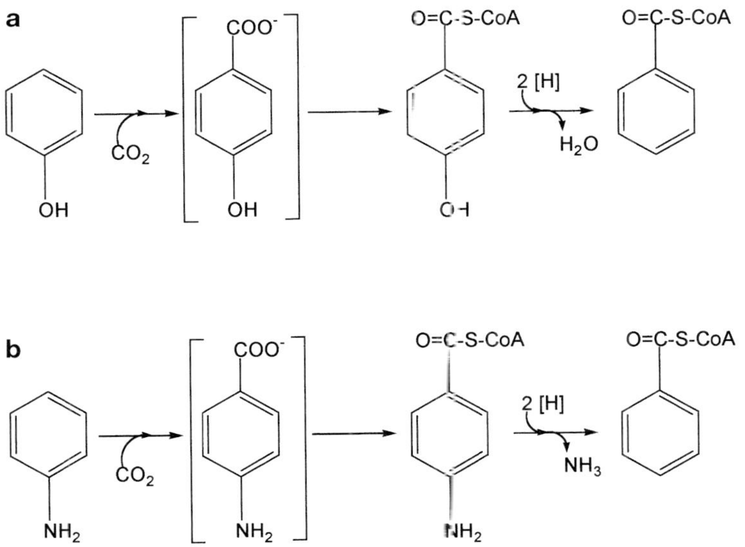

**Fig. 8.6** Anaerobic degradation of (a) phenol and (b) aniline by anaerobic bacteria. The compounds in brackets have not been identified.

The 4-hydroxybenzoate formed is activated through a ligase reaction, analogous to benzoyl-CoA formation from benzoate, to form 4-hydroxybenzoyl-CoA, which is subsequently reductively dehydroxylated to benzoyl-CoA (Fig. 8.6a).

4-Hydroxybenzoate may be degraded anaerobically through the same pathway. However, in cultures 4-hydroxybenzoate is not stable but is slowly decarboxylated to phenol, perhaps by an enzyme activity related to phenyl phosphate carboxylase.

3-Hydroxybenzoate is comparably stable and does not decarboxylate spontaneously. Instead, the hydroxyl group is reductively eliminated by fermenting bacteria to allow further degradation through the benzoyl-CoA pathway, as shown with *Sporotomaculum hydroxybenzoicum* (Müller and Schink, 2000).

Aniline is degraded anaerobically through a pathway analogous to that for phenol degradation. The initial activation is accomplished through carboxylation to 4-aminobenzoate, which is subsequently activated to 4-aminobenzoyl-CoA and undergoes reductive deamination to benzoyl-CoA (Fig. 8.6b; Schnell and Schink, 1991). The initial carboxylation reaction has not been studied in cell-free extracts so far, and nothing is known about an activated intermediate that can provide the carboxylation reaction with the necessary energy.

Aminobenzoates, diaminobenzenes, and aminohydroxybenzenes are degraded very slowly in anaerobic enrichment cultures, and nothing is known about the degradation pathways (Schnell and Schink, unpublished data).

### 8.2.6.3 Cresols

Cresols (methylphenols) are anaerobically degraded through three different pathways, depending on the type of substitution. p-Cresol is hydroxylated at the methyl group by an oxygen-independent reaction, probably through a quinomethide intermediate as suggested earlier for an aerobic *Pseudomonas* strain (Fig. 8.7a; Hopper, 1978). The redox potential of this oxidation reaction is in the range of +100 mV (calculated after Thauer et al., 1977), and the reaction is therefore easy to do for a nitrate-reducing bacterium that couples this oxidation, e.g., with the reduction of a *c*–type cytochrome at +235 mV (Hopper et al., 1991). Sulfate-reducing or fermenting bacteria, on the other hand, have difficulty in disposing of these electrons. o-Cresol can be carboxylated to 3-methyl-4-hydroxybenzoate and further degraded as such (Fig. 8.7b; Bisaillon et al., 1991; Rudolphi et al., 1991). An alternative pathway could lead through methyl group hydroxylation, analogous to p-cresol, to form salicylic acid as an intermediate (Suflita et al., 1989; Schink et al., 1992), but this pathway has only been hypothesized so far. The pathway of anaerobic m-cresol degradation has been elucidated recently in the sulfate-reducing bacterium *Desulfobacterium cetonicum*. This degradation follows a strategy analogous to anaerobic toluene degradation by nitrate-reducing bacteria (Müller et al., 1999): the methyl group of m-cresol adds to fumarate to form 3-hydroxybenzylsuccinate. Activation and beta oxidation lead to

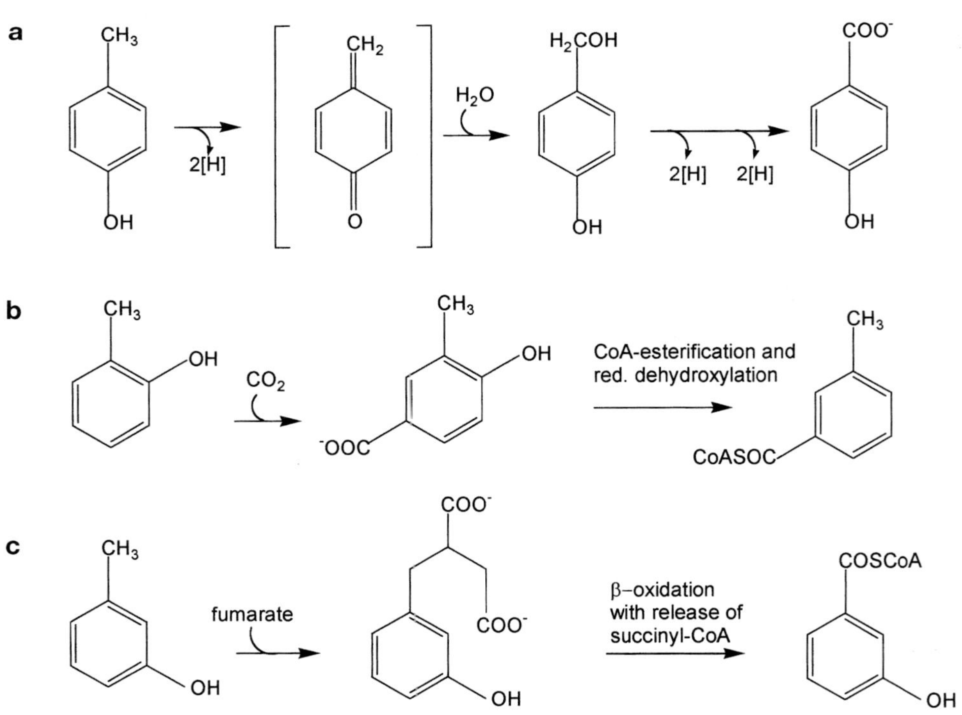

**Fig. 8.7** Anaerobic degradation of cresols. (a) Degradation of p-cresol, (b) degradation of o–cresol, (c) degradation of m-cresol.

succinyl-CoA and benzoyl-CoA (Fig. 8.7c). Thus, the new type of methyl group activation by addition to fumarate appears not to be restricted to the activation of hydrocarbons (see below).

#### 8.2.6.4 Hydroquinone and Catechol

Hydroquinone is degraded by sulfate-reducing and fermenting bacteria. The degradation pathway has been studied in a sulfate-reducing *Desulfococcus* strain (Gorny
and Schink, 1994a) and a fermenting bacterium that was later described as *Syntrophus gentianae* (Gorny and Schink, 1994b). In both species, hydroquinone is carboxylated to gentisic acid; again, this carboxylation could not be studied in cell-free extracts and the way of energization of this reaction is unknown. Gentisate is activated
to gentisyl-CoA through a CoA-ligase reaction and reductively dehydroxylated to
benzoyl-CoA, which enters the modified benzoyl-CoA pathway. The dehydroxylation
of both hydroxyl groups appears to proceed in a single step. Alternatively, gentisic
acid is utilized by the fermenting bacterium.

Catechol, the key intermediate in aerobic breakdown of aromatic compounds, is
by far the slowest phenolic compound to be degraded under anoxic conditions. The
biochemistry of catechol degradation has been studied so far only in a sulfate-reducing *Desulfobacterium* strain, which carboxylates catechol to protocatechuate (Gorny
and Schink, 1994c). Protocatechuate is activated to form protocatechuyl-CoA, which
is subsequently dehydroxylated to benzoyl-CoA. Efforts to isolate nitrate-reducing or
fermenting bacteria with hydroquinone or catechol as substrate have failed so far.

#### 8.2.6.5 Resorcinol

An entirely different strategy is used in the anaerobic degradation of resorcinol and
its derivatives. The two hydroxyl groups in resorcinol are in positions relative to each
other that allow tautomerization to a cyclohexene dione derivative having three isolated double bonds (Fig. 8.8a). Cell-free extracts of a fermenting *Clostridium* strain
convert resorcinol to dihydroresorcinol (cyclohexanedione; Tschech and Schink,
1985; Kluge et al., 1990), which is further hydrolyzed to 5-oxohexanoate, probably
through a nucleophilic attack on one of the carbonyl carbon atoms (Fig. 8.8a).

The resorcinol carboxylates β- and γ-resorcylate are degraded by the same fermenting bacterium after decarboxylation to resorcinol. These decarboxylations are
chemically easy, because in these compounds the carboxylic group is located ortho
or para to electron-withdrawing hydroxyl groups.

In cultures of nitrate-reducing bacteria growing with resorcinol as the sole substrate, no resorcinol-reducing activity could be identified (Kluge et al., 1990; Gorny
et al., 1992). The nitrate reducer *Azoarcus anaerobius* destabilizes the ring by intro-
ducing an additional hydroxyl group to form hydroxyhydroquinone (Fig. 8.8b; Philipp and Schink, 1998). The enzyme involved is membrane-bound, and the hydroxylation is coupled to reduction of nitrate to nitrite. In a later oxidation step, hydroxyhydroquinone is oxidized to hydroxybenzoquinone (Philipp and Schink, 1998). The
ring fission reaction has not been resolved yet.

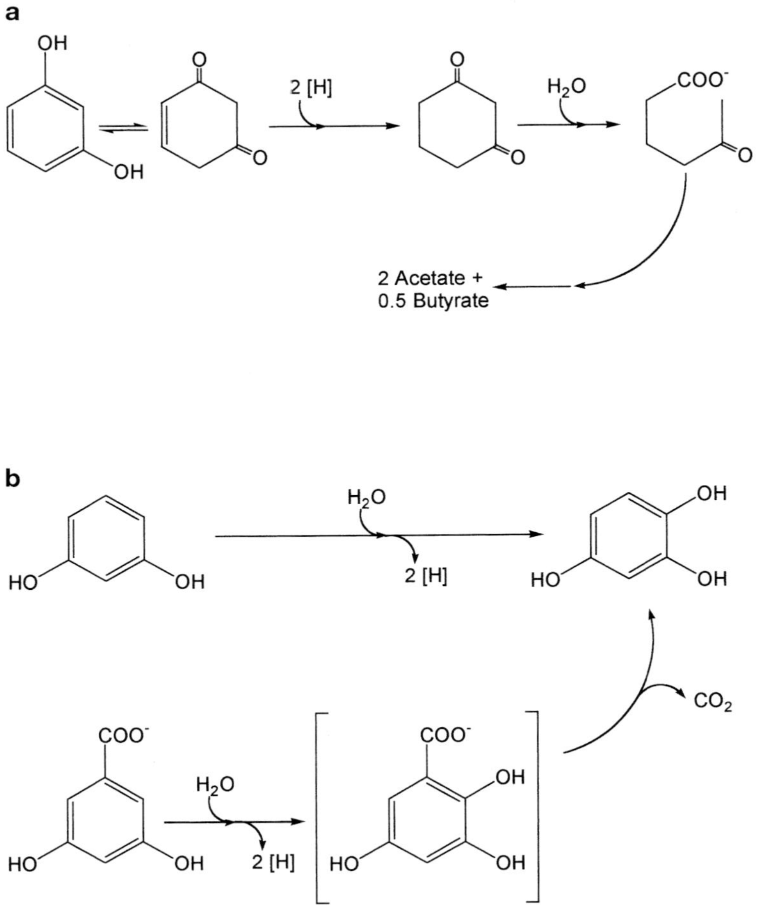

**a**

2 Acetate +
0.5 Butyrate

**b**

**Fig. 8.8** Anaerobic degradation of resorcinol and α-resorcylate. (a) Resorcinol degradation by a fermenting bacterium, *Clostridium* strain KN245, (b) degradation of resorcinol and α–resorcylate by nitrate-reducing bacteria.

### 8.2.6.6 Trihydroxybenzenes and Trihydroxybenzoates

Among the three trihydroxybenzene isomers, pyrogallol and phloroglucinol are degraded quickly by fermenting bacteria (Schink and Pfennig, 1982). Phloroglucinol degradation has been studied in detail with *Eubacterium oxidoreducens* and *Pelobacter acidigallici*. Phloroglucinol is reduced by an NADPH-dependent reductase to dihydrophloroglucinol (Fig. 8.9; Haddock and Ferry, 1989; Brune and Schink, 1992), and the same strategy is followed by *Holophaga foetida* strain TMBS4 (Kreft and Schink, 1993). Hydrolytic ring cleavage leads to 3-hydroxy-5-oxohexanoic acid, which is thiolytically cleaved to three acetate residues (Brune and Schink, 1992). This pathway is easy to conceive, because the 1,3,5 arrangement of three hydroxyl groups on the aromatic ring allows tautomerization to 1,3,5-trioxocyclohexane to an extent that makes the molecule susceptible to a nucleophilic attack on the oxocarbon groups. The second trihydroxybenzene isomer, pyrogallol, cannot be hydrolyzed directly but is isomerized to phloroglucinol through a transhydroxylation reaction (Fig. 8.9; Krumholz and Bryant, 1988; Brune and Schink, 1990). The reaction requires 1,3,4,5-tetrahydroxybenzene as a cosubstrate, and the enzyme transfers a hydroxyl group from the tetrahydroxybenzene to pyrogallol, thus releasing phloroglucinol as product and the tetrahydroxybenzene as coproduct (Brune and Schink, 1990; Reichenbecher et al., 1996).

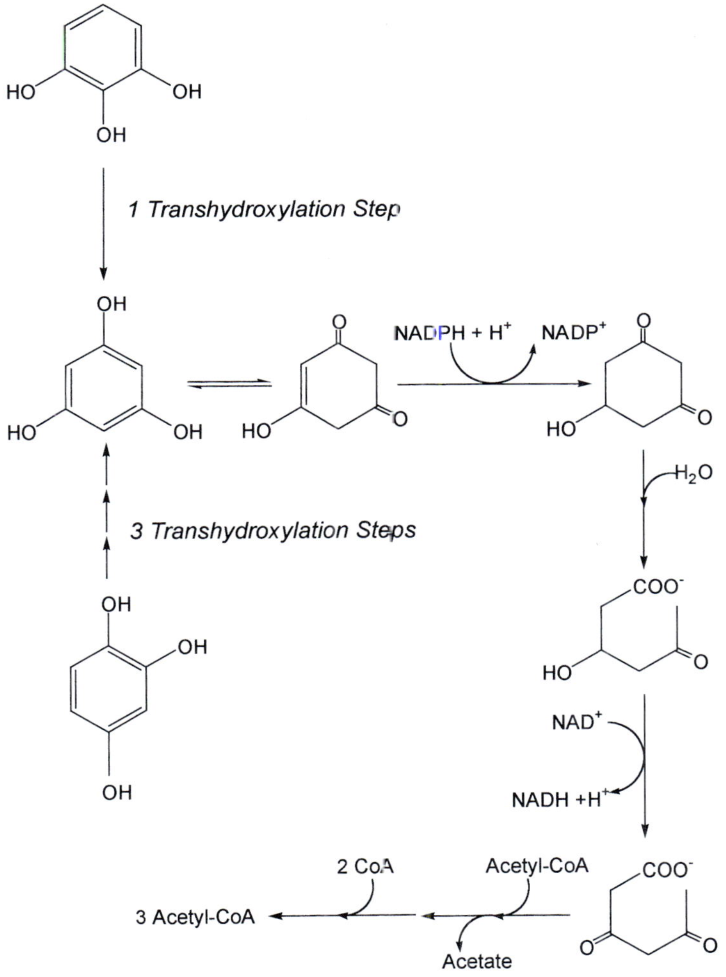

**Fig. 8.9** Degradation of trihydroxybenzenes by fermenting bacteria.

### 8.2.6.7  **Hydroxyhydroquinone, a New Important Intermediate**

The third trihydroxybenzene isomer, hydroxyhydroquinone, is converted by the fermenting bacterium *Pelobacter massiliensis* to three acetates via three subsequent transhydroxylation reactions, analogous to the pyrogallol–phloroglucinol transhydroxylation (Schnell et al., 1991; Brune et al., 1992). Alternative pathways of hydroxyhydroquinone degradation were found in nitrate-reducing and sulfate-reducing bacteria. Hydroxyhydroquinone degradation by nitrate-reducing bacteria was mentioned above in the context of nitrate-dependent resorcinol degradation (Philipp and Schink, 1998). The reaction sequence leads to an acetate and a succinate residue, suggesting that the hydroxyhydroquinone intermediate is cleaved between carbon atoms 1 and 2 and between 3 and 4 (Fig. 8.10). Another alternative pathway of hydroxyhydroquinone degradation was found in the sulfate-reducing bacterium *Desulfovibrio inopinatus*. This bacterium destabilizes hydroxyhydroquinone by reduction to dihydrohydroxyhydroquinone, to form acetate and an as-yet unidentified 4-carbon derivative (Reichenbecher et al., unpublished data). Since *D. inopinatus* is unable to

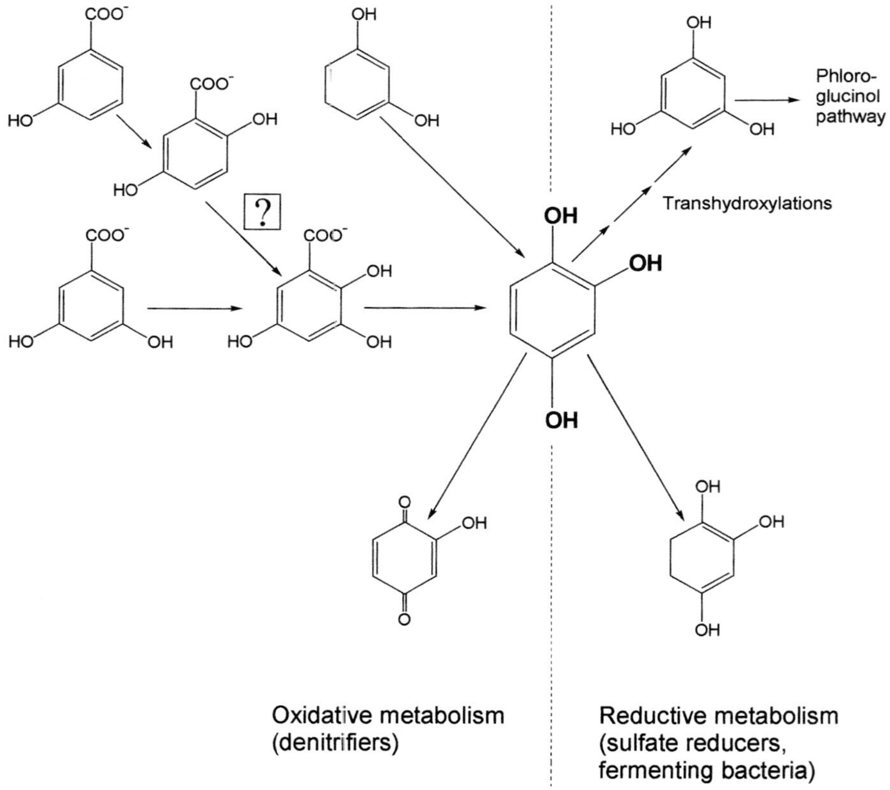

**Fig. 8.10** Hydroxyhydroquinone as a new intermediate in anaerobic degradation of various aromatic compounds.

oxidize acetate, the final products are two acetate and two $CO_2$, and one mole of sulfate is reduced concomitantly to sulfide.

Hydroxyhydroquinone has gained additional interest recently because it was found to be an intermediate in the nitrate-dependent degradation of resorcinol, 3-hydroxybenzoate, 3,5-dihydroxybenzoate, and perhaps also gentisate. The strategy of oxidative destabilization of aromatic compounds through hydroxylation, which these nitrate-reducing bacteria use, resembles to some extent the strategy of aerobic bacteria, and nitrate-dependent degradation of phenolic compounds thus follows a strategy that is somewhat of a mix between the oxidative aerobic strategy and the typical reductive strategy followed by strictly anaerobic bacteria.

### 8.2.6.8  Aromatic Hydrocarbons

Anaerobic degradation of aromatic hydrocarbons has been a matter of dispute for several years until reliable conversion balances with fast growing enrichment cultures or pure cultures were provided. Today, several pure cultures of nitrate-reducing or sulfate-reducing bacteria are available which oxidize toluene (methylbenzene) and have been characterized in detail (Dolfing et al., 1990; Schocher et al., 1991; Evans et al., 1991; Altenschmidt and Fuchs, 1991; Seyfried et al., 1994; Rabus et al., 1993; Rabus and Widdel, 1995; Beller et al., 1996; Spormann and Widdel, 2000). Initial experiments indicated that toluene degradation proceeded through oxidation of the methyl group (Altenschmidt and Fuchs, 1991), with benzoyl-CoA as central intermediate. Nonetheless, the anticipated methyl hydroxylation reaction transforming toluene to benzylalcohol could never be observed in vitro. Labeling experiments with intact cells provided evidence that toluene was activated by addition of fumarate, through a radical intermediate, to form benzyl succinate (Fig. 8.11a; Biegert et al., 1996), and this mechanism was confirmed in nitrate-reducing and sulfate-reducing bacteria (Beller and Spormann, 1997a, b, 1998). Benzylsuccinate synthase has been characterized in great detail in the meantime (Achong et al., 2001; Verfurth et al., 2004). Benzylsuccinate releases succinyl-CoA through beta oxidation, leading again to benzoyl-CoA as key intermediate. This new type of methyl group activation is also employed in anaerobic degradation of alkanes and of *m*–cresol (see above).

Anaerobic degradation of *o*-, *m*-, and *p*-xylene has been documented mostly by tracer experiments with sediment samples or in enrichment cultures. Pure cultures of sulfate-reducing xylene degraders are now available (Harms et al., 1999), and evidence is available that at least *m*-xylene is activated by addition to fumarate (Krieger et al., 1999; Verfurth et al., 2004).

Ethylbenzene is oxidized by denitrifying and sulfate-reducing bacteria (Rabus and Widdel, 1995; Ball et al., 1996; Rabus and Heider, 1998). Here, the sidechain is hydroxylated to form 1-phenylethanol as the first oxidation product (Fig. 8.11b; Ball et al., 1996; Rabus and Heider, 1998; Johnson and Spormann, 1999). The oxidizing enzyme is a novel molybdenum enzyme (Johnson et al., 2001). The subsequent pathway leads via oxidation to acetophenone and carboxylation to benzoylacetate and then through thiolytic cleavage to an acetyl residue plus benzoyl-CoA (Heider et al., 1999).

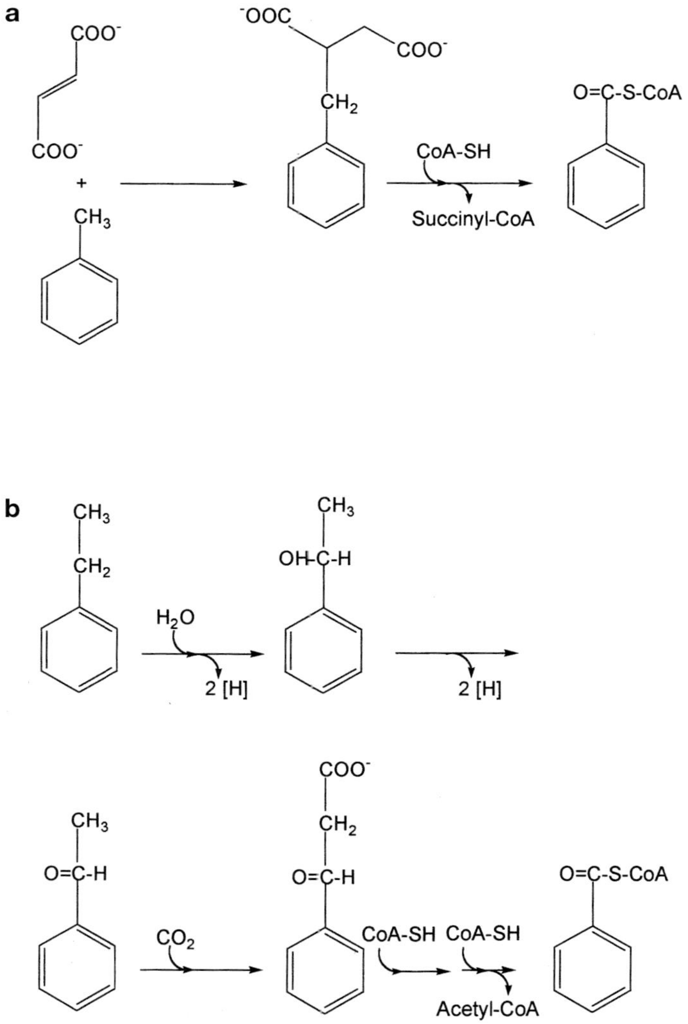

**Fig. 8.11** Anaerobic degradation of (a) toluene and (b) ethylbenzene by nitrate-reducing or sulfate-reducing bacteria.

Much less is known about anaerobic benzene oxidation. Oxygen-independent hydroxylation and degradation of benzene was observed in methanogenic enrichment cultures (Vogel and Grbić-Galić, 1986; Edwards and Grbić-Galić, 1994) and in sulfate-reducing (Lovley et al., 1995), nitrate-reducing (Burland and Edwards, 1999), or iron(III)-reducing enrichments (Lovley et al., 1996). Isolation of a pure culture of nitrate-reducing benzene degraders has been reported recently (Coates et al., 2001), but the biochemistry of benzene activation still remains an open question.

Naphthalene is degraded anaerobically by various enrichments and also by pure cultures of sulfate-reducing bacteria (Galushko et al., 1999). Its degradation is probably initiated by a carboxylation reaction (Zhang and Young, 1997) and subsequent reductive dearomatization (Annweiler et al., 2002), but the details of this pathway have still to be worked out.

## 8.2.7
### Degradation of Halogenated Organics

Halogenated organics are widespread in nature and are formed especially as secondary metabolites by plants, marine algae, fungi, and certain bacteria at low rates (for a review, see Hutzinger, 1982). It is not surprising, therefore, that a broad variety of bacteria and fungi can degrade such compounds and that this capacity has grown in the past to include also the majority of synthetic halogenated compounds. Dehalogenation can proceed basically through an oxidative, a hydrolytic, or a reductive reaction (Fetzner and Lingens, 1994; El Fantroussi et al., 1998). Among anaerobic bacteria, the reductive elimination of halogen substituents is the most common type of reaction and was first observed in enrichment cultures with 3,5-dihydroxybenzoate (Suflita et al., 1982). Later, several other aliphatic and aromatic compounds were found to be dehalogenated (mainly dechlorinated) in anoxic incubation experiments (Bouwer and McCarty, 1983), and today nearly all chlorinated organics can be dehalogenated and further degraded in strictly anaerobic microbial cultures. As a rule, reductive dechlorination is the preferred process for degradation of compounds with higher degrees of halogenation. The reaction is an excellent electron sink ($E_h$ +250 to +580 mV), and highly halogenated compounds are much more amenable to a nucleophilic attack on the respective carbon atom than to an oxidative reaction.

The overall reaction of reductive dehalogenation can be diagrammed as in Figure 8.12 for a chlorinated compound: electrons derived from molecular hydrogen, formate, or more complex organic compounds are transferred to the halogenated substrate to release the organic residue in a reduced form, together with chloride. It has been shown in several instances that this redox process can yield metabolic energy through a respiratory mechanism, which implies that the process may establish a net translocation of protons across the cytoplasmic membrane and that the proton gradient drives ATP synthesis through a membrane-bound ATP synthase complex. The biochemistry of the dechlorination reaction has been studied best with bacteria converting tetrachloroethene via trichloroethene to dichloroethene (Holliger et al., 1999). The dechlorination reaction employs a corrinoid as cofactor, which attacks the carbon-chlorine linkage in its reduced Co(I) form and converts it to the Co(III) form, probably through two subsequent one-electron transfer reactions (Schumacher et al., 1997; Neumann et al., 1998; Neumann et al., 2002; Maillard et al., 2003). Whether this reaction mechanism is used also in the dehalogenation of aromatic compounds has still to be elucidated. There appear to be numerous differences with respect to the electron carriers involved and the spatial arrangement of the enzyme components in the cytoplasmic membrane.

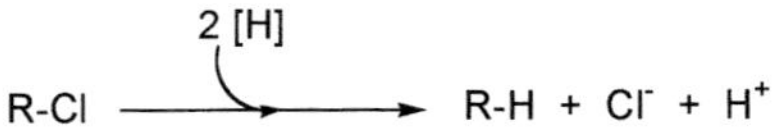

$$R\text{-}Cl \xrightarrow{\quad 2\,[H] \quad} R\text{-}H + Cl^- + H^+$$

**Fig. 8.12** Reductive dehalogenation of a chloro-organic compound.

Halo-organics can also be reductively dehalogenated by reduced Fe(II) phases (Pecher et al., 2002; Elsner et al., 2004) that are formed on the surface of Fe(III) minerals.

### 8.2.8
### Degradation of Sulfonates

Sulfonated organics are rare in nature: only taurine, coenzyme M, cysteate, and a few secondary metabolites are known to contain sulfonate substituents. Aerobic degradation of such compounds typically requires an oxygenase reaction, which hydroxylates the neighboring carbon atom and releases sulfite (Cook et al., 1999). Some sulfonates can also be partly degraded in the absence of molecular oxygen and can serve as sulfur sources under conditions of sulfur limitation (Chien et al., 1995; Cook et al., 1999). The biochemistry of anaerobic sulfur release from sulfonates has been studied extensively with taurine and related compounds, which are metabolized through sulfoacetaldehyde as key intermediate (Laue et al., 1997a, b). The sulfono group is released from sulfoacetaldehyde probably through a thiamine pyrophosphate-dependent reaction that forms sulfite and an acetyl residue, analogous to the reaction elucidated earlier with an aerobic bacterium (Kondo and Ishimoto, 1975). Since this desulfonation reaction requires an oxo group positioned β to the sulfur atom for linkage to the coenzyme, it remains questionable whether this concept can also be applied to desulfonation of commercial sulfonates, such as alkyl sulfonates or alkylbenzene sulfonates, which so far resist anaerobic degradation.

### 8.2.9
### Degradation of Nitroorganics

Among the nitroaromatic compounds, trinitrotoluene (TNT) is of major importance as a soil pollutant because it has accumulated at old ammunition factory sites over several decades. The electron-withdrawing effect of the nitro substituents makes an oxidative attack on nitroaromatics rather difficult. Indeed, trinitrotoluene is attacked by aerobic bacteria primarily through a reductive reaction that transforms nitroaromatics to the corresponding amino derivatives (Naumova et al., 1989) or reductively eliminates the nitro groups via a so-called Meisenheimer complex (Vorbeck et al., 1994). In the presence of oxygen, the partly reduced derivatives can react with each other to form a rather inert polymer (Heiss and Knackmuss, 2002). A reductive approach is taken also by strictly anaerobic bacteria, e.g., sulfate-reducing bacteria: trinitrotoluene is converted via diaminonitrotoluene to triaminotoluene. Whereas the first step can also be catalyzed purely chemically without the participation of microbial cells or enzymes, reduction of diaminonitrotoluene to triaminotoluene requires the participation of microbial cells or enzyme fractions (Preuss et al., 1993). The further fate of triaminotoluene is unclear. It is partly utilized as a nitrogen source by a sulfate-reducing bacterium; the remnant product possibly polymerizes, especially in the presence of traces of oxygen (Preuss et al., 1993). The same is likely to happen in contaminated soils; so far, there is no reliable proof of complete degradation of TNT by anaerobic bacteria or by anaerobic and aerobic bacteria cooperating in a two-step process (see Chapter 11, this volume).

## 8.3
## Concluding Remarks

Anaerobic degradation can be applied in technical devices for treatment of waste material, often leading to $CH_4$ and $CO_2$ as products, which can be exploited as energy source or as a basis for biosynthetic processes. Moreover, anaerobic degradation proceeds in many anoxic habitats such as the intestinal tracts of humans and animals, sediments, and oxygen-deprived microenvironments in soil, sewage sludge, etc. Knowledge of the capacities, strategies and limits of anaerobic degradation processes is therefore needed to assess the potential risk of synthetic compounds to health or to the environment, no matter whether such synthetics are released intentionally (as with plant protection agents), inadvertently through wastewater treatment, or accidentally through spills.

This overview shows that the degradative potential of anaerobic microbial communities is much greater than assumed only a few years ago: a broad variety of compounds can be subject to anaerobic degradation, most often down to methane and carbon dioxide as final products. Aliphatic hydrocarbons are degraded if they contain unsaturated bonds, preferentially if these are located terminally, but saturated long-chain aliphatic hydrocarbons are also anaerobically degradable. These processes are slow and can be applied only in long-term incubations, if at all. Ether compounds are degraded anaerobically if they are methyl ethers or if they can be transformed into hemiacetals through, e.g., hydroxyl shift reactions. In the anaerobic degradation of ketones, the primary activation reaction is a carboxylation rather than an oxidation step.

Mononuclear aromatic compounds can be degraded anaerobically rather efficiently if they carry at least one carboxy, hydroxy, methoxy, amino, or methyl substituent, and four major degradation pathways have been elucidated in the recent past, which differ basically from the well known aerobic oxygenase-dependent pathways. The degradation kinetics differ considerably, depending on the sites of substitution.

Halogenated aliphatics and aromatics are reductively dehalogenated, more efficiently and better than by aerobes, the higher the degree of halogenation. Anaerobic degradation of sulfonates appears to be restricted to only a few compounds, whereas the majority of synthetic sulfonates (detergents) are degraded efficiently only in the presence of oxygen. Nitro-substituted compounds are preferentially attacked through reduction, and anaerobic processes therefore appear to be advantageous over aerobic ones. The same applies to azo compounds, which are not discussed here.

Several types of reactions were identified which activate or destabilize comparably inert substrates in the absence of oxygen. Among these are carboxylations, addition to fumarate, reductions and reductive eliminations, rearrangements of aliphatic carbon skeletons, cobalamin-dependent nucleophilic substitutions, and oxygen-independent hydroxylations. Numerous reactions proceed through radical mechanisms, and the diversity of radical chemistry in the absence of oxygen appears to be considerably greater than in its presence.

Transformation of polymeric compounds is restricted in the anaerobic world to extracellular hydrolysis reactions unless the polymer can be taken up into the cell, as occurs with polyethylene glycol. There is no equivalent in the anoxic world to the fungal lignin-degrading enzyme apparatus. Therefore, polynuclear aromatics (lignin, other polyphenols) remain comparably recalcitrant in anoxic environments and represent barriers to microbial attack in the absence of molecular oxygen.

## Acknowledgments

Experimental work in my laboratory has been funded mainly by the Deutsche Forschungsgemeinschaft, the Bundesministerium für Forschung und Technologie, the Fonds der Chemischen Industrie, and the University of Konstanz. Bodo Philipp and Jochen Müller designed the figures. Thanks are also due to the dedicated and talented graduate students who carried out the experimental work and contributed to the concepts with their own ideas. This overview is based on two review articles from our group (Schink, 1995; Schink et al., 2000), which cover special aspects of this survey.

## References

Achong, G. R., Rodriguez, A. M., Spormann, A. M., Benzylsuccinate synthase of *Azoarcus* sp. strain T: cloning, sequencing transcriptional organization, and its role in anaerobic toluene and *m*-xylene mineralization, *J. Bacteriol.* **2001**, *183*, 6763–5770.

Aeckersberg, F., Bak, F., Widdel, F. Anaerobic oxidation of saturated hydrocarbons to CO₂ by a new type of sulfate-reducing bacterium, *Arch. Microbiol.* **1991**, *156*, 5–14.

Aeckersberg, F., Rainey, F. A., Widdel, F., Growth, natural relationships, cell fatty acids and metabolic adaptation of sulfate-reducing bacteria utilizing long-chain alkanes under anoxic conditions, *Arch. Microbiol.* **1998**, *170*, 361–369.

Altenschmidt, U., Fuchs, G., Anaerobic degradation of toluene in denitrifying *Pseudomonas* sp.: indication of toluene methylhydroxylation and benzoyl-CoA as central aromatic intermediate, *Arch. Microbiol.* **1991**, *156*, 152–158.

Annweiler, E., Michaelis, W., Meckenstock, R. U., Identical ring cleavage products during anaerobic degradation of naphthalene, 2-methylnaphthalene, and tetralin indicate a new metabolic pathway, *Appl. Environ. Microbiol.* **2002**, *68*, 852–858.

Bache, R., Pfennig, N., Selective isolation of *Acetobacterium woodii* on methoxylated aromatic acids and determination of growth yields, *Arch. Microbiol.* **1981**, *130*, 255–261.

Ball, H. D., Johnson, H. A., Reinhard, M., Spormann, A. M., Initial reactions in anaerobic ethylbenzene oxidation by a denitrifying bacterium, strain EB1, *J. Bacteriol.* **1996**, *178*, 5755– 5761.

Beller, H. R., Spormann, A. M., Anaerobic activation of toluene and *o*-xylene by addition to fumarate in denitrifying strain T, *J. Bacteriol.* **1997a**, *179*, 670–676.

Beller, H. R., Spormann, A. M., Benzylsuccinate formation as a means of anaerobic toluene activation by sulfate-reducing strain PRTOL1, *Appl. Environ. Microbiol.* **1997b**, *63*, 3729–3731.

Beller, H. R., Spormann, A. M., Analysis of the novel benzyl succinate synthase reaction for anaerobic toluene activation based on structural studies of the product, *J. Bacteriol.* **1998**, *180*, 5454– 5457.

Beller, H. R., Spormann, A. M., Sharma, P. K., Cole, J. R., Reinhard, M., Isolation and characterization of a novel toluene-degrading sulfate-reducing bacterium, *Appl. Environ. Microbiol.* **1996**, *62*, 1188–1196.

Berman, M. H., Frazer, A. Z., Importance of tetrahydrofolate and ATP in the anaerobic O-demethylation reaction for phenylmethylethers, *Appl. Environ. Microbiol.* **1992**, *58*, 925–931.

Bernhardt, F. H., Staudinger, H., Ullrich, V., Eigenschaften einer *p*-Anisat-O-Demethylase im zellfreien Extrakt von *Pseudomonas* species, *Hoppe-Seyler's Z. Physiol. Chem.* **1970**, *351*, 467–478.

Biegert, T., Fuchs, G., Heider, J., Evidence that oxidation of toluene in the denitrifying bacterium *Thauera aromatica* is initiated by formation of benzylsuccinate from toluene and fumarate, *Eur. J. Biochem.* **1996**, *238*, 661–668.

Birks, S. J., Kelly, D. J., Assay and properties of acetone carboxylase, a novel enzyme involved in acetone-dependent growth and $CO_2$ fixation in *Rhodobacter capsulatus* and other photosynthetic and denitrifying bacteria. *Microbiology* **1997**, *143*, 755–766.

Bisaillon, J. G., Lépine, F., Beaudet, R., Sylvestre, M., Carboxylation of *o*-cresol by an anaerobic consortium under methanogenic conditions, *Appl. Environ. Microbiol.* **1991**, *57*, 2131–2134.

Boetius, A., Ravenschlag, K., Schubert, C. J., Rickert, D., Widdel, F., Gieseke, A., Amann, R., Jorgensen, B. B., Witte, U., Pfannkuche, O., A marine microbial consortium apparently mediating anaerobic oxidation of methane, *Nature* **2000**, *407*, 623–626.

Boll, M., Fuchs, G., Benzoyl-CoA reductase (dearomatizing), a key enzyme of anaerobic aromatic metabolism: ATP dependence of the reaction, purification and some properties of the enzyme from *Thauera aromatica* strain K172, *Eur. J. Biochem.* **1995**, *234*, 921–933.

Boll, M., Fuchs, G., Tilley, G., Armstrong F. A., Lowe, D. J., Unusual spectroscopic and electrochemical properties of the 2[4Fe-4S] ferredoxin of *Thauera aromatica*, *Biochemistry* **2000**, *39*, 4929–4938.

Boll, M., Laempe, D., Eisenreich, W., Bacher, A., Mittelberger, T., Heinze J., Fuchs, G., Nonaromatic products from anoxic conversion of benzoyl-CoA with benzoyl-CoA reductase and cyclohexa-1,5-diene-1-carbonyl-CoA hydratase, *J. Biol. Chem.* **2000**, *275*, 21889–21895.

Bonnet-Smits, E. M., Robertson, L. A., Van Dijken, J. P., Senior, E., Kuenen, J. G., Carbon dioxide fixation as the initial step in the metabolism of acetone by *Thiosphaera pantotropha*, *J. Gen. Microbiol.* **1988**, *134*, 2231–2289.

Bouwer, E. J., McCarty, P. L., Transformation of halogenated organic compounds under denitrification conditions, *Appl. Environ. Microbiol.* **1983**, *45*, 1295–1299.

Breinig, S., Schiltz, E., Fuchs, G., Genes involved in anaerobic metabolism of phenol in the bacterium *Thauera aromatica*, *J. Bacteriol.* **2000**, *182*, 5849–5863.

Brune, A., Schink, B., Pyrogallol-to-phloroglucinol conversion and other hydroxyl-transfer reactions catalyzed by cell extracts of *Pelobacter acidigallici*, *J. Bacteriol.* **1990**, *172*, 1070–1076.

Brune, A., Schink, B., Phloroglucinol pathway in the strictly anaerobic *Pelobacter acidigallici*: fermentation of trihydroxybenzenes to acetate via triacetic acid, *Arch. Microbiol.* **1992**, *157*, 417–424.

Brune, A., Schnell, S., Schink, B., Sequential transhydroxylations converting hydroxy hydroquinone to phloroglucinol in the strictly anaerobic fermenting bacterium, *Pelobacter massiliensis*, *Appl. Environ. Microbiol.* **1992**, *58*, 1861–1868.

Burland, S. M., Edwards, E. A., Anaerobic benzene biodegradation linked to nitrate reduction, *Appl. Environ. Microbiol.* **1999**, *65*, 529–533.

Chien, C.-C., Leadbetter, E. R., Godchaux III, W., Sulfonate sulfur can be assimilated for fermentative growth, *FEMS Microbiol. Lett.* **1995**, *129*, 189–194.

Coates, J. D., Chakraborty, R., Lack, J. G., O'Connor, S. M., Cole, K. A., Bender, K. S., Achenbach, L. A., Anaerobic benzene oxidation coupled to nitrate reduction in pure culture by two strains of *Dechloromonas*, *Nature* **2001**, *411*, 1039–1043.

Cook, A. M., Laue, H., Junker, F., Microbial desulfonation, *FEMS Microbiol. Rev.* **1999**, *22*, 399–419.

Dagley, S., Catabolism of aromatic compounds by microorganisms, *Adv. Microb. Physiol.* **1971**, *6*, 1–46.

Dolfing, J., Zeyer, P., Binder-Eicher, P., Schwarzenbach, R. P., Isolation and characterization of a bacterium that mineralizes toluene in the absence of molecular oxygen, *Arch. Microbiol.* **1990**, *154*, 336–341.

Edwards, E. A., Grbić-Galić, D., Anaerobic degradation of toluene and *o*-xylene by a methanogenic consortium, *Appl. Environ. Microbiol.* **1994**, *60*, 313–322.

Egli, T., Bally, M., Uetz, T., Microbial degradation of chelating agents used in detergents with special reference to nitrilotriacetic acid (NTA), *Biodegradation* **1990**, *1*, 121–132.

El Fantroussi, S., Naveau, H., Agathos, S. N., Anaerobic dechlorinating bacteria, *Biotechnol. Prog.* **1998**, *14*, 167–188.

Engelmann, T., Kaufmann, F., Diekert, G., Isolation and characterization of a veratrol:corrinoid protein methyl transferase from *Acetobacterium dehalogenans*, *Arch. Microbiol.* **2001**, *175*, 376–383.

Elsner, M., Schwarzenbach, R. P., Haderlein, S. B., Reactivity of Fe(II)-bearing minerals toward reductive transformation of organic contaminants, *Environ. Sci. Technol.* **2004**, *38*, 799–807.

Evans, P. J., Mang, D. T., Kim, K. S., Young, L. Y., Anaerobic degradation of toluene by a denitrifying bacterium, *Appl. Environ. Microbiol.* **1991**, *57*, 1139–1145.

Evans, W. C., Biochemistry of the bacterial catabolism of aromatic compounds in anaerobic environments, *Nature* **1977**, *270*, 17–22.

Evans, W. C., Fuchs, G., Anaerobic degradation of aromatic compounds, *Annu. Rev. Microbiol.* **1988**, *42*, 289–317.

Fetzner, S., Lingens, F., Bacterial dehalogenases: biochemistry, genetics, and biotechnological applications, *Microbiol. Rev.* **1994**, *58*, 641–685.

Foß, S., Harder, J., *Thauera linaloolentis* sp. nov. and *Thauera terpenica* sp. nov., isolated on oxygen-containing monoterpenes (linalool, menthol, and eucalyptol) and nitrate, *Syst. Appl. Microbiol.* **1998**, *21*, 365–373.

Foß, S., Heyen, U., Harder, J., *Alcaligenes defragrans* sp. nov., description of four strains isolated on alkenoic monoterpenes (((*c*)-menthene, α-pinene, 2-carene and α-phellandrene) and nitrate, *Syst. Appl. Microbiol.* **1998**, *21*, 237–244.

Frings, J., Schramm, E., Schink, B., Enzymes involved in anaerobic polyethylene glycol degradation by *Pelobacter venetianus* and *Bacteroides* strain PG1, *Appl. Environ. Microbiol.* **1992**, *58*, 2164–2167.

Fuchs, G., El-Said, M., M., Altenschmidt, U., Koch, J., Lack, A. et al., Biochemistry of anaerobic biodegradation of aromatic compounds, in: *Biochemistry of Microbial Degradation* (Ratledge, C., ed.), pp. 513–553. Dordrecht **1994**: Kluwer.

Gallert, C., Knoll, G., Winter, J., Anaerobic carboxylation of phenol to benzoate: use of deuterated phenols revealed carboxylation exclusively in the C4 position, *Appl. Microbiol. Biotechnol.* **1991**, *36*, 124–129.

Galushko, A., Minz, D., Schink, B., Widdel, F., Anaerobic degradation of naphthalene by a pure culture of a novel type of marine sulfate-reducing bacterium, *Environ. Microbiol.* **1999**, *1*, 415–420.

Gorny, N., Schink, B., Hydroquinone degradation via reductive dehydroxylation of gentisyl-CoA by a strictly anaerobic fermenting bacterium, *Arch. Microbiol.* **1994a**, *161*, 25–32.

Gorny, N., Schink, B., Complete anaerobic oxidation of hydroquinone by *Desulfococcus* sp. strain Hy5: indications of hydroquinone carboxylation to gentisate, *Arch. Microbiol.* **1994b**, *162*, 131–135.

Gorny, N., Schink, B., Anaerobic degradation of catechol by *Desulfobacterium* sp. strain Cat2 proceeds via carboxylation to protocatechuate, *Appl. Environ. Microbiol.* **1994c**, *60*, 3396–3340.

Gorny, N., Wahl, G., Brune, A., Schink, B., A strictly anaerobic nitrate-reducing bacterium growing with resorcinol and other aromatic compounds, *Arch. Microbiol.* **1992**, *158*, 48–53.

Haddock, J. D., Ferry, J. G., Purification and properties of phloroglucinol reductase from *Eubacterium oxidoreducens* G-41, *J. Biol. Chem.* **1989**, *264*, 4423–4427.

Harder, J., Anaerobic methane oxidation by bacteria employing $^{14}$C-methane uncontaminated with $^{14}$C-carbon monoxide, *Mar. Geol.* **1997**, *137*, 13–23.

Harder, M., Probian, C., Microbial degradation of monoterpenes in the absence of molecular oxygen, *Appl. Environ. Microbiol.* **1995**, *61*, 3804–3808.

Harms, G., Zengler, K., Rabus, R., Aeckersberg, F., Minz, D. et al., Anaerobic oxidation of *o*-xylene, *m*-xylene, and homologous alkyl benzenes by new types of sulfate-reducing bacteria, *Appl. Environ. Microbiol.* **1999**, *65*, 999–1004.

Harwood, C. S., Burchhardt, G., Herrmann, H., Fuchs, G., Anaerobic metabolism of aromatic compounds via the benzoyl-CoA pathway, *FEMS Microbiol. Rev.* **1999**, *22*, 439–458.

Heider, J., Fuchs, G., Anaerobic metabolism of aromatic compounds, *Eur. J. Biochem.* **1997**, *243*, 577–596.

Heider, J., Spormann, A. M., Beller, H. R., Widdel, F., Anaerobic bacterial metabolism of hydrocarbons, *FEMS Microbiol. Rev.* **1999**, *22*, 459–473.

Heiss, G., Knackmuss, H. J., Bioelimination of trinitroaromatic compounds: immobilization versus mineralization, *Curr. Opin. Microbiol.* **2002**, *5*, 282–287.

Hippe, H., Caspari, D., Fiebig, K., Gottschalk, G., Utilization of trimethylamine and other *N*-methyl compounds for growth and methane formation by *Methanosarcina barkeri*, *Proc. Natl. Acad. Sci. USA* **1979**, *76*, 494–498.

Holliger, C., Wohlfarth, G., Diekert, G., Reductive dechlorination in the energy metabolism of anaerobic bacteria, *FEMS Microbiol. Rev.* **1999**, *22*, 383–398.

Hopper, D. J., Incorporation of [$^{18}$O]water in the formation of *p*-hydroxybenzyl alcohol by the *p*-cresol methylhydroxylase from *Pseudomonas putida*, *Biochem. J.* **1978**, *175*, 345–347.

Hopper, D. J., Bossert, I. D., Rhodes-Roberts, M. E., *p*-Cresol methylhydroxylase from a denitrifying bacterium involved in anaerobic degradation of *p*-cresol, *J. Bacteriol.* **1991**, *173*, 1298–1301.

Hutzinger, O., *The Handbook of Environmental Chemistry*. Berlin **1982**: Springer-Verlag.

Hylemon, P. B., Harder, J., Biotransformation of monoterpenes, bile acids, and other isoprenoids in anaerobic ecosystems, *FEMS Microbiol. Rev.* **1999**, *22*, 475–488.

Janssen, P. H., Schink, B., $^{14}$CO$_2$ exchange with acetoacetate catalyzed by dialyzed cell-free extracts of the bacterial strain BunN grown with acetone and nitrate, *Eur. J. Biochem.* **1995**, *228*, 677–682.

Johnson, H. A., Spormann, A. M., In vitro studies on the initial reactions of anaerobic ethylbenzene mineralization, *J. Bacteriol.* **1999**, *181*, 5662–5668.

Johnson, H. A., Pelletier, D. A., Spormann, A. M., Isolation and characterization of anaerobic ethylbenzene dehydrogenase, a novel Mo-Fe-S enzyme, *J. Bacteriol.* **2001**, *183*, 4536–4542.

Kaufmann, F., Wohlfahrt, G., Diekert, G., Isolation of O-demethylase, an ether-cleaving enzyme system of the homoacetogenic strain MC, *Arch. Microbiol.* **1997**, *168*, 136–142.

Kaufmann, F., Wohlfarth, G., Diekert, G., O-demethylase from *Acetobacterium dehalogenans*: cloning, sequencing, and active expression of the gene encoding the corrinoid protein, *Eur. J. Biochem.* **1998**, *257*, 515–521.

Keltjens, J. T., Vogels, G. D., Conversion of methanol and methylamines to methane and carbon dioxide, in: *Methanogenesis* (Ferry, J. G., ed.), pp. 253–303. New York **1993**: Chapman & Hall.

Kerem, Z., Bao, W., Hammel, K. E., Rapid polyether cleavage via extracellular one-electron oxidation by a brown-rot basidiomycete, *Proc. Natl. Acad. Sci. USA* **1998**, *95*, 10373–10377.

Kiene, R. P., Oremland, R. S., Catena, A., Miller, L. G., Capone, D. G., Metabolism of reduced methylated sulfur compounds in anaerobic sediments and by a pure culture of an estuarine methanogen, *Appl. Environ. Microbiol.* **1986**, *52*, 1037–1045.

Kluge, C., Tschech, A., Fuchs, G., Anaerobic metabolism of resorcylic acids (*m*–dihydrox–ybenzoic acids) and resorcinol (1,3-benzenediol) in a fermenting and in a denitrifying bacterium, *Arch. Microbiol.* **1990**, *155*, 68–74.

Koch, J., Eisenreich, W., Bacher, A., Fuchs, G., Products of enzymatic reduction of benzoyl-CoA, a key reaction in anaerobic aromatic metabolism, *Eur. J. Biochem.* **1993**, *211*, 649–661.

Kondo, H., Ishimoto, M., Purification and properties of sulfoacetaldehyde sulfolyase, a thiamine pyrophosphate-dependent enzyme forming sulfite and acetate, *J. Biochem.* **1975**, *78*, 317–325.

Kreft, J.-U., Schink, B., Demethylation and further degradation of phenyl methylethers by the sulfide-methylating homoacetogenic bacterium strain TMBS4, *Arch. Microbiol.* **1993**, *159*, 308–315.

Kreft, J., Schink, B., O-Demethylation by the homoacetogenic anaerobe *Holophaga foetida* studied by a new photometric methylation assay using electrochemically produced cob(I)alamin, *Eur. J. Biochem.* **1994**, *226*, 945–951.

Krieger, C. J., Beller, H. R., Reinhard, M., Spormann, A. M., Initial reactions in anaerobic oxidation of *m*-xylene by the denitrifying bacterium *Azoarcus* sp. strain T, *J. Bacteriol.* **1999**, *181*, 6403–6410.

Krumholz, L. R., Bryant, M. P., Characterization of the pyrogallol–phloroglucinol isomerase of *Eubacterium oxidoreducens*, *J. Bacteriol.* **1988**, *170*, 2472–2479.

Lack, A., Fuchs, G., Carboxylation of phenylphosphate by phenol caboxylase, an enzyme system of anaerobic phenol metabolism in a denitrifying *Pseudomonas* sp., *J. Bacteriol.* **1992**, *174*, 3629–3636.

Lack, A., Fuchs, G., Evidence that phenol phosphorylation to phenylphosphate is the first step in anaerobic phenol metabolism in a denitrifying *Pseudomonas* sp , *Arch. Microbiol.* **1994**, *161*, 306–311.

Laue, H., Denger, K., Cook, A., Taurine reduction in anaerobic respiration of *Bilophila wadsworthia* RZA-TAU, *Appl. Environ. Microbiol.* **1997a**) *63*, 2016–2021.

Laue, H., Denger, K., Cook, A., Fermentation of cysteine by a sulfate-reducing bacterium, *Arch. Microbiol.* **1997b**) *168*, 210–214.

Lengeler, J. W., Drews, G., Schlegel, H. G. (eds.), *Biology of the Prokaryotes*, Stuttgart **1999**: Georg Thieme Verlag.

Lovley, D. R., Coates, J. D., Woodward, J. C., Phillips, E. J. P., Benzene oxidation coupled to sulfate reduction, *Appl. Environ. Microbiol.* **1995**, *61*, 953–958.

Lovley, D. R., Woodward, J. C., Chapelle, F. H., Rapid anaerobic benzene oxidation with a variety of chelated Fe(III) forms. *Appl. Environ. Microbiol.* **1996**, *62*, 288–291.

Lukins, H. B., Foster, J. W., Methylketone metabolism in hydrocarbon utilizing mycobacteria, *J. Bacteriol.* **1963**, *85*, 1074–1087.

Maillard, J., Schumacher, W., Vazquez, F., Regeard, C., Hagen, W. R., Holliger, C., Characterization of the corrinoid iron-sulfur protein tetrachloroethene reductive dehalogenase of *Dehalobacter restrictus*, *Appl. Environ. Microbiol.* **2003**, *69*, 4628–4638.

Meckenstock, R. U., Krieger, R., Ensign, S., Kroneck, P. M. H., Schink, B., Acetylene hydratase of *Pelobacter acetylenicus*: molecular and spectroscopic properties of the tungsten iron-sulfur enzyme, *Eur. J. Biochem.* **1999**, *264*, 176–182.

Müller, J. A., Galushko, A. S., Kappler, A., Schink, B., Anaerobic degradation of *m*-cresol by *Desulfobacterium cetonicum* is initiated by formation of 3-hydroxybenzylsuccinate, *Arch. Microbiol.* **1999**, *172*, 287–294.

Müller, J. A., Schink, B., Initial steps in the fermentation of 3-hydroxybenzoate by *Sporotomaculum hydroxybenzoicum*, *Arch. Microbiol.* **2000**, *173*, 288–295.

Nauhaus, K., Boetius, A., Kruger, M., Widdel, F., In vitro demonstration of anaerobic oxidation of methane coupled to sulphate reduction in sediment from a marine gas hydrate area, *Environ. Microbiol.* **2002**, *4*, 296–305.

Naumova, R. P., Selivanovskaya, S. Y., Cherepneva, I. E., Conversion of 2,4,6-trinitrotoluene under conditions of oxygen and nitrate respiration of *Pseudomonas fluorescens*, *Appl. Biochem. Microbiol.* **1989**, *24*, 409–413.

Neumann, A., Siebert, A., Trescher, T., Reinhardt, S., Wohlfarth, G., Diekert, G., Tetrachloroethene reductive dehalogenase of *Dehalospirillum multivorans*: substrate specificity of the native enzyme and its corrinoid cofactor, *Arch. Microbiol.* **2002**, *177*, 420–426.

Neumann, A., Wohlfarth, G., Diekert, G., Tetrachloroethene dehalogenase from *Dehalospirillum multivorans*: cloning, sequencing of the encoding genes, and expression of the *pceA* gene in *Escherichia coli*, *J. Bacteriol.* **1998**, *180*, 4140–4145.

Nörtemann, B., Total degradation of EDTA by mixed cultures and a bacterial isolate, *Appl. Environ. Microbiol.* **1992**, *58*, 671–676.

Oremland, R. S., Kiene, R. P., Mathrani, I., Whiticar, M. J., Boone, D. R., Description of an estuarine methylotrophic methanogen which grows on dimethylsulfide, *Appl. Environ. Microbiol.* **1989**, *55*, 994–1002.

Pecher, K., Haderlein, S. B., Schwarzenbach, R. P., Reduction of polyhalogenated methanes by surface-bound Fe(II) in aqueous suspensions of iron oxides, *Environ. Sci. Technol.* **2002**) *36*, 1734–1741.

Peters, F., Rother, M., Boll, M., Selenocysteine-containing proteins in anaerobic benzoate metabolism of *Desulfococcus multivorans*, *J. Bacteriol.* **2004**, *186*, 2156–2163.

Philipp, B., Schink, B., Evidence of two oxidative reaction steps initiating anaerobic degradation of resorcinol (1,3-dihydroxybenzene) by the denitrifying bacterium *Azoarcus anaerobius*, *J. Bacteriol.* **1998**, *180*, 3644–3649.

Platen, H., Schink, B., Methanogenic degradation of acetone by an enrichment culture, *Arch. Microbiol.* **1987**, *149*, 136–141.

Platen, H., Schink, B., Anaerobic degradation of acetone and higher ketones via carboxylation by newly isolated denitrizing bacteria, *J. Gen. Microbiol.* **1989**, *135*, 883–891.

Platen, H., Schink, B., Enzymes involved in anaerobic degradation of acetone by a denitrifying bacterium, *Biodegradation* **1991**, *1*, 243–251.

Platen, H., Temmes, A., Schink, B., Anaerobic degradation of acetone by *Desulfococcus biacutus* sp. nov., *Arch. Microbiol.* **1990**, *154*, 355–361.

Preuss, A., Fimpel, J., Diekert, G., Anaerobic transformation of 2,4,6-trinitrotoluene (TNT), *Arch. Microbiol.* **1993**, *159*, 345–355.

Rabus, R., Heider, J., Initial reactions of anaerobic metabolism of alkyl benzenes in denitrifying and sulfate-reducing bacteria, *Arch. Microbiol.* **1998**, *170*, 337–384.

Rabus, R., Widdel, F., Anaerobic degradation of ethyl benzene and other aromatic hydrocarbons by new denitrifying bacteria, *Arch. Microbiol.* **1995**, *163*, 96–103.

Rabus, R., Nordhaus, R., Ludwig, W., Widdel, F., Complete oxidation of toluene under strictly anaerobic conditions by a new sulfate-reducing bacterium, *Appl. Environ. Microbiol.* **1993**, *59*, 1444–1451.

Rabus, R., Wilkes, H., Behrends, A., Armstroff, A., Fischer, T., Pierik, A. J., Widdel, F., Anaerobic initial reaction of *n*-alkanes in a denitrifying bacterium: evidence for (1-methylpentyl)succinate as initial product and for involvement of an organic radical in *n*-hexane metabolism, *J. Bacteriol.* **2001**, *183*, 1707–1715.

Reichenbecher, W., Brune, A., Schink, B., Transhydroxylase of *Pelobacter acidigallici*: a molybdoenzyme catalyzing the conversion of pyrogallol to phloroglucinol, *Biochim. Biophys. Acta* **1994**, *1204*, 217–224.

Reichenbecher, W., Rüdiger, A., Kroneck, P. M. H., Schink, B., One molecule of molybdopterin guanine dinucleotide is associated with each subunit of the heterodimeric Mo-Fe-S protein transhydroxylase of *Pelobacter acidigallici* as determined by SDS/PAGE and mass spectrometry, *Eur. J. Biochem.* **1996**, *237*, 406–413.

Rosner, B., Schink, B., Purification and characterization of acetylene hydratase of *Pelobacter acetylenicus*, a tungsten iron-sulfur protein, *J. Bacteriol.* **1995**, *177*, 5767–5772.

Rudolphi, A., Tschech, A. Fuchs, G., Anaerobic degradation of cresols by denitrifying bacteria, *Arch. Microbiol.* **1991**, *155*, 238–248.

Rueter, P., Rabus, R., Wilkes, H., Aeckersberg, F., Rainey, F. A. et al., Anaerobic oxidation of hydrocarbons in crude oil by denitrifying bacteria, *Nature* **1994**, *372*, 445–458.

Schink, B., Degradation of unsaturated hydrocarbons by methanogenic enrichment cultures, *FEMS Microbiol. Ecol.* **1985a**, *31*, 69–77.

Schink, B., Fermentation of acetylene by an obligate anaerobe, *Pelobacter acetylenicus* sp. nov., *Arch. Microbiol.* **1985b**, *142*, 295–301.

Schink, B., Principles and limits of anaerobic degradation: environmental and technological aspects, in: *Biology of Anaerobic Microorganisms* (Zehnder, A. J. B., ed.), pp. 771–846. New York **1988**: Wiley.

Schink, B., Chances and limits of anaerobic degradation of organic compounds, in: *Mikrobielle Eliminierung chlororganischer Verbindungen* (Cuno, M., ed.), Schriftenreihe Biologische Abwasserreinigung Vol. 6, pp. 57–67. Berlin **1995**: Technische Universität.

Schink, B., Energetics of syntrophic cooperations in methanogenic degradation, *Microbiol. Mol. Biol. Rev.* **1997**, *61*, 262–280.

Schink, B., Pfennig, N., Fermentation of trihydroxybenzenes by *Pelobacter acidigallici* gen. nov. sp. nov., a new strictly anaerobic nonspore forming bacterium, *Arch. Microbiol.* **1982**, *133*, 195–201.

Schink, B., Brune, A., Schnell, S., Anaerobic degradation of aromatic compounds, in: *Microbial Degradation of Natural Compounds* (Winkelmann, G., ed.), pp. 219–242. Weinheim **1992**: VCH.

Schink, B., Philipp, B., Müller, J., Anaerobic degradation of phenolic compounds. *Naturwissenschaften* **2000**, *87*, 12–23.

Schnell, S., Schink, B., Anaerobic aniline degradation via reductive deamination of 4-aminobenzoyl CoA in *Desulfobacterium anilini*, *Arch. Microbiol.* **1991**, *155*, 183–190.

Schnell, S., Brune, A., Schink, B., Degradation of hydroxy hydroquinone by the strictly anaerobic fermenting bacterium *Pelobacter massiliensis* sp. nov., *Arch. Microbiol.* **1991**, *155*, 511–516.

Schocher, R. J., Seyfried, B., Vazquez, F., Zeyer, J., Anaerobic degradation of toluene by pure cultures of denitrifying bacteria, *Arch. Microbiol.* **1991**, *157*, 7–12.

Schöcke, L., Schink, B., Biochemistry and energetics of fermentative benzoate degradation by *Syntrophus gentianae*, *Arch. Microbiol.* **1999**, *171*, 331–337.

Schramm, E., Schink, B., Ether-cleaving enzyme and diol dehydratase involved in anaerobic polyethylene glycol degradation by an *Acetobacterium* sp., *Biodegradation* **1991**, *2*, 71–79.

Schumacher, W., Holliger, C., Zehnder, A. J., Hagen, W. R., Redox chemistry of cobalamin and iron-sulfur cofactors in the tetrachloroethene reductase of *Dehalobacter restrictus*, *FEBS Lett.* **1997**, *409*, 421–425.

Seyfried, B., Glod, G., Schocher, R., Tschech, A., Zeyer, J., Anaerobic degradation of toluene by pure cultures of denitrifying bacteria, *Appl. Environ. Microbiol.* **1994**, *60*, 4047–4052.

Sluis, M. K., Ensign, S. A., Purification and characterization of acetone carboxylase in a CO₂-dependent pathway of acetone metabolism by *Xanthobacter* strain Py2, *Proc. Natl. Acad. Sci. USA* **1997**, *94*, 8456–8461.

Sluis, M. K., Small, F. J., Allen, J. R., Ensign, S. A., Involvement of an ATP-dependent carboxylase in a CO₂-dependent pathway of acetone metabolism by *Xanthobacter* strain Py2, *J. Bacteriol.* **1996**, *178*, 4020–4026.

Sluis, M. K., Larsen, R. A., Krum, J. G., Anderson, R., Metcalf, W. W., Ensign, S. A., Biochemical, molecular, and genetic analyses of the acetone carboxylases from *Xanthobacter autotrophicus* strain Py2 and *Rhodobacter capsulatus* strain B10, *J. Bacteriol.* **2002**, *184*, 2969–2977.

Spormann, A., M., Widdel, F., Metabolism of alkylbenzenes, alkanes, and other hydrocarbons in anaerobic bacteria, *Biodegradation* **2000**, *11*, 85–105.

Stupperich, E., Konle, R., Corrinoid-dependent methyl transfer reactions are involved in methanol and 3,4-dimethoxybenzoate metabolism by *Sporomusa ovata*, *Appl. Environ. Microbiol.* **1993**, *59*, 3110–3116.

Suflita, J. M., Horowitz, A., Shelton, D. R., Tiedje, J. M., Dehalogenation: a novel pathway for the anaerobic biodegradation of haloaromatic compounds, *Science* **1982**, *218*, 1115–1117.

Suflita, J. M., Liang, L.-N., Saxena, A., The anaerobic degradation of *o*-, *m*-, and *p*-cresol by sulfate-reducing bacterial enrichment cultures obtained from a shallow anoxic aquifer, *J. Ind. Microbiol.* **1989**, *4*, 255–266.

Tarvin, D., Buswell, A. M., The methane fermentation of organic acids and carbohydrates, *J. Am. Chem. Soc.* **1934**, *56*, 1751–1755.

Taylor, D. G., Trudgill, P. W., Gripps, R. E., Harris, P. R., The microbial metabolism of acetone, *J. Gen. Microbiol.* **1980**, *118*, 159–170.

Thauer, R. K., Jungermann, K., Decker, K., Energy conservation in chemotrophic anaerobic bacteria, *Bacteriol. Rev.* **1977**, *41*, 100–180.

Tschech, A., Fuchs, G., Anaerobic degradation of phenol via carboxylation to 4-hydroxybenzoate: in vitro study of isotope exchange between ¹⁴CO₂ and 4-hydroxybenzoate, *Arch. Microbiol.* **1989**, *152*, 594–599.

Tschech, A., Schink, B., Fermentative degradation of resorcinol and resorcylic acids, *Arch. Microbiol.* **1985**, *143*, 52–59.

Verfurth, K., Pierik, A. J., Leutwein, C., Zorn, S., Heider, J., Substrate specificities and electron paramagnetic resonance properties of benzylsuccinate synthases in anaerobic toluene and *m*-xylene metabolism, *Arch. Microbiol.* **2004**, *181*, 155–162.

Vogel, T. M., Grbić-Galić, D., Incorporation of oxygen from water into toluene and benzene during anaerobic fermentative transformation, *Appl. Environ. Microbiol.* **1986**, *51*, 200–202.

Vorbeck, C., Lenke, H., Fischer, P., Knackmuss, H. J., Identification of a hydride–Meisenheimer complex as a metabolite of 2,4,6-trinitrotoluene by a *Mycobacterium* strain, *J. Bacteriol.* **1994**, *176*, 932–934.

Wagener, S., Schink, B., Anaerobic degradation of non-ionic and anionic surfactants in enrichment cultures and fixed-bed reactors, *Water Res.* **1987**, *21*, 615–622.

Wagener, S., Schink, B., Fermentative degradation of non-ionic surfactants and polyethylene glycol by enrichment cultures and by pure cultures of homoacetogenic and propionate-forming bacteria, *Appl. Environ. Microbiol.* **1988**, *54*, 561–565.

Wilkes, H., Rabus, R., Fischer, T., Armstroff, A., Behrends, A., Widdel, F., Anaerobic degradation of *n*-hexane in a denitrifying bacterium: further degradation of the initial intermediate (1-methylpentyl)succinate via skeleton rearrangement, *Arch. Microbiol.* **2002**, *177*, 235–243.

Zehnder, A. J. B., Brock, T. D., Anaerobic methane oxidation: occurrence and ecology, *Appl. Environ. Microbiol.* **1980**, *39*, 194–204.

Zehnder, A. J. B., Stumm, W., Geochemistry and biogeochemistry of anaerobic habitats, in: *Biology of Anaerobic Microorganisms* (Zehnder, A. J. B., ed.), pp. 1–38. New York **1988**: Wiley.

Zhang, X., Young, L. Y., Carboxylation as an initial reaction in the anaerobic metabolism of naphthalene and phenanthrene by sulfidogenic consortia, *Appl. Environ. Microbiol.* **1997**, *63*, 4759–4764.

# 9
# Soil Remediation and Disposal

Michael Koning, Karsten Hupe, and Rainer Stegmann

## 9.1
## Introduction

For the treatment of contaminated sites, securing as well as remediation methods
are applied. While remediation achieves decontamination or reduction of pollutants,
securing sets up technical barriers for the protection of the environment. Since the
source of contamination remains in place and the technical barriers are subject to
aging and environmental influences, securing measures often represent only a time
restricted solution and future remediation activities become necessary.

Remediation methods are classified according to their operation location as well
as to processing aspects. Thus, ex situ and in situ processes are available on the one
hand, and thermal, chemical, physical, and biological processes on the other. The ex
situ processes require excavation of the contaminated soil and soil treatment either
at the site (on-site remediation) or at an external soil treatment plant (off-site remediation). In contrast, in situ treatment takes place at the site in the contaminated soil
itself, without any soil excavation.

Thermal processes are used for the treatment of highly concentrated organic pollutants, but they are suitable only to a small extent for the elimination of heavy metals. With soil scrubbing, the coarse-grain fraction >63 μm is purified, transferring
the pollutants into the water phase and/or into the fine-grain fraction. This fine fraction is highly loaded with pollutants and thus has to be treated and disposed of afterwards. The biopile process is applied on a large scale as a state-of-the-art technology.
It is an effective process for the treatment of biologically degradable pollutants such
as mineral oil and its derivatives, aliphatic hydrocarbons, phenols, formaldehyde,
and other soil contaminants.

In remediation practice, there is a trend toward actions with a minimum of requirements. Contaminated sites are remediated or secured, depending on the intended after-use, e.g., housing, development of commercial or industrial facilities,
or as recreational areas. Depending on the kind of use, specific target values have to
be met after treatment. If the contaminated site is not being used and so far no major dangerous contamination of the groundwater, surface water, etc. has occurred,

*Environmental Biotechnology. Concepts and Applications.* Edited by H.-J. Jördening and J. Winter
Copyright © 2005 WILEY-VCH Verlag GmbH & Co. KGaA, Weinheim
ISBN: 3-527-30585-8

one relies more and more on natural attenuation processes in soil and groundwater. Economical active (e.g., selective biostimulation) and passive in situ measures (reactive walls, funnel and gate systems) are developed and need to be investigated with regard to their long-term effectiveness.

For the ex situ remediation of soils that are nearly not amenable to treatment (silty soils contaminated with chlorinated hydrocarbons, polycyclic aromatic hydrocarbons, or tar oil) combined processes with soil scrubbing as basic process are increasingly being considered. For the subsequent treatment of problematic polluted residues (fine-grain fraction, process waters, etc.), a variety of processes are available, e.g., suspension bioreactors, oxidation with ozone or $H_2O_2$, and thermal treatment.

Experiences in dealing with hazardous old sites have basically shown that balancing pre-investigations are of essential importance regarding the process evolution as well as the evaluation of remediation measures. Far-reaching investigations have been carried out in the field of microbiological (ex situ) treatment, and standardizations have been defined (Dechema, 1992; GDCh, 1996). In the future, emphasis should be placed on process evaluation for the in situ treatment so as to predict the processes of natural attenuation by balancing as far as possible the fate of pollutants.

## 9.2
## Thermal Processes

Thermal soil purification is based mainly on transfer of the pollutants from the soil matrix into the gas phase by thermal energy input. The pollutants are released from the soil by vaporization and then burned. The polluted gas is purified further.

The different processing concepts for the thermal purification of contaminated soils are characterized by variations of the process parameters (e.g., temperature range, retention time for solids and waste gas in certain temperature zones, supply of oxygen, supply of reactive gases for gasification, supply of inert gas, kind of heat input, and optimum heat utilization, etc.). The large variability has led to a multitude of ex situ and in situ process combinations.

### 9.2.1
### Thermal Ex Situ Processes

The basic principle of a thermal soil purification plant includes the following processing steps (Figure 9.1):
1. soil conditioning
2. thermal treatment
3. waste gas purification

In soil conditioning, the contaminated soil is freed of interfering foreign matter (e.g., scrap, plastics) broken, sieved, and homogenized by mechanical preparation processes to be of a particle size consistent with the technical requirements of the subsequent thermal treatment [<20 mm (fluidized bed) to 80 mm]. During thermal

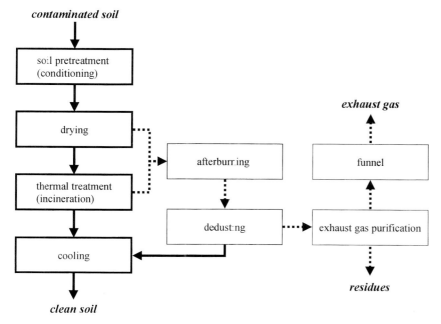

**contaminated soil**

**exhaust gas**

**residues**

**clean soil**

**Fig. 9.1** The thermal ex situ treatment process.

treatment the soil is heated so that volatile pollutants are stripped from it. In the gas phase above the soil, the combustion of pollutants takes place, but in this phase complete destruction of volatile pollutants cannot be achieved. For this reason the gases are burned in an afterburner chamber at high temperature (~1200 °C) for a certain retention time. Under these conditions dioxins are also destroyed.

Two different technologies are mainly used in thermal soil treatment:

- processes with an exclusively thermal effect: directly and indirectly heated processes (pyrolysis) with subsequent gas treatment (afterburning, condensation) (rarely used)
- processes with a thermal effect and additional measures (common practice)

Processes having an exclusively thermal effect (directly and indirectly heated) are conventional processes available on a large scale, using, e.g., rotary kiln plants, fluidized-bed plants, and sintering strand plants, for which long-term experience in practice has been achieved. Most of the processes work in the low-temperature range of 100 to 550 °C (residence time of the soil in the furnace 10–20 min; throughput 20 to 30 Mg (megagramms) $h^{-1}$ (VDI, 2002)) at which the structure of the soil is not fundamentally changed and humic components are only partly destroyed. In the early days of thermal soil treatment, high-temperature processes were practiced in which the soil was heated to temperatures of 800 to 1100 °C (residence time of the soil in the furnace 20–45 min; throughput 2 to 7 Mg $h^{-1}$ (VDI, 2002)). Under these conditions, partial liquefying or sintering of the soil particles is possible. At the same time, the organic components and the clay minerals of the soil are destroyed to a large extent, hydroxides are changed into oxides, and primary minerals are crushed by gritting. Furthermore, the quantity of $NO_x$ increases rapidly at temperatures >1000 °C, so that special equipment could be necessary for decreasing $NO_x$ in the waste gas.

Depending on the contents of pollutant, water, and fine soil particles, thermal treatment with indirect heating is carried out under reducing conditions (pyrolysis) in a temperature range between 400 and 600 °C (residence time of the soil in the furnace approx. 45 min; throughput 5 to 13 Mg h$^{-1}$ (VDI, 2002)).

The processes using an exclusively thermal effect show significant differences regarding the waste gas purification systems used. The selection of the waste gas purification system used, is decisive influenced by the local regulations on air emissions. Usually, aggregates are used, which are normally used for flue gas purification in large-scale power plants and in waste incineration plants. The waste gas purification equipment in these processes mainly contains three partial units:

1. high-temperature afterburners
2. dedusters
3. flue gas purifiers

The separation of fine soil particles from the gas stream (dedusting) is done by means of cyclones, hot gas filters, or combinations of different filter techniques specifically adapted to the process. Although the organic pollutants (hydrocarbons) are completely oxidized during the afterburning step (high temperature afterburning) at 900–1300 °C, flue gas purification minimizes the amounts of inorganic pollutants such as hydrogen chloride, hydrogen fluoride, and sulfur dioxide and brings the levels of dust and heavy metal emissions down to target values. Fundamentally, flue gas purification processes can be divided into a dry sorption process, a wet cleaning process (wet scrubbers), and a combination of both (semidry process). If reduction of nitrogen oxides is required, special measures are necessary. The efficiency of the various processes increases from the dry sorption through the semidry up to the wet scrubber process – as do the total costs. But at the same time, the degree of pollutant separation effected by these processes increases so that – especially regarding the wet cleaning processes – high waste gas quality requirements can be met. To guarantee the target values of the waste gas, activated carbon filters are used as a final step in almost all processes. An operation without excess water can be achieved by treatment of the washing water (heavy metal precipitation and stabilization) and subsequent recirculation for semidry and wet cleansing processes (Fortmann and Jahns, 1996).

Processes involving thermal effects and additional measures include various designs that use a temperature range of 60–350 °C, at which the pollutants (especially low-boiling and medium-boiling hydrocarbons) are stripped from the soil by the influence of heat and additional measures (e.g., steam stripping, vacuum). The efficiency of the process depends not only on the effect of the temperature, but also on the physical properties of the pollutants.

For these processes, the waste gas treatment normally includes no high-temperature afterburning of the waste gas. The purification is carried out via condensation and/or absorption (Fortmann and Jahns, 1996; ITVA, 1997).

## 9.2.2
## Thermal In Situ Processes

The use of thermal in situ processes is still not state-of-the-art. Thermal in situ processes differ mainly in the kind of energy input for heating the soil matrix to transfer the pollutants into the gas phase. To capture and treat the gas phase, a combination of soil vapor extraction (SVE) and subsequent gas treatment is necessary. Compared with pure soil vapor extraction, the required treatment period can be reduced.

In the steam-injection process, a hot steam–air mixture is passed into the unsaturated soil zone by steam–air injection (60–100 °C). In consequence, the volatile as well as the semivolatile compounds (NAPL) pass into the gas phase. The soil gas phase is extracted by means of gas extraction systems and is treated afterwards. Transport of the mobilized pollutants toward groundwater must be prevented by specific temperature control and by specific adjustment of the steam–air mixture. This process has limitations for soils with low permeability and for soils with very high inhomogeneities, because these soils require long periods of time for heating (Betz et al., 1998). As an alternative, the temperature within the unsaturated soil zone can be increased by imposing high-frequency electromagnetic fields by using electrodes (Jütterschenke, 1999).

## 9.2.3
## Application of Thermal Processes

In principle, all kinds of pollutants that can be stripped from the soil under the influence of thermal energy can be treated by means of thermal processes. The operation temperatures and retention times depend on the type and concentration of the pollutants as well as on the intended use of the treated soil material.

Thermal ex situ processes are preferably used when high initial concentrations of organic compounds are found and a high degree of purification is required. They can remove mainly petroleum hydrocarbons (TPH), polycyclic aromatic hydrocarbons (PAH), volatile organic hydrocarbons (benzene, toluene, ethylbenzene, xylenes (BTEX)), phenolic compounds, cyanides, and chlorinated compounds such as polychlorinated biphenyls (PCB), pentachlorophenol (PCP), volatile halogenated hydrocarbons, chlorinated pesticides, polychlorinated dibenzodioxins (PCDD), and polychlorinated dibenzofurans (PCDF). Furthermore, thermal ex situ processes can be used as a pre- or post-treatment step in conjunction with other ex situ processes.

Compared to thermal ex situ processes that also use additional measures, processes having an exclusively thermal effect are mainly used for soils characterized by single substance class contamination with volatile pollutants. Because of their comparatively low generation of heat, in situ processes are suitable only for pollutants that can be stripped in the lower temperature range (e.g., BTEX).

Basically, soil materials of all particle size distributions can be purified by thermal ex situ processes. A limitation on the proportion of silt (<30% to 50%) can become necessary for economic reasons. The use of thermal in situ processes can also be restricted because of inhomogeneities or unsuitable soil water content.

The efficiency of thermal treatment processes for removing organic pollutants from contaminated soils approaches almost 100% and is usually higher than the efficiency of biological or chemical/physical ex situ processes. However, to evaluate the application of various treatment processes, additional aspects, e.g., the necessary energy input, technical requirements, treatment costs, possibilities for reuse of the treated soils, and other aspects have to be considered. Usually, the costs of thermal soil treatment are higher than the costs of biological or chemical/physical processes.

## 9.3
## Chemical/Physical Processes

Chemical/physical soil treatment processes are mainly extraction and/or wet classification processes. The principle of ex situ soil scrubbing technologies is to concentrate the contaminants in a small residual fraction by separation. In general, water (with or without additives) is used as an extracting agent. For the transfer of contaminants from the soil to the extracting agent, two mechanisms are of importance:
- strong shearing forces induced by pumping, mixing, vibration, high-pressure water jets (to break up agglomerates of polluted and nonpolluted particles and disperse contaminants into the extracting phase)
- dissolution of contaminants by extracting agents

In situ extraction basically consists of percolation of an aqueous extracting agent into the contaminated soil. Percolation can be achieved by means of surface trenches, horizontal drains, or vertical deep wells. Soluble contaminants present in the soil dissolve in the percolate, which is pumped up and treated on-site.

### 9.3.1
### Chemical/Physical Ex Situ Processes

During soil scrubbing, the pollutants are detached from the soil particles by means of mechanical energy and/or solubilizing effects, often supported by surfactants. In consequence, pollutants become concentrated in the liquid phase and in the solid fine fraction of the soil containing pollutants sorbed onto the surface. Water, optionally enhanced with additives, serves as a dissolving agent and as a medium for transport. In general, soil scrubbing consists of the following steps (Figure 9.2):
1. soil pretreatment
2. soil washing
3. separation by gravity (classification)
4. separation of dispersed particles
5. separation of process water and rinsing of the purified soil fraction

6. process water recirculation
7. wastewater purification
8. waste gas purification

First, the contaminated soil is mechanically prepared to separate coarse substances that might disturb subsequent process steps. The soil, which includes rubble, is granulated to < 30–100 mm in diameter by crushing and sieving. This also serves as a homogenization step. In the next step, the solids are dispersed in the liquid phase and strongly agitated, so as to detach the pollutants and to separate the fine particles from the coarse particles. Sometimes chemical additives are added to the water (acids, bases, surface-active substances/surfactants) to overcome the binding between pollutants and soil particles. The conventional equipment for wet separation is high-pressure water jet pipes, centrifugal impact equipment, washing drums, vibration screws, attrition cells, blade washers, and fluidized-bed reactors. Pollutants are frequently situated in and on light matter (e.g., charcoal, slag, wood). It is not possible to transfer this contamination into the washwater. Constituents are removed in suitable sorting equipment: sedimenting machines, up-current separators, helical separators, flotation equipment, etc. For successful sorting, classification by particle size must be provided upstream of the sorting equipment (VDI, 2002). Depending on the process, the forces necessary for the detachment of the contaminants are restricted to a certain range of particle size, for economical reasons. The fine particles

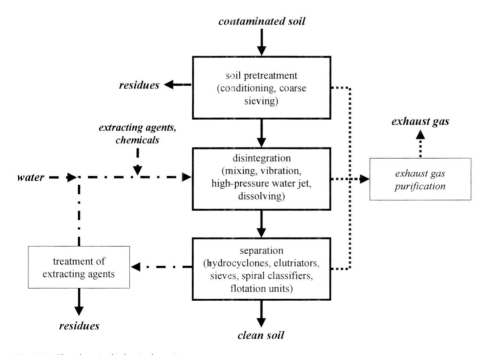

**Fig. 9.2**  The chemical/physical ex situ process.

have to be removed and disposed of or be treated by other processes. The separation of the fine particles from the soil takes place in three successive steps:

1. separation of the highly polluted fine particles from coarser components that are polluted at only a low level, using hydrocyclones (separation down to 0.1 to 0.01 mm)
2. separation of the fine particles from the process water by means of coagulation, flotation, and sedimentation
3. dewatering of the fine particles using screens and filter systems, e.g., filter presses

After separation of the polluted fine particles, the purified coarse fraction has to be separated from the process water. For this, drain screens or vacuum band filters are used, depending on the particle size. For the removal of the remaining process water, rinsing with uncontaminated water is done, followed by another dewatering step. To reduce the consumption of rinse water, the water is recirculated (process water recirculation). To keep the concentration of the pollutants in the process water at a certain level, some of the water – usually up to 10% – is separated and treated in a wastewater purification plant. This water is treated just enough so that the water can be reused as rinsing water instead of taking fresh water. Depending on the kind of pollution, various processes or process combinations of wastewater treatment technology are used (e.g., oxidation, reduction, neutralization, emulsion breaking, heavy metal precipitation, gravel/sand filtration, active carbon adsorption, extraction, membrane separation, ion exchange). The temperature during the treatment may be high enough that – supported by heavy mixing –organic compounds having a low vapor pressure volatilize. For this reason, the relevant areas are enclosed, and the air is captured and subsequently purified by means of separators for solids, drop separators, or activated carbon filters (ITVA, 1994a; Heimhard et al., 1996; VDI, 2002).

The particle size separation cut – below which the soil particles of the contaminated fine grain fraction are separated – is at 25 to 63 μm. There are technical and economical limitations for treating the fine grain portion. The treatment of high-silt fractions is uneconomical, because the portion of highly polluted residues increases drastically and consequently the disposal costs increase. Depending on the plant configuration, the treatable fine grain portion is between 25% and 40% of the soil. Relative to the plant input, the average amount of residual material is between 2% and 30%. If the percentage of fines exceeds approximately 30%, the disposal costs become the main cost factor for soil washing (Wilichowski, 2001).

## 9.3.2
### Chemical/Physical In Situ Processes

During pump-and-treat processes, water is supplied to the soil so as to leach out the contaminants. The contaminated water is pumped to the surface and treated. Surfactants that increase the solubility of the pollutants may be added to the water. The extracted washing water is treated with standard wastewater treatment technologies.

The possible applications of chemical/physical in situ processes are especially limited by the permeability of the soil. For hydraulic in situ measures, the soil permeability needs to have a permeability factor $k_t$ of at least $5 \times 10^{-4}$ m s$^{-1}$. Side effects such as bioclogging can be responsible for further decreasing the natural permeability of the soil. The problem always remains that the entire amount of the injected water is not recaptured by pumping, so that there is a risk of pollutant transport into the groundwater. In addition, the added surfactants may be a source of secondary pollution.

Soil vapor extraction (SVE) is an effective and economical process for decreasing highly volatile pollutants (e.g., BTEX) in the unsaturated zone of permeable soils ($k_f < 10^{-3}$ m s$^{-1}$). Perforated pipes are placed in the contaminated soil area. The volatile pollutants are sucked out of the soil by using low vacuum blowers. The extracted pollutants and condensates are treated on-site using activated carbon filters, compost filters, etc. The kind of treatment system used depends on the amount of air to be treated as well as the kind of pollutants. The time needed for extraction of the pollutants to an acceptable degree lasts from months to years. Often, complete decontamination of the soil is not achieved (ITVA, 1997).

The efficiency of the soil vapor extraction process is influenced by the characteristics of the soil (permeability, moisture content, temperature, homogeneity) and the kind of pollutants (vapor pressure). Often, the pollutants are in the liquid phase in the soil, so that volatilization takes place at the border of the liquid plume, which prolongs the extraction process. Sometimes extraction can be enhanced by increasing the temperature in the soil (e.g., by adding steam).

### 9.3.3
### Application of Chemical/Physical Processes

There is no limit to the kind of pollutants that can be treated by soil washing processes, as long as they can be detached from soil particles and solubilized in the washing water. Therefore, all kinds of pollutant groups have been treated with the ex situ soil scrubbing process: BTEX, TPH, PAH, PCB, heavy metals, and PCDD/PCDF (VDI, 2002). A systematic approach to estimating the prospects of recycling contaminated soils by soil-washing processes was described by Feil et al. (1997). The actual purification of the polluted liquid phase and of the fine particle fraction can take place outside the soil washing plant, in separate treatment facilities. But the polluted fine fraction is usually dumped in a landfill, meaning that no actual treatment, but only separation, is achieved.

### 9.4
### Biological Processes

During biological treatment soil microorganisms convert organic pollutants (e.g., hydrocarbons) into mainly $CO_2$, water, and biomass. Some of the pollutants can also be immobilized by binding to the humic substance fraction. Degradation may take place under aerobic as well as under anaerobic conditions. The aerobic process is predominantly used in soil remediation. For efficient biological treatment of contaminated soils, it is important to optimize the environmental conditions for the microorganisms (oxygen supply, water content, pH value, etc.). By appropriate adjustment of the conditions, the degradation processes are enhanced and the degree of degradation is improved, especially for high concentrations of readily degradable pollutants. To stimulate biological activity, soil homogenization, active aeration, moistening or drying, heating, addition of nutrients and substrates, or inoculation with microorganisms can be done. However, in comparison to thermal or chemical/physical treatment processes, less energy input but longer treatment periods are generally required.

### 9.4.1
### Biological Ex Situ Processes

Biological ex situ processes usually follow the following treatment steps (Figure 9.3):
1. mechanical pre-treatment
2. addition of water, nutrients, substrates and microorganisms
3. biological treatment

After excavation, the contaminated soil is mechanically broken up and sieved to remove disruptive material, homogenize the soil material, and loosen the soil structure. In addition, oxygen supply to the soil particles is improved. Mineral components separated during the sieving process can be crushed and later added to the contaminated soil. To activate biological degradation of the contaminants, water, nu-

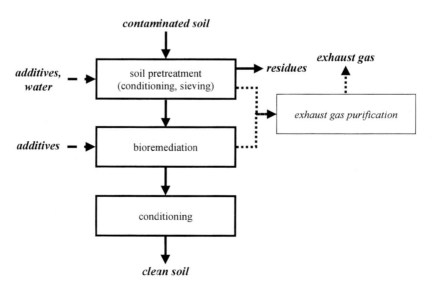

**Fig. 9.3** The biological ex situ process.

trients, and substrates are added. Optionally, substances that improve the soil structure or, in very special situations, microorganisms are added. Organic additives like compost, bark, wood chips, or straw serve as cosubstrates or nutrient sources for the microorganisms and as structural material. Biological treatment of contaminated soils is done in thin soil layers (landfarming process), in biopiles, or in bioreactors (dry and slurry reactors).

In the landfarming process, the contaminated soil is treated in thin layers of up to 0.4 m thickness. Therefore, large treatment areas are required if large amounts of contaminated soil need to be treated. The pretreated soil is placed on foil, concrete, or a clay layer. Enhanced oxygen supply as well as mixing are done by plowing, harrowing, or milling at regular intervals (Cookson, 1995).

Biopiles for soil remediation are constructed similar to biopiles for composting organic wastes. Rectangular, oblong, or pyramidal forms are used. The height of the biopiles is usually between 0.8 and 3.0 m. The biopile process can be carried out at water contents below and above the maximum water holding capacity (dry and wet systems). Although dry systems can be operated with or without agitation (dynamic or static biopile process), wet systems are all static. In contrast to the static biopile process, the principle of the dynamic biopile process is decomposition of the soil by repeatedly plowing and turning the biopiles. If necessary, water and nutrients are added during the turning process. This increases the bioavailability of the pollutants, and the contamination is brought into close contact with microorganisms, nutrients, water, and air. Today, the biopile process is predominantly done dry and is used mainly for aerobic degradation processes. To achieve unrestricted aerobic degradation processes, oxygen contents >1 vol.% have to be ensured in all parts of the biopiles. Theoretical calculations and field investigations have shown that passive aeration of biopiles made of sandy soil material and having a water content optimal for the aerobic degradation of organic pollutants (~95% of the plastic limit) requires piles limited to approx. 2.0 m in height. If the use of taller biopiles is required, active aeration systems become necessary to achieve an economical biopile process (Koning et al., 2001).

In the bioreactor process, soils are treated in the solid or slurry phase. The principle of solid-phase reactors is mechanical decomposition of the soil by attrition and by intensive mixture of the components in a closed container. This ensures that contamination, microorganisms, nutrients, water, and air are brought into permanent contact. Additives to improve the soil structure are usually not necessary. The soil can be aerated with an active aeration system or via exhaustion. However, the exhaust air has to be purified, for example, by the use of activated carbon filters or biofilters. Soils that are not suitable for solid bioreactors (clayey and silty soils) can be treated as a slurry in suspension bioreactors. Suspension bioreactors are also suitable for treating the residual fine-particle fractions from the soil scrubbing process, which are highly loaded with contaminants. After treatment, the slurry is dewatered, and the contaminated water fraction is purified. Usually much of the water is recirculated (ITVA, 1994b). In comparison to landfarming and biopile processes, bioreactors offer better conditions for process control; therefore, usually shorter treatment periods are required. However, compared to the biopile process, only a few bioreactors are used, since the advantages gained often do not justify the relatively high technical effort.

## 9.4.2
## Biological In Situ Processes

The ability of microorganisms to degrade organic pollutants under environmental conditions naturally present in the field is the basis for intrinsic bioremediation processes. As part of the natural attenuation processes, these biological degradation processes can contribute to decreasing the levels of organic pollutants in soil over long-term periods (Newman and Barr, 1997). However, although under suitable environmental conditions natural biodegradation processes can be used as a passive remedial measure, the natural biodegradation processes can come to a standstill under limiting environmental conditions and in the absence of additional biostimulating measures. Therefore, environmental conditions must be monitored in a natural attenuation program (ITVA, 2003).

In biological in situ treatment, the environmental conditions for the biological degradation of organic pollutants are optimized as far as possible. Oxygen usually has to be supplied, which can be done by artificial aeration or by addition of electron acceptors such as nitrate or oxygen-releasing compounds. Sometimes also $H_2O_2$ or $O_3$ dissolved in water is added. By these means the organic contaminants are degraded. If oxygen is supplied via the water phase, nutrients and, sometimes bacteria, are also added. However, usually the authochtonous microflora is adapted to the present contaminants and addition of cultured microorganisms is not necessary.

The supply of oxygen to the saturated respectively into the unsaturated soil zone is also the basis for active aeration processes like bioventing (pressure aeration of the unsaturated soil zone), air sparging (pressure aeration of the saturated soil zone), and bioslurping (combined soil atmosphere and groundwater exhaustion) (Gidarakos and Schachtebeck, 1996).

## 9.4.3
## Application of Biological Processes

The biological turnover of organic pollutants depends mainly on the bioavailability and biodegradability of the contaminants, as well as on the environmental conditions for the degrading microorganisms. Therefore, the degree of biological degradation achieved in a technical process is influenced by many factors (e.g., type, concentration, and physical state of the contaminants, soil type, content of organic substances, adjustment of the environmental conditions) and can be limited for biological, physicochemical, or technical reasons (Dechema, 1991).

Bioremediation methods are used for several different contaminants. Biological treatment of TPH-contaminated soils can be considered state-of-the-art technology. Biological degradation can also be expected for BTEX and phenols, although release of the volatile substances has to be taken into consideration. PAHs are biodegradable only to a certain extent (up to 4-ring PAHs) and the rate of degradation is relatively slow (Kästner, 2000). Furthermore, PAH can be located within coal particles and thus be not bioavailable (Weibenfels et al., 1992). Regarding soils contaminated by TNT, an irreversible fixing of the conversion product TAT (2,4,6-triaminotoluene) in the soil components can be achieved by means of anaerobic biological treatment with very low redox potentials (Reis and Held, 1996). PCBs are relatively inert. Nevertheless, their biological degradation by anaerobic dechlorination and aerobic oxidation is possible (Abramovicz, 1990). However, PCBs are in general not treated biologically due to their low biodegradation rates.

Statements in the literature about possible biodegradation endpoints vary for different types of contamination and treatment methods. Basically, remaining residues from the contamination are expected.

## 9.5
## Disposal

Due to economical reasons, the excavation and disposal of contaminated soil at landfills became a frequent used option in redevelopment projects for contaminated sites. Although, the contaminated soil is removed from the original site, the contaminated soil remains usually untreated and therefore represents a potential source of environmental risk at the landfill site. The same applies to residues from ex situ soil treatment processes (e.g. sludges from wet soil scrubbing) which are disposed of at landfill sites. From a technical point of view, the landfilling of contaminated soil can be regarded as a securing measure with future remediation activities required. In order to save landfill capacity for other wastes and to promote the utilization of decontaminated soil, preference should be given to remediation measures.

## 9.6
## Utilization of Decontaminated Soil

A major aspect in ex situ soil remediation is the reuse of decontaminated soil. During the various treatment processes, the soil materials change their chemical and physical properties in different ways. Residual concentrations of contaminants and of organic materials originating from organic additives (e.g., compost, bark, wood chips) can restrict the reuse of biologically treated soil. This soil is not normally suitable for reuse as filling material or in agriculture. Therefore, it is often used in landscaping. Thermally treated soil can be used as filling material (i.e., refill where excavated) but is not suitable for vegetation due to its inert nature. Soil from wet scrubbers can be used in a similar way. It is often not easy to find adequate possibilities for utilization of the treated soil.

A crucial factor for the reuse of decontaminated soils is toxicological/ecotoxicological assessment. For this purpose bioassays are conducted to measure the possible impacts of treated soils. Bioassays should be an appropriate tool if treated soils have to be tested with regard to their hazard potential. They integrate the effects of all relevant substances, including those not considered or recorded in chemical analyses (Dechema, 1995; Klein, 1999).

**9.7**
**Conclusions**

The selection of a suitable remediation process depends on the kind and concentration of pollutants, the soil type, the local availability of remediation processes, and economical aspects. Adequate treatment processes are available for all kinds of situations: biologically degradable pollutants should preferably be treated biologically. Soils contaminated with non-biodegradable organic pollutants can be treated by thermal processes. Heavy metals can be concentrated in the fine soil fraction by means of wet scrubbers (soil washing). Wet scrubbers can also be used on organically polluted soils as a concentration step; the fine fraction may subsequently be treated thermally or biologically (e.g., as a slurry) (Mann et al., 1995; Kleijntjens, 1999; Koning, 2002).

To treat soils under controlled conditions, ex situ treatment, in which the soil is excavated and treated in specialized plants, should be preferred. Of course, these processes are costly, since, in addition to the actual treatment costs, excavation, transport, and pre-treatment of the soil (sorting out bulky materials, homogenization, etc.) must be considered. Today the costs are very much influenced by strong competition and are often not real costs.

In situ treatment avoids excavation and is therefore less costly, but it is often less effective and less controllable due to ubiquitous soil inhomogeneities. Additionally, it has to be assured that during in situ remediation no secondary pollution takes place and uncontrolled movement of the pollutants into uncontaminated areas is prevented. Therefore, extensive monitoring and securing measures may be necessary. This is especially true when natural attenuation processes are taken into consideration (ITVA, 2003).

The excavation of significantly polluted soil and its disposal in landfills should be abolished. The possibilities for reuse of treated soils should be improved. It is essential that, as a first step in soil treatment, pre-investigations be performed to predict as far as possible the efficiency of the selected treatment process. This is especially true for biological soil treatment (Dechema, 1992; Hupe et al., 2001).

**References**

Abramovicz, D. A., Aerobic and anaerobic biodegradation of PCBs: a review. *Crit. Rev, Biotechnol.* **1990**, *10*, 241–251.

Betz, C, Farber, A., Green, C. M., Koschitzky, H.-E, Schmidt, R., Removing volatile and semi-volatile contaminants from the unsaturated zone by injection of steam/air mixture, in: *Contaminated Soil '98*, pp. 575–584. London **1998**: Telford.

Cookson, J. T. Jr., *Bioremediation Engineering: Design and Application.* New York **1995**: McGraw-Hill.

Dechema E.V., *Einsatzmöglichkeiten und Grenzen mikrobiologischer Verfahren zur Bodensanierung.* Frankfurt/Main **1991**: Dechema.

Dechema E.V., *Labormethoden zur Beurteilung der Biologischen Bodensanierung.* Frankfurt/Main **1992**: Dechema.

Dechema E.V., *Bioassays for Soils.* Frankfurt/Main **1995**: Dechema.

Feil, A., Neeße, T., Hoberg, H., Washability of contaminated soil, *AufbereitungsTechnik* **1997**, *38*, 399–409.

Fortmann, J., Jahns, P., Thermische Boden-reinigung, in: *Altlasten: Erkennen, Bewerten, Sanieren* 3rd edn. (Neumaier, H., Weber, H. H., eds.), pp. 272–303. Berlin **1996**: Springer-Verlag.

GDCh (Gesellschaft Deutscher Chemiker), Leitfaden: Erfolgskontrolle bei der Boden-reinigung; Arbeitskreis 'Bodenchemie und Boden-Okologie' der Fachgruppe Umwelt-chemie und Okotoxikologie, *GDCh Mono-graphien* Vol. 4. Frankfurt/Main **1996**: GDCh.

Gidarakos, E., Schachtebeck, G., In-situ Sa-nierung von Mineralölschäden durch Bio-venting, Bioslurping und Air-Sparging, *TerraTech* **1996**, 3, 50–54.

Heimhard, H.-J., Fell, H. I, Weilandt, E., Waschen, in: *Altlasten: Erkennen, Bewerten, Sanieren,* 3rd edn. (Neumaier, H., Weber, H. H., eds.), pp. 303–339. Berlin **1996**: Springer-Verlag.

Hupe, K., Koning, M., Lüth, J.-C., Chors, I., Heerenklage, J, Stegmann, R., Application of test systems for a balance-based exam-ination of biodegradation of contaminants in soil, in: *Treatment of Contaminated Soil: Fundamentals, Analysis, Applications* (Steg-mann, R., Brunner, G., Calmano, W., Matz, G., eds), pp. 637–649, Berlin **2001**: Spring-er-Verlag.

ITVA, Arbeitshilfe H 1-1, *Dekontamination durch Bodenwaschverfahren.* Berlin **1994a**: IVTA.

ITVA, Arbeitshilfe H1-3, *Mikrobiologische Verfahren zur Bodendekontamination.* Berlin **1994b**: IVTA.

ITVA, Arbeitshilfe H 1-6, *Thermische Verfah-ren zur Bodendekontamination.* Berlin **1997**: IVTA.

ITVA, Arbeitshilfe H 1-12, *Monitored Natural Attenuation* (Entwurf). Berlin **2003**: IVTA.

Jütterschenke, P., Thermische In-situ-Boden-sanierung unter Einsatz hochfrequenter elektromagnetischer Felder: eine innovative Methode zur Reinigung kontaminierter Böden, in: *Innovative Techniken der Boden-sanierung: ein Beitrag zur Nachhaltigkeit,* Deutsche Bundesstiftung Umwelt, (S. Heiden ed.), pp. 153–171. Heidelberg **1999**: Spektrum Akademischer Verlag.

Kästner, M., Degradation of aromatic and polyaromatic compounds, in: *Biotechnology, Vol. 11b: Environmental Processes II* (Rehm,

H.-J., Reed, G., Pühler, A., Stadler, P., eds.), pp. 211–239. Weinheim **2000**: Wiley-VCH.

Kleijntjens, R., The slurry decontamination process: bioprocessing of contaminated solid waste streams, *Proc. 9th Eur. Congr. Biotech-nol. ECB9,* July 11–15, Brussels, Belgium **1999**.

Klein, J., Biological soil treatment: status, de-velopment and perspectives. *Bioremediation 1999: State of the art and future perspectives, Proc. 9th Eur. Congr. Biotechnol ECB9,* July 11–15, Brussels, Belgium **1999**.

Koning, M., Cohrs, I., Stegmann, R., Develop-ment and application of an oxygen-controlled high-pressure aeration system for the treat-ment of TPH-contaminated soils in high bio-piles (a case study), in: *Treatment of Contami-nated Soil: Fundamentals, Analysis, Applica-tions* (Stegmann, R., Brunner, G., Calmano, W., Matz, G., eds), pp. 399–414, Berlin **2001**: Springer-Verlag.

Koning, M., Optimierung in der biologischen ex situ Bodenreinigung. Hamburger Berichte 20, (Stegmann, R. ed.). Stuttgart **2002**: Verlag Abfall Aktuell.

Mann, V, Klein, I, Pfeiffer, E, Sinder, C, Nitschke, V, Hempel, D. C., Bioreaktorver-fahren zur Reinigung feinkörniger, mit PAK kontaminierter Böden, *TerraTech* **1995**, 3,69–72.

Newman, A. W., Barr, K. D., Assessment of natural rates of unsaturated zone hydrocar-bon bioattenuation, in: *In Situ and On-site Bioremediation* Vol. 1 (Alleman, B. C, Leeson, A., eds.), pp. 1–5. Columbus, OH **1997**: Bat-telle.

Reis, K.-H,, Held, T., Mikrobiologische In-situ-Verfahren zur Dekontaminierung 2,4,6-Tri-nitrotoluol (TNT)-kontaminierter Böden mit-tels kontrollierter Humifizierung; In-situ-Sa-nierung von Böden; Resumee und Beiträge des 11. Dechema-Fachgespräches Umwelt-schutz, pp. 323–328. Frankfurt/Main **1996**: Dechema.

VDI, Emission control: plants for physical and chemical, thermal and biological soil treat-ment; immobilisation methods (VDI Guide-line 3898). *VDI/DIN-Handbuch Reinhaltung der Luft (Air Pollution Prevention),* Düsseldorf **2002**: VDI.

Weibenfels, W. D., Klewer, H.-J., Langhoff, J., Adsorption of PAHs by soil particles: influ-ence on biodegradability and biotoxicity,

*Appl. Microbiol. Biotechnol.* **1992**, *36*, 689–696.

Wilichowski, M., Remediation of soils by washing processes: a historical overview, in: *Treatment of Contaminated Soil: Fundamentals, Analysis, Applications* (Stegmann, R., Brunner, G., Calmano, W., Matz, G., eds), pp. 417–433, Berlin **2001**: Springer-Verlag.

# 10
# Bioremediation by the Heap Technique

Volker Schulz-Berendt

## 10.1
## Introduction

Although the potential of microorganisms to degrade contaminants like petroleum hydrocarbons has been known for more than 100 years, the technical application of this knowledge has a history of only about 15 years. During this time biological soil remediation has made strong development, marked by great efforts in research and development, manifold conceptual and technical innovations, as well as economical ups and downs.

Today biological treatment of contaminated soil is the most-used technology for large-scale soil remediation (Schmitz and Andel, 1997), with global proliferation and an expanding international market (Cookson, 1995). The heap technique has an especially high potential for widespread use, because this technology is easy to handle and needs only a low technical and monetary input.

A large number of investigations and case studies all over the world have shown the potentials and limits of soil remediation. Biological treatment of contaminated soil by the heap technique is considered to be the most effective and competitive technology for dealing with pollution by petroleum hydrocarbons (Schulz-Berendt, 1999). Nevertheless, there is some need for further development, especially for technical solutions to enhance the height of soil heaps or to establish thermophilic conditions to use the high metabolic potential of extremophile microorganisms (Sorkoh et al., 1993; Feitkenhauer, 1998).

This chapter describes the principles of the heap technique and the different approaches and technological solutions. It also shows the advantages and limits of this technique and the research being done to overcome these problems. Additionally, it discusses some economical and legal considerations.

*Environmental Biotechnology. Concepts and Applications.* Edited by H.-J. Jördening and J. Winter
Copyright © 2005 WILEY-VCH Verlag GmbH & Co. KGaA, Weinheim
ISBN: 3-527-30585-8

## 10.2
## Principles of the Heap Technique

The heap technique for biological soil treatment is an ex situ technology, that is, the contaminated soil is excavated and separated from the uncontaminated material. In contrast to so-called 'landfarming', where the contaminated material is spread over a large area in a comparably thin layer and mixed with the existing soil cover, in the heap technique the contaminated soil is prepared by homogenizing and mixing with additives and then piled up in heaps. Figure 10.1 shows the different steps of treatment, as implemented in the Terraferm system (Henke, 1989).

Before starting the treatment, samples of the excavated contaminated soil are tested for biological degradation of the pollutants by standardized laboratory procedures (Dechema, 1992). After a positive result is obtained in these tests, the first step in soil preparation is the separation of non-soil material like plastics, metals, etc., as well as stones having a particle size of more than 40–60 mm. Although non-soil material has to be disposed of or treated by other methods, the stones can be crushed and added back to the soil. Often the stones are not contaminated to the same extent as the fine particles, because the pollutants did not penetrate into them. If so, the separated stones can be reused directly. Therefore, separation is not only important for the degradation process but is also a way to reduce the volume of material that has to be treated.

The most important steps in pretreatment are homogenization of soil material and mixing with additives. Homogenization means that the normally inhomogene-

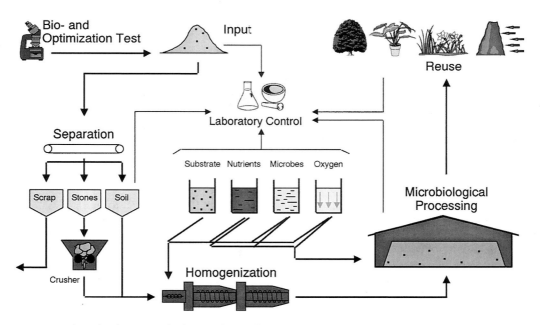

**Fig. 10.1** Biological soil treatment by the Terraferm technology.

ous distribution of the pollutants is changed into an average concentration of contaminants in the total volume of soil. Statistical evaluation of 35 large-scale remediation projects using the heap technique (Krass et al., 1998) shows that the standard deviation of the hydrocarbon concentration is reduced significantly after homogenization and that the average level of concentration decreases to about 50% of that calculated for the original material. This second effect is due to overestimation by the analysis of the original material because samples are normally taken from more-contaminated parts of the soil.

Determination of requirements for and addition of additives of suitable quality and quantity is another important step for successful soil treatment with the heap technique. As shown in Figure 10.1, additives can be divided into different categories. Here, 'substrate' indicates all additives that improve the physical and chemical soil structure. Depending on soil quality parameters like particle size, pH, organic matter, etc., and the results of laboratory testing, materials such as compost, bark, lime, tensides, etc. are added to create optimal environmental conditions for the degrading microflora. 'Substrate' can also be used to enhance the soil temperature by including a high amount of easily degradable organics, although with the risk of a high level of carbon sources that are competitive to the contaminant. Otherwise, the added carbon source can be used as cosubstrate for energy supply or an inducer for degrading enzymes; this use has been investigated for the degradation of chlorinated hydrocarbons like tetrachloroethylene and trichloroethylene (Ewers et al., 1990; Koziollek et al., 1999; Meyer et al., 1993).

The second class of additives are the nutrients for the degrading microflora. Because contaminated soil often comes from industrial sites and is often excavated from layers several meters deep, it usually has no significant content of nutrients like nitrogen, phosphorous, or potassium. Of course, degrading microorganisms need these substances for growth and metabolism. Most often, mineral fertilizers are used as liquids or in granular form to supplement the soil with these compounds. Because nutrients normally remain in the soil after treatment, it is important not to overdose with these additives. The level of fertilizer should not exceed that used in conventional agricultural practice.

As a third element, specialized microorganisms can be added to the soil during the mixing and homogenization procedure. Inoculation with bacteria, fungi, or enzymes to enhance the degradation process is controversial in the scientific community. In contrast to many studies in the United States, investigations in Germany have shown no significant effects of added cultures of specialized microorganisms on petroleum hydrocarbon degradation (Dott and Becker, 1995), which is the prior application of the heap technique. The potential of the autochthonous microflora is normally sufficient for effective degradation.

Good results have been obtained by adding complex mixtures of 'substrates', nutrients, and microorganisms, e.g., compost (Hupe et al., 1998) or activated sludge from wastewater treatment plants. The overall goal of the mixing and homogenization process is to obtain optimum conditions for aerobic metabolism of the contaminating substances.

For sufficient homogenization and mixing of contaminated soil, special machines and aggregates have been developed, which have to combine powerful homogenization and mixing units with controlled and sophisticated dosing of the various additives. Today, specialized equipment is available (Fig. 10.2) with capacities of 50 t h$^{-1}$ or more and which have been adapted to different soil qualities from sand to clay. Crushing units can be integrated directly in this machinery.

After pretreatment, the soil is transferred to the degradation area and piled in heaps. According to the environmental regulations for treating hazardous wastes in Germany, the heaps must be located in a closed space. Depending on local climatic conditions, locating the process in a closed system is not only necessary for complying with environmental regulations but is also an important tool for controlling the degradation process, especially the water content and temperature.

The area must be prepared below the ground surface to prevent contaminated seepage water from penetrating into the subsoil. This can be done by compacting the soil or by installation of an area sealed with concrete or asphalt. Often, layers of 1.0–1.5-mm-thick high-density polyethylene (HDPE) are used to ensure safe and sustainable enclosure of the contaminated material.

To minimize emissions of volatile compounds to the air, the heaps are set up in structures such as tents or sheds or covered with plastic sheeting or membranes. With this measure it is also possible to protect the heaps from unsuitable weather conditions like rainfall or extreme temperature. The design, construction, and material of the cover depend on the kind of heap technique that is used. In the past five years in Germany, soil treatment has changed from on-site to off-site installations. Therefore, the heap technique is used mainly in stationary treatment centers, which are permanent installations and normally equipped with a treatment shed or a similar building in which the heaps are set up.

During the process of degradation, the soil is monitored continually by analyzing samples from different parts of the heap. The main control parameters are:

• concentration of the contaminants
• water content
• concentration of available nutrients
• biological activity (soil respiration)

Depending on the monitoring results, the degradation conditions are optimized by aeration, addition of water or nutrients, and further homogenization. The treatment

**Fig. 10.2** Soil preparation unit type 'mole'.

ends after reaching the target values, which are sufficient for reuse of the cleaned soil. The time of treatment differs greatly, depending on the kind and concentration of the contaminants, the target values that have to be reached, and the soil quality. The normal residence time is in the range of several months. For example, Krass et al. (1998) detected average halftimes of 85 d with a 95% confidence interval in a range from 75.4 to 94.6 d.

After treatment by the heap technique, the soil quality is adequate for it to be used as topsoil for landscaping or as dumpsite cover. Because of the treatment procedure the soil is free of larger stones, very homogeneous, and enriched in nutrients and humic substances.

## 10.3
## Different Heap Techniques

Besides the different technical solutions for below-surface preparation and heap cover, various treatment systems have been developed to establish and maintain suitable conditions for microbiological degradation of toxic compounds. Figure 10.3 shows several approaches concerning humidity, agitation, aeration, and temperature and their impact on the area needed for treatment.

The first technologies used for large-scale biological soil remediation were open-air heap installations with facilities for water recycling (Altmann et al., 1988). The moisture in the heaps is above the maximum water holding capacity (whc), and the seepage water is collected via a drainage system in a pond at one end of the heap (Fig. 10.4).

From the collecting pond, the water is pumped to the surface of the heap and spread over the heap. Nutrients and other soluble additives are mixed with the water, and the microorganisms are supplied with the added substances and oxygen via the water phase.

In contrast to these 'wet' systems, comparable 'dry' technologies have been developed and are mainly used today. The moisture in the dry heaps is below the maximum whc, so that water seepage is prevented and the soil pores are filled with water and air.

Systems with low water content can be operated with (dynamic) and without (static) agitation, but the wet systems are all static. By agitation, i.e., turning and mixing the soil at time intervals of days to weeks depending on the level of biological activity, the heap is aerated and the water or nutrient content can be readjusted. With the static and dry systems, no additional supply of additives is possible, so all ingredients must be added during pretreatment. Therefore, special additives such as slow-release fertilizers must be used with these technologies. It is also necessary to install some kind of aeration system to supply the microorganisms with oxygen.

One advantage of the static and dry system is the height of heaps, which can be up to 3–5 m, whereas in dynamic systems the heaps are no higher than about 2 m, because the special turning machines (Fig. 10.5) have a limited turning depth.

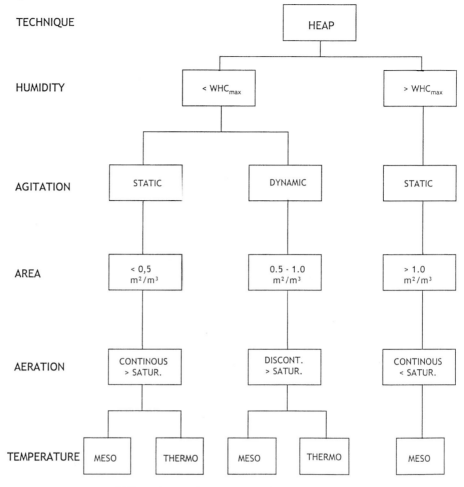

TECHNIQUE

HEAP

HUMIDITY

< WHC$_{max}$

> WHC$_{max}$

AGITATION

STATIC

DYNAMIC

STATIC

AREA

< 0,5
m$^2$/m$^3$

0.5 - 1.0
m$^2$/m$^3$

> 1.0
m$^2$/m$^3$

AERATION

CONTINOUS
> SATUR.

DISCONT.
> SATUR.

CONTINOUS
< SATUR.

TEMPERATURE

MESO

THERMO

MESO

THERMO

MESO

**Fig. 10.3** Comparison of different heap techniques.

**Fig. 10.4** Open-air heap.

**Fig. 10.5** Turning machine on the surface of a dynamic heap.

The 'wet' systems require the greatest area, because the limited capacity of water to transport oxygen leads to anaerobic layers if the height of the heaps is >0.5 m.

Supply with oxygen is the most critical factor in biological soil remediation by the heap technique. Because of the high oxygen consumption during aerobic degradation of hydrocarbons, an aeration system must be installed in all static heaps. The 'wet' technologies supply oxygen through the water phase. Therefore, the efficiency of oxygen transport to the different parts of the heap is limited by the concentration of dissolved oxygen and cannot exceed saturation. Investigations of changes in oxygen concentrations in different parts of the heaps show a rapid decrease during the first phase of degradation, with creation of anaerobic zones and methane production inside the heaps (Koning et al., 1999). Recent results of large-scale experiments show that high-pressure injection of air can solve this problem and may be an efficient alternative to dynamic treatment technologies.

If oxygen supply is sufficient, the temperature in the heaps increases by biological activity to a level of 30–35 °C. Especially in closed systems this temperature level can be maintained independent of outside conditions, so that optimum mesophilic conditions can be established by the heap technique. Research on extremophile microorganisms and their practical biotechnological application indicates a high potential of thermophilic bacteria for hydrocarbon degradation (Sorkoh et al., 1993). The establishment of thermophilic conditions in heaps is one approach to increasing degradation efficiency. Besides using the existing climatic conditions, in some regions of the world, such as Arabia, Africa, and South America, it is possible to increase the temperature by adding easily degradable organic matter. This leads to enhanced oxygen consumption, which must also be ensured by the technical design.

Today the dry and dynamic solution is the common technology for treating petroleum hydrocarbons by the heap technique. 'Wet' solutions have failed because of their long degradation times of (1–2 years) and the large demand for space. If the dry and static approach could overcome problems with limited oxygen supply, it may be the technology of the future because of the small area needed for the installation and the possibility of thermophilic process design, which together can enhance the efficiency of the heap technique.

**10.4**
**Efficiency and Economy**

As mentioned in Sections 10.1–10.3, the heap technique has been used in large-scale bioremediation for more than 10 years and is a standard technology for clean-up of soil contaminated with petroleum hydrocarbons. Depending on their origin, petroleum hydrocarbons can differ greatly. Successful bioremediation with the heap technology has been described for contamination with gasoline, light and heavy heating oil, and crude oil from different drilling areas. For successful bioremediation, not only the quality of the pollutant is important, but also its concentration and the target value that has to be reached. With the heap technique it is possible to achieve degradation rates of about 80%–90% in a reasonable time depending on the quality of the pollutant. This means that starting concentrations should not exceed $10\,000$–$20\,000$ mg kg$^{-1}$ to reach end concentrations of $1000$–$2000$ mg kg$^{-1}$, which is sufficient for reuse on industrial sites. As mentioned in Section 10.2, the average concentration in the heap is significantly lower than that calculated from the average value from single soil samples; therefore, concentrations of up to $100\,000$ mg kg$^{-1}$ can be treated after homogenization.

In addition to petroleum hydrocarbons, the following classes of pollutants can be treated by the heap technique:
- BTEX aromatics
- phenols
- polycyclic aromatic hydrocarbons (PAH) having up to four aromatic rings
- explosives (TNT, RDX)

Each of the different compounds requires some modifications to the basic technique. Especially if volatile substances have to be treated, emissions must be controlled and waste air has to be treated. An example of how to manage this problem is shown in Figure 10.6 for the remediation of a soil contaminated with phenols and aromatics.

For efficient extraction of volatiles from the soil, extraction pipes are installed in the heap for soil air extraction. During setup of the heap and the turning procedure in the dynamic system, the air of the tent is exhausted and treated, instead of extracting the soil air. The treatment system consists of four biofilter units and two activated-carbon filters. In addition to the usual tent material, another plastic cover having a thin layer of aluminum is installed inside the tent as an effective barrier to the volatile compounds. This example demonstrates that the heap technique is very flexible and can be modified from a very simple installation to a high-performance technology.

Another modification has been used for the bioconversion of explosives, especially TNT and its derivates. For biological detoxification, TNT (trinitrotoluene) is converted to TAT (triaminotoluene) under anaerobic conditions (Lenke et al., 1997). Under aerobic conditions TAT is fixed to the soil matrix. To establish strictly anaerobic conditions in the heap, organic material is added to the soil to a high extent. During degradation of the organic material, all oxygen is consumed and the temper-

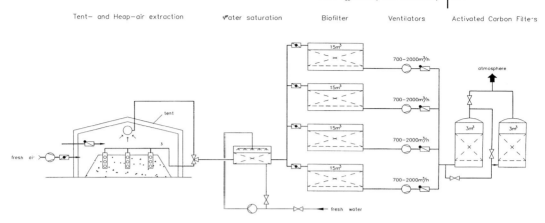

**Fig. 10.6** Schematic representation of soil air extraction from heaps.

ature reaches levels of about 60 °C. By turning the soil, aerobic conditions appear for a short time so that TAT is bound to the soil matrix and cannot be detected after treatment. With this technology, the treatment time can be reduced to less than four weeks to achieve target values (Fig. 10.7). The concentration of PAH is also reduced in the heaps. This example shows that the heap technology can be used for thermophilic degradation processes and that, under thermophilic conditions, degradation times can be reduced significantly.

Efficiency has to be discussed in relation to the economics of the process. Because of the low investment costs for installation and the simple operation procedures, the heap technique is of course a low-cost technology compared with other biological technologies like bioreactors or soil washing and incineration. Therefore, about 80% of the soil treatment plants installed in Germany are biological treatment plants (Schmitz and Andel, 1997), most of them working with the heap technique. However, because of the high number of treatment plants in Germany, with an annual capacity of about 2 500 000 t the competition is extremely strong. During the last decade prices for soil remediation have decreased from nearly 100 US$ per ton to 30–50 US$ per ton. Even for the simple heap technique, this is the absolutely lowest limit for any reliable operation.

On the other hand, legislative regulations regarding soil handling and plant operation have become stricter, so that nearly half the costs for soil bioremediation by the heap technique are due to measures set by the authorities. In contrast to this high level of environmental-safety and health-care concerns regarding treatment plants that destroy contaminants and lead to recycling of soil for different uses, dumping contaminated soil and leaving the problem for the next generation is still allowed.

Nevertheless, soil treatment by the heap technique is suitable for many different situations involving contaminants, soil quality, or climatic conditions. It is adjustable by technical modifications to meet any requirement of the degradation process. Therefore, although it was the first large-scale remediation technique, it still needs further development to become the biological treatment technology of the future.

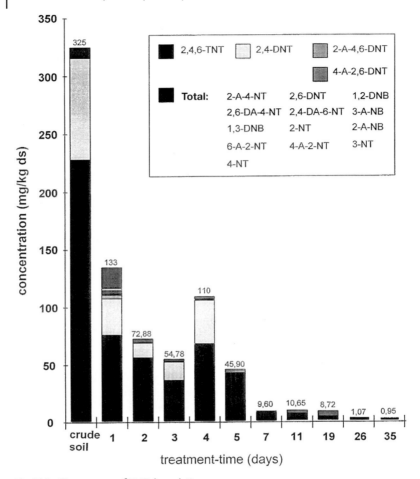

**Fig. 10.7** Time course of TNT degradation.

## References

Altmann, B. R. et al., *DGMK Forschungsbericht 396-02* – Erfahrungsbericht über die biologische Ex-situ-Sanierung ölverunreinigter Böden. Hamburg **1988**: DGMK.

Cookson, J. T. Jr., *Bioremediation Engineering: Design and Application*. New York **1995**: McGraw-Hill.

Dechema, *Labormethoden zur Beurteilung der biologischen Bodensanierung*. Frankfurt/Main **1992**: Dechema.

Dott, W., Becker, P. M., Functional analysis of communities of aerobic heterotrophic

bacteria from hydrocarbon-contaminated soils, *Microb. Ecol.* **1995**, *30*, 285–296.

Ewers, J., Freier-Schröder, D., Knackmuss, H.-J., Selection of trichloroethene (TCE) degrading bacteria that resist inactivation by TCE, *Arch. Microbiol.* **1990**, *154*, 410–413.

Feitkenhauer, H., *Biodegradation of Aliphatic and Aromatic Hydrocarbons at High Temperatures: Kinetics and Applications*, Thesis, Technical University Hamburg, Harburg **1998**.

Henke, G. A., Experience reports about on-site bioremediation of oil-polluted soils, in: *Recy-*

*cling International* (Thomé-Kozmiensky, K. J., ed.), pp. 2178–2183. Berlin **1989**: EF-Verlag.

Hupe, K., Koning, M., Lemke, A., Lüth, J.-C., Stegmann, R., Steigerung der Reinigungsleistung bei MKW durch die Zugabe von Kompost, *TerraTech* **1998**, *1*, 49–52.

Koning, M., Brauckmeier, J., Lüth, J.-C., Ruiz-Saucedo, U., Stegmann, R. et al., Optimization of the biological treatment of TPH-contaminated soils in biopiles, *Proc. 5th Int. Symp. In Situ and On Site Bioremediation*, San Diego, CA **1999**.

Koziollek, P., Bryniok, D., Knackmuss, H.-J., Ethene as an auxiliary substrate for co-oxidation of *cis*-1,2-dichloroethene and vinyl chloride, *Arch. Microbiol.* **1999**, *172*, 240–246.

Krass, J. D., Mathes, K., Schulz-Berendt, V., Scale up of biological remediation processes: evaluating the quality of laboratory derived prognoses for the degradation of petroleum hydrocarbons in clumps, in: *Proc SECOTOX 99*, 5th Eur. Conf. Ecotoxicol. Environ. Safety (Kettrup, A., Schramm, K.-W., eds.). March 15–17, 1998, Munich **1998**.

Lenke, H., Warrelmann, J., Daun, G., Walter, U., Sieglen, U., Knackmuss, H.-J., Bioremediation of TNT contaminated soil by an anaerobic/aerobic process, in: *In situ and Onsite Bioremediation* Vol. 2 (Alleman, B. C., Leeson, A., Eds.), pp. 1–2. Columbus, OH **1997**: Battelle Press.

Meyer, O., Refae, R. I., Warrelmann, J., Reis, H. von, Development of techniques for the bioremediation of soil, air and groundwater polluted with chlorinated hydrocarbons: the demonstration project at the model site in Eppelheim, *Microb. Releases* **1993**, *2*, 2–11.

Schmitz, H.-J., Andel, P., Die Jagd nach dem Boden wird härter, *TerraTech* **1997**, *5*, 17–31.

Schulz-Berendt, V., Biologische Bodensanierung: Praxis und Defizite, in: *Bödenökologie: interdisziplinäre Aspekte* (Köhler, H., Mathes, K., Breckling, B., eds.), Berlin **1999**: Springer-Verlag.

Sorkoh, N. A., Ibrahim, A. S., Ghanoum, M. A., Radwan, S. S., High-temperature hydrocarbon degradation by *Bacillus stearothermophilus* from oil polluted Kuwait desert, *Appl. Microbiol. Biotechnol.* **1993**, *39*, 123–126.

# 11
# Bioreactors

René H. Kleijntjens and Karel Ch. A. M  Luyben

## 11.1
## Introduction

### 11.1.1
### Contaminated Solid Waste Streams (Soils, Sediments, and Sludges)

Waste recycling plays a key role in the development of a sustainable economy (Suzuki, 1992). The classical approach, remediation without the production of recycled materials, does not contribute to durable material flows. Moreover, the production of reusable materials is a necessity to make waste treatment an attractive economic solution. Recycling, however, cannot be done without regard to the effort and costs needed. The overall environmental benefits should be positive and fit within the local economic and legal framework.

A practical way to qualify these benefits is found in three issues:
- the quality of the recycled products
- the amount of energy required per ton
- the cost per ton

Solid waste streams (contaminated soils, sediments, and sludges) can be recycled. The solids have to be transformed into usable products while the contaminants are removed or destroyed. If the contaminants are organic (such as mineral oil, PAH, solvents, BTEX, PCB), the use of bioreactors can result in environmental benefits (Riser-Roberts, 1998). In bioreactors populations of soil organisms degrade the contaminants to yield carbon dioxide, water, and harmless byproducts (Schlegel, 1986).

A prerequisite for the implementation of new technologies such as bioprocessing is the definition of the goals of recycling and treatment. In the Netherlands legal targets for recycled materials were set in the Dutch building materials law (Building Material Act, 1995). Depending on the content and the leaching of components, two different ways for using recycled materials in plants are defined:
- category 1 products, needing no further isolation
- category 2 products, needing further isolation and monitoring

*Environmental Biotechnology. Concepts and Applications.* Edited by H.-J. Jördening and J. Winter
Copyright © 2005 WILEY-VCH Verlag GmbH & Co. KGaA, Weinheim
ISBN: 3-527-30585-8

In the Dutch practice, treatment processes are aimed at the production of category 1 recycled products. Although each country still has its own standards, and standardization is far away (Northcliff et al., 1998), it is clear that practical recycling standards accelerate the development needed to create sustainable technologies.

11.1.2
**Characteristics of Contaminated Solids**

In soils the solid matrix is frequently dominated by sand, while the water content may be <25%; various levels of debris can be found depending on the history of the site, but rarely exceed 10%. River, harbor, and canal sediments contain mostly water (frequently >60% to 70%), while the fine fraction (<63 µm) dominates the solids if dredging takes place in the upper sediment layers. Industrial and municipal sludges are mostly very moist (>95% water) and have a large content of organics (>60%) (INES, 1997). Industrial sludges mostly originate directly from corrosion and wear from equipment or water treatment units on the site.

Disregarding the heterogeneous nature of the waste, the contaminant behavior is largely determined by the fines (Werther and Wilichowski, 1990), because submicron particles such as humic–clay structures and clay agglomerates have an extremely high adsorption capacity (Brady, 1984). The solid waste, therefore, basically contains a contaminated fine fraction, a less contaminated sand–gravel fraction, cleaner debris, and a contaminated water phase.

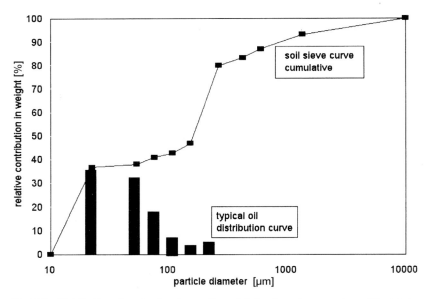

Fig. 11.1 Distribution of contaminants over the solids fractions (in percentage of the total amount of oil present) and particle size distribution (cumulative).

In Figure 11.1 the line shows a typical particle size distribution for soils having a dominant fraction of fines (the step between 10 and 20 µm) and a dominant sand fraction (the step between 150 and 250 µm).

The bars show the measured mineral oil concentration as distributed over the fractions. For efficient processing of this waste, only the fractions <150–200 µm should be treated.

In general one can state that, to develop an appropriate recycling technology, the particle features of the solids have to match the type of process operation. For bio-processing this implies integration of the separation technology and bioreactors in a sequence of operation in which the clean fractions are removed from the feed before entering the reactor (Kleijntjens et al. 1999).

## 11.2
## Bioreactors

### 11.2.1
### Reactor Configurations

The aerated bioreactor for solids processing is a three-phase (solid–liquid–gas) multiphase system. The solid phase contains the adsorbed contaminants, the liquid phase (process water) provides the medium for microbial growth, and aeration complicates the system. Nutrients and adapted biomass may be added to enhance breakdown. Furthermore, processing conditions (temperature, pH, $O_2$ level, etc.) can be monitored and to some extent controlled.

Regarding the bioreactor configuration there are two major topics:

- physical state of the multiphase system
  - bioreactors with a restricted solids holdup: slurry reactors (typical solids holdup <40 wt%)
  - bioreactor with restricted humidity: solid-state fermentation (solids content >50 wt%)
- operation mode
  - batch operation: no fresh material is introduced to the bioreactor during processing; the composition of the content changes continuously
  - continuous operation (plug flow): fresh material is introduced and treated material removed during processing, the composition in the reactor remains unchanged with time (Levenspiel, 1972); in practice, semicontinuous operation is often used (batch-wise feeding and removal giving small fluctuations in the reactor)

Three basic reactor configurations exist:
- slurry bioreactors
- solid-state fixed-bed bioreactors
- rotating-drum dry solid bioreactors

Characteristic of all types of slurry bioreactors (Fig. 11.2) is the need for energy input to sustain a three-phase system in which the solid particles are suspended; the force of gravity acting on the solids has to be compensated for by the drag forces executed by the liquid motion (Hinze, 1959). In a properly designed slurry system, the energy input is used to establish three phenomena:

- suspension
- aeration
- mixing

A slurry bioreactor can work properly only if these three measures are balanced. For each reactor configuration, the appropriate processing conditions depend on parameters such as the reactor scale, particle size distribution, slurry density, slurry viscosity, oxygen demand of the biomass, and the solids holdup (Kleijntjens, 1991).

For solid-state fermentations there is no need to maintain a solids–liquid suspension; a compact, moist, solid phase determines the system. Both the fixed-bed reactor and the rotating-drum bioreactor are suited to solid-state fermentation (Fig. 11.3). In

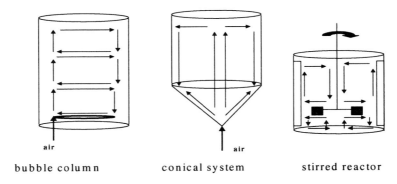

bubble column            conical system            stirred reactor

**Fig. 11.2**  Common configurations for slurry bioreactors.

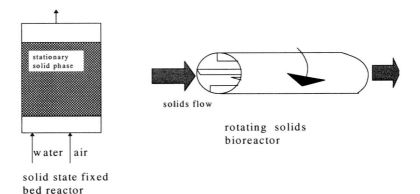

solid state fixed
bed reactor

**Fig. 11.3**  Bioreactors for solid-state processing: (a) fixed-bed reactor, (b) rotating-drum reactor.

the fixed-bed reactor the contaminated solids rest on a drained bottom as a stationary phase. Forced aeration and the supply of water are mostly applied as a continuous phase (Riser-Roberts, 1998). Fixed-bed reactors are mostly batch operated. Although landfarming might be considered a form of solid-state batch treatment under fixed-bed conditions (Harmsen, 1991), this technique offers limited control options (in comparison to other solid-state treatment) and is therefore not considered to be a bioreactor within the present context.

Continuous solid-state processing is possible in the rotating system. Here, the solid phase (as a compact moist material) is 'screwed and pushed' through the reactor. In line with slurry processing, energy is required to transport the solids through the system.

## 11.2.2
### Diffusion of Contaminants out of Solid Particles

Regardless of the bioreactor type, the presence of sufficient water is crucial. Not only is a water activity around 100% a necessity for microbial action (van Balen, 1991), but biodegradation fully depends on the availability of the components in the water phase. On the microlevel, diffusion of the adsorbed contaminants into the bulk phase is the rate-determining step (Fig. 11.4). Mathematical models focus on capturing of the different physical processes into single diffusion parameters (Wu and Geschwend, 1988). The overall diffusion process depends also on the flow conditions around the particles (Crank, 1975).

An illustration of the importance of the flow around the particles was given by Koning et al. (1998), who analyzed microbial breakdown of a petroleum-contaminat-

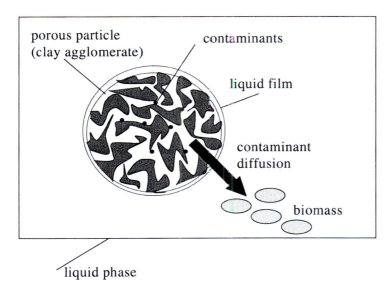

**Fig. 11.4**  Diffusion of contaminants.

ed soil. A conversion level of 80% was reached within about 10 d in a slurry reactor, but under fixed-bed conditions some 150 d were required to reach the same conversion level. They concluded that "the enhanced breakdown is based on an increasing bioavailability of the contaminant due to suspending of and the mechanical strain on the soil material in the slurry reactors". In another slurry experiment, it was shown that manipulation of reactor parameters such as the energy input result in faster breakdown (Reynaarts et al., 1990).

The reactor type, scale, energy input, and aeration rate each have an impact on the diffusion. The better these parameters can be optimized, the faster the breakdown proceeds.

Having established proper diffusion conditions, the presence of adapted biomass capable of degrading the desorbed contaminants is essential. Organisms that are already present in the waste are almost always the most practical source of adapted biomass.

Of the three different bioreactors, the slurry system offers the best features to substantially influence the diffusion rate. Intense multiphase mixing allows for the exchange of desorbed contaminants, nutrients, and biomass through the medium. In addition, particle deagglomeration has been measured (Oostenbrink et al., 1995); this results in smaller particles and a faster diffusion rate (smaller particles mean shorter diffusion distances).

For solid-state processes in fixed-bed reactors there are fewer factors to influence the microenvironment of the pollution. Only air and liquid (nutrients) can be forced into the system while the solids are packed. On the microlevel, the rotating solids bioreactor is intermediate between the fixed-bed and slurry systems, and deagglomeration of the moist mass and phase exchange can take place.

## 11.3
## Slurry Bioreactors

### 11.3.1
### Slurry Processing

A slurry bioreactor functions only with a pretreated feedstock; therefore, the bioreactor is necessarily integrated with washing–separation operations and includes a dewatering operation at the end of the process (Kleijntjens, 1991; Robra et al., 1998).

A typical setup of an integrated (slurry) bioprocess is shown in Figure 11.5. First, the feedstock is screened using a wet vibrating screen to remove the debris (typical size >2–6 mm). Second, sand fractions are removed by one or more separation techniques such as sieves, hydrocyclones, Humphrey spirals, flotation cells, jigs, and upflow columns; a typical separation diameter (the so-called cutpoint) for the hydrocyclone depicted is 63 µm (Cullinane et al., 1990). In the cyclone the slurry flow is split into a sand fraction (particle size >63 µm) at the bottom and a fine fraction at the top (<63 µm).

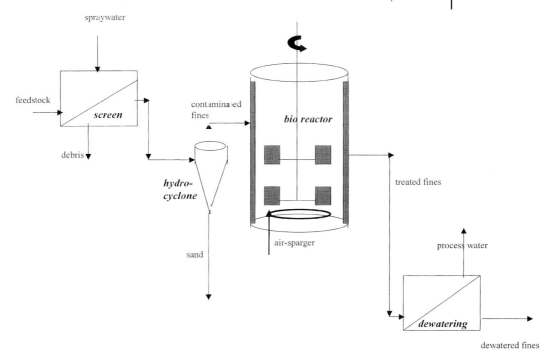

**Fig. 11.5** Typical setup for a slurry bioprocess using a batch-operated aerated stirred-tank reactor (typical solids holdup is 20 wt%).

The top flow of the cyclone, containing the contaminated fines, is fed to the bioreactor (a stirred tank is shown, but any of the three types in Figure 11.2 can be chosen). The final operation results in a dewatered product containing the fines and a flow of process water.

## 11.3.2
### Batch Operation

A slurry bioreactor is often designed as a standard continuously stirred tank reactor (CSTR) having a mechanical stirrer, baffles, and a sparger at the bottom (Fig. 11.5). In the system a three-phase suspension of contaminated solids, water, and air is maintained. Batch degradation experiments in aerated, stirred, slurry bioreactors have been carried out frequently.

Figure 11.6a shows the experimental results from a batch experiment on a 4-L scale. Using a conventional stirred bioreactor with baffles (stirrer speed 600 rpm, aeration rate 1.5–3 L min$^{-1}$, temperature kept at 30 °C), adapted biomass and nutrients were added (ammonium, phosphate, and potassium) while the pH was kept at 7–8. The soil was an oil-contaminated sandy soil having an initial oil concentration

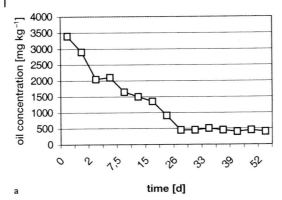

a

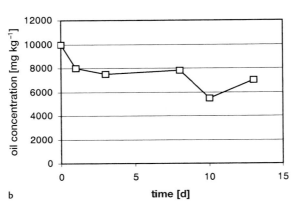

b

**Fig. 11.6** Decontamination time course in (a) batch processing of soil (4 L), (b) batch processing of sediment (650 L).

of 3500 mg kg⁻¹. Within 30 d the concentration dropped to about 500 mg kg⁻¹, but during the following period (up to day 56) no significant degradation was measured.

Figure 11.6b shows a batch degradation curve for a heavily polluted harbor sediment, as measured on the pilot scale (650 L) in an air-agitated slurry reactor (the dual injected turbulent separation (DITS) reactor). The DITS reactor was originally the first reactor unit in the Slurry Decontamination Process (Fig. 11.7). But for this experiment the DITS reactor was disconnected and used as a standalone batch reactor. The batch process showed a decrease from 10 000 mg kg⁻¹ of mineral oil contaminant to 6000–7000 mg kg⁻¹.

Other batch processes have been carried out (see also Section 11.3.3). Generalizing the results of these batch experiments enables a 'typical' batch curve to be constructed (Fig. 11.8): extensive breakdown in the first few days is followed by slower degradation during the second stage. In this stage only a small percentage of the contaminant is degraded over a relatively long time period.

The exact shape of the curve and the conversion level at which breakdown stops vary with the composition and age of the solids and the type of contaminant.

To understand the typical batch curve, three major phenomena have to be considered:

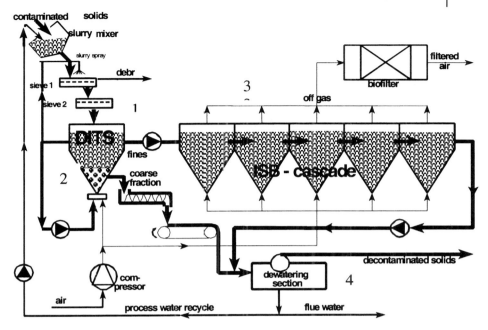

**Fig. 11.7** Slurry decontamination process (SDP), including dual injected turbulent separation reactor (DITS) and a cascade of interconnected suspension bioreactors (ISB cascade).

- In batch processing the biomass, growing on a complex substrate, has to adapt continuously to different components, since the 'easy' parts are degraded first (Teschner and Wehner, 1985). This pattern inevitably leads to increasing difficulties for the resident microbial population. The adaptation rate of the population to the available contaminants is insufficient.
- The contaminant desorption kinetics are of such an order that at lower concentrations there is a limited 'driving force' for the adsorbed contaminants to be released

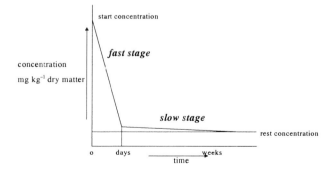

**Fig. 11.8** Generalized curve for decontamination during batch operation.

from the solids (Di Toro and Horzempa, 1982). Since microbial breakdown proceeds only if the contaminant is dissolved in the water phase, this type of kinetics does not allow for easy breakdown at low concentrations of contaminant.

- During batch processing, inhibitory side products of the microbial breakdown may be increasingly released into the medium; in addition, the physical condition of the solids may change (e.g., attrition), which may lead to unfavorable conditions for the microorganisms. An example would be the drop in pH due to humification processes, which can accompany contaminant breakdown.

### 11.3.3
### Full Scale Batch Processes

#### 11.3.3.1 The DMT–Biodyn Process

A full scale batch process has been implemented as the DMT–Biodyn process (Sinder et al., 1999). The process consists of a fluidized-bed slurry reactor in which fined-grained contaminated soils are treated. For aeration, the slurry circulation loop is treated in an external bubble column (Fig. 11.9). The reactor configuration allows a high solids loading, up to 50 wt%. The DMT–Biodyn process was designed by the Deutsche Montan Technologie GmbH to treat PAH-polluted sites at former coal facilities in the Federal Republic of Germany (Nitschke, 1994).

On a pilot scale (a 1.2 m³ reactor) various technological parameters were investigated and the hydraulic feed system at the bottom was optimized. Experiments showed that PAH degraded rapidly without a noticeable lag phase; the measured oxygen consumption rate correlated with the PAH degradation rate. For this specific soil (starting concentration of PAH about 250 mg kg⁻¹), the target levels were reached after 6 d. Comparison of the pilot results with results on a laboratory scale using a respirometer showed that the degradation of this test soil in each of these systems was similar. With a higher level of PAH contamination (starting at 1100 mg kg⁻¹), the target levels could not be reached (treatment time, 30 d). The researchers concluded that, for this specific soil, "adsorption of the PAH to the organic matrix results in a reduced bioavailability". Later experiments on this contaminated soil, in

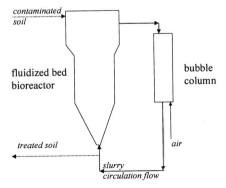

**Fig. 11.9** DMT–Biodyn process configuration.

which bioprocessing and soil washing was combined, resulted in final contaminant levels below target values.

In Sweden a full scale (55 m³) DMT–Biodyn processor was installed to treat a contaminated gas plant area. The first trials on this scale have been successfully completed (Sinder et al., 1999).

### 11.3.3.2  The FORTEC Process

In the Netherlands the FORTEC process (fast organic removal technology) for treating contaminated sediments combines several separation technologies with bioreactor treatment (ten Brummelen et al., 1997). The first step in the process is the use of hydrocyclones to remove the sand fractions. The efficiency of the separation step is optimized by a multistage hydrocyclone configuration, and separation down to a particle size of 20 μm is achieved. After the sand separation, the contaminated fine fraction is fed to the bioreactor section. Based on initial experiments with aerated stirred-tank bioreactors, a full scale bioreactor section with a total volume of 300 m³ (several reactors) was constructed (RIZA, 1997).

A heavily contaminated harbor sediment (port of Amsterdam) was treated with the FORTEC process. After sand removal the fine fraction (<20 μm) was treated batchwise in full scale bioreactors. On this scale 85% of the PAH and 78% of the mineral oil were removed after 15 d of batch processing (ten Brummelen et al., 1997).

### 11.3.3.3  The OMH Process

In the United States a large scale slurry bioreactor (750 m³) was used to treat creosote-contaminated lagoon solids stabilized with fly ash (total PAH was 11 g kg⁻¹). An extensive pretreatment to classify the material was combined with an aerated and stirred (900 rpm) bioreactor (20% solids load). Remediation with respect to the PAH contaminants was determined to be 82%–99% conversion for the 3- and 4-ring PAHs and 34%–78% for PAHs having more rings (Jerger et al., 1993).

### 11.3.3.4 The Huber Process

On a scale of 30 m³, a batch process for treating the fine fraction (<200 μm) of a diesel oil-contaminated soil was carried out in an air-lift bioreactor. The oil concentration dropped from 12 000 to 2000 mg kg⁻¹ after 2 weeks of residence time (Blank-Huber et al., 1992). Stripping effects of the diesel oil contaminant were determined at about 20% by laboratory experiments.

### 11.3.4
### Sequential Batch Operation (Semicontinuous)

To overcome some of the limitations of batch processing, continuous operation in a plug-flow system is considered. As the contaminated solids travel through the system, specific conditions in the successive reactors develop. A practical way of achieving a plug flow system is to use a sequence of batch reactors (Fig. 11.10). In an experimental cascade of three stirred-batch slurry reactors, the contaminated solids (oil contamination) are periodically transferred to the next step (only part of the reactor slurry content is transferred, the so-called 'slug') (Apitz et al., 1994).

Figure 11.11 shows the breakdown pattern in the transferred slug during its residence in each individual reactor. The authors explained "that each successive stage of the cascade maintains a microbial consortium that is optimized to consume organic compounds of increasing complexity. When compared to the batch process the biocascade was shown to be more effective, both in terms of the rate and degree of degradation" (Apitz et al., 1994).

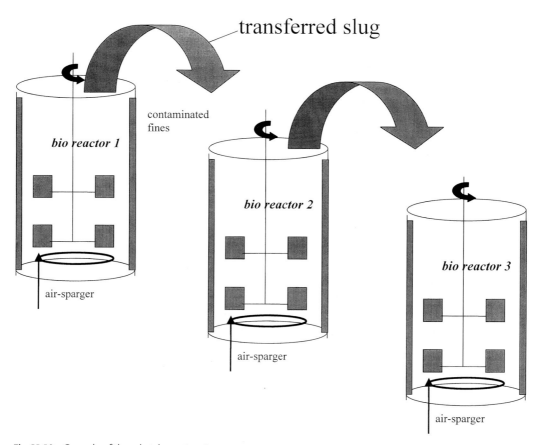

**Fig. 11.10** Cascade of three batch reactors in sequence.

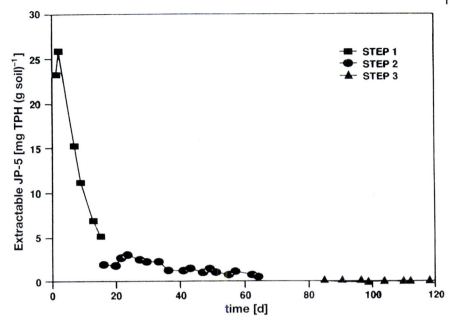

**Fig. 11.11** Level of contamination in one slug of material traveling through the sequenced batch process (three-step biocascade). Successive steps reflect the degradation of more recalcitrant components (Apistz et al., 1994; with permission).

### 11.3.5
### Continuous Operation

In the Netherlands a continuous plug-flow system has been achieved with the Slurry Decontamination Process (SDP) (Fig. 11.7). This process contains four major unit operations (Kleijntjens, 1991).

1. The contaminated solids are mixed with (process) water to form a slurry and are sized by passage over a vibrating screen. In this wet sieving step, debris is removed and a slurry having the proper density (about 30 w/w%) is prepared.
2. In the first reactor/separator (a tapered air-lifted bioreactor: the DITS reactor), the sand fractions are removed by means of a fluidized bed. Extensive organic material is removed by fine screening of the light material. In addition, agglomerates of contaminated fines are demolished owing to the power input and are thereby opened to biological breakdown (inoculation with the active biomass also takes place here).
3. In a second reactor stage, the contaminated fine fraction is treated. The second stage consists of a cascade of interconnected bioreactors (ISB cascade).
4. A dewatering stage completes the process; the water released is partly recirculated as process water to mix fresh solids into a slurry.

Figure 11.12 shows the configuration of the DITS bioreactor (Luyben and Kleijntjens, 1988). In the tapered bioreactor system, energy is introduced at the bottom by the simultaneous injection of compressed air and slurry that was recycled from the reactor content itself. Shown are the dual injectors, the settlers (used in the recycle flow), and the tapered vessel. This system has been built on scales of 400 L, 800 L, and 4 m³. It has been operated for 2.5 years to test various solid waste streams. The integral process was operated semicontinuously on a pilot scale (3 m³ working reactor volume).

Figure 11.13 shows the experimental results for a heavily polluted harbor sediment during a steady-state period of six weeks. Nutrients (nitrogen, phosphorous, and potassium) were added and the temperature was kept at 30 °C. The steady-state PAH concentration in the solids is shown as a function of time. The upper symbols show the feed concentrations in the slurry mill, which have an average of about 350 mg kg⁻¹ (the input data are scattered because of the heterogeneous feedstock). In the DITS reactor the steady-state concentration dropped to around 100 mg kg⁻¹ (first part of the microbial breakdown).

In the ISB cascade the average concentration dropped to 30–40 mg kg⁻¹. After dewatering, the final concentration increased somewhat in the filter cake. The overall

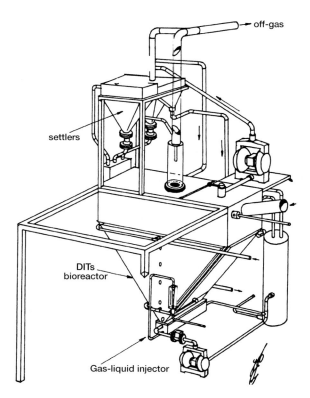

**Fig. 11.12** Technical schema of the DITS bioreactor.

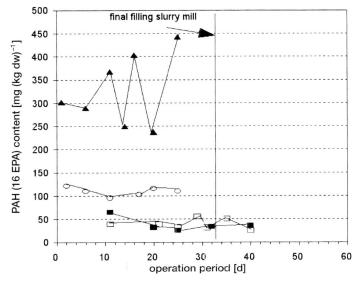

**Fig. 11.13** PAH breakdown pattern during steady-state treatment of sediment by the SDP process: ▲ slurry mill, ○ DITS reactor, ■ ISB reactor, □ press cake.

PAH degradation level was about 92% for an overall residence time of 16 d. No significant evaporation was measured in the off-gas. In this sediment, the degradation of mineral oil showed a similar pattern (Fig. 11.14); however, an overall degradation level of only 65% was achieved. Final concentrations after dewatering were around 3000 mg kg$^{-1}$. In this sediment, mineral oil seems to be less available for microbial breakdown than the PAH. Recycling standards for PAH could be achieved, but levels of oil contamination remained too high.

A contaminated soil treated by the slurry process showed a steady-state breakdown pattern (Fig. 11.15). Starting at moderate contamination levels of 1200–1300 mg kg$^{-1}$ mineral oil, final concentrations of <50 mg kg$^{-1}$ were achieved, which is well below the recycling standards.

Comparing the steady-state continuous results (Fig. 11.14) with the batch results (Fig. 11.6b) for the same sediment and contaminant, we can clearly see that the continuous mode results in much better conversion (final concentration in continuous mode is ±3000 mg kg$^{-1}$ vs. 7000 mg kg$^{-1}$ in batch mode). This improvement due to continuity is in accordance with observations on cascade breakdown (Apitz et al., 1994). However, despite the improvement shown in Figure 11.14, the Dutch recycling standard (mineral oil = 500 mg kg$^{-1}$) for this specific sediment was not met.

For a less contaminated solid waste (Fig. 11.15), the recycling standard was met without difficulty: a mineral oil starting concentration of 900 mg kg$^{-1}$ was decreased to an output concentration <50 mg kg$^{-1}$. If this oil conversion is compared with the results in Figure 11.14 (8000 mg kg$^{-1}$ of mineral oil input was decreased to 3000 mg kg$^{-1}$ output), we can see that, at higher concentrations, limited bioavailability seriously hinders microbial breakdown. As a consequence, it is often difficult to

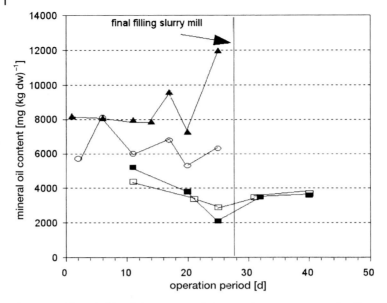

**Fig. 11.14** Mineral oil breakdown pattern during steady-state treatment of sediment by the SDP process: ▲ slurry mill, ○ DITS reactor, ■ ISB reactor, □ press cake.

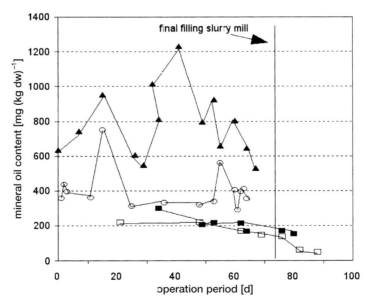

**Fig. 11.15** Mineral oil breakdown pattern during steady-state treatment of soil by the SDP process: ▲ slurry mill, ○ DITS reactor, ■ cascade, □ press cake.

achieve complete conversion and achieve low end concentrations when starting with high input concentrations.

We should note that analytical difficulties (especially for oil) in the handling of wet samples can easily lead to incorrect data. For example, in addition to a standard deviation of 20% within one laboratory, sample-exchange assays between various laboratories led to results that had a variation of more than 75% (Warbout and Ouboter, 1988). Bearing in mind these variations, analytical results should be regarded with care. This holds even more when samples of processed solids are compared to recycling standards.

## 11.4
## Solid-State Bioreactors

### 11.4.1
### Process Configuration

Systems operating under conditions of restricted moisture can be referred to as solid-state fermentations. A solid-state fermentation includes the growth of microorganisms, both fungi and bacteria, on a moist solid. In contrast to the slurry systems, solid-state systems operate with:
- limited pretreatment
- restricted energy input (no intense multiphase mixing)
- a limited amount of unit operations (the solids remain as a 'solid phase')

In a solid-state fermenter, process conditions are maintained by controlling the temperature, humidity, and aeration. In contrast to slurry systems, a compact, moist, solid phase dominates the reactor physics. Because handling the moist mass is tedious, solid-state fermentation typically takes place as a batch operation. Continuous operation depends on the possibility of practical and smooth loading and unloading procedures.

### 11.4.2
### Batch Operation: Composting

In line with its traditional role as a soil fertilizer, compost has been used as an additive in soil remediation. Composting mostly takes place in fixed-bed reactors (Fig. 11.3a). Basically, the compost is added to stimulate microbial breakdown. In experiments, soil contaminated with hydrocarbons has been mixed with compost in various ratios (soil–compost ratios of $2:1$, $3:1$, and $4:1$). In 3-L test batch reactors, the hydrocarbon degradation was >90% after a period of 44 d. Compared with the results in the absence of added compost, soil–compost systems had a much faster degradation rate and a lower end concentration (Lotter et al., 1990). In addition to composting, experiments focusing on the use of white-rot fungi have also been carried out (Schaeffer et al., 1995).

Another example of composting contaminated soil (in a column) was presented by Gorostiza et al. (1998). In a soil column the degradation of pentachlorophenol with and without added compost was tested; compost addition significantly enhanced the degradation rate (residence time about 60 d).

On a larger scale (in Finland), composting systems have been implemented by mixing contaminated soil with spruce bark chips as a bulking agent, lime, and nutrients beforehand. The bed is constructed as a biopile and can be turned by a tractor-drawn screw mixer. Degradation temperature rose during composting to as high as 44 °C. During five months the mineral oil concentration dropped from 2400 to 700 mg kg$^{-1}$ (Puustinen et al., 1995).

In another experiment, four biopiles of 10 m$^3$ each were created to treat chlorophenol-contaminated soil. Chalk, commercial fertilizer (NPK), and bark chips (as a bulk aeration agent) were added. After two months 80% of the contaminant was removed (Laine and Jorgensen, 1995).

## 11.4.3
### (Semi)Continuous Operation: The Rotating-drum Bioreactor

In a rotating-drum bioreactor (or revolving tubular reactor), the reactor itself acts as a rotating conveyer in which the contaminated solids are mixed (Fig. 11.3). In a (semi)continuous operation the solids are transported (screwed) through the system. Loading at the head of the process is combined with unloading at the back. The solids residence time in the system is determined by the shape of the screw blades, the rotation speed, the scale, and the amount of material present. Experiments have been carried out in a bioreactor with a length of 25 m and a diameter of 3.5 m. A warm air blower provided an appropriate temperature and oxygen, a sprinkler system was installed and nutrients added; the residence time was 14 d. For starting concentrations of mineral oil contaminant ranging from 1000 to 6000 mg kg$^{-1}$, at a temperature of 22 °C, the contaminant concentration decreased to 50–350 mg kg$^{-1}$. Most breakdown activity took place in the first few days (Munckhof and Veul, 1990).

In a revolving tubular reactor developed by BioteCon in Germany, having a length of 18 m and a diameter of 2.5 m, between 0.5 and 1 t h$^{-1}$ of soil can be treated (Kiehne et al., 1995). Model drums with diameters of 0.5 m and 1 m were tested. The typical moisture content was 15%, the revolution speed was 1 h$^{-1}$, and the aeration rate 0.125 VVM. In the trials the breakdown of motor fuel, kerosene, and phenol was investigated. Under batch conditions kerosene was degraded from 8300 mg kg$^{-1}$ to 500 mg kg$^{-1}$ in 300 h. When the same soil was treated in a repeated batch (25% of the treated soil was returned and mixed with new soil), within 180 h end concentrations of 180 mg kg$^{-1}$ were reached. The increase in microbial activity due to the reuse of biomass was considered responsible for the enhanced breakdown.

In an experiment a combination between heap leaching and reactor processing was tested. After treatment in the reactor for 120 h, kerosene-contaminated soil was deposited in a heap for four weeks. The treatment was finalized by a second reactor treatment for 100 h. With a total residence time in the reactor of 220 h, an overall conversion of 98.5% was achieved. It was concluded that for "contaminated soils in the order of 500–3000 t the revolving tubular reactor especially demonstrates its advantage" (Kiehne et al., 1995). The technology has been developed to operate on-site.

Another solid-state process was developed by Umweltschutz–Nord (Bremen, Germany). Using a steel tube having a length of 45 m and an diameter of 3 m, the contaminated soil is screwed through the Terranox system. For mineral oil-contaminated soils a treatment period of 6–8 weeks is used; the system can also be used for a combination of anaerobic and aerobic treatments.

On a smaller scale (3, 6, and 90 L) a blade mixing bioreactor was operated as a solid-state process (Hupe et al., 1995). The authors concluded that, in addition to the benefit of adding compost, moisture enhances the process; however, the water content should be <65% of the maximum moisture content to obtain solid-state processing conditions. The aeration rate is also a key parameter.

## 11.5
## Comparison of Bioreactors

Table 11.1 shows a technological and economical characterization of the three bioreactor systems. It is clear that composting is the 'low technology' type of bioreactor, requiring only limited technological infrastructure and investment. Slurry process-

**Table 11.1** Comparison of bioreactors.

| Technology | Economics |
|---|---|
| **Biological slurry processing** | |
| • Batch and (semi)continuous | • Substantial investment (equipment) |
| • Water addition | • Complex plant or trained operators required |
| • Energy input | • Large scale needed (economy of scale) |
| • Addition of nutrients and/or biomass | • Use as recycling technology |
| • Controlled conditions in closed systems | • Cost per ton determined by surface, machinery, energy, and labor costs |
| **Composting** | |
| • Batch operation | • Low investment |
| • Limited use of technology and machinery | • Large surface required |
| • Low energy input | • Use as decontamination technology only |
| • Addition of compost to the solids | • May be used on small scale |
| • Limited control options | • Cost per ton determined by surface and labor costs |
| • Longer treatment times | |
| **Rotating solids bioreactor** | |
| • (Semi)continuous | • Substantial investment (equipment) |
| • Limited control options | • Mobile plant on-site |
| • Moderate scale required | • Small volumes |
| • Moderate energy input | • Cost per ton similar to that of slurry systems |

ing requires a technological infrastructure, and rotating-drum systems have an intermediate position.

Experiments show that during batch processing, the microbial breakdown slows down at low concentrations and full conversion is not reached. Continuous processing significantly improves the results.

The economic feasibility of bioprocessing depends largely on the nature of the waste streams to be treated. Basically, waste streams having a large water content and a considerable fine fraction (which is predominantly contaminated with organic components) are suitable for bioprocessing.

From the economic point of view, continuous slurry processing in an off-site installation can be beneficial only on a large scale. Capacities >40 000–60 000 t a$^{-1}$ (depending on the local conditions) are needed to benefit from the economy of scale and thus to perform at acceptable cost levels. For batch-operated composting processes, smaller volumes, handled close to or on the site, may offer solutions if sufficient area is available and time is not limited. Rotating-drum systems in which solid-state processing is achieved are considered to perform well for smaller volumes (up to 3000 t) in a mobile plant used on-site.

## 11.6
## Conclusions and Outlook

### 11.6.1
### Conclusions

- In terms of product quality, energy consumption, and costs, bioprocessing has the capability to function as a useful waste recycling technology.
- Comparing the three bioreactor types considered, we can conclude that
  - slurry processing offers the best option for rapid, controlled integral treatment
  - composting can be used as a batch approach (biopiles) for smaller amounts
  - rotating-drum bioreactors are a solid-state alternative to slurry processing
- In the decontamination market, bioreactor processing should focus on wet waste streams having a large content of contaminated fines; the contamination should be essentially organic in character.
- Unambiguous analytical techniques and procedures and clear (international) recycling standards are needed to properly evaluate the various technologies, the end products, and calibration standards.
- Research has to identify the reasons for the hampered breakdown as measured in various field experiments. Especially for large input concentrations, the bioavailability (or even toxicity) of the contaminants may seriously interfere with optimal microbial breakdown conditions. Ways have to be found to overcome these limitations.
- To fully explore the possibilities of bioprocessing, notice should be taken of the benefits of (semi)continuous processing.

## 11.6.2
## Outlook

- The trend toward increasing use of environmental biotechnology will result in the availability of organisms able to degrade almost any organic contaminant. To transform the potentials of these organisms into feasible bioprocesses, the proper combination of bioreactors and separation techniques (to remove the clean fractions such as sand) is crucial.

- Various options focusing on the integration of bioreactors with landfarming, ripening, or phytoremediation will be explored. In addition, in situ treatment based on bioreactor research, such as biorestoration or bioscreens will be fully developed.

- Biotechnological processes will be explored and used on a larger scale. Owners of contaminated sites that might be treated by bioprocesses have to benefit from legal and financial stimuli such as eco taxes on waste disposal.

## References

Apitz, S. E., Pickwell, G. V., Meyer-Schulte, K. J., Kirtay, V., Douglass, E., A slurry bio-cascade for the enhanced degradation of fuels in soils, *Fed. Environ Restoration and Waste Minimization Conf.* Vol. 2, New Orleans **1994**, pp. 1288– 1299.

Balen van, A., Influence of the water activity on *Zymomonas mobils*, Thesis, Technical University of Delft, The Netherlands **1991**.

Blank-Huber, M., Huber, E., Huber, S., Hutter, J., Heiss, R., Development of a mobile plant, in: *Proc. Soil Decontamination Using Biological Processes*, Karlsruhe. Frankfurt/Main **1992**: Dechema.

Brady, N. C., *Nature and Properties of Soils*. New York **1984**: Macmillan.

Brummelen ten, E., Oostra, R., Pruijn, M., Weller, B., Bioremediation with the Fortec process: a safe harbour for sediment, *Int. Conf. Contaminated Sediments*, preprints Vol. 1, p. 405, Rotterdam, The Netherlands **1997**.

Building Material Act, *Bouwstoffenbesluit boden: en oppervlaktewaterbescherming*, TK 22683, Staatsdrukkerij, Den Haag, The Netherlands **1995**.

Crank, J., *The Mathematics of Diffusion*, 2nd edn. Oxford **1975**: Clarendon.

Cullinane, M. J., Averett, D. E., Shafer, R. A., Male, J. M., Truitt, C. L., Bradbury, M. R., *Contaminated Dredged Material*, New Jersey **1990**: Noyes Data Corporation.

Gorostiza, I., Susaeta, I., Bibao, V., Diaz, A. I., San Vicente, A. I., Salas, O., Biological removal of wood preservative (PCP) waste in soil, autochtonous microflora and effect of compost addition, in: *Proc. Contaminated Soil*, p. 1149. London **1998**: Thomas Telford.

Harmsen, J., in: *Possibilities and Limitations of Landfarming, On-Site Bioreclamation* (Hincheeve, Olfenbuttel, eds.). London **1991**: Butterworth Heinemann.

Hinze, J. O., *Turbulence*. New York **1959**: McGraw-Hill.

Hupe, K., Heerenklage, J., Luth, J., Stegmann, R., Enhancement of the biological degradation processes in contaminated soils, in: *Proc. Contaminated Soil* (Van de Brink, W. J., Bosman, R., Arendt, F., eds.), p. 853. Dordrecht **1995**: Kluwer.

INES, *Project Bioslib*, Stichting Europort Botlek Belangen, The Netherlands **1997**.

Jerger, D., Cady, D. J., Bentjen, S., Exner, J., Full scale bioslurry reactor treatment of creosote-contaminated material at south eastern wood preserving Superfund site, in: *Speaker Abstract* of *In-situ and On-site Bioreclamation*, 2nd Int. Symp. San Diego, CA **1993**.

Kiehne, M., Berghof, K., Muller-Kuhrt, L., Buchholz, R., Mobile revolving tubular reactor for continuous microbial soil decontamination, in: *Proc. Contaminated Soil* (Van de Brink, W. J., Bosman, R., Arendt, V., eds.), p. 873. Dordrecht **1995**: Kluwer.

Kleijntjens, R. H., *Biotechnological slurry process for the decontamination of excavated polluted soils*, Thesis, Technical University of Delft, The Netherlands **1991**.

Kleijntjens, R. H., Kerkhof, L., Schutter, A. J., Luyben, K. C. A. M., The slurry decontamination process, bioprocessing of contaminated solid waste streams, *Abstract* 2489, *Abstract Book of the 9th European Congress of Biotechnology*, July **1999**, Brussels.

Koning, M., Hupe, K., Luth, C., Cohrs, C., Stegmann, R., Comparative investigation into biological degradation in fixed bed and slurry reactors, in: *Proc. Contaminated Soil*, p. 531. London **1998**: Thomas Telford.

Laine, M., Jorgensen, S., Pilot scale composting of chlorophenol-contaminated saw mill soil, in: *Proc. Contaminated Soil* (Van de Brink, W. J., Bosman, R., Arendt, F., eds.), p. 1273. Dordrecht **1995**: Kluwer.

Levenspiel, O., *Chemical Reaction Engineering.* New York **1972**: Wiley.

Lotter, S., Stregmann, R., Heerenklage, J., Basic investigation on the optimization of biological treatment of oil contaminated soils, in: *Proc. Contaminated Soil* (Van de Brink, W. J., Bosman, R., Arendt, F., eds.), p. 967. Dordrecht **1990**: Kluwer.

Luyben, K., Kleijntjens, R., Werkwijze en inrichting voor het scheiden van vaste stoffen, *Dutch Patent* 8802728, Den Haag **1988**.

Munckhof, P., Veul, F., Production scale trials on the decontamination of oil polluted soil in a rotating bioreactor at field capacity, in: *Proc. Contaminated Soil* (Van de Brink, W. J., Bosman, R., Arendt, F., eds.). Dordrecht **1990**: Kluwer.

Nitschke, V., *Entwicklung eines Verfahrens zur mikrobiologischen Reinigung feinkörniger mit PAH belasteter Boden*, Thesis, Universität Gesamthochschule Paderborn, Germany **1994**.

Northcliff, S., Bannick, C., Paetz, A., International standardization for soil quality, in: *Proc. Contaminated Soil*, May 1998, Edinburgh, pp. 83–91. London **1998**: Thomas Telford.

Oostenbrink, I., Kleijntjens, R., Mijnbeek, G., Kerkhof, L., Vetter, P., Luben, K., Biotechnological decontamination using a 4 m³ pilot plant of the slurry decontamination process, in: *Proc. Contaminated Soil* (Van de Brink, W. J., Bosman, R., Arendt F., eds.). Dordrecht1995: Kluwer.

Puustinen, K., Jorgensen, S., Strandberg, T., Suortti, M., Bioremediation of oil contaminated soil from service stations: evaluations of biological treatment, in: *Proc. Contaminated Soil* (Van de Brink, W. J., Bosman, R., Arendt, F., eds.), p. 1325. Dordrecht **1995**: Kluwer.

Reynaarts, H., Bachmann, A., Jumelet, J., Zehnder, A., Effect of desorption and mass transfer on the aerobic mineralization of α-HCH in contaminated soils, *Environ. Sci. Technol.* **1990**, *24*, 1493.

Riser-Roberts, E., *Remediation of Petroleum Contaminated Soils*. Boca Raton, FL **1998**: CRC Press.

RIZA Report 97-067, *Hoofdrapport pilot sanering Petroleumhaven Amsterdam: monitoring en evaluatie.* SDU-Den Haag, The Netherlands **1997**.

Robra, K., Somitsch, W., Becker, J., Jernej, J., Schneider, M., Battisti, A., Off-site bioremediation of contaminated soil and direct reutilization of all oil fractions, in: *Proc. Contaminated Soil*. London **1998**: Thomas Telford.

Schaeffer, G., Hattwig, S., Unterste, M., Hupe, K., Heerenklage, J. et al., PAH degradation in soil: microbial activity or inoculation, in: *Proc. Contaminated Soil* (Van de Brink, W. J., Bosman, R., Arendt, F., eds.), p. 415. Dordrecht **1995**: Kluwer.

Schlegel, H. G., *Allgemeine Mikrobiologie*. Stuttgart **1986**: Thieme.

Sinder, C., Klein, J., Pfeifer, F., The DMT–Biodyn process, a suspension reactor for biological treatment of fine grained soil contaminated with PAH, Abstract 2812, *Abstract Book of the 9th European Congress of Biotechnology*, July 1999, Brussels **1999**.

Suzuki, M., Waste management according to Japanese experience, *4th World Congr. Chemical Engineering*, June 1991, Dechema, Frankfurt, Germany **1992**.

Teschner, M., Wehner, H., Chromatographic investigations on biodegraded crude oils, *Chromatographia* **1985**, *2*, 407–416.

Toro, Di D. M., Horzempa, L., Reversible and resistant components of PCB adsorption–desorption isotherms, *Environ. Sci. Technol.* **1982**, *16*, 594–602.

Warbout, J., Ouboter, P. S. H., Een vergelijking tussen verschillende methode om het gehalte aan minerale olie in grond te bepalen, *H₂O* **1988**, *21*, 1.

Werther, J., Wilichowski, M., Investigations in the physical mechanisms involved in washing processes, in: *Proc. Contaminated Soil* (Arendt, F., Hinsenveld, M., Van de Brink, W. J., eds.), pp. 907–920. Dordrecht **1990**: Kluwer.

Wu, S., Geschwend, P., Numerical modelling of sorption kinetics of organic compounds to soil and sediment particles, *Water Res. Des.* **1988**, *24*, 1373–1383.

# 12
# In-situ Remediation

T. Held and H. Dörr

## 12.1
## Introduction

The input of contaminants into soil and groundwater may lead to a persistent pollution. The methods for remediation of the environmental compartments contaminated with organic contaminants comprise, in addition to physical and chemical methods, also biological technologies. The technologies are subdivided into ex situ and in situ methods. Ex situ methods include excavation of the soil and subsequent treatment at the site (on-site) or elsewhere (off-site). In situ means that the soil remains in its natural condition during treatment. Generally, in situ technologies also include ex situ components, e.g., water- or gas-treatment plants.

The goal of in situ technologies is to mineralize the contaminants microbiologically to form harmless end products. Many biodegradation processes, especially of nonchlorinated compounds (e.g., mineral oil hydrocarbons), require aerobic conditions or denitrification. Although these contaminants may also be degraded under other conditions (e.g., iron reduction, sulfate reduction), these bioprocesses are not very effective and are therefore not used in enhanced bioremediation. These processes are productive, i.e., the contaminants serve as sources of carbon and energy. Chlorinated hydrocarbons usually require anaerobic conditions. They may be degraded productively (during dehalorespiration, chlorinated hydrocarbons serve as obligate electron acceptors) or co-metabolically. In the latter, microorganisms cannot grow on these contaminants but need an additional substrate for growth and biodegradation. Additional nutrient elements such as nitrogen and phosphate as well as electron acceptors are often lacking. In addition to such mineralization, other biochemical processes that result in a reduction in toxicity can also be applied in in situ remediation processes. These are
- cometabolic aerobic or anaerobic transformation
- humification
- precipitation
- solubilization
- volatilization

*Environmental Biotechnology. Concepts and Applications.* Edited by H.-J. Jördening and J. Winter
Copyright © 2005 WILEY-VCH Verlag GmbH & Co. KGaA, Weinheim
ISBN: 3-527-30585-8

With the use of these processes, a wide variety of contaminants seem to be treatable in situ, including

- mineral oil hydrocarbons
- monoaromatic and polyaromatic compounds
- chlorinated or nitrated aliphatics and aromatics
- inorganic ions, including simple and complex cyanides
- heavy metals

However, until now only a limited number of possible techniques have been developed for practical application. The main reason for the actual state of development is that the in situ degradability of contaminants is limited by numerous factors (Fig. 12.1). Most of these factors, like low solubility, strong sorption on solids, sequestration with high molecular weight matrices, diffusion into macropores of soils and sediments, scavenging by insoluble and lipophilic phases, lead to a limited mass transfer. This lack of bioavailability may prevent sufficient degradation and result in persistence of the contaminants.

While planning an in situ remediation process, one has to consider that the technology covers processes working on various scales (from nanometers to kilometers) (Fig. 12.2). Import and export of contaminants by bacterial cells and induction of the degradative enzymes occurs on the nanometer scale. Surface processes occur on the micrometer scale. In the range of micrometers to millimeters diffusion processes (soil micro pores) as well as micro-inhomogeneities occur. The meter scale represents small inhomogeneities, e.g., silt lenses in a sandy aquifer. Finally, the kilome-

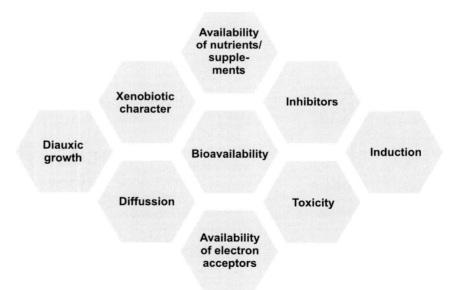

**Fig. 12.1** Limiting factors influencing the degradability of contaminants.

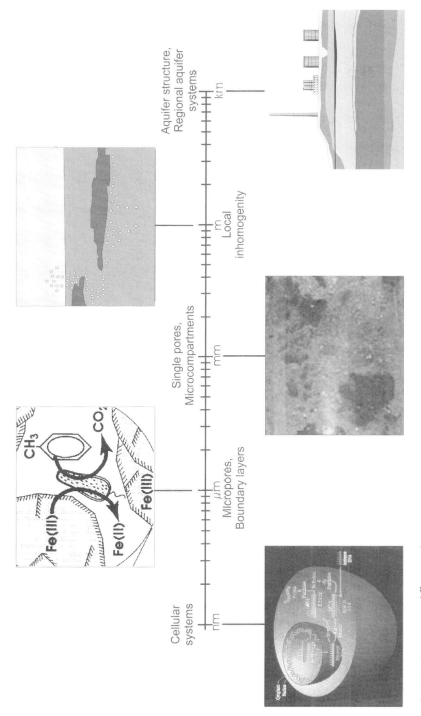

**Fig. 12.2** Processes on different scales.

ter scale covers local aquifer systems. From scale to scale, different constraints limit the processes. In general, the constraints of the largest scale, i.e., hydraulic site conditions, might be the overall limiting processes. Hydraulically favorable conditions are offered by course sediments with a hydraulic conductivity of $>10^{-4}$ m s$^{-1}$.

Some of the main features of in situ technologies are to transport supplements (nutrient salts, electron acceptors or donors) to the contaminants and to remove reaction products, as well as to induce proper environmental conditions for contaminant degradation. This may, e.g., be implemented by pumping of reaction-product–enriched groundwater, water cleaning, and reinfiltration of water enriched with nutrients.

The efficiency of in situ biodegradation depends importantly on the distribution of the nutrients in suitable concentrations within the aquifer. Biodegradation processes that require aerobic conditions may be limited by the lack of oxygen. The maximum $O_2$ concentration dissolved in water that can be infiltrated is given by its solubility of about 50 mg L$^{-1}$. The oxygen supply can be increased by using hydrogen peroxide ($H_2O_2$), which is decomposed to $O_2$ and water enzymatically or by geogenic components. However, the concentrations of $H_2O_2$ in the infiltration water are limited by its toxicity to bacteria at concentrations above 1000 mg L$^{-1}$ and the possibility of $O_2$ bubble formation, which would lead to a decrease in hydraulic conductivity and thus to decreased nutrient and oxygen supply.

The demand for oxygen or other electron acceptors decreases during the course of the remediation. In the beginning, electron acceptor consumption is usually limited by the supply rate. In a later phase, when all readily available compounds have been degraded, the electron acceptor consumption is limited by the rate of contaminant dissolution or diffusion out of micropores. Especially in this phase is it important, if $O_2$ is added, to use proper $O_2$ infiltration concentrations to avoid bubble formation.

The use of denitrifying conditions for contaminant degradation by the infiltration of nitrate may also lead to the formation of bubbles, because the end product of the denitrification is nitrogen ($N_2$), which has a comparably low solubility. If nitrate is overdosed or the exfiltration rate is too slow, this may result in $N_2$ degassing and bubble formation.

## 12.2
## Investigations

Each site has to be investigated not only for its geological, hydrogeological, and contaminant situation, but also for the site-specific degradability. The following description refers to laboratory methods for investigation of the microbiological degradability of contaminants. These methods are divided into two stages [1] (Fig. 12.3). During the first stage the general degradability is determined, i.e., whether the contaminants at the given site are sufficiently degradable. Conclusive demonstration of the fate of the contaminants can be obtained only by using tracer (e.g., $^{14}$C-labeled substances). However, this is possible only in closed laboratory systems and not on a

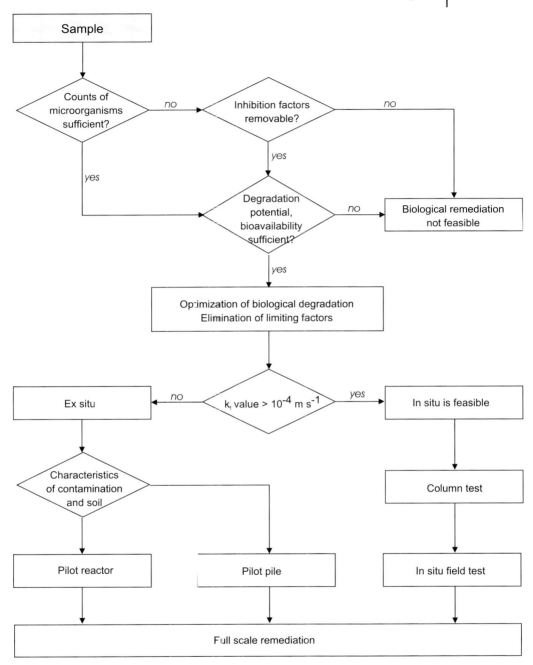

**Fig. 12.3** Laboratory test methods for investigating and optimizing microbial degradation of contaminants (according to [1]).

technical scale. Hence, it is necessary to use indirect parameters to demonstrate biodegradation.

Investigations of numerous sites have shown that the indigenous microflora often include specific contaminant degraders. Therefore, in principle, the degradation potential is given, unless the contamination reaches toxic concentrations. Environmental conditions are usually unfavorable for rapid degradation. In the first stage of investigation the maximum degradation rate is determined, instead of a realistic rate.

During the second stage, basic data for planning the remediation, such as nutrient demand, remediation duration, and achievable end concentrations, are determined by test methods that simulate the remediation technology (benchtop scale).

The results of all site investigations and of risk assessments are considered when planning the remediation. Today, sufficient experience is available to scale up the results of bench-top scale investigations to a technical scale. Pilot scale investigations are carried out when new technologies without sufficient practical experience are being considered, recalcitrant contaminants need to be treated, or the site exhibits special complex conditions.

The data collected for a specific site are used to describe the site (conceptual site model). From this site model, laboratory investigations are designed. The laboratory results are used to plan the technical scale remediation. For in situ remediation, it is reasonable to use an additional planning instrument – modeling. These models are fed with parameters determined during site and laboratory investigations.

## 12.3
## Remediation Technologies

### 12.3.1
### General Considerations

Figure 12.4 shows the typical characteristics of long-term pollution. Contaminants enter the soil and migrate downwards. The amount of contaminants remaining in the unsaturated zone is controlled by sorption, diffusion into soil pores, and retention by capillary forces. Generally, most contaminants have low solubility; hence, high amounts of contaminants may appear – when they reach the groundwater table. Depending on their buoyant density, a separate phase on top of the groundwater, called LNAPL (light nonaqueous-phase liquid), or at the base of the aquifer (groundwater zone), called DNAPL (dense nonaqueous-phase liquid) may occur. Minor amounts of contaminants are dissolved in the groundwater and are transported with the natural groundwater flow, forming a contaminant plume. The spatial extent of the plume and the contaminant concentrations depend on the duration of the pollution, the sorption and transport characteristics, and the efficiency of the natural biotic and abiotic degradation processes.

Depending on the characteristics of the contaminants and on the site conditions, different technologies have to be chosen. The technologies are divided into technol-

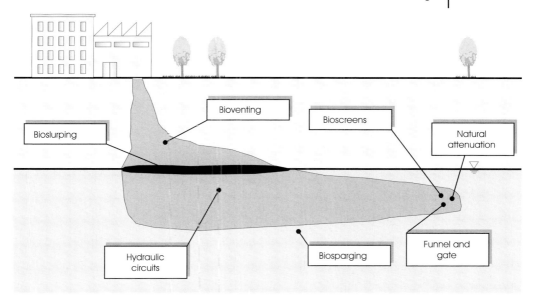

**Fig. 12.4** Localization of different microbial in situ technologies.

ogies for treatment of the unsaturated and saturated soils. A further subdivision comprises active and passive technologies. All these technologies can be combined with bioaugmentation, which is the addition (i.e., infiltration) of specific contaminant-degrading bacteria previously isolated and propagated in the laboratory. However, bioaugmentation is still controversial, because the majority of the infiltrated cells become attached to the soil within a few centimeters. Furthermore, establishing added microflora in an environmental compartment requires highly specific conditions [2]. Usually, the infiltration of nutrients during the remediation causes the added bacteria to be overgrown by the autochthonous microflora. Most technologies include not only degradation of the contaminants in situ but also physical removal of the contaminants (e.g., together with exhausted soil vapor or pumped groundwater), which requires an additional treatment step.

## 12.3.2
## Treatment of Unsaturated Soil (Bioventing)

The only biotechnological in situ technology available for treating unsaturated soil is bioventing. The process scheme is shown in Figure 12.5 [3] and is based on vacuum-enhanced soil vapor extraction. The pressure difference in the subsurface causes an inflow of atmospheric air and therefore an oxygen supply, as needed for aerobic contaminant degradation. Depending on site conditions, nutrients may need to be added, e.g., by sprinkling nutrient solutions on top of the soil or by installing horizontal infiltration drainage above the contaminant soil zone.

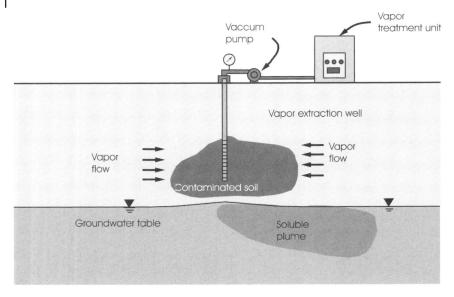

**Fig. 12.5** Process scheme of bioventing.

One of the most important tasks of process design is to ensure a sufficient air flow regime within the soil. In particular, the geometry of the exfiltration wells, the necessity of active or passive air injection wells, and the need for ground sealing need to be considered. High contaminant concentrations may clog the soil pores, leading to reduced efficiency of oxygen supply. If this occurs, a pulsed soil vapor extraction method may be of advantage.

If the contaminants to be treated are volatile, the extracted soil vapor has to be treated, e.g., by sorption of the contaminants on activated carbon or by biodegradation within a biofilter. Bioventing is applicable to treating petroleum hydrocarbons, aromatic hydrocarbons, and other comparable contaminants. Because soil vapor extraction dries the soil, the water budget of the soil can be adjusted with the nutrient addition. Optimum biodegradation requires a water content of 40%–60% of the maximum water holding capacity. A lower water content reduces the biodegradation rates, a higher water content leads to water-saturated zones in which air flow is not possible and aerobic biodegradation is prevented.

Bioventing is easy to monitor (Section 12.4). Under optimal conditions degradation rates of about 0.2–20 mg kg$^{-1}$ d$^{-1}$ for petroleum products in soils of medium permeability can be achieved. The degradation rates in the processes for the treatment of the saturated soil are influenced by many more parameters; hence; these rates diverge to a much higher extent.

## 12.3.3
## Treatment of Saturated Soil

### 12.3.3.1 Hydraulic Circuits

Hydraulic circuits comprise groundwater pumping, cleaning, addition of nutrients, and reinfiltration. The contaminants are degraded in the subsurface (in situ) or are removed with the groundwater and eliminated in the groundwater treatment plant. This technology is the first to be applied in situ and thus is the most understood. The technique has mainly been used to treat mineral oil contamination. Mineral oil hydrocarbons (MHC) are extractable to only a very small extent. During complete remediation less than 1% of the MHC can be removed with the exfiltrated groundwater; the rest has to be biodegraded in situ. However, monoaromatic hydrocarbons, a cocontaminant of mineral oil products, can be washed out to a greater extent. The rate of in situ degradation and exfiltration depends on the solubility of the contaminants, the kinetics of biodegradation, and the process technology. Because pumping and treatment of groundwater are expensive, the process may be designed to minimize the amount of groundwater to be pumped. Both the process and the complete in situ infrastructure, including the positioning and size of pumps and infiltration wells, can be chosen on the basis of process modeling.

Below, a specific hydraulic circuit design is described. At the site a gravel filter (0.5 m) was installed at the level of the groundwater table after and before refilling of the excavation on-site treated of unsaturated soil. The gravel filles contained drainage pipes that collected and transported the groundwater to the pumping wells. The pumped groundwater was sent to a water treatment plant, where Fe, Mn (to prevent clogging of the plant by hydroxides), and the contaminants were removed. The cleaned water was supplemented with nutrients (urea, phosphoric acid) and the electron acceptors hydrogen peroxide ($H_2O_2$) and nitrate ($NO_3^-$). The supplemented water was infiltrated at the bottom of the aquifer and pumped from the groundwater table. The resulting vertical groundwater flow direction improved uniform distribution of the nutrients. The locations of the pumping wells have to be chosen in so that no contaminants can escape from the site along with the natural groundwater flow. This may also be achieved by enclosing the site within a slurry wall reaching down to the aquiclude (groundwater-impermeable layer).

### 12.3.3.2 Special Groundwater Wells

A variety of special groundwater wells have been developed, which have two common features. They cause groundwater circulation and stripping within the well, resulting in an intensive throughput of groundwater and, therefore, an efficient supply of nutrients and electron acceptors (air oxygen or others, e.g., $H_2O_2$). Furthermore, volatile compounds are stripped within the well. The waste air is extracted and cleaned on-site.

The individual techniques are called groundwater circulation wells, in-well stripping, and BioAirliftT. The wells consist of a combined system of groundwater removal and infiltration within the wells. To allow circulation of groundwater the well is screened at the bottom and at the groundwater table. Both areas are separated by a cover pipe and bentonite sealing. A pipe is used to inject atmospheric air at the bottom of the well. This causes upstreaming of the water according to the principle of a mammoth pump and simultaneously a stripping of volatile compounds. The elevation of the water table within the well leads to infiltration of the oxygen-enriched groundwater at the top of the groundwater level. After circulating within the aquifer, the water enters the well again at the bottom.

Additional elements can include nutrient infiltration pipes within the well. An electric water pump may be installed instead of the mammoth pump. If so, no stripping and no oxygen enrichment of the groundwater occur. An electric pump may also be installed in addition to a mammoth pump. Furthermore, a permeable bioreactor containing immobilized contaminant-degrading bacteria can be installed between the points of water input and output. However, the water flow velocity is usually too high to allow significant degradation of the contaminants within the residence time in the bioreactor. Therefore, the reactor material also contains activated carbon. The contaminants are sorbed onto the carbon, which is reactivated by biodegradation of the contaminants within the reactor. The reactor may also contain other materials such as an ion exchanger (to remove heavy metals) or only activated carbon, if the contaminants are not sufficiently biodegradable (e.g., PAH).

If there is additional contamination of the unsaturated soil, the outlet of the groundwater may be installed above the groundwater table. With this type of operation the unsaturated zone (at least the groundwater fluctuation zone) can be flushed with nutrient-enriched water. Alternatively, the well can be combined with a vapor-extraction system. Here, bioventing of the unsaturated soil may also be induced. These special wells have been used to remediate sites polluted with petroleum hydrocarbons, aromatic hydrocarbons, or volatile chlorinated hydrocarbons.

### 12.3.3.3 Biosparging and Bioslurping

Both biosparging and bioslurping are not restricted to treatment of the saturated zone, but are also used for treatment of the unsaturated zone. Biosparging is the injection of atmospheric air into the aquifer (Fig. 12.6) [4, 5], which results in the formation of small branched channels through which air moves to the unsaturated zone. Outside these channels all processes are limited by diffusion. Therefore, highly branched channels are desired, which can be achieved by pulsing the air injection. Biosparging enhances the in situ stripping of volatile contaminants, desorption of contaminants, and their degradation by enriching the groundwater with oxygen. Because the contaminants are transported to the unsaturated zone, biosparging is usually combined with soil vapor extraction. Biosparging is applicable if the sparging point can be installed below the zone of contamination, because air flows upward, forming a cone. The angle of the cone and the degree of branching of the channels in a given soil depend mainly on the injection pressure, which should be only a little higher than the pressure of the water column. Usually the radius of influence of a biosparging well is determined by a pilot test at the site. At various distances from the sparging point, additional wells are installed to monitor groundwater level and

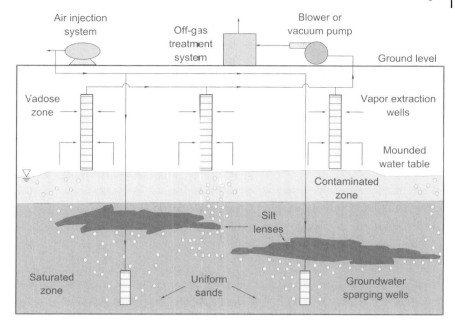

**Fig. 12.6** Process scheme of biosparging.

oxygen saturation. An increase of the groundwater table is observed only in the beginning of the treatment, and after the air channels are formed the groundwater returns to its original level. Biosparging is very sensitive to soil inhomogeneities. Zones of lower permeability may deflect the air channels. Zones of high permeability may act like open pipes channeling the air flow. In both cases the soil above these zones remains untreated. Additional nutrients to enhance microbiological degradation may be infiltrated with the same biosparging wells or with separate infiltration wells.

Bioslurping (also known as vacuum-enhanced recovery) is the only technology that also treats free product phases floating on top of the groundwater [3, 6]. Bioslurping wells (gas-tight at the well head) are mainly screened within the groundwater fluctuation zone. A suction pipe is directed within the free phase. By applying a vacuum, a mixture of free product, soil vapor, and groundwater is extracted. Above ground all three media (free product, waste air, and water) are separated. Free product is collected, and water and waste air are cleaned separately. The treatment plant can be small and run at comparatively low cost because only a small amount of groundwater and soil air is extracted. The main advantage of bioslurping is the horizontal flow of the free product. Compared to conventional recovery systems, e.g., by hydraulic gradients toward a pumping well, bioslurping does not enhance the smearing of the free phase to greater depths of the aquifer. The accompanying extraction of soil vapor leads to enrichment of the soil with oxygen, comparable to bioventing. Hence biodegradation in this zone is enhanced. Bioslurping can be combined with infiltration of nutrient salts into the vadose zone or into the saturated zone.

### 12.3.3.4 Passive Technologies

Experience shows that all active technologies require homogenous geological conditions. If these conditions do not pertain, passive technologies may be of advantage. Furthermore, the low solubility of hydrophobic organic contaminants and the slow diffusion of contaminants that have been within soil micropores for decades may result in a low efficiency of remediation technologies based on induced groundwater flow. Passive technologies are used at or near the end of the contaminant plume. They consist of constructed zones (reactors) in which the contaminants are degraded. If the zones cover the complete cross section of the plume, the technologies are called activated zone, bioscreen, reactive wall, or reactive trench. The main difference between these techniques is the need for soil excavation. Whereas activated zones or bioscreens are arranged without any soil management, the reactive wall and reactive trench techniques require construction of a subsurface bioreactor. Activated zones can be arranged, for example, as a line of narrow wells perpendicular to the direction of groundwater flow (Fig. 12.7). Incompletely passive technologies comprise alternating pumping and reinfiltration of groundwater in closed, directly linked loops, combined with an in-line nutrient amendment system consisting of a

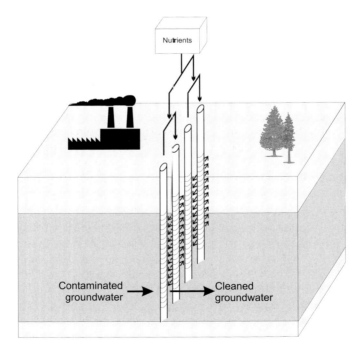

**Fig. 12.7** Process scheme of activated zones.

water-driven proportional feeder and a reservoir [7]. In this system the autochthonous microbial population is stimulated to adapt to a new and suitable redox situation and to develop the appropriate contaminant-degrading activity [8]. In such systems the hydraulic conductivity of the activated zone is the same as in the surrounding aquifer. Any alteration, e.g., by iron hydroxide precipitation, may lead to a reduction of hydraulic conductivity and a changed groundwater flow regime, which may result in deficient contaminant treatment.

Completely passive systems were developed because such exfiltration–infiltration loops consume significant amounts of energy. The advantage of passive systems is the highest when the energy demand is the lowest. This is true of completely passive systems, in which the remediation time is long and the contaminant load to be treated is usually low. In these completely passive systems, the wells are, for example, charged with solid cylinders consisting of highly permeable structural material (sand/cement) and so-called oxygen-release compounds (ORC®), which represent a proprietary $MgO_2$ formulation. This compound releases oxygen over a period of about 300 d. If the ORC is exhausted, the cylinders can be easily replaced with fresh ones.

Because degradation of numerous contaminants requires anaerobic conditions, it is necessary to supply electron donors (e.g., hydrogen) in completely passive systems. Hence, hydrogen-release compounds (HRC®) were developed. This product is used especially to enhance the in situ transformation of volatile highly chlorinated hydrocarbons. These viscous compounds must be supplied via high-pressure injections.

At present, many different substances are used for injection into the groundwater to construct so-called in situ reactive zones (IRZ), including $H_2O_2$ for aerobic biodegradation and molasses, whey, or chitin for anaerobic biodegradation. Current investigations are concerned with the electrochemical generation of hydrogen as electron donor ($2\ H_2O \rightarrow O_2 + 2\ H_2$). The cathode where $H_2$ is generated can be located within an anaerobic treatment zone, but the anode where $O_2$ is generated is located within an aerobic treatment zone.

Reactive walls or comparable systems are local zones in a natural porous medium exhibiting high contaminant retention capacity and increased bioactivity. For reactive walls, systems with high longevity and without significant maintenance or the necessity of nutrient replenishment are desirable. Reactive walls can be composed of a mixture of organic waste (compost, wood chips, sewage sludge, etc.) and of, e.g., limestone for pH correction. The organic waste serves as a nutrient source, a structural material to establish high permeability, and a source of bacteria. A carrier (e.g., activated carbon) coated with specific contaminant-degrading microorganisms can also be used. Reactive walls can be in place for the complete duration of the treatment. If so, the necessary amount of nutrients is calculated on the basis of mass balance. However, it is difficult to estimate the fraction of nutrient mass that will be available for contaminant removal. Alternatively, reactive walls can be constructed so that the wall material is exchangeable or restorable. The materials have to be homogenized prior to installation to avoid channeling within the wall. Online monitoring of wall permeability is necessary to avoid changes in the predicted groundwater flow regime. The thickness of the wall depends on the groundwater flow velocity within the wall, contaminant concentration, degradation rates, and the required concentrations at the downgradient side of the reactive wall. Long-time changes in the values of these parameters have to be considered. The use of a numeric groundwater flow model for designing the wall is helpful. At present, several types of reactive walls have been developed; however, experience on a technical scale is limited. Reactive walls are implemented as denitrification zones [9] or metal barriers with biprecipitation [10] (see Section 12.3.3.6). An adverse effect during metal precipitation is high concentrations of Fe(II) and Mn(II), because they consume most of the precipitation capacity. Furthermore, because the metals stay within the reactive wall, mainly as metal sulfides, the permeability of the wall may decrease with time. If environmental conditions will not change and the long-time stability of the insoluble metal sulfides is known, they may remain in the subsurface; otherwise, the wall material has to be removed.

In general, reactive walls can be used with all biochemical processes that eliminate pollutants. However, site-specific conditions always need to be considered to show which technology is feasible and also economical.

Funnel-and-Gate™ is a system that channels contaminated groundwater, usually at the front of a plume, by means of impermeable walls (the funnel) toward gates within the wall, where a reactor is located (Fig. 12.8). To design such systems, use of a groundwater model that also considers inhomogeneities of the subsurface is essential. Within the gates, the same bioprocesses may be installed as in reactive walls. However, the smaller width of the bioreactors is favorable, because constructive measures to exchange the reactor material (e.g. reactor material filled cassette) are easier to implement. However, the groundwater flow velocity is increased within the gates and the groundwater table may rise. Full scale experience with Funnel-and-Gate technology is still rare in Europe. In particular, the longevity can be calculated only by extrapolation of short-term monitored processes. The advantage of these passive systems is low operation costs over long periods, since only monitoring is necessary.

### 12.3.3.5 Natural Attenuation

Contaminant transport in the groundwater results in dilution of the contaminants. In a certain distance from the contaminant source, concentration ranges can be reached in which available nutrients and electron acceptors are sufficient to allow complete degradation of the residual contaminants, resulting in a steady state of contaminant spreading and degradation or even in shrinkage of the plume if the source is removed or exhausted. The use of this site feature as 'remediation' is possible only if the extent of the plume in time and space can be forecasted reliably. Furthermore, it is necessary that sensitive receptors (targets) are not affected.

An essential part of natural attenuation is detailed knowledge of the hydrogeology and the use of reactive groundwater modeling. The geological and hydrogeological parameters of the model (e.g., sorption coefficients, hydraulic conductivity, ground-

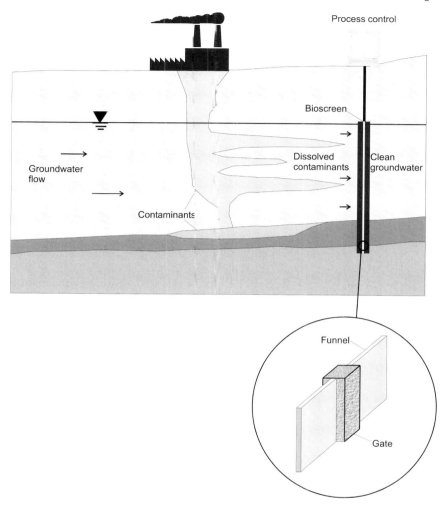

**Fig. 12.8** Process scheme of funnel-and-gate technology.

water flow velocity), as well as the biological degradation rates within individual redox zones, have to be determined by site-specific and/or laboratory investigations. These parameters and also the determined distribution of the contaminants are fed into the model. With the model the future extension of the plume can be estimated, allowing one to decide whether natural attenuation will be sufficiently effective. Experience has shown that natural attenuation is applicable in only a limited number of situations. However, it might be suitable as a post-treatment phase after active remediation. Natural attenuation already has been used to remove chlorinated ethenes, monoaromatic hydrocarbons (BTEX), and petroleum hydrocarbons [11].

### 12.3.3.6 Evolving Technologies

Most of the evolving technologies have been investigated only on the laboratory scale, but some have already been tested on a pilot or even technical scale. However, detailed experience and wide commercial use are not available. Until now heavy metals have been treated with physicochemical technologies. Although these elements are not 'degradable', they are not biochemically inert. Several microbiological transformations are known that mainly alter the physicochemical behavior of metals [12], including

- solubilization and sorption (bioleaching, biosorption, phytoremediation)
- precipitation (bioprecipitation)
- volatilization by alkylation

Solubilization was originally developed to support the mining of metals (Me) by leaching (bioleaching) [13]. Various metabolites play an important role in bioleaching, e.g., surfactants, chelators, and organic acids. Many bacteria, e.g., *Thiobacillus* sp. and *Leptospirillum ferrooxidans* are reported to be able to leach metals. The metals, which occur mainly in solid ore deposits ($MS_2$), are solubilized by two indirect mechanisms: via thiosulfate or via polysulfides and sulfur. Some metal sulfides are chemically attacked by Fe(III) hexahydrate ions, resulting in the formation of thiosulfate, which is oxidized to sulfuric acid. Other metal sulfides are attacked by Fe(III) and protons, which leads to the formation of polysulfides as intermediates and finally to elemental sulfur, which is biooxidized to sulfuric acid. The two mechanisms can be simplified by the following equations [14]:

- thiosulfate mechanism ($FeS_2$, $MoS_2$, $WS_2$):

$$FeS_2 + 6\ Fe^{3+} + 3\ H_2O \rightarrow S_2O_3^{2-} + 7\ Fe^{2+} + 6\ H^+ \tag{1}$$

$$S_2O_3^{2-} + 8\ Fe^{3+} + 5\ H_2O \rightarrow 2\ SO_4^{2-} + 8\ Fe^{2+} + 10\ H^+ \tag{2}$$

- polysulfide mechanism (e.g., ZnS, $CuFeS_2$, PbS):

$$MeS + Fe^{3+} + H^+ \rightarrow Me^{2+} + 0.5\ H_2S_n + Fe^{2+}\ (n \geq 2) \tag{3}$$

$$0.5\ H_2S_n + Fe^{3+} \rightarrow 0.125\ S_8 + Fe^{2+} + H^+ \tag{4}$$

$$0.125\ S_8 + 1.5\ O_2 + H_2O \rightarrow SO_4^{2-} + 2\ H^+ \tag{5}$$

The bioleaching method requires pumping of the groundwater and removal of the solubilized metals in a groundwater treatment plant. For treatment, biosorption may be used. In summary, although bioleaching is a pump-and-treat process, the higher solubility of the 'contaminants', i.e., the solubilized metals, may result in enhanced removal of the contaminants compared to classical pump-and-treat procedures.

Another technology for removing dissolved contaminants from soil and groundwater is phytoremediation: contaminants are taken up through the roots of plants and trees. The choice of plants depends on the characteristics of the contaminants and the soil as well as on the three-dimensional distribution of the contaminants, because the efficiency of phytoremediation is restricted to the depth of root growth. Some contaminants, e.g., nitroaromatics, are taken up but are not transported within the plant. Transformation of the contaminants is usually not effective enough for final elimination and is therefore of minor importance. For example, heavy metals can be accumulated without significant transformation by some plants to a very high extent. Hence, for final removal of the contaminants, the plants have to be harvested and eliminated. Because of the high water demand of some trees, phytoremediation can also be used for a hydraulic limitation of a pollution plume.

A main problem in the mining industry is acidification and contamination of groundwater with dissolved metal ions together with sulfate. With bioprecipitation, both problems can be solved together. Addition of an organic substrate (represented by 'CH$_2$O') leads to the formation of sulfide and an increase in the pH. The metals (Me$^{2+}$) are precipitated as a nontoxic metal sulfide in an abiotic reaction:

$$SO_4{}^{2-} + CH_2O \rightarrow H_2S + HCO^{3-} \tag{6}$$
$$Me^{2+} + H^2S \rightarrow MeS\downarrow + H^+ \tag{7}$$

Many metals can also be precipitated as metal carbonates or hydroxides. The reactions are disturbed by the presence of oxygen. However, O$_2$ is readily consumed when an organic substrate is added. Most formed metal sulfides are very stable even if the redox milieu recovers after finalization of the organic substrate addition. Remobilization can occur when the pH drops to low values. If excess formation of H$_2$S occurs, a second, aerobic treatment phase is required, in which surplus H$_2$S is reoxidized to sulfate. With bioprecipitation, contamination with at least the following metals can be treated: Pb, Zn, Cu, Cd, Ni The mobile, very toxic, hexavalent chromium ion (Cr$^{6+}$) has been reported to serve as an alternative electron acceptor for dissimilatory Fe(III)-reducing bacteria, which use reduction of Cr$^{6+}$ to the less mobile, nontoxic, trivalent chromium (Cr$^{3+}$), which precipitates as chromium hydroxide to gain energy for growth. Chromium hydroxide is stable also in aerobic environments. Comparable reactions are reported for uranium (U$^{6+}$). Fe(II) and S$^{2-}$ can also reduce Cr(VI) in nonenzymatic reactions.

In addition to these redox reactions, microorganisms can form and degrade organometals. Alkylation and dealkylation are carried out by a wide range of microorganisms, which changes several important parameters such as toxicity, volatility, and water solubility of the contaminants [15]. For example, arsenic can be removed from the vadose zone by this bioprocess. For microorganisms this process is linked to detoxification of the environment, because the volatile transformation products evaporate readily into the atmosphere. However, the volatile products are rather toxic. Therefore, using this bioprocess as a remediation technology requires extraction of the volatile products (e.g., by soil vapor extraction). The extracted contaminated soil vapor can be treated chemically: the compound is dealkylated and the product is removed in a gas scrubber. The following microbial pathways for reducing inorganic arsenic to volatile organic compounds can occur. Arsenate (As$^{5+}$) and arsenite (As$^{3+}$) are transformed to arsine (AsH$_3$) by bacteria, preferably under anaerobic conditions.

Fungi and some bacteria transform arsenate and arsenite to di- and trimethylarsine in an aerobic environment. The process can take place only when an appropriate C source is available.

$$As^{5+}, As^{3+} + C_{org} \rightarrow (CH_3)_2As + (CH_3)_3As \tag{8}$$

In addition, several other biotransformation reactions forming a variety of other products can occur. For example, $As^{5+}$ can be consumed as an electron acceptor during the oxidation of organic matter, resulting in its reduction to the more soluble and more toxic $As^{3+}$ [2]. Until now, the necessary environmental conditions for a controlled microbial arsenic volatilization have not been investigated. Although such reduction processes (leading to gaseous $As^{3-}$ compounds) are well documented from laboratory experiments to occur with significant transformation rates, substantial natural arsenic losses in the field have not yet been observed [16]. The bioprocess that adds methyl groups to metal ions, forming volatile organometals that volatilize from the environmental compartment, is a general detoxification process used by bacteria. It is well documented, e.g., for mercury (Hg) [17].

Concerning radioactive metals, we should note that a bacterium, *Deinococcus radiodurans*, was isolated that transforms, e.g., toluene, even under high doses of γ radiation. Only the organic pollutants were degraded; the radionuclide of course remained unaffected. The survival of this microorganism in a radioactive environment led to speculation that probably some as-yet unknown bacteria exist which will allow complete removal of radionuclides from the environment by bioleaching. Future progress may also be achieved by constructing genetically engineered microorganisms (GEM) that contain, e.g., a complete degradation sequence for final mineralization of contaminants, specific genes to resist unfavorable environments, or other improved biochemical features, like a deficiency in adhesion, which will be useful for bioaugmentation.

## 12.4
## Monitoring

The conditions for the success of a remediation measure are the controllability of its bioprocesses and the technological skills of the operators. The reactions occurring within an aquifer during in situ microbiological remediation are manifold, including both abiotic and biotic processes. In situ bioremediation can be successful only if the transport of nutrients and electron acceptors and donors to the contaminants and the removal of metabolic end products (e.g., $CH_4$, $N_2$) is sufficient. Otherwise, gas bubbles can form, resulting in a change in the transport process. For this reason, hydrogeological parameters have to be included in the monitoring. Remediation can be successful only if it is accompanied by close monitoring. The tasks of monitoring are therefore to obtain information for the following purposes:
- controlling the addition of electron acceptors, donors, and nutrient salts
- controlling biogeochemical site conditions

- demonstrating that remediation processes are occurring
- indicating when the remediation target has been reached

In a continuous remediation process, the monitoring data are used to optimize the bioprocess and to correct malfunctions.

When unsaturated soil is treated, the most important monitoring instrument is the in situ respiration detector, which determines the consumption of oxygen and formation of carbon dioxide in the soil vapor without soil vapor extraction. Because $CO_2$ formation can be influenced by abiotic $CO_2$ fixation within the soil matrix (e.g., as carbonates), only $O_2$ consumption allows good determination of the actual degradation rate with the help of stoichiometric factors. That is, we know that the mineralization of 1 g of mineral oil hydrocarbons requires the consumption of 3.5 g of $O_2$ (neglecting biomass formation). The change in the degradation rate with time allows assessment of the quality of the installed remediation measures. A decrease in the degradation rate may indicate a lack of water, nutrients, or degradable contaminants.

Monitoring remediation measures in saturated soil is often more complex. Usually, groundwater samples are taken and the following parameters are determined:
- contaminants
- metabolites (specific or as dissolved organic carbon, DOC)
- degradation end products ($CO_2$, $CH_4$, ethene, ethane)
- nutrient salts (e.g., $NH_4^+$, $PO_4^{3-}$)
- electron acceptors ($O_2$, $NO_3^-$, Fe(III), Mn(IV), $SO_4^{2-}$)
- reduced electron acceptor species ($NO_2^-$, Fe(II), Mn(II), $S^{2-}$)
- electron donors
- field parameters (pH, redox potential, electrical conductivity, temperature)
- optional bacterial counts (total counts, contaminant degraders, denitrifiers)

Because of the variety of possible metabolites, it is usually not possible to determine specific compounds separately; they are usually summarized as DOC. Sometimes (e.g., when water treatment plants are involved in the remediation process), more detailed information on the character of the DOC is required. This information can be supplied by LC–OCD (liquid chromatography with organic carbon detection) analysis, by which the DOC is divided into fractions of humic substances, building blocks, low molecular weight acids, amphiphilic substances, and polysaccharides. Monitoring the degradation end products is necessary if balancing of the remediation is required and complete biodegradation must be shown. However, until now no reliable methods for balancing technical-scale in situ processes are available. Usually a more pragmatically approach is chosen, i.e., only a limited number of parameters are determined. These parameters include monitoring the nutrients and electron acceptors or electron donors (in anaerobic processes) to ensure a sufficient supply of these supplements. The field parameters are monitored to ensure that the remediation measures have led to environmental conditions optimal for contaminant degradation. Here, an increase in the counts of specific contaminant-degrading bacteria is expected. However, investigations have shown that remediation measures strongly interfere with the indigenous microflora, changing the biodiversity in complex ways. The effects are not yet well understood.

Because many in situ technologies are based on controlled hydraulic conditions, it is evident that maintenance of the designed infiltration rate or groundwater flow rate is of utmost importance. Most pollutants exhibit low solubility and a high tendency to bind to the soil matrix, resulting in strong retardation of their transport. Elimination of the contaminants is enhanced by biodegradation at the place where the contaminants are located. Hence, transport of nutrients and electron acceptors and donors to the contaminants is necessary. Once the pollutants are solubilized, they are degraded prior to resorption. Thus, biodegradation increases the concentration gradient between sorbed/solid and dissolved substances, which accelerates desorption, dissolution, and diffusion of pollutants out of micropores (where they are not available for biodegradation). Changing the hydraulic conductivity of the aquifer strongly affects the progress of remediation. Hence, monitoring the hydraulic conductivity is necessary. A decrease in hydraulic conductivity can be caused by formation of gas bubbles (due to excess infiltration of $H_2O_2$ or high denitrification rates), precipitation of Fe oxide (during changing redox conditions), clogging with produced biomass, or shifting of fine soil particles caused by fast infiltration. By correlating an observed decrease in hydraulic conductivity with actual remediation measures, the influence of the remediation on the hydraulic conductivity can be minimized.

Generally, owing to the partitioning equilibrium of soluble gases between groundwater and soil vapor, monitoring of volatile contaminants, degradation end products ($CO_2$, $CH_4$), electron acceptor ($O_2$), and natural tracer radon ($^{222}Rn$) can be used to control the remediation. Degradation end products are monitored to balance the in situ processes. Monitoring $O_2$ can prevent oversaturation with respect to this electron acceptor. By monitoring the natural tracer radon, transport from groundwater to soil vapor and from soil vapor to the atmosphere and also dilution of soil vapor with atmospheric air can be determined.

## 12.5
## Outlook

Until now various in situ remediation technologies have been developed, and the growing experience with in situ technologies as well as ongoing research today allows many processes that occur in the subsurface to be predicted during remediation planning. This results in the application of economic remediation processes. Recent developments have considered natural limitations in a much better way than before, resulting in increased use of slower processes like the passive technologies. There is still a large potential for development of in situ technologies. Although today more efforts are made in environmental protection management, the current environmental situation shows that end-of-the-pipe technologies will still be needed for decades.

# References

1 Klein, J., Dott, W., *Laboratory Methods for the Evaluation of Biological Soil Cleanup Processes.* Frankfurt **1992**: Dechema.

2 Anderson, R. T., Lovley, D. R., Ecology and biogeochemistry of in situ groundwater bioremediation, in: *Advances in Microbial Ecology* Vol. 15 (Jones, ed.), pp. 289–350. New York **1997**: Plenum.

3 Nyer E. K., Palmer, P. L., Carman, E. P., Boettcher, G., Bedessem, J. M., Lenzo, F., Crossman, T. L., Rorech, G. J., Kidd, D., *In Situ Treatment Technology.* Boca Raton, FL **2001**: CRC Lewis.

4 Rogers, S. W., Ong, S. K., Influence of porous media, airflow rate and air channel spacing on benzene NAPL removal during air sparging. *Environ. Sci. Technol.* **2000**, *34*, 764–770.

5 Reddy, K. R., Kosgi, S., Zhou, J., A review of in situ air sparging for the remediation of VOC-contaminated saturated soils and groundwater, *Hazardous Waste – Hazardous Materials* **1995**, *12*, 97–117.

6 Rosansky, S. H., Kramer, J., Coonfare, C., Wickramanayake, G. B., Moring, B. K., Maughon, M. J., Casley, M. J., Optimization and monitoring of bioslurping systems for free-product recovery, in: *Risk, Regulatory, and Monitoring Considerations: Remediation of Chlorinated and Recalcitrant Compounds* (Wickramanayake, G. B., Gavaskar, A. R., Kelley, M. E., Nehring, K. W., eds.), pp. 339–348. Columbus, OH **2000**: Battelle Press.

7 Spuij, F., Alphenaar, A., de Wit, H., Lubbers, R., v/d Brink, K. et al., Full-scale application of in situ bioremediation of PCE-contaminated soil, in: *Pap. 4th Int. In Situ On-Site Biorem. Symp. 4*, pp. 431–437. Columbus, OH **1997**: Battelle Press.

8 Rijnaarts, H. H. M., Brunia, A., Van Aalst, M., In situ bioscreens, in: *Pap. 4th Int. In Situ On-Site Biorem. Symp. 4*, pp. 203–208. Columbus, OH **1997**: Battelle Press.

9 Robertson, W. D., Cherry, J. A., Long-term performance of the Waterloo denitrification barrier, *Land Contam. Reclam.* **1997**, *5*, 183–188.

10 Benner, S. G., Blowes, D. W., Ptacek, C. J., A full-scale porous reactive wall for prevention of acid mine drainage, *Ground Water Monit. Rem.* **1997**, *17*, 99–107.

11 Wiedemeier, T. H., Wilson, J. T., Kampbell, D. H., Miller, R. N., Hansen, J. E., Technical protocol for implementing intrinsic remediation with long-term monitoring for natural attenuation of fuel contamination dissolved in groundwater, *AFC Environ. Excellence Books.* **1995**: AFB, San Antonio, Texas.

12 Singh, B., Wilson, M. J., Geochemistry of acid mine waters and the role of microorganisms in such environments: a review, *Adv. Geoecol.* **1997**, *30*, 159–192.

13 Rawlings, D. E., Silver, S., Mining with microbes. *Biotechnology* **1995**, *13*, 773–778.

14 Schippers, A., Sand, W. Bacterial leaching of metal sulfides proceeds by two indirect mechanisms: via thiosulfate or via polysulfides and sulfur, *Appl. Environ. Microbiol.* **1999**, *65*, 319–321.

15 White, C., Gadd, G. M., Reduction of metal cations and oxyanions by anaerobic and metal-resistant microorganisms: chemistry, physiology, and potential for the control and bioremediation of toxic metal pollution, in: *Extremophiles: Microbial Life in Extreme Environments* (Horikoshi, K., Grant, W. D., eds.), pp. 233–254. New York **1998**: Wiley-Liss.

16 Stoeppler, M., Arsenic, in: *Elements and Their Compounds in the Environment. Occurrence, Analysis and Biological Relevance*, 2nd Ed. (Merian, E., Anke, M., Ihnat, M., Stoeppler, M. eds.), pp. 1321–1364. Weinheim **2004**: Wiley-VCH.

17 Matilainen, T., Involvement of bacteria in methyl mercury formation in anaerobic lake waters, *Water Air Soil Pollut.* **1995**, *80*, 757–764.

# 13
# Composting of Organic Waste

Frank Schuchardt

## 13.1
## Introduction

Composting is the biological decomposition of the organic compounds of wastes under controlled aerobic conditions. In contrast to uncontrolled natural decomposition of organic compounds, the temperature in waste heaps can increase by self-heating to the ranges of mesophilic (25–40 °C) and thermophilic microorganisms (50–70 °C). The end product of composting is a biologically stable humus-like product for use as a soil conditioner, fertilizer, biofilter material, or fuel.

The objectives of composting can be stabilization, volume and mass reduction, drying, elimination of phytotoxic substances and undesired seeds and plant parts, and sanitation. Composting is also a method for decontamination of polluted soils. Almost any organic waste can be treated by this method. The pretreatment of organic waste by composting before landfilling can reduce the emissions of greenhouse gases.

In any event, composting of wastes is conducted with the objective of high economic effectiveness and has the goal of compost production with the lowest input of work and expenditure. The consequence of this approach is the effort to optimize the biological, technical, and organizational factors and elements that influence the composting process. The factors that influence the composting process are well known and have been published in numerous reviews and monographs. The period since 1970 has been characterized by the development of new strategies, composting processes, and technologies and the optimization of existing processes against the background of an expanding market for composting technology. Among others, reasons for these developments are rising costs for sanitary landfills, improved environmental protection requirements, as well as new laws, ordinances, and regulations. The realization that resources are limited and the idea of recycling refuse back to soil have also provided important impetuses for developments in this field.

Numerous publications and reviews on composting are published in specific journals such as *Compost Science and Utilization*, *Bioresource Technology*, and *Biosystems Engineering*. Some monographs give an overview of the field of composting [1–8].

*Environmental Biotechnology. Concepts and Applications*. Edited by H.-J. Jördening and J. Winter
Copyright © 2005 WILEY-VCH Verlag GmbH & Co. KGaA, Weinheim
ISBN: 3-527-30585-8

## 13.2
## Waste Materials for Composting

The origins of organic waste for composting are households, industry, wastewater treatment plants, agriculture, horticulture, landscapes, and forestry (Table 13.1). The amount, composition, and physical characteristics of plant wastes are influenced by numerous factors such as the origin, production process, preparation, season, collecting system, social structure, and culture. The wide range in the amount and composition of waste requires analyses for planning a composting plant and for estimating the compost quality in each individual situation.

**Table 13.1** Nutrient content per dry weight of some wastes for composting [3; own analyses].

| Waste | $VS^a$ [%] | C/N [-] | N [%] | $P_2O_5$ [%] | $K_2O$ [%] | CaO [%] | MgO [%] |
|---|---|---|---|---|---|---|---|
| Kitchen waste | 20–80 | 12–20 | 0.6–2.2 | 0.3–1.5 | 0.4–1.8 | 0.5–4.8 | 0.5–2.1 |
| Biowaste | 30–70 | 10–25 | 0.6–2.7 | 0.4–1.4 | 0.5–1.6 | 0.5–5.5 | 0.5–2.0 |
| Garden and green waste | 15–75 | 20–60 | 0.3–2.0 | 0.1–2.3 | 0.4–3.4 | 0.4–12 | 0.2–1.5 |
| Garbage | 25–50 | 30–40 | 0.8-1.1 | 0.6–0.8 | 0.5–0.6 | 4.4-5.6 | 0.8 |
| Feces (human) | 15–25 | 6–10 | 2 | 1.8 | 0.4 | 5.4 | 2.1 |
| Wastewater sludge (raw) | 20–70 | 15 | 4.5 | 2.3 | 0.5 | 2.7 | 0.6 |
| Wastewater sludge (anaerobic stabilized) | 15–30 | 15 | 2.3 | 1.5 | 0.5 | 5.7 | 1.0 |
| Dung | | | | | | | |
|   Cattle | 20.3 | 20 | 0.6 | 0.4 | 0.7 | 0.6 | 0.2 |
|   Horses | 25.4 | 25 | 0.7 | 0.3 | 0.8 | 0.4 | 0.2 |
|   Sheep | 31.8 | 15–18 | 0.9 | 0.3 | 0.8 | 0.4 | 0.2 |
|   Pigs | 18.0 | 15–20 | 0.8 | 0.9 | 0.5 | 0.8 | 0.2 |
| Liquid manure | | | | | | | |
|   Cattle | 10–16 | 8–13 | 3.2 | 1.7 | 3.9 | 1.8 | 0.6 |
|   Pigs | 10–20 | 5–7 | 5.7 | 3.9 | 3.3 | 3.7 | 1.2 |
|   Chickens | 10–15 | 5–7 | 9.8 | 8.3 | 4.8 | 17.3 | 1.7 |
| Beet leaves | 70 | 15 | 2.3 | 0.6 | 4.2 | 1.6 | 1.2 |
| Straw | 90 | 100 | 0.4 | 2.3 | 2.1 | 0.4 | 0.2 |
| Bark, fresh | 90–93 | 85–180 | 0.5–1.0 | 0.02–0.06 | 0.03–0.06 | 0.5–1 | 0.04–0.1 |
| Bark mulch | 60–85 | 100–130 | 0.2–0.6 | 0.1–0.2 | 0.3–1.5 | 0.4–1.3 | 0.1–0.2 |
| Wood chips | 65–85 | 400–500 | 0.1–0.4 | 1.0 | 0.3–0.5 | 0.5–1.0 | 0.1–0.15 |
| Leaves | 80 | 20–60 | 0.2–0.5 | — | — | — | — |
| Reed | 75 | 20–50 | 0.4 | — | — | — | — |
| Peat | 95–99 | 30–100 | 0.6 | 0.1 | 0.03 | 0.25 | 0.1 |
| Paunch manure | 8.5–17 | 15–18 | 1.4 | 0.6 | 0.9 | 2.0 | 0.6 |
| Grape marc | 81 | 50 | 1.5–2.5 | 1.0–1.7 | 3.4–5.3 | 1.4–2.4 | 0.2 |
| Fruit marc | 90–95 | 35 | 1.1 | 0.6 | 1.6 | 1.1 | 0.2 |
| Tobacco | 85–88 | 50 | 2.0–2.4 | 0.5–6.6 | 5.1–6.0 | 5.0 | 0.1–0.4 |
| Paper | 75 | 170–180 | 0.2–1.5 | 0.2–0.6 | 0.02–0.1 | 0.5–1.5 | 0.1–0.4 |

[a] VS Volatile solids.

**Table 13.2** Heavy metal content per dry weight of some wastes for composting [4].

| Waste | Zn [mg kg⁻¹] | Cu [mg kg⁻¹] | Cd [mg kg⁻¹] | Cr [mg kg⁻¹] | Pb [mg kg⁻¹] | Ni [mg kg⁻¹] | Hg [mg kg⁻¹] |
|---|---|---|---|---|---|---|---|
| Biowaste | 50–470 | 8–81 | 0.1–1 | 5–130 | 10–183 | 6–59 | 0.01–0.8 |
| Green waste | 30–138 | 5–31 | 0.2–0.9 | 28–86 | 24–138 | 9–27 | 0.1–3.5 |
| Paper | 93 | 60 | 0.2 | 4 | 20 | 1 | 0.08 |
| Paper, printed | 112 | 66 | 0.2 | 31 | 78 | 3 | 0.04 |
| Paper, collected | 40 | 21 | 0.2 | 3 | 12 | 1 | 0.06 |
| Paper sludge | 150–1500 | 15–100 | 0.1–1.5 | 30–300 | 70–90 | 5–15 | 0.2–0.5 |
| Bark | 150–300 | 40–60 | 0.6–2.1 | 30–63 | 20–57 | 12–20 | 0.1–0.5 |
| Bark mulch | 40–500 | 10–30 | 0.1–2 | 500–1000 | 50–100 | 30–60 | 0.1–1 |
| Wood chips | 58–137 | 8–11 | 0.1–0.2 | 6–8 | 13–53 | 4 | 0.1 |
| Grapes marc | 60–80 | 100–200 | 0.5 | 2.5–7.4 | 10 | 1–4 | 0.02–0.04 |
| Fruit marc | 20–30 | 9.5 | 0.2 | 0.02–1 | 0.3–1 | 2–4 | 0.03 |
| Brewers' grains | 13 | 6 | 0.3 | 16 | 10 | 16 | 0.04 |
| Oil seed residues | 4 | 1 | 0.03–0.05 | 0.1 | 0.1–0.4 | 1–3 | 0.01 |
| Cacao hulls | 89 | 7–12 | 0.25 | 0.5 | 0.4 | 0.3 | 0.02 |

The content of heavy metals and organic compounds in the waste is of great importance, particularly with regard to the use of compost as soil conditioner and fertilizer (Table 13.2). Ways to reduce the heavy metal content in compost from biowaste are to collect the waste separately and to obtain detailed information of the producers of the waste. The fact that the concentrations of heavy metals increase as the organic compounds are degraded also must be taken into consideration.

Analyses of biowaste show a content of impurities (e.g., plastics, glass, metals, rocks) between 0.5% and 5.0%, depending on the social structure, buildings, and public relations work in the area [2]. In densely built-up areas the contamination is higher than in others. More than 90% of the impurities have a size >60 mm, and 90% of the biowaste materials have a size <60 mm [1].

Depending on climatic and cultural conditions (e.g., growing and harvest times, holidays, traditions) some plant wastes, like leaves or branches, are not available during the whole year. An important factor for the operation of a composting plant can be the fluctuation in the composition of the wastes, in particular, the water content [5].

## 13.3
## Fundamentals of Composting Process

Degradation of the organic compounds in waste during composting is initiated predominately by a very diverse community of microorganisms: bacteria, actinomyctes, and fungi [7, 9, 10, 18, 23, 24, 28, 33, 36]. Just as in biological wastewater treatment, an additional inoculum for the composting process is not generally necessary, because of the high number of microorganisms in the waste itself and their short generation time. Invertebrate animals play no role in the rotting process during the first phase at a high temperature level. Nevertheless, earthworms are sometimes used in waste management and to produce a high-value compost [1, 8, 20, 27]

Rotting waste material, even during well aerated composting, is characterized by aerobic and anaerobic microbial processes at the same time (Fig. 13.1) [13]. The relation between aerobic and anaerobic metabolism depends on the physical properties of the waste/compost [31], including the structure of the heap, its porosity, its water content and capacity, its free air space, and the availability of nutrients.

The aerobic microorganisms in the rotting material need free water and oxygen for their activity. End products of their metabolism are water, carbon dioxide, $NH_4$ (or, at higher temperature and pH >7, $NH_3$), nitrate, nitrite (nitrous oxide as a product of nitrification), heat, and humus or humus-like products. The waste air from the aerobic metabolism in compost heaps contains evaporated water, carbon dioxide, ammonia, and nitrous oxide. The end products of the anaerobic microorganisms are methane, carbon dioxide, hydrogen, hydrogen sulfide, ammonia, nitrous oxide, nitrogen gas (both from denitrification) and water as liquid [5, 12, 14–17, 19, 21, 25, 26, 30, 34]

Mature compost consists of components that are difficult to digest or undegradable components (lignin, lignocellulosics, minerals), humus, microorganisms, water, and mineral nitrogen compounds. The organisms that take part in the composting process are microorganisms (bacteria, actinomyces, mildews) in the first phase

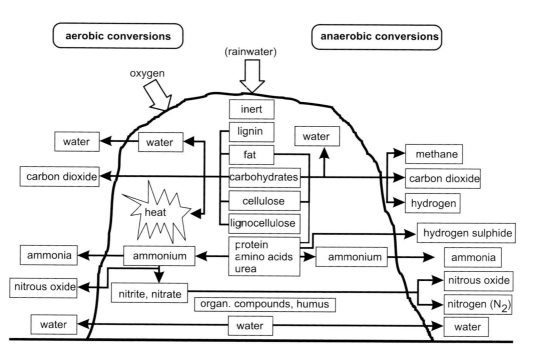

**Fig. 13.1** Substrates and products of microbial activity in a compost heap.

of composting. They each have optimal growing conditions at different temperatures: psychrophilics between 15 and 20 °C, mesophilics between 25 and 35 °C, and thermophilics between 55 and 65 °C. In mature compost at temperatures below 30–35 °C, other organisms such as protozoa, collembolans, mites, and earthworms join in the biodegradation.

A pile of organic wastes consists of solid, liquid, and gaseous phases, and the microorganisms depend on free water for their metabolism (Fig. 13.2). Dissolved oxygen, from the gas phase in the heap, must be available for the activity of aerobic microorganisms. To make sure that oxygen transfer from the gas phase to the liquid phase and carbon dioxide transfer from the liquid phase to the gas phase occur, a permanent partial-pressure gradient must be maintained, which is possible only by a permanent exchange of the gas phase by forced or natural aeration [2, 22]. Specifications about the optimal water content for composting are meaningful only in combination with the knowledge of the specific type of waste to be composted, its structure and volume of air pores (Table 13.3) [35]. In general, the water content can be higher when the waste structure, air pore volume, and water capacity are higher and more stable (also during the rotting process). Theoretically, the water content for composting can be 100%, provided the oxygen supply is sufficient for microbial activity.

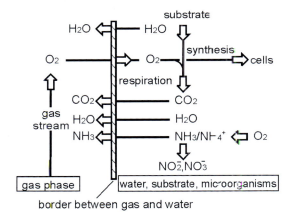

**Fig. 13.2** Metabolism of aerobic microorganisms at the gas/water interface.

Table 13.3   Optimal water content and structure of wastes for composting.

| Waste | Water Content [%] | Structure | Air Pore Volume [%] |
|---|---|---|---|
| Woodchips, cut trees, and brushwood | 75–90 | good | >70 |
| Straw, hay, cut grass | 75–85 | good | >60 |
| Paper | 55–65 | middle | <30 |
| Kitchen waste | 50–55 | middle/bad | 25–45 |
| Sewage sludge | 45–55 | bad | 20–40 |

In addition to a sufficient content of free water, the microorganisms need a C/N ratio in the substrate of 25–30 for optimal development and fast enough rotting process, and the carbon should be readily available. At C/N ratios below the optimum, the danger of nitrogen loss as ammonia gas increases (especially when the temperature rises and the pH is >7). If the C/N ratio is higher than optimum, the composting process needs a longer time to stabilize the waste material.

Figure 13.3 shows the relations between the factors influencing the rotting process. The structure of the waste, i.e., its consistency and the configuration and geometry of the solids, determines the pore volume (whether filled with water or air) and the air flow resistance of a compost heap. These in turn influence the gas exchange and the oxygen and carbon dioxide concentrations in the air pores and liquid phase. When these factors are optimum, exothermic microbial activity is rapid, leading to increasing temperature by build up of heat within the heap. Microbial activity is affected by the water content, nutrients (C/N ratio, availability), and pH. The mass and volume of the heap influence the temperature according to the heat capacity and heat losses by irradiation. Heat convection within the heap, which is conditioned by the temperature difference between the material and the atmosphere, affects the gas exchange. The gas exchange and the temperature influence the evaporation of water and thus also the proportion of water-filled pores [3, 4, 6, 11, 22, 29].

One effect of the activity of the different microbial groups is a characteristic temperature curve during composting (Fig. 13.4). After a short lag, the temperature increases exponentially to 70–75 °C. At 40 °C there is often a lag during the changeover from mesophilic to thermophilic microorganisms. After reaching a maximum, the temperature declines slowly to the level of the atmosphere. The progression of the temperature curve depends on numerous factors such as the kind and preparation

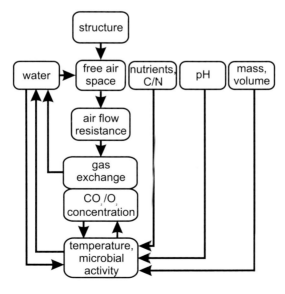

**Fig. 13.3** Factors influencing the composting process.

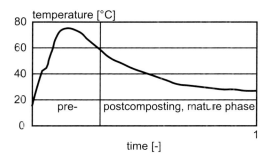

**Fig. 13.4** Characteristic temperature curve during composting process.

of the waste, the surface/volume ratio of the heap, air temperature, wind velocity, aeration rate, C/N ratio, processing technique, and mixing frequency.

The first phase of the composting process, up to a temperature up to about 60 °C, is called the pre- and main composting; the second phase is called the post-composting or mature phase. Both phases are characterized by different processes (Table 13.4).

Frequently, the designers of a composting facility must consider both phases by dividing the entire composting process into different technical stages, especially when the wastes have a risk of strong odor emissions:

• Pre- and main composting occurs in closed reactors or in roofed facilities, and in frequently-mixed or forced-aerated windrows.
• The post-composting/mature phase is done in windrows.

The consequences for the composting process are basically to optimize the factors that influence the rotting process. The most important factor is, for a given composition of waste, to ensure gas exchange in the heap. This can be done by taking the following measures:

• adapting the height of the heap to the structure, water content, and oxygen demand (high during pre- and main composting, low during the mature phase)
• turning (mixing, loosening) the windrows
• constructing windrows in thin, ventilatable layers

**Table 13.4** Phases and characteristics of the composting process.

| Pre- and Main Composting | Postcomposting, Mature Phase |
|---|---|
| Degradation of easily degradable compounds: sugar, starch, pectin, protein | Degradation of difficult-to-decay degradable compounds: hemicellulose, wax, fat, oil, cellulose, lignin |
| Inactivation of pathogenic microorganisms and weed seeds | Composition of high molecular weight compounds (humus) |
| High oxygen demand | Low oxygen demand |
| Emissions of odor and drainage water | Low emissions |
| Time: 1–6 weeks | Time: 3 weeks to 1 year |

- mixing and loosening the rotting material in reactors (in rotating drums, with tools)
- using forced aeration
- decreasing the streaming resistance by adding bulking material having a rough structure or in the form of pellets

**13.4**
**Composting Technologies**

The production of compost consists of preparing and conditioning the raw material, followed by the actual composting (Fig. 13.5). To produce a marketable product it is necessary to convert the compost to an end product. The aim of raw material preparation and conditioning is to optimize conditions for the following composting process, to remove impurities so as to protect the technical equipment, to reduce the input of heavy metals and hazardous organic components (if the impurities contain these components), and to meet quality requirements for the finished compost. The basic steps of raw material preparation and conditioning are:

- disintegration of rough wastes (e.g., wood scraps, trees, brush, long grass) by chopping, crushing, or grinding to increase the surface area available for microbial activity
- dehydration or (partial) drying of water-rich, structureless wastes (e.g., sludge, fruit remains) if they are too wet for the composting process
- addition of water (fresh water, wastewater, sludge) if the wastes are too dry for the composting process

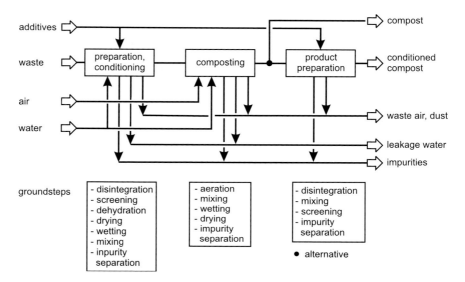

**Fig. 13.5**  Basic flow sheet of compost production.

- mixing of components (e.g., wet and dry wastes, N-rich and C-rich wastes, wastes with rough and fine structure)
- manual or automatic separation of impurities (glass, metals, plastics)

The products of preparation and conditioning of the wastes are waste air (depending on the composition and the conditions of storage, it may include bad smells and dust) and possibly drainage water beneath the raw material. The basic steps of the subsequent composting process may be:

- aeration to exchange the respiration gases oxygen and carbon dioxide and to remove water (the only essential step during composting)
- mixing to compensate for irregularities in the compost heap (e.g., dry zones at the surface, wet zones at the bottom, cool zones, hot zones) and to renew the structure for better aeration
- moistening of dry material to improve microbial activity
- drying of wet material by aeration or/and mixing to increase the free air pore space for microbial activity or to improve the structure of the compost for packaging
- manual removal of impurities

The products of the composting process are a biologically stabilized compost, waste air, and drainage water (when the material is very wet). It may be necessary to prepare the compost for transport, storage, sale, and its application. When post-preparation is needed, the basic steps can be:

- sieving the compost to obtain different fractions for marketing or to remove impurities
- manually or automatically removing impurities
- drying wet compost to prevent formation of a clumpy, muddy product and drainage of water during storage
- disintegrating clumps in the compost by crushing or grinding to prevent problems that may occur when the fertilizer is packaged
- mixing the compost with additives (soil, mineral fertilizer) to produce potting mixes or gardening soils.

Disintegration (crushing, chopping, grinding), especially of bulky wastes containing wood pieces, is necessary to increase the surface area available for the microorganisms and to ensure the functioning of the machines and equipment used in subsequent stages of the process (e.g., turning machine or tools, screens, belt conveyor). The intensity of disintegration depends on the velocity of the biodegradation of the waste, the composting process, the dimensions of the heap, the composting time, and the intended application of the final product. For disintegration of organic wastes, chopping machines or various kinds of mills (cutting, cracking, hammer, screw) are mainly used [1].

The raw waste or compost is screened to separate particles with a required granule size. These particles can be the organic raw material for composting, the compost itself, or impurities. In practice, drum- and plain-screens (with hole plates, wire grates, stars, or profile iron) are usually used. The size of the sieve holes depends on the subsequent use of the compost or on whether impurities are being removed (>80 mm: removal of impurities; 80 to 40 mm: production of mulching material; 10 to 25 mm, production of compost for landscaping, agriculture, and gardening [1].

**13.5**
**Composting Systems**

Composting systems can be classified into nonreactor systems and reactor or vessel systems (Fig. 13.6) [1, 2, 4–6]

13.5.1
**Nonreactor Composting**

Figure 13.7 shows the types of nonreactor composting systems.

**Field composting:** During field composting, which is the simplest way of composting organic wastes, all microbial activity takes place in a thin layer at the soil surface or within a few centimeters of the soil surface (arable land or grassland). This system is useful for treating both sludge and green wastes (grass, straw, brushwood). To ensure rapid and uniform decomposition, green wastes need to be chopped. Mulching machines can be used if the wastes are growing in the same area (e.g., vineyard prunings); otherwise, collected wastes are spread out with a manure spreader after chopping. Because the waste material surface exposed to the atmosphere is large,

**composting systems**

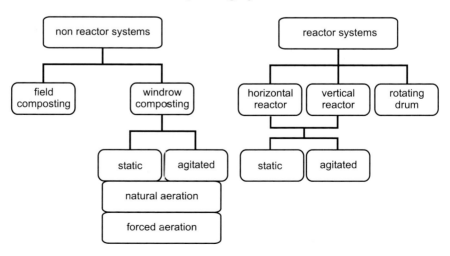

**Fig. 13.6** Classification of composting systems.

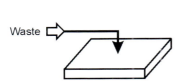

① Field "composting"

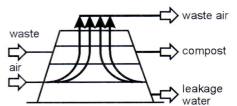

④ Windrow composting; natural aeration; vertically growing heap

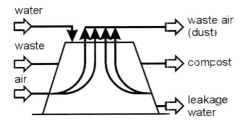

② Windrow composting; natural aeration; one step build-up

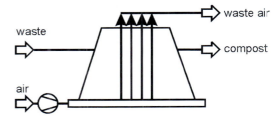

⑤ Windrow composting; forced aeration; ventilation

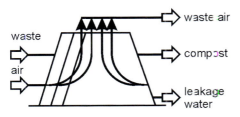

③ Windrow composting; natural aeration; horizontally growing heap

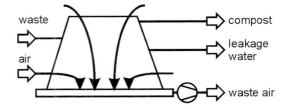

⑥ Windrow composting; forced aeration; deaeration

**Fig. 13.7** Classification of nonreactor composting systems.

self-heating does not occur, and therefore neither do thermal disinfecting or killing of weed seeds. Therefore, only wastes without problems of hygiene or weed seeds can be utilized in this kind of composting. In the narrower sense of the definition of composting, field composting is not composting, because there is no self-heating and no real process control.

**Windrow composting:** The main characteristic of nonreactor windrow composting is direct contact between the waste material and the atmosphere and, therefore, interdependence between the two. The composting process influences the atmosphere by emitting odors, greenhouse gases, spores, germs, and dust. The atmosphere, which carries the respiration gas oxygen, can influence the composting process by

- supplying rain water
  - advantage: adds water, which is needed if the material for composting is or has become too dry, thus resulting in more rapid biodegradation
  - disadvantages: blocks free airspace, favors anaerobic conditions and associated odor emissions, decreases compost quality, increases drainage water
- changes in air temperature
  - advantages: high air temperatures can increase the evaporation rate of very wet wastes, increasing the amount of free air space; high temperatures can shorten the lag phase at the start of the process
  - disadvantages: high air temperatures can increase the evaporation rate, leading to insufficient moisture; low air temperatures can delay or inhibit self-heating
- changes in air humidity
  - advantages: low air humidity can increase the evaporation rate of very wet wastes; high air humidity reduces the evaporation rate
  - disadvantages: low air humidity can increase the evaporation rate, leading to insufficient moisture; high air humidity can decrease the evaporation rate, leading to too much moisture
- supplying wind
  - advantages and disadvantages: wind can intensify the effects of air temperature and humidity

The extent of contact between waste material and atmosphere can be influenced by covering the piles with mature compost material, straw, or special textile or fleece materials that allow gas exchange but reduce the infiltration of rain water. The cross section shape of a windrow compost pile can be triangular or trapezoidal. The height, width, and shape of a windrow depend on the waste material, climatic conditions, and the turning equipment.

Natural aeration in windrows can be supported by (1) addition of bulking material to the waste, (2) using bulking material as an aeration layer at the bottom of the windrow (20–30 cm), (3) aeration pipes from the bottom of the windrow, and (4) perforated floor (Fig. 13.8).

To ensure a high quality of the compost, windrows are disturbed from time to time by turning. The effects of turning are (1) mixing of the material for homogenization (dry or wet zones at the surface, wet zones at the bottom) and for killing pathogenic microorganisms and weed seeds, (2) renewing the structure and free airspace, and (3) increasing evaporation to dry the waste material or the mature compost. The turning frequency depends on the kind and structure of the waste and the quality requirements of the finished compost. It can vary from several times a day (at the start of the process when the oxygen demand is high or for drying mature compost) to once every several weeks.

Machines and equipment for turning include tractor mounted front-end loaders, wheel loader shovels, manure spreaders, tractor-driven windrow turning machines,

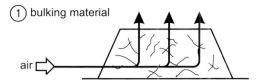

(1) bulking material

air ⟹

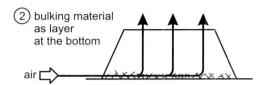

(2) bulking material
as layer
at the bottom

air ⟹

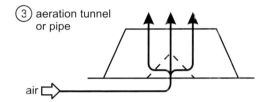

(3) aeration tunnel
or pipe

air ⟹

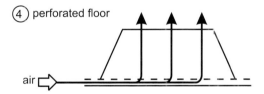

(4) perforated floor

air ⟹

**Fig. 13.8** Ways to improve natural aeration in compost windrows.

and self-driven windrow turning machines (Fig. 13.9). The mixing quality of front-end loaders and wheel loaders is relatively poor and requires an experienced driver. Compression of the (wet) wastes by the weight of the machinery can be a disadvantage.

An example of a simple open-windrow composting plant with a tractor-driven turning machine is shown in Figures 13.10 and 13.11. It consists of a concrete or asphalt floor area with an open shed for storing the finished compost. All the drainage and rain water are collected in a tank or basin. Large pieces of waste (branches, trees) are chopped periodically by a machine that is driven from one place to another. After separating the impurities from the biowaste, windrows are formed from both components with a wheel loader. The windrows are turned frequently with a tractor-driven turning machine (or a self-driven machine). The finished compost is screened with a mobile screening device. The oversize fractions from the screening are

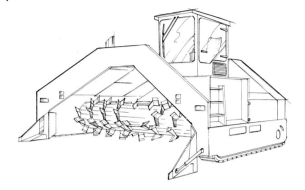

**Fig. 13.9** Self-driven windrow turning machine (drawing: Backhus GmbH, Edewecht).

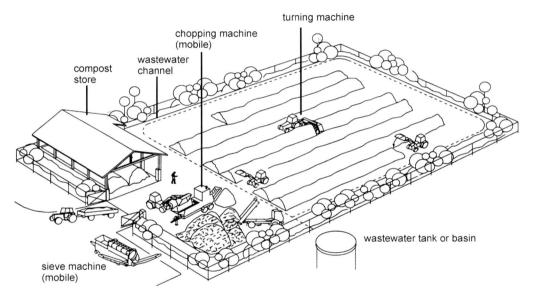

**Fig. 13.10** Open windrow composting plant with turning machine [3].

reused in the compost plant, and the finished compost is stored under a roof until it is sold.

## 13.5.2
### Reactor Composting

Every method of composting in an enclosed space (e.g., container, box, bin, tunnel, shed) with forced exchange of respiration gases is a type of reactor composting. Composting reactors (Fig. 13.12) can be classified according to the manner of material flow as

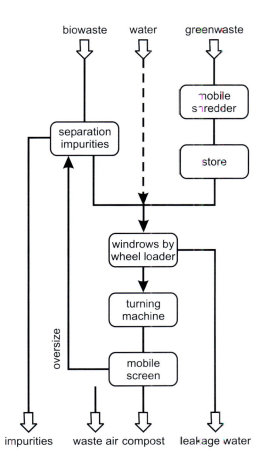

biowaste    water    greenwaste

separation impurities

mobile shredder

store

windrows by wheel loader

turning machine

oversize

mobile screen

impurities    waste air    compost    leakage water

**Fig. 13.11** Flow sheet of an open windrow composting plant.

- horizontal-flow reactors having
  - a static solids bed or
  - an agitated solids bed
- vertical-flow reactors
- rotating-drum reactors

With few exceptions, all reactors have a means of controlled forced aeration and enable the waste air and the drainage water to be collected and treated. Addition of water or other additives is possible only when the waste material is mixed in the reactor. If possible, composting processes are used only for precomposting, because of the high costs of reactor composting compared to windrow composting.

**Horizontal-flow reactors with static solids bed:** This is a batch system in which the waste material is loaded with a wheel loader or transport devices into a horizontal reactor or is covered by a foil or textile material. The forced aeration, in positive or neg-

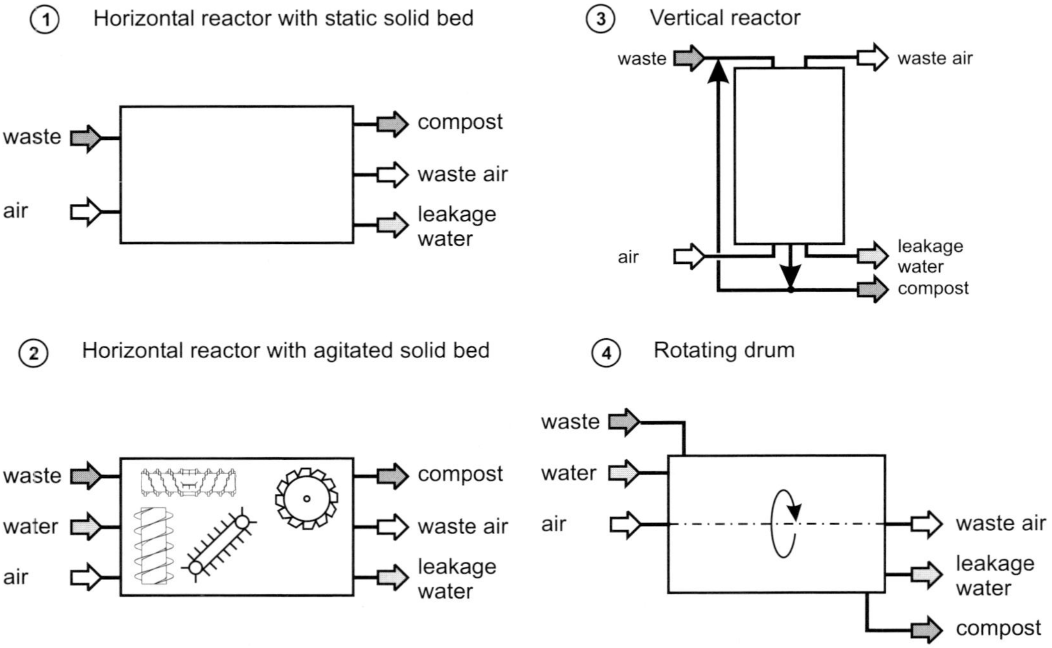

① Horizontal reactor with static solid bed

③ Vertical reactor

**Fig. 13.12** Reactor composting systems.

ative mode or alternating, issues from pipes at the bottom of the material or from holes in the floor. The waste air, with its odor components and water, can be treated in a biofilter or biowasher. In some systems some of the waste air is recycled. The air flow rate can be controlled according to the temperature in the material or the oxygen/carbon dioxide concentration in the air. The retention time in the reactor is between several days (precomposting) and several weeks (mature compost). The end product can be inhomogeneous (partially too dry) and not biologically stabilized, because there is no turning/mixing and no water addition to the waste material and is forced aeration from only one direction.

**Horizontal-flow reactors with agitated solids bed:** In this reactor type the waste material can be turned mechanically and water can be added, in contrast to a horizontal-flow reactor with a static solids bed. The devices for mixing the wastes can be horizontally or vertically operating rotors or screws, scraper conveyors, or shovel wheels. Fully automated functioning of the whole process is possible.

**Vertical-flow reactors:** In this reactor type the waste material flows vertically, with or without stages, with mass flow from the top to the bottom as a plug flow system or, if the material from the outlet at the bottom is loaded back in at the top, as a mixed system. Forced aeration occurs from the bottom or from vertical pipes in the material. This process can also be run in a fully automated way.

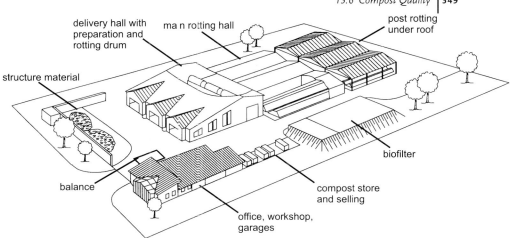

**Fig. 13.13**   Composting plant with rotting drum in closed shed [3].

**Rotating-drum reactor:** The waste material is loaded into a horizontal slowly rotating drum with forced aeration. The filling capacity is approximately 50%. The material is transported (plug flow) in the helical pathway from one end of the drum to the other and is mixed intensively. Self-heating starts after a short time. Water addition is possible. Rotating-drum reactors can also be used as mixing equipment.

An example of a composting plant with a rotating-drum reactor is shown in Figures 13.13 and 13.14. This plant is characterized by an enclosed building for receiving the biowaste, preparing it for the rotting process (disintegration, separation of impurities), pre-rotting in a drum, and main-rotting in turned windrows (composting I, Figure 13.14). All highly contaminated waste air can be collected and treated with a biofilter. Only the maturing phase of the compost (composting II, Figure 13.14) occurs under natural climatic conditions in windrows under a roof. In a fully enclosed composting plant, even the maturing phase takes place in a closed shed.

## 13.6
## Compost Quality

To be used as a fertilizer and soil conditioner, compost must meet certain quality requirements, such as (1) optimal maturity, (2) favorable contents of nutrients and organic matter, (3) favorable C/N ratio, (4) neutral or alkaline pH, (5) low contents of heavy metals and organic contaminants, (6) no components that interfere with plant growth, (7) mostly free from impurities, (8) mostly free from germinatable seeds and living plant parts, (9) low content of rocks, (10) typical smell of forest soil, and (11) dark brown to black.

'Maturity' and 'stability' are different properties of compost. Stability is defined in terms of the bioavailability of organic matter, which relates to the rate of decompo-

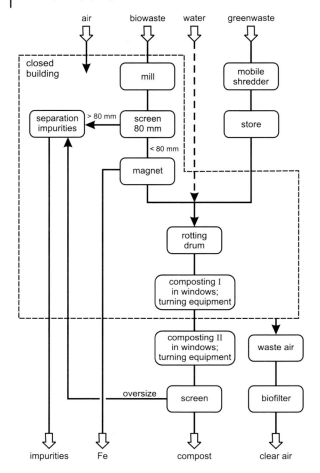

**Fig. 13.14** Flow sheet of a composting plant with rotting drum in closed shed.

sition. Maturity describes the suitability of the compost for plant growth and has been associated with the degree of humification. Several methods are used in practice to determine maturity and stability: e.g., simple field tests in Dewar flasks, tests on plants, respiration activity, chemical analyses, nuclear magnetic resonance (NMR) [1 to 26]. The various procedures for determining the maturity and stability of compost are discussed in the literature; general, final tests have not been agreed upon.

Concerning the compost quality factors that relate to human health, such as the contents of heavy metals, hazardous organic compounds, parasites, and other disease organisms, several countries have passed laws, ordinances, regulations, and norms.

# References

## Section 13.1: Introduction

1 DeBertoldi M (ed.), *Compost: Production, Quality and Use*. London **1987**: Elsevier.
2 Haug RT, *The Practical Handbook of Composting*. Boca Raton, FL **1993**: Lewis.
3 Thomé-Kozmiensky KJ (ed.), *Biologische Abfallbehandlung*. Berlin **1995**: Erich Freitag.
4 White P, Franke M, Hindle P, *Integrated Solid Waste Management: A Lifecycle Inventory*. London **1995**: Blackie.
5 DeBertoldi M, Sequi P, Lemmes B, Papi T (eds.), *The Science of Composting*. London **1996**: Blackie.
6 Epstein E, *The Science of Composting*. Lancaster, Basel **1997**: Technomic.
7 Bidlingmaier W (ed.), *Biologische Abfallverwertung*. Stuttgart **2000**: Ulmer.
8 Insam H, Riddech N, Klammer S (eds.), *Microbiology of Composting*. Berlin/Heidelberg **2002**: Springer.

## Section 13.2: Waste Materials for Composting

1 Fricke K, Stellenwert der biologischen Abfallbehandlung in integrierten Entsorgungskonzepten, in: *Biologische Verfahren der Abfallbehandlung* (Dott W, Fricke K, Oetjen R, eds.), pp. 1–58. Berlin **1990**: Erich Freitag Verlag.
2 Fricke K, Turk T, Stand und Stellenwert der Kompostierung in der Abfallwirtschaft, in: *Bioabfallkompostierung – Flächendeckende Einführung* (Wiemer K, Kern M, eds.), pp. 13–98. Witzenhausen **1991**: M.I.C. Baeza.
3 Bidlingmaier W, Anlageninput und erzeugte Kompostqualität. in: *Biologische Abfallbehandlung II* (Wiemer K, Kern M., eds.), pp. 109–120. Witzenhausen **1995**: M.I.C. Baeza
4 Thomé-Kozmiensky KJ (ed.), *Biologische Abfallbehandlung*. Berlin **1995**: Erich Freitag
5 Schuchardt F, Composting of plant residues and waste plant materials. in: *Biotechnology* (Rehm HJ, Reed G, eds.) Vol. 11c: *Environmental Processes III*, pp. 101–125, Weinheim **2000**: Wiley-VCH

## Section 13.3: Fundamentals of Composting Process

1 Edwards CA, Neuhauser EF (eds.), *Earthworms in Waste and Environmental Management*. The Hague **1988**: SPB Academic
2 Michel FC, Reddy CA, Effect of oxygenation level on yard trimmings composting rate, odor production, and compost quality in bench-scale reactors. *Compost Sci Util* **1998**, *6(4)*, 6–14
3 Schaub-Szabo SM, Leonard JJ, *Compost Sci Util* **1999**, *7(4)*, 15–24
4 Shi W, Norton JM, Miller BE, et al., *Appl Soil Ecol* **1999**, *11(1)*, 17–28
5 Sommer SG, Dahl P, *J Agric Eng Res* **1999**, *74(2)*, 145–153
6 Van Ginkel JT, Raats PAC, Van Haneghem IA, *Neth J Agric Sci* **1999**, *47(2)*, 105–121
7 Chamuris GP, Koziol-Kotch S, Brouse TM, *Compost Sci Util* **2000**, *8(1)*, 6–11
8 Dominguez J, Edwards CA, Webster M, Vermicomposting of sewage sludge: effect of bulking materials on the growth and reproduction of the earthworm *Eisenia andrei*. *Pedobiologia* **2000**, *44(1)*, 24–32
9 El-Din SMSB, Attia M, Abo-Sedera SA, Field assessment of composts produced by highly effective cellulolytic microorganisms. *Biol Fert Soils* **2000**, *32(1)*, 35–40
10 Kutzner HJ, Microbiology of composting. in: *Biotechnology* (Rehm HJ, Reed G. eds.) Vol. 11c: *Environmental Processes III*, pp 35–100, Weinheim **2000**: Wiley-VCH
11 Larney FJ, Olson AF, Carcamo AA et al., Physical changes during active and passive composting of beef feedlot manure in winter and summer. *Bioresource Technol* **2000**, *75(2)*, 139–148
12 Moller HB, Sommer SG, Andersen BH, Nitrogen mass balance in deep litter during the pig fattening cycle and during composting. *J Agric Sci* **2000**, *135*, 287–296
13 Schuchardt F, Composting of plant residues and waste plant materials. in: *Biotechnology* (Rehm HJ, Reed G eds.) Vol. 11c: *Environmental Processes III*, pp 101–125, Weinheim **2000**: Wiley-VCH
14 Amon B, Amon T, Boxberger J, et al., Emissions of $NH_3$, $N_2O$ and $CH_4$ from dairy cows housed in a farmyard manure tying stall (housing, manure storage, ma-

nure spreading). *Nutr Cycl Agroecosys* **2001**, *60*, 103–113

15  Beck-Friis B, Smars S, Jonsson H, et al., Gaseous emissions of carbon dioxide, ammonia and nitrous oxide from organic household waste in a compost reactor under different temperature regimes *J Agric Eng Res* **2001**, *78*, 423–430

16  Elwell DL, Keener HM, Wiles MC, et al., *T ASAE* **2001**, *44*, 1307–1316

17  Hao XY, Chang C, Larney FJ, et al., *J Environ Qual* **2001**, *30*, 376–386

18  Hassen A, Belguith K, Jedidi N, et al., Microbial characterization during composting of municipal solid waste. *Bioresource Technol* **2001**, *80*, 217–225

19  Huang GF, Wu QT, Li FB, et al., Nitrogen transformations during pig manure composting. *J Environ Sci China* **2001**, *13*, 401–405

20  Ndegwa PM, Thompson SA, Integrating composting and vermicomposting in the treatment and bioconversion of biosolids. *Bioresource Technol* **2001**, *76*, 107–112

21  Barrington S, Choiniere D, Trigu. M, et al., Effect of carbon source on compost nitrogen and carbon losses. *Bioresource Technol* **2002**, *83(3)*, 189–194

22  (Barrington S, Choiniere D, Trigui M, et al., Compost airflow resistance. *Biosyst Eng* **2002**, *81*, 433–441

23  Hart TD, De Leij FAAM, Kinsey G, et al., Strategies for the isolation of cellulolytic fungi for composting of wheat straw. *World J Microb Biot* **2002**, *18*, 471–480

24  Jensen HEK, Leth M, Iversen JJL, Effect of compost age and concentration of pig slurry on plant growth. *Compost Sci Util* **2002**, *10(2)*, 129–141

25  Noble, R, Hobbs, PJ, Mead, A, et al., Influence of straw types and nitrogen sources on mushroom composting emissions and compost productivity. *J Ind Microbiol Biot* **2002**, *29(3)*, 99–110

26  Richard TL, Hamelers HVM, Veeken A, et al., Moisture relationships in composting processes. *Compost Sci Util* **2002**, *10(4)*, 286–302

27  Singh A, Sharma S, Composting of a crop residue through treatment with microorganisms and subsequent vermicomposting. *Bioresource Technol* **2002**, *85(2)*, 107–111

28  Tiquia SM, Tam NFY, Characterization and composting of poultry litter in forced-aeration piles. *Process Biochem* **2002**, *37*, 869–880

29  Veeken A, de Wilde V, Hamelers B, Passively aerated composting of straw-rich pig manure: effect of compost bed porosity. *Compost Sci Util* **2002**, *10(2)*, 114–128

30  Wolter M, Prayitno S, Schuchardt F, Comparison of greenhouse gas emissions from solid pig manure during storage versus during composting with respect to different dry matter contents. *Landbauforsch Volk* **2002**, *52(3)*, 167–174

31  Agnew JM, Leonard JJ, The physical properties of compost. *Compost Sci Util* **2003**, *11*, 238–264

32  Barrington S, Choiniere D, Trigui M, Knight W, Compost convective airflow under passive aeration *Bioresource Technol* **2003**, *86*, 259–266

33  Bolta SV, Mihelic R, Lobnik F, et al., Microbial community structure during composting with and without mass inocula. *Compost Sci Util* **2003**, *11*, 6–15

34  Fukumoto Y, Osada T, Hanajima D, Haga K, Patterns and quantities of $NH_3$, $N_2O$ and $CH_4$ emissions during swine manure composting without forced aeration: effect of compost pile scale. *Bioresource Technol* **2003**, *89(2)*, 109–114

35  Liang C, Das KC, McClendon RW, The influence of temperature and moisture contents regimes on the aerobic microbial activity of a biosolids composting blend. *Bioresource Technol* **2003**, *86(2)*, 131–137

36  Principi P, Ranalli G, da Borso F, et al., Microbiological aspects of humid husk composting. *J Environ Sci Heal B* **2003**, *38*, 645–661

## Section 13.4: Composting Technologies

1  Schuchardt F, Composting of plant residues and waste plant materials. in: *Biotechnology* (Rehm HJ, Reed G, eds.) Vol. 11c: *Environmental Processes III*, 101–125, Weinheim **2000**: Wiley-VCH

## Section 13.5: Composting Systems

1  Haug RT, *The Practical Handbook of Composting*. Boca Raton, FL **1993**: Lewis
2  Thomé-Kozmiensky KJ (ed.), *Biologische Abfallbehandlung*. Berlin **1995**: Erich Freitag Verlag.
3  Gronauer A, Claassen N, Ebertseder T et al., *Bioabfallkompostierung*. BayLfU 139 (**1997**)
4  Bidlingmaier W (Ed.), *Biologische Abfallverwertung*. Stuttgart **2000**: Ulmer Verlag
5  Krogmann U, Körner I, Technology and strategies of composting. in: *Biotechnology* (Rehm HJ, Reed G, eds.) Vol. 11c: Environmental Processes III, pp 127–150, Weinheim **2000**: Wiley-VCH
6  Schuchardt F, Composting of plant residues and waste plant materials. in: *Biotechnology* (Rehm HJ, Reed G, eds.) Vol. 11c: Environmental Processes III, 101–125, Weinheim **2000**: Wiley-VCH


## Section 13.6: Compost Quality

1  Adani F, Genevini PL, Gasperi F, Zorzi G, Organic matter evolution index (OMEI) as a measure of composting efficiency. *Compost Sci Util* **1997**, *5(2)*, 53–62
2  Popp L, Fischer P. Claassen N, Biologisch-biochemische Methoden zur Reifebestimmung von Komposten. *Agrobiol Res* **1998**, *51(3)*, 201–212
3  Saharinen MH, Evaluation of changes in CEC during composting. *Compost Sci Util* **1998**, *6(4)*, 29–37
4  Fauci MF, Bezdicek DF, Caldwell D, et al., *Compost Sci Util* **1999**, *7(2)*, 17–29
5  Namkoong W, Hwang EY, Cheong JG, et al., *Compost Sci Util* **1999**, *7(2)*, 55–62
6  Warman PR, *Compost Sci Util* **1999**, *7(3)*, 33–37
7  Koenig A, Bari QH, Application of self-heating test for indirect estimation of respirometric activity of compost: theory and practice. *Compost Sci Util* **2000**, *8(2)*, 99–107
8  Ouatmane A, Provenzano MR, Hafidi M, et al., *Compost Sci Util* **2000**, *8(2)*, 124–134
9  Wu L, Ma LQ, Martinez GA, *J Environ Qual* **2000**,. *29*, 424–429

10  Butler TA, Sikora LJ, Steinhilber PM, et al., Compost age and sample storage effects on maturity indicators of biosolids compost. *J Environ Qual* **2001**, *30*, 2141–2148
11  Eggen T, Vethe O, Stability indices for different composts. *Compost Sci Util* **2001**, *9(1)*, 19–26
12  Provenzano MR, de Oliveira SC, Silva MRS, et al., Assessment of maturity degree of composts from domestic solid wastes by fluorescence and Fourier transform infrared spectroscopies. *J Agric Food Chem* **2001**, *49*, 5874–5879
13  Smith DC, Hughes JC, A simple test to determine cellulolytic activity as indicator of compost maturity. *Commun Soil Sci Plan* **2001**, *32*, 1735–1749
14  Levanon D, Pluda D, Chemical, physical and biological criteria for maturity in composts for organic farming. *Compost Sci Util* **2002**, *10*, 339–346
15  Weppen P, Determining compost maturity: evaluation of analytical properties. *Compost Sci Util* **2002**, *10(1)*, 6–15
16  Wu L, Ma LQ, Relationship between compost stability and extractable organic carbon. *J Environ Qual* **2002**, *31*, 1323–1328
17  Adani F, Gigliotti G, Valentini F, Laraia R, Respiration index determination: a comparative study of different methods. *Compost Sci Util* **2003**, *11*, 144–151
18  Benito M, Masaguer A, Moliner A, Arrigo N, Palma RM, Chemical and microbiological parameters for the characterisation of the stability and maturity of pruning waste compost. *Biol Fert Soils* **2003**, *37*, 184–189
19  Brewer LJ, Sullivan DM, Maturity and stability evaluation of composted yard trimmings. *Compost Sci Util* **2003**, *11*, 96–112
20  Changa CM, Wang P, Watson ME, et al., Assessment of the reliability of a commercial maturity test kit for composted manures. *Compost Sci Util* **2003**, *11*, 125–143
21  Chen YN, Nuclear magnetic resonance, infra-red and pyrolysis: application of spectroscopic methodologies to maturity determination of composts. *Compost Sci Util* **2003**, *11*, 152–168
22  Chica A, Mohedo JJ, Martin MA, Martin A., Determination of the stability of MSW compost using a respirometric technique. *Compost Sci Util* **2003**, *11*, 169–175

**23** Cooperband LR, Stone AG, Fryda MR, Ravet JL., Relating compost measures of stability and maturity to plant growth. *Compost Sci Util* **2003**, *11*, 113–124

**24** Korner, I, Braukmeier, J, Herrenklage, J, et al., Investigation and optimization of composting processes: test systems and practical examples. *Waste Manage* **2003**, *23*, 17–26

**25** Rynk R, The art in the science of compost maturity. *Compost Sci Util* **2003**, *11*, 94–95

**26** Zubillaga MS, Lavado RS, Stability indexes of sewage sludge compost obtained with different proportions of a bulking agent. *Commun Soil Sci Plan* **2003**, *34*, 581–591

# 14
# Anaerobic Fermentation of Wet and Semidry Garbage Waste Fractions

Norbert Rilling

## 14.1
## Introduction

During the last 30 years the amount of solid wastes has rapidly increased. Now one of the primary aims in waste management is to reduce the amount of waste to be disposed by prevention, reduction, and utilization (KrW-/AbfG, 1994).

By means of separate collection and biological treatment of biowaste, the amount of municipal solid waste (MSW) to be incinerated or landfilled will be significantly reduced.

Biological treatment of garbage waste fractions can be carried out aerobically (composting) or anaerobically (anaerobic digestion). Each technique is appropriate for a certain spectrum of wastes. Today, most biological waste is composted because this technology is already well developed, with quite a lot of experience at hand, but anaerobic processes advance in their importance for the utilization of solid organic waste.

Anaerobic digestion, which is typically conducted inside a closed vessel in which the temperature and moisture are controlled, is particularly suited to wastes with a high moisture content and a high amount of readily biodegradable components. In Sections 14.2 and 14.3 the characteristics of the anaerobic process are discussed with respect to their ecological and economical aspects.

## 14.2
## Basic Aspects of Biological Waste Treatment

The leading aim of separate collection and treatment of biological wastes is stabilization of the waste by microbial degradation. The product is a compost that, if its pollutant content is low, can be used as a fertilizer or soil conditioner and thus be fed back into the natural cycle.

In contrast to the commonly established composting processes, the technique of anaerobic fermentation of waste is relatively young and dynamic. With great scien-

*Environmental Biotechnology. Concepts and Applications.* Edited by H.-J. Jördening and J. Winter
Copyright © 2005 WILEY-VCH Verlag GmbH & Co. KGaA, Weinheim
ISBN: 3-527-30585-8

tific expenditure, process developments and optimizations are being pursued, so it may be assumed that the technological potential of biowaste fermentation has not yet been fully exhausted.

## 14.2.1
### Biochemical Fundamentals of Anaerobic Fermentation

Biogas is produced whenever organic matter is microbially degraded in the absence of oxygen. In nature this process can be observed in marshlands, in marine sediments, in flooded rice fields, in the rumen of ruminants, and in landfill sites (Maurer and Winkler, 1982).

Anaerobic degradation is effected by various specialized groups of bacteria in several successive steps, each step depending on the preceding one. For industrial-scale application of anaerobic fermentation processes it is necessary to have a thorough knowledge of these interactions, to avoid substrate limitation and product inhibition.

The entire anaerobic fermentation process can be divided into three steps (Fig. 14.1):
1. hydrolysis
2. acidification
3. methane formation

At least three groups of bacteria are involved in the anaerobic fermentation process.

First, during hydrolysis, the mostly water-insoluble biopolymers such as carbohydrates, proteins, and fats are decomposed by extracellular enzymes to water-soluble monomers (e.g., monosaccharides, amino acids, glycerin, fatty acids) and thus made accessible to further degradation.

In the second step (acidification) the intermediates of hydrolysis are converted into acetic acid ($CH_3COOH$), hydrogen ($H_2$), carbon dioxide ($CO_2$), organic acids,

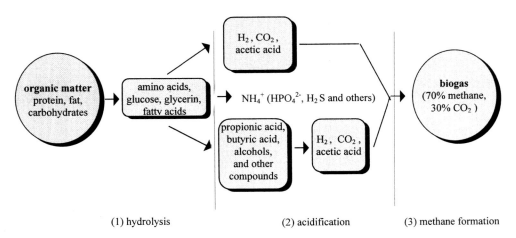

(1) hydrolysis        (2) acidification        (3) methane formation

**Fig. 14.1**   The three stages of anaerobic processing of organic matter (Sahm, 1981).

amino acids, and alcohols by different groups of bacteria. Some of these intermediate products (acetic acid, hydrogen, and carbon dioxide) can be directly used by methanogenic bacteria, but most of the organic acids and alcohol are decomposed into acetic acid, hydrogen, and carbon dioxide during acidogenesis. Only these products, as well as methanol, methylamine, and formate, can be transformed into carbon dioxide and methane ($CH_4$) by methanogenic bacteria during the third and last step, methane formation.

### 14.2.1.1 Hydrolytic and Acid-forming (Fermentative) Bacteria

The first group of bacteria is very heterogeneous: besides obligate anaerobic bacteria, also facultative anaerobic bacterial strains occur. The high molecular weight compounds of the waste biomass (proteins, polysaccharides, fats) are decomposed into low molecular weight components by enzymes that are excreted by fermentative bacteria. This first step is inhibited by lignocellulose-containing materials, which are degraded only very slowly or incompletely. Subsequently, acid-forming bacteria transform the hydrolysis products into hydrogen, carbon dioxide, alcohols, and organic acids such as acetic, propionic, butyric, lactic, and valeric acids. The formation of the acids decreases the pH.

### 14.2.1.2 Acetic Acid- and Hydrogen-forming (Acetogenic) Bacteria

The group of acetogenic microorganisms represents the link between fermentative and methanogenic bacteria. They decompose alcohols and long-chain fatty acids into acetic acid, hydrogen, and carbon dioxide. It is a characteristic of acetogenic bacteria that they can grow only at a very low hydrogen partial pressure. For this reason they live in close symbiosis with methanogenic and sulfidogenic bacteria, which use hydrogen as an energy source (Sahm, 1981).

### 14.2.1.3 Methane-forming (Methanogenic) Bacteria

The group of methanogenic bacteria is formed of extreme obligate anaerobic microorganisms which are very sensitive to environmental changes. They transform the final products of acidic and acetogenic fermentation into methane and carbon dioxide. About 70% of the methane is produced by the degradation of acetic acid and about 30% by a redox reaction from hydrogen and carbon dioxide (Roediger et al., 1990).

The slowest step is rate-determining for the whole process of anaerobic fermentation. Although methanogenesis of acetic acids is the rate-limiting step in the anaerobic fermentation of easily degradable substances, hydrolysis can be rate-limiting when sparingly degradable substances occur. Because of the complexity of the anaerobic degradation mechanisms and the stringent requirements of the microorganisms, process operation is very important for fermentation processes. To achieve optimized, undisturbed anaerobic degradation, the speed of decomposition in the consecutive steps should be equal.

14.2.2
**Influence of Processing Conditions on Fermentation**

The activity of microorganisms in anaerobic fermentation processes depends mainly on water content, temperature, pH, redox potential, and the presence of inhibitory factors.

### 14.2.2.1  Water Content

Bacteria take up the available substrates in dissolved form. Therefore, biogas production and the water content of the initial material are interdependent. When the water content is below 20% by weight, hardly any biogas is produced. With increasing water content biogas production is enhanced, reaching its optimum at 91%–98% water by weight (Kaltwasser, 1980).

### 14.2.2.2  Temperature

The process of biomethanation is very sensitive to changes in temperature, and the degree of sensitivity depends on the temperature range. For most methane bacteria, the optimum temperature range is between 30 and 37 °C (Maurer and Winkler, 1982). Here, temperature variations of ±3 °C have minor effect on the fermentation (Winter, 1985). In the thermophilic range, however, i.e., at temperatures between 55 and 65 °C, a fairly constant temperature has to be maintained, since deviations by only a few degrees cause a drastic reduction of the degradation rates and thus of biogas production.

### 14.2.2.3  pH Level

The pH optimum for methane fermentation is between pH 6.7 and 7.4. If the pH of the medium drops below 6, because the balance is disturbed and the acid-producing dominate the acid-consuming bacteria, the medium becomes inhibitory or toxic to the methanogenic bacteria. In addition, strong ammonia production during the degradation of proteins may inhibit methane formation if the pH exceeds 8. Normally, acid and ammonia production vary only slightly due to the buffering effect of carbon dioxide/bicarbonate ($CO_2/HCO^{3-}$) and ammonia/ammonium ($NH_3/NH_4^+$), which are formed during fermentation, and the pH normally stays constant between 7 and 8.

### 14.2.2.4  Redox Potential and Oxygen

Methane bacteria are very sensitive to oxygen and have lower activity in the presence of oxygen. The anaerobic process, however, shows a certain tolerance to small quantities of oxygen. Even continuous, but limited, oxygen introduction is normally tolerated (Mudrak and Kunst, 1991). The redox potential can be used as an indicator of the process of methane fermentation. Methanogenic bacterial growth requires a relatively low redox potential. Hungate (1966; cited by Braun, 1982) found −300 mV to be the minimum value. Changes in redox potential during the fermentation process are caused by a decrease in oxygen content as well as by the formation of metabolites like formate or acetate.

### 14.2.2.5 Inhibitory Factors

The presence of heavy metals, antibiotics, and detergents can inhibit the process of biomethanation. With reference to investigations of Konzeli-Katsiri and Kartsonas (1986), Table 14.1 lists the limit concentrations (mg $L^{-1}$) for inhibition and toxicity of heavy metals in anaerobic digestion.

### 14.2.3
### Gas Quantity and Composition

Biogas is a mixture of various gases. Independent of the fermentation temperature, a biogas is produced which consists of 60%–70% methane and 30%–40% carbon dioxide. Trace components of ammonia ($NH_3$) and hydrogen sulfide ($H_2S$) can be detected. The caloric value of the biogas is about 5.5–6.0 kWh $m^{-3}$. This corresponds to about 0.5 L of diesel oil.

If the chemical composition of the substrate is known, the yield and composition of the biogas can be estimated from Eq. (1) (with reference to Symons and Buswell, 1933):

$$C_n H_a O_b + \left( n - \frac{a}{4} - \frac{b}{2} \right) H_2O \rightarrow \left( \frac{n}{2} + \frac{a}{8} - \frac{b}{4} \right) CH_4 + \left( \frac{n}{2} - \frac{a}{8} + \frac{b}{4} \right) CO_2 \qquad (1)$$

Table 14.2 shows the mean composition and specific quantity of biogas as dependent on the kind of degraded substances.

For anaerobic digestion of the organic fraction of municipal solid waste, an average biogas yield of 100 $m^3$ $t^{-1}$ moist biowaste and having a methane content of about 60% by volume may be assumed.

**Table 14.1** Inhibition of anaerobic digestion by heavy metals (Konzell-Katsiri and Kartsonas, 1986).

| Heavy Metal | Inhibition (mg $L^{-1}$) | Toxicity (mg $L^{-1}$) |
|---|---|---|
| Copper (Cu) | 40–250 | 170–300 |
| Cadmium (Cd) | – | 20–600 |
| Zinc (Zn) | 150–400 | 250–600 |
| Nickel (Ni) | 10–300 | 30–1000 |
| Lead (Pb) | 300–340 | 340 |
| Chromium III (Cr) | 120–300 | 200–500 |
| Chromium VI (Cr) | 100–110 | 200–420 |

**Table 14.2** Mean composition and specific yields of biogas in relation to the kind of substances degraded.

| Substance | Gas Yield (m³ kg⁻¹ TS) | CH₄ Methane Content (Vol. %) | CO₂ Carbon Dioxide Content (Vol. %) |
|---|---|---|---|
| Carbohydrates | 0.79 | 50 | 50 |
| Fats | 1.27 | 68 | 32 |
| Proteins | 0.70 | 71 | 29 |
| Municipal solid waste (MSW) | 0.1–0.2 | 55–65 | 35–45 |
| Biowaste | 0.2–0.3 | 55–65 | 35–45 |
| Sewage sludge | 0.2–0.4 | 60–70 | 30–40 |
| Manure | 0.1–0.3 | 60–65 | 35–40 |

## 14.2.4
### Comparison of Aerobic and Anaerobic Waste Treatment

Professional biological treatment must include the separation of interfering matter, sanitation, and microbial degradation of the readily and moderately degradable substances so that the final product is biologically stable, compatible with plant roots, and, as far as possible, free of pollutants and can be applied as a soil improver in horticulture and agriculture. Aerobic composting and anaerobic fermentation are, in principle, available as processes for the biological treatment of organic residues.

Composting is suitable for the stabilization of rather dry solid waste, and anaerobic processes are used for very moist waste (e.g., kitchen garbage) which is easy to degrade. Both systems have advantages and disadvantages. Their general properties are listed and compared in Table 14.3.

A great advantage of anaerobic fermentation is the production of biogas that can be used as a source of energy. Either local users can be found for the gas recovered from the process, or it can be cleaned and upgraded for inclusion in a gas supply network. By way of comparison, during composting all the energy is released as heat and cannot be used. In addition, intensive composting requires a lot of energy for artificial aeration of the waste material.

The technical expenditures for anaerobic fermentation are higher than for composting, but if standards, especially concerning the reduction of emissions (odor, germs, noise, dust), are raised, the technical expenditures for composting can be expected to grow as well.

The duration of anaerobic and aerobic treatment depends very much on the substrate and the process used, so that the times required cannot be compared in general. The same is true for the floor space required.

A major advantage of anaerobic digestion in comparison with aerobic composting is the ability of engineers to have total control over gaseous and liquid emissions, as well as having the potential to recover and use methane gas generated as the wastes degrade. For composting, the problem of odor control has not yet been sufficiently

**Table 14.3**  Comparison of aerobic and anaerobic waste treatment (according to Rilling, 1994a).

| Characteristics | Anaerobic Digestion | Aerobic Composting |
|---|---|---|
| Phases | solids, liquid | solids, liquid, gas |
| Degradation rate | up to 80% volatile solids | up to 50% volatile solids |
| Energy consumption | excess produced | demands input |
| Technical expenditure | in the same range | |
| Duration of the process | 1–4 weeks (only anaerobic stage) | 4–16 weeks (depending on the process) |
| Post-treatment | generally necessary (post-composting about 2–8 weeks) | generally none |
| Floor space required | comparatively low | comparatively low or high (depending on the process) |
| Odor emission | comparatively low | comparatively high |
| Stage of development | little experience but increasing | much experience |
| Costs | in the same range | |
| Sanitation | external process step | integrated |
| Suitability of wastes | wide (wet and dry wastes) | narrow (dry wastes) |

solved. Large volumes of odiferous waste air have to be treated by costly means to avoid complaints from surround communities. In contrast, hardly any odorous emissions occur in anaerobic fermentation, because this biological treatment takes place in closed reactors. Malodorous waste air is produced only during loading and unloading the reactor.

No general statement can be made as to whether anaerobic or aerobic fermentation is the more favorable process, as in each individual case many factors must be considered. This means that in the field of waste treatment, as well as in the field of wastewater treatment, several different processes with similar goals can have certain advantages – depending on the operational area.

## 14.3
## Processes of Anaerobic Waste Treatment

Anaerobic fermentation has been used successfully for many years as a treatment for wastewater, sewage sludge, and manure. However, anaerobic digestion of municipal solid waste is a relatively new technique which has been developed in the past 10–15 years.

Only biodegradable household wastes – i.e. those of organic or vegetable origin – can be processed in anaerobic digestion plants. Garden waste, or green wastes as they are often called, can also be included, but woodier materials (like branches) are less suitable because of their relatively longer decomposition time under anaerobic conditions.

As a result of anaerobic fermentation combined with an additional post-composting step, a material is produced that is usually similar to the compost produced by aerobic processes. It can be used as fertilizer, soil conditioner, or peat substitute.

Although composting is widely used for wastes containing high amounts of dry matter, anaerobic digestion has turned out to be a good alternative for treating wet organic wastes (Fig. 14.2). At present, the anaerobic fermentation technique is mainly used in Western Europe, where more than 30 companies offer anaerobic treatment plants commercially for the digestion of putrefiable solid waste.

### 14.3.1
### Procedures of Anaerobic Waste Fermentation

Generally, the following steps are required for the anaerobic treatment of organic waste (Rilling, 1994a):
1. delivery and storage of the biological waste
2. preprocessing of the incoming biological waste
3. anaerobic fermentation
4. storage and treatment of the digester gas
5. treatment of the process water
6. post-processing of the digested material

Figure 14.3 shows the possible treatment steps used in biowaste fermentation. In principle, all fermentation processes can be described as a combination of a selection of these treatment steps. The process technology demanded for the implementation of the different steps of the treatment differs very much, depending on the anaerobic process chosen. In general, the gas production increases and the detention time decreases with increasing energy input for preparation of the material and the fermentation itself (mesophilic/thermophilic).

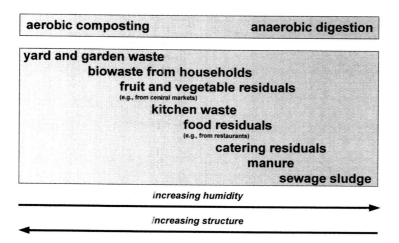

**Fig. 14.2** Suitability of wastes for aerobic composting and anaerobic digestion (according to Kern et al., 1996).

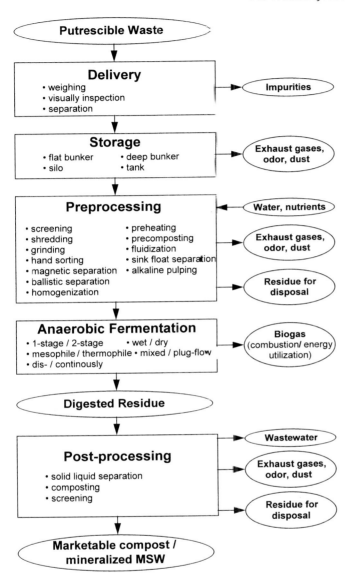

**Fig. 14.3** Possible treatment steps used in biowaste fermentation.

### 14.3.1.1 Delivery and Storage

For both composting and fermentation the wastes are pretreated before the actual biological stabilization occurs. The supplied biowastes are quantitatively and qualitatively recorded by weighing, are visually inspected at an acceptance station, and are unloaded into a flat or deep bunker or a collecting tank that serves as a short-term intermediate storage place and permits continuous feeding to the subsequent pretreatment plant.

### 14.3.1.2 **Preprocessing**

The purpose of pretreatment is to remove pollutants and interfering matter as well as to homogenize and condition the biowaste. The kind of pretreatment depends on the specific system of the anaerobic fermentation process.

Dry fermentation processes use dry preprocessing, where sieves, shredders, grinders, homogenization drums, metal separators, ballistic separators, and hand sorting sections can be combined. In wet fermentation processes the biowastes are additionally mixed with water, homogenized, and shredded. By means of sink–float separation other foreign substances can be removed.

### 14.3.1.3 **Anaerobic Fermentation**

Having separated any recyclable or unwanted materials from the incoming wastes, the organic material is shredded and fed into the digester. Shredding results in a material that can be handled more easily. In addition, material having a larger surface area is more easily broken down by the bacteria. Shredding in a drum can be combined with a precomposting step. However, shredding causes the material to lose structure. The desired particle size after shredding may be between 5 and 40 mm, in certain cases up to 80 mm. Yard and garden waste (especially branches, etc.) have to be shredded separately before composting.

If very wet wastes, like sewage sludge, are included, the addition of further water may not be necessary, but for household organic wastes, water is usually added to dilute the solids.

Wastes with low structure and a high moisture content are best for fermentation, but wastes rich in structure can also be degraded anaerobically by means of dry fermentation processes. Heat is needed to adjust the required process temperatures to about 35 °C (mesophilic process) or 55 °C (thermophilic process), and sometimes water must be added. During fermentation, organic degradation takes place anaerobically, i.e., under the exclusion of oxygen, in closed, temperature-regulated containers. Depending on the process operation, the material consistency may vary between well-structured matter and thick sludge or fluid suspension. The optimum pH value is in the neutral range.

The output of the fermentation reactor is a wet, organically stabilized, fermentation residue and biogas. After dewatering, a soil-improving product comparable to compost can be obtained from the fermentation residue by aerobic post-composting. The wastewater generated during draining can in part be recirculated into the pretreatment unit to adjust the water content. Surplus wastewater has to be treated and discharged. With only minor energy loss, the biogas can be used in decentralized fuel-burning power stations to produce electrical power and heat so that, in general, the fermentation process can be operated in an energy-neutral manner, and the surplus can be marketed by feeding it into the public power and heat supply mains.

When the fermentation is confined only to easily degradable organic waste components, energy can be produced with minimal technical expenditure, and the odor and energy-intensive prefermentation steps can be omitted. In a subsequent composting step the medium and difficult-to-degrade organic substances, which can be degraded anaerobically only to a limited extent, are aerobically decomposed at low cost. Thus, when investigating the question of 'fermentation *or* composting' the answer may often suggest the demand for fermentation *and* composting.

#### 14.3.1.4 Post-processing

To complete the stabilization and disinfection of the digested residue, some kind of refining process is needed before it can be used for agriculture or horticulture. After possible dewatering and/or drying, the anaerobically fermented waste is generally transferred to aerobic biological post-treatment and matured for about 2–4 weeks to become a good, marketable compost.

After drying and, if required, purification, the biogas can be used as an energy source.

Depending on local regulations, the excess process water is transferred to a wastewater treatment plant or may be applied directly to farmland as a liquid fertilizer.

### 14.3.2
### Process Engineering of Anaerobic Fermentation of Biowastes

Anaerobic fermentation processes are generally suitable for the biological treatment of readily degradable substances having low structure and high water content, e.g., kitchen waste. At present, several processes for the anaerobic fermentation of organic solid wastes are under development. The processes differ in the number of biodegradation stages (one- or two-stage processes), separation of liquid and solids (one- or two-phase system), water content (dry or wet fermentation), feed method (continuous or discontinuous), and the means of agitation. The most important characteristics of anaerobic fermentations are compiled in Table 14.4.

Anaerobic fermentation of biowaste can be achieved by one-stage or two-stage fermentation. In the one-stage process (Table 14.5) hydrolysis, acidification, and methane formation take place in one reactor, so it is not possible to achieve optimum re-

**Table 14.4** Characteristics of anaerobic waste treatments (according to Rilling, 1994a).

| | *Characteristics* | |
|---|---|---|
| Stages of biodegradation | one-stage | two-stage |
| Separation of liquid and solids | one-phase dry fermentation | two-phase wet fermentation |
| Total solids content | 25%–45% | <15% |
| Water content | 55%–75% | >85% |
| Feed method | discontinuous | continuous |
| Agitation | none | stirring, mixing, percolation |
| Temperature | mesophilic (30–37 °C) | thermophilic (55–65 °C) |

**Table 14.5** Comparison of one- and two-stage processes.

| Process Operation | One-Stage | Two-Stage |
|---|---|---|
| Operational reliability | in the same range | |
| Technical equipment | relatively simple | very complex |
| Process control | compromise solution | optimal |
| Risk of process instability | high | minimal |
| Retention time | long | short |
| Degradation rate | reduced | increased |

action conditions for the overall process, due to slightly different environmental requirements during the different stages of the fermentation. Therefore, the degradation rate is reduced and the retention time increases. The basic advantage of one-stage process operation is the relatively simple technical installation of the anaerobic digestion plant and lower costs.

In two-stage processes (Table 14.5), hydrolysis and acidification on the one hand and methane formation on the other hand take place in different reactors so that, e.g., mixing and adjustment of the pH can be optimized separately, permitting higher degradation degrees and loading rates. As a result, the detention time of the material can be decreased significantly. This, of course, involves a more sophisticated technical design and operation, resulting in higher costs.

During the first stage the organic fraction is hydrolyzed. As a result, dissolved organics and mainly organic acids, as well as $CO_2$ and low concentrations of hydrogen, are produced. In the second stage the highly concentrated water is supplied to an anaerobic fixed-film reactor, sludge blanket reactor, or other appropriate system where methane and $CO_2$ are produced as final products. Reciprocal inhibiting effects are excluded so that high process stability with a better methane yield is obtained.

### 14.3.2.1 Dry and Wet Fermentation

Different anaerobic digestion systems can handle wastes with different moisture contents; they are classified as dry fermentation processes (water content between 55% and 75%) and wet fermentation processes (water content >85%.) Table 14.6 shows advantages and disadvantages of dry and wet fermentation.

With the dry fermentation process, little or no water is added to the biowaste. As a consequence, the material streams to be treated are minimized. The resulting advantages are smaller reactor volumes and easier dewatering of the digested residue. On the other hand, operating with high contents of dry matter places higher requirements on mechanical pretreatment and conveyance, on the gas-tightness of charge and discharge equipment, and, if planned, on mixing in the reactors. Bridging of the material and the possibility of clogging have to be avoided. Because of the low mobility is dry fermentation, a defined residence time can be reached by approximating plug flow, which is particularly important under the aspect of product hygiene in the thermophilic operation mode. The degradation rates in dry fermentation processes

**Table 14.6** Comparison of wet and dry fermentations.

| Process Mode | Dry | Wet |
|---|---|---|
| Total solids content | high<br>25–45% | low<br>2–15% |
| Reactor volume | minimized | increased |
| Conveyance technique | expensive | simple |
| Agitation | difficult | easy |
| Scumming | little risk | high risk |
| Short circuit flow | little risk | high risk |
| Solid–liquid separation | simple | expensive |
| Variety of waste components | small | great |

are lower than in wet fermentation, due to the larger particle size and reduced substrate availability.

When wet fermentation processes are used, the organic wastes are ground to a small particle size and mixed with large quantities of water so that sludges or suspensions are obtained. This allows the use of simple, established mechanical conveyance techniques (pumping) and the removal of interfering substances by sink–float separation. At the same time, the reactor contents can be easily mixed, which permits controlled degassing and defined concentration equalization in the fermenter. As a consequence, the degradation performance of the microorganisms is optimized. The mean substrate concentrations and thus also the related degradation rates are lower than in plug flow systems, since, for completely mixed systems, the concentrations in the system are equal to the outlet concentrations. Mixing is limited by the shear-sensitivity of methane bacteria; however, too-low a degree of mixing may result in floating and sinking layers. Homogeneity and a fluid consistency permit easier process control. By fluidizing the biowaste, the mass to be treated increases until the 5-fold, depending on the total solids content of the substrate with the consequence that the aggregates and reactors have to be made much larger. Fluidization and dewatering of the fermentation suspension especially require considerable technical and energetic expenditures. But if the degrees of degradation are the same, recycling the liquid phase from the dewatering step to the fluidization of the input material, makes it possible to reduce the wastewater quantity to an amount comparable to that used in dry fermentation and to keep a considerable part of the required thermal energy within the system.

### 14.3.2.2 Continuous and Discontinuous Operation

When an anaerobic process is run in continuous operation mode, the reactor is fed and discharged regularly. Completely mixed and plug flow systems are available. Enough substrate is fed into the reactor to replace the putrefied material as it is discharged. Therefore, the substrate must be flowable and uniform. Steady provision of nutrients in the form of raw biodegradable waste enables stable process operation and constant biogas yield. Depending on the reactor design and the means of mixing, short circuits may occur, and the retention time therefore cannot be guaranteed for each part of the substrate in completely mixed systems.

In the discontinuous operation mode (batch process), the fermentation vessel is completely filled with raw garbage mixed with inoculum (e.g., digestate from another reactor) and then completely discharged after a fixed detention time. Batch digesters are easy to design, comparatively low in cost, and suitable for all wet and dry organic wastes (Table 14.7).

### 14.3.2.3 Thermophilic and Mesophilic Operation

The optimal process temperatures for methane fermentation are in the mesophilic temperature range (about 35 °C) and in the thermophilic temperature range (about 55 °C).

Reactors designed for mesophilic operation are heated to 30 to 40 °C. According to the experience gained with this way of operation, the process stability is high. Minor temperature variations have only a small effect on mesophilic bacteria.

The advantages of mesophilic process operation result from the lower amount of heat to be supplied and the related higher net energy yield. In addition, higher process stability is achieved, since a broad spectrum of mesophilic methane bacteria that show low sensitivity to temperature changes exist.

The thermophilic range requires temperatures between 50 and 60 °C. Under certain circumstances thermophilic process operation allows faster substrate turnover, so that the residence times can be shorter. The higher expenditure of energy to maintain the process temperature is a disadvantage.

When the process is run under thermophilic conditions for a defined residence time, sanitation in the reactor is possible,; otherwise, sanitation has to be achieved in a separate treatment step or by aerobic after-composting. On the other hand, the net energy yield is lower, due to the higher heat requirement, and the temperature sensitivity of the microorganisms reduces the process stability. Table 14.8 lists the advantages and disadvantages of mesophilic and thermophilic process operations.

### 14.3.2.4 Agitation

For a high degradation activity of the bacteria, it is necessary to provide the active biomass with sufficient degradable substrate. Simultaneously, the metabolic products

**Table 14.7** Comparison of continuous and discontinuous feed.

| *Process Operation* | *Continuous* | *Discontinuous* |
|---|---|---|
| Retention time | shorter | longer |
| Technical equipment | complex | simple |

**Table 14.8** Comparison of mesophilic and thermophilic process operation.

| Process Operation | Mesophilic (35 °C) | Thermophilic (55 °C) |
|---|---|---|
| Process stability | higher | lower |
| Temperature sensitivity | low | high |
| Energy demand | low | high |
| Degradation rate | decreased | increased |
| Detention time | longer or the same | shorter or the same |
| Sanitation | no | possible |

of the organisms have to be removed (Dauber, 1993). These requirements can be met by mechanical mixing or other agitation of the reactor contents. Another possibility is to install a water recirculation system, by which the process water, which ensures nutrient provision and the removal of metabolic products, trickles through the biowaste in the reactor (Rilling and Stegmann, 1992). Other processes use compressed biogas for total or partial mixing of the material.

### 14.3.3
### Survey of Anaerobic Fermentation Processes

During the past few years several anaerobic processes for the utilization of solid waste have been developed. Each has its own benefits. Currently, the European market offers at least 30 different processes or process variants. Figure 14.4 shows (without claiming to be complete) the processes organized according to water content and number of fermentation stages and indicates that most of the offered processes are one- or two-stage wet fermentation systems. The number of dry fermentation processes is comparatively small. One primary reason for this situation is the higher technical effort required for dry fermentation. Although wet fermentation can be based on the well known and successful technology used in sewage sludge treatment and the digestion of manure, dry fermentation requires the employment of new, innovative technologies, especially in the field of gas-tight filling and emptying, as well as in conveyance systems.

An example of a one-step wet fermentation system is the DBA–Wabio process (Fig. 14.5); an example of a two-step wet fermentation system is the BTA process (Fig. 14.6); and examples of one-step dry fermentation systems are the ATF process (Fig. 14.7) and the Dranco process (Fig. 14.8).

### 14.3.4
### Feedstock for Anaerobic Digestion

The feedstock for an anaerobic digestion plant can be organic wastes that have been separately collected and delivered to the plant ready for processing or, alternatively, municipal solid waste (MSW) or a fraction of MSW (e.g., <100 mm) from a mechanical sorting plant in which the other fraction is a kind of refuse-derived fuel. A fur-

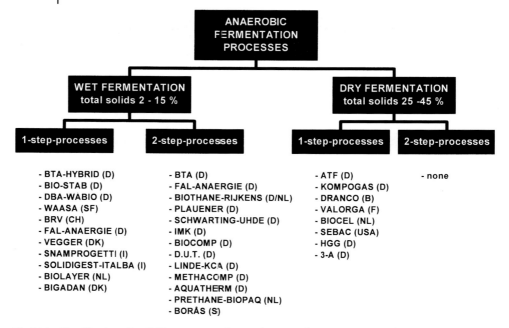

**Fig. 14.4** Classification of available processes of anaerobic waste fermentation (according to Rilling, 1994a, updated).

ther source of organic waste is 'green wastes' collected at centralized collection points.

At least the purity of the raw material fed into the anaerobic digestion process dictates the quality of the horticultural product coming out at the end of the process.

The range of application of the anaerobic digestion process is very broad. In principle, any organic material can be digested, i.e. (Rilling, 1994b):

- organic municipal solid waste
- waste from central markets (e.g., fruit, vegetable, and flower residuals)
- slaughterhouse waste (paunch manure)
- residues from the fish processing industry
- food wastes from hotels, restaurants, and canteens
- bleaching soil
- drift materials such as seaweed or algae
- agricultural waste
- manure
- beer draff
- fruit or wine marc
- sewage sludge

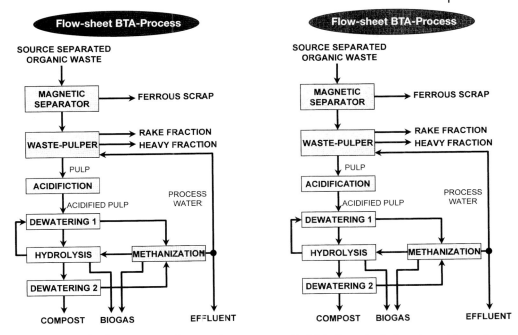

**Fig. 14.5** Example of a one-step wet fermentation plant (DBA–Wabio).

**Fig. 14.6** Example of a two-step wet fermentation plant (Biotechnische Abfallbehandlung, BTA).

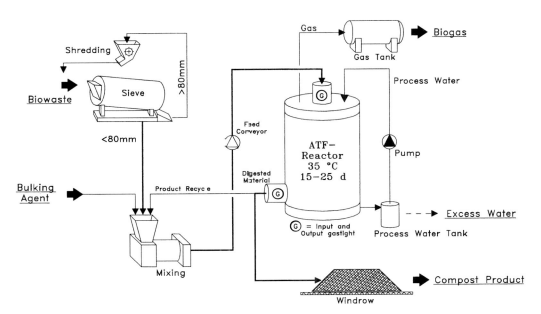

**Fig. 14.7** Example of a one-step dry fermentation plant (ATF process) (Rilling et al., 1996).

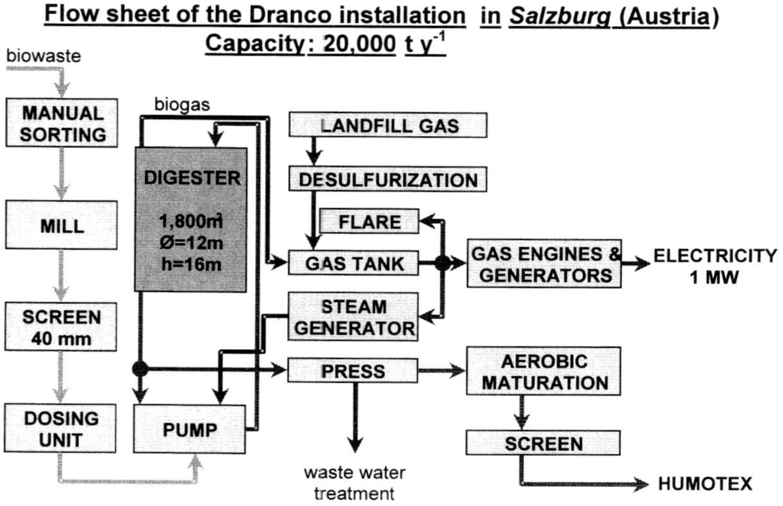

**Flow sheet of the Dranco installation** in *Salzburg* **(Austria)**
**Capacity: 20,000 t y⁻¹**

**Fig. 14.8** Example of a one-step dry fermentation plant (Dranco process) (Six et al., 1995)

## 14.4
## Conclusions

The separate collection of biological wastes and their biological treatment are an important part of waste utilization. There is no universal process for the biological treatment of biological waste, nor will there be. Nevertheless, composting anaerobic digestion is turning out to be a useful technology in both its economic and ecological aspects.

The basic aims of anaerobic biological waste treatment are

- decrease of waste volume
- mass reduction of organic substance for stabilization of waste
- conservation of nutrients and recycling of organic substances (fertilizer or soil improvement)
- energy recovery by generation of biogas
- odor reduction
- sanitation

Anaerobic digestion is a sustainable treatment technique for many kinds of organic wastes. Based on the expectations of its future rising market share, the increasing number of processes offered shows the dynamic and innovative force of anaerobic fermentation technology.

# References

Braun, R., *Biogas: Methangärung organischer Abfallstoffe*, Berlin **1982**: Springer-Verlag.

Dauber, S., Einflußfaktoren auf die anaeroben biologischen Abbauvorgänge, in: *Anaerobtechnik: Handbuch der anaeroben Behandlung von Abwasser und Schlamm* (Böhnke, B., Bischofsberger, W., Seyfried, C. F., eds.), pp. 62–93. Berlin **1993**: Springer-Verlag.

Hungate, R. E., *The Rumen and Its Microbes*. New York **1966**: Academic.

Kaltwasser, B. J., *Biogas: Regenerative Energieerzeugung durch anaerobe Fermentation organischer Abfälle in Biogasanlagen*. Wiesbaden **1980**: Bauverlag.

Kern, M., Mayer, M., Wiemer, K., Systematik und Vergleich von Anlagen zur anaeroben Abfallbehandlung, in: *Biologische Abfallbehandlung III* (Wiemer, K., Kern, M., eds.) pp. 409–437. Witzenhausen **1996**: M.I.C. Baeza.

Konzeli-Katsiri, A., Kartsonas, N., Inhibition of anaerobic digestion by heavy metals, in: *Anaerobic Digestion of Sewage Sludge and Organic Agricultural Wastes* (Bruce, A. M., Konzeli-Katsiri, A., Newman, P. J., eds.), pp. 104–119. London **1986**: Elsevier Applied Science.

KrW-/AbfG, *Gesetz zur Förderung der Kreislaufwirtschaft und Sicherung der umweltverträglichen Beseitigung von Abfällen* (**1994**) BGBl I S. 2705.

Maurer, M., Winkler, J.-P., *Biogas: Theoretische Grundlagen, Bau und Betrieb von Anlagen*. Karlsruhe **1982**: C. V. Müller.

Mudrack, K., Kunst, S., *Biologie der Abwasserreinigung*. Stuttgart **1991**: Gustav Fischer.

Rilling, N., Anaerobe Trockenfermentation für Bioabfall, *Ber. Abwassertechn. Ver.* **1994a**, *44*, 985–1002.

Rilling, N., Untersuchungen zur Vergärung organischer Sonderabfälle, in: *Anaerobe Behandlung von festen und flüssigen Rückständen*, Dechema Monographien, Bd. 130 (Maerkl, H., Stegmann, R., eds.), pp. 185–205. Weinheim **1994b**: VCH.

Rilling, N., Stegmann, R., High solid content anaerobic digestion of biowaste, in: *Proc. 6th Int. Solid Wastes Congr. ISWA 92*, Madrid, Spain **1992**, Ategrus, Bilbao.

Rilling, N., Arndt, M., Stegmann, R., Anaerobic fermentation of biowaste at high total solids content: experiences with ATF system, in: *Management of Urban Biodegradable Wastes* (Hansen, J. A., ed.), pp. 172–180. London **1996**: James & James.

Roediger, H., Roediger, M., Kapp, H., *Anaerobe alkalische Schlammfaulung*. München **1990**: Oldenbourg.

Sahm, H., Biologie der Methanbildung, *Chem. Ing. Technol.* **1981**, *53*, 854–863.

Symons, G. E., Buswell, A. M., The methane fermentation of carbohydrates, *J. Am. Chem. Soc.* **1933**, *55*, 2028.

Six, W., Kaendler, C., De Baere, L., The Salzburg plant: a case study for the biomethanization of biowaste, in: *Proc. 1st Int. Symp. Biol. Waste Manag.: A Wasted Chance?* Bochum, Germany **1995**, BWM, Oelde.

Winter, J., Mikrobiologische Grundlagen der anaeroben Schlammfaulung, *gwf Wasser Abwasser* **1985**, *126*, 51–56.

# 15
# Landfill Systems, Sanitary Landfilling of Solid Wastes, and Long-term Problems with Leachate

Kai-Uwe Heyer and Rainer Stegmann

## 15.1
## Introduction

In most countries the sanitary landfill plays a most important role in the context of solid waste disposal and will remain an integral part of solid waste management. The quality of landfill design, according to technical, social, and economic developments, has improved dramatically in recent years. Design concepts are mainly devoted towards ensuring minimal environmental impact in accordance with observations made concerning the operation of old landfills. The major environmental concern associated with landfills is related to the discharge of leachate into the environment, and current landfill technology is primarily determined by the need to prevent and control leachate problems. To reduce the emissions of gases that cause global warming, the control of landfill gas gains more and more importance.

To reduce these landfill emissions significantly in the future, the amount of biodegradable organics prior to deposition has to be reduced (see EC Landfill Directive, 1999). Moreover, waste pretreatments may be required, for example, mechanical–biological pretreatment or thermal treatment (ABFABLV, 2001).

## 15.2
## Biochemical Processes in Sanitary Landfills

To promote understanding of the requirements for landfill systems and sanitary landfilling of municipal solid wastes, the main chemical, physical, and biological factors that influence the processes in landfills and the leachate quality are described.

The mechanisms that regulate mass transfer from wastes to leaching water, from which leachate originates, can be divided into three categories (Andreottola, 1992):
- hydrolysis of solid waste and biological degradation
- solubilization of soluble salts contained in the waste
- wash out of fines

*Environmental Biotechnology. Concepts and Applications.* Edited by H.-J. Jördening and J. Winter
Copyright © 2005 WILEY-VCH Verlag GmbH & Co. KGaA, Weinheim
ISBN: 3-527-30585-8

The first two categories of mechanisms, which have greater influence on the quality of leachate produced, are included in the more general concept of waste stabilization in landfills.

## 15.2.1
### Aerobic Degradation Phases

The first phase of aerobic degradation of organic substances is generally of limited duration, due to the high oxygen demand of waste relative to the limited quantity of oxygen present inside a landfill (Phase I, Figure 15.1). The only layer of a landfill involved in aerobic metabolism is the upper layer, where oxygen is trapped in fresh waste and is supplied by diffusion and rainwater. Usually the aerobic phase is quite short, and no substantial leachate generation takes place.

In very old landfills, for which only the more refractory organic carbon material remains in the landfilled wastes, a second aerobic phase may appear in the upper layer of the landfill. In this phase the methane production rate is very low and air stars diffusing from the atmosphere, giving rise to aerobic zones and zones with redox potentials too high for methane formation (Christensen and Kjeldsen, 1989).

## 15.2.2
### Anaerobic Degradation Phases

Three different phases can be identified in the anaerobic decomposition of waste:
- acid fermentation
- intermediate anaerobic phase
- methanogenic fermentation

The first phase of anaerobic degradation is acid fermentation, which causes a decrease in leachate pH, high concentrations of volatile acids, and considerable concentrations of inorganic ions (e.g., $Cl^-$, $SO_4^{2-}$, $Ca^{2+}$, $Mg^{2+}$, $Na^+$). The initial high content of sulfates may be slowly reduced as the redox potential drops and metal sulfides are gradually generated that are of low solubility and precipitate iron, manganese, and other heavy metals that were dissolved by the acid fermentation (Christensen and Kjeldsen, 1989). The decrease in pH is caused by the high production of volatile fatty acids and the high partial pressure of $CO_2$. The increased concentration of anions and cations is due to leaching of easily soluble materials, consisting of original waste components and degradation products of organic substances. Initial anaerobic processes are elicited by a population of mixed anaerobic microbes, composed of strictly anaerobic bacteria and facultative anaerobic bacteria. The facultative anaerobic bacteria reduce the redox potential with the result that methanogenic bacteria can grow. In fact, the latter are sensitive to the presence of oxygen and require a redox potential below $-330$ mV. Leachate from this phase is characterized by high $BOD_5$ values (commonly >10 000 mg $O_2$ $L^{-1}$), high $BOD_5/COD$ ratios (commonly >0.7), acidic pH values (typically 5–6), and ammonia (often 500–1000 mg $L^{-1}$) (Ro-

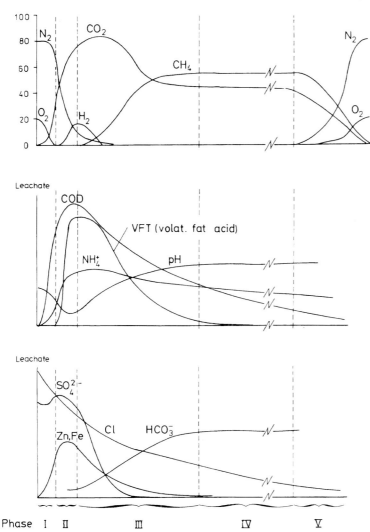

**Fig. 15.1** Time courses of compositions of leachate and gas in a landfill cell (Christensen and Kjeldsen, 1989).

binson, 1989), the latter due to hydrolysis and fermentation of proteinaceous compounds in particular.

The second intermediate anaerobic phase (Phase III, Figure 15.1) starts with slow growth of methanogenic bacteria. This growth may be inhibited by an excess of organic volatile acids, which are toxic to methanogenic bacteria at concentrations of 6000–16 000 mg L$^{-1}$ (Stegmann and Spendlin, 1989). The methane concentration in the gas increases, while hydrogen, carbon dioxide, and volatile fatty acids decrease.

Moreover, the concentration of sulfate decreases owing to biologic reduction. Conversion of fatty acids causes an increase in pH values and alkalinity, with a consequent decrease in solubility of calcium, iron, manganese, and heavy metals. The latter are probably precipitated as sulfides. Ammonia is released and is not converted in the anaerobic environment.

The third phase of anaerobic degradation (Phase IV, Figure 15.1) is characterized by methanogenic fermentation elicited by methanogenic bacteria. At this stage, the composition of leachate is characterized by almost neutral pH values and low concentrations of volatile acids and total dissolved solids; the biogas generally consists of more than 50% methane. This confirms that solubilization of the majority of organic components has decreased at this stage of landfill operation, although the process of waste stabilization will continue for several decades.

Leachates produced during this phase are characterized by relatively low $BOD_5$ values and low ratios of $BOD_5/COD$. Ammonia continues to be released by the first-stage acetogenic process.

Table 15.1 shows the ranges of leachate concentrations depending on the degradation phase for some relevant parameters. Ehrig (1990) compiled leachate concentrations from German landfills from the 1970s and 1980s. According to his evaluation the organics (COD, $BOD_5$, TOC), as well as AOX, $SO_4$, Ca, Mg, Fe, Mn, Zn, and Cr, are determined by the biochemical processes in the landfill, and there are striking differences between the acid phase and the methanogenic phase.

Kruse (1994) investigated 33 landfills in Northern Germany, the leachate concentrations mainly derive from the late 1980s and early 1990s. He defined three characteristic periods according to the $BOD_5/COD$ ratio:

- acid phase:          $BOD_5/COD \geq 0.4$
- intermediate phase:  $0.4 > BOD_5/COD > 0.2$
- methanogenic phase:  $BOD_5/COD \leq 0.2$

Between the two investigations there are significant differences concerning the quantities of organic parameters. In the younger landfills (Kruse, 1994), leachate concentrations of COD, $BOD_5$, and TOC are lower than those determined by Ehrig (1990) some 10 years earlier. This can be explained by developments in the technology of waste landfilling. In many younger landfills waste deposition and compaction in thin layers, in combination with an aerobic pretreated bottom layer, were carried out. This led to a reduction in the period for the acid phase and to an accelerated conversion of organic leachate components into the gaseous phase, as well as the degradation of organics to methane and carbon dioxide.

### 15.2.3
### Factors Affecting Leachate Composition

The chemical composition of leachate depends on several parameters, including those concerning waste mass and site localization and those deriving from design and management of the landfill. Of the former. the main factors influencing leachate quality are discussed in this section.

Table 15.1  Constituents in leachates from MSW landfills (Ehrig, 1990; Kruse, 1994).

| Parameter | Unit | Leachate from MSW Landfills (Ehrig, 1990) | | | | Leachate from MSW Landfills (Kruse, 1994) | | | | | |
| | | Acid Phase | | Methanogenic Phase | | Acid Phase | | Intermediate Phase | | Methanogenic Phase | |
| | | Range | Medium | Range | Medium | Range | Medium | Range | Medium | Range | Medium |
|---|---|---|---|---|---|---|---|---|---|---|---|
| pH | – | 4.5–7 | 6 | 7.5–9 | 8 | 6.2–7.8 | 7.4 | 6.7–8.3 | 7.5 | 7.0–8.3 | 7.6 |
| COD | mg $O_2$ $L^{-1}$ | 6000–60000 | 22000 | 500–4500 | 3000 | 950–40000 | 9500 | 700–28000 | 3400 | 460–8300 | 2500 |
| $BOD_5$ | mg $O_2$ $L^{-1}$ | 4000–40000 | 13000 | 20–550 | 180 | 600–27000 | 6300 | 200–10000 | 1200 | 20–700 | 230 |
| TOC | mg $L^{-1}$ | 1500–25000 | 7000 | 200–5000 | 1300 | 350–12000[b] | 2600[b] | 300–1500[b] | 880[b] | 150–1600[b] | 660[b] |
| AOX | µg $L^{-1}$ | 540–3450 | 1674 | 524–2010 | 1040 | 260–6200 | 2400 | 260–3900 | 1545 | 195–3500 | 1725 |
| Organic N[a] | mg $L^{-1}$ | 10–4250 | 600 | 10–4250 | 600 | | | | | | |
| $NH_4$-N[a] | mg $L^{-1}$ | 30–3000 | 750 | 30–3000 | 750 | 17–1650 | 740 | 17–1650 | 740 | 17–1650 | 740 |
| TKN[a] | mg $L^{-1}$ | 40–3425 | 1350 | 40–3425 | 1350 | 250–2000 | 920 | 250–2000 | 920 | 250–2000 | 920 |
| $NO_2$-N[a] | mg $L^{-1}$ | 0–25 | 0.5 | 0–25 | 0.5 | | | | | | |
| $NO_3$-N[a] | mg $L^{-1}$ | 0.1–50 | 3 | 0.1–50 | 3 | | | | | | |
| $SO_4$ | mg $L^{-1}$ | 70–1750 | 500 | 10–420 | 80 | 35–925 | 200 | 20–230 | 90 | 25–2500 | 240 |
| Cl | mg $L^{-1}$ | 100–5000 | 2100 | 100–5000 | 2100 | 315–12400 | 2150 | 315–12400 | 2150 | 315–12400 | 2150 |
| Na[a] | mg $L^{-1}$ | 50–4000 | 1350 | 50–4000 | 1350 | 1–6800 | 1150 | 1–6800 | 1150 | 1–6800 | 1150 |
| K[a] | mg $L^{-1}$ | 10–2500 | 1100 | 10–2500 | 1100 | 170–1750 | 880 | 170–1750 | 880 | 170–1750 | 880 |
| Mg | mg $L^{-1}$ | 50–1150 | 470 | 40–350 | 180 | 30–600 | 285 | 90–350 | 200 | 25–300 | 150 |
| Ca | mg $L^{-1}$ | 10–2500 | 1200 | 20–600 | 60 | 80–2300 | 650 | 40–310 | 150 | 50–1100 | 200 |
| Total P[a] | mg $L^{-1}$ | 0.1–30 | 6 | 0.1–30 | 6 | 0.3–54 | 6.8 | 0.3–54 | 6.8 | 0.3–54 | 6.8 |
| Cr[a] | mg $L^{-1}$ | 0.03–1.6 | 0.3 | 0.3–1.6 | 0.3 | 0.002–0.52 | 0.155 | 0.002–0.52 | 0.155 | 0.002–0.52 | 0.155 |
| Fe | mg $L^{-1}$ | 20–2100 | 780 | 3–280 | 15 | 3–500 | 135 | 2–120 | 36 | 4–125 | 25 |
| Ni[a] | mg $L^{-1}$ | 0.02–2.05 | 0.2 | 0.02–2.05 | 0.2 | 0.01–1 | 0.19 | 0.01–1 | 0.19 | 0.01–1 | 0.19 |
| Cu[a] | mg $L^{-1}$ | 0.004–1.4 | 0.08 | 0.004–1.4 | 0.08 | 0.005–0.56 | 0.09 | 0.005–0.56 | 0.09 | 0.005–0.56 | 0.09 |
| Zn | mg $L^{-1}$ | 0.1–120 | 5 | 0.03–4 | 0.6 | 0.05–16 | 2.2 | 0.06–1.7 | 0.6 | 0.09–3.5 | 0.6 |
| As[a] | mg $L^{-1}$ | 0.005–1.6 | 0.16 | 0.005–1.6 | 0.16 | 0.0053–0.11 | 0.0255 | 0.0053–0.11 | 0.0255 | 0.0053–0.11 | 0.0255 |
| Cd[a] | mg $L^{-1}$ | 0.0005–0.14 | 0.006 | 0.0005–0.14 | 0.006 | 0.0007–0.525 | 0.0375 | 0.0007–0.525 | 0.0375 | 0.0007–0.525 | 0.0375 |
| Hg[a] | mg $L^{-1}$ | 0.0002–0.01 | 0.01 | 0.0002–0.01 | 0.01 | 0.000002–0.025 | 0.0015 | 0.000002–0.025 | 0.0015 | 0.000002–0.025 | 0.0015 |
| Pb[a] | mg $L^{-1}$ | 0.008–1.02 | 0.09 | 0.008–1.02 | 0.09 | 0.008–0.4 | 0.16 | 0.008–0.4 | 0.16 | 0.008–0.4 | 0.16 |

[a] Parameter more or less independent from the biochemical degradation phase.
[b] DOC.

### 15.2.3.1 Waste Composition

The nature of the waste organic fraction considerably influences the degradation of waste in the landfill and thus also the quality of the leachate produced. The inorganic content of the leachate depends on the contact between waste and leaching water, as well as on the pH and the chemical balance at the solid–liquid interface. In particular, the majority of metals are released from the waste mass under acid conditions.

### 15.2.3.2 Water Balance

In Figure 15.2, the main factors affecting the water balance of landfills are represented schematically. The water balance of a landfill may be described by means of the water balance equation, where $L_F$ describes the proportion of precipitation which, after decreases due to evaporation and surface effluent, is actually introduced into the landfill body:

$$P - V_E - V_T - E_S - S \pm R \pm W_D + W_C = E_B$$

where:

$P$ precipitation, controlled water addition, if required
$V_E$ evaporation
$V_T$ transpiration
$E_S$ effluent surface
$S$ storage
$L_F$ climatic leachate formation
  $L_F = P - V_E - V_T - E_S$
$R$ retention
$W_D$ water demand or release by biological conversion
$W_C$ consolidation
$E_B$ leachate effluent at the landfill base (into a drainage system, or underground if no bottom sealing exists)

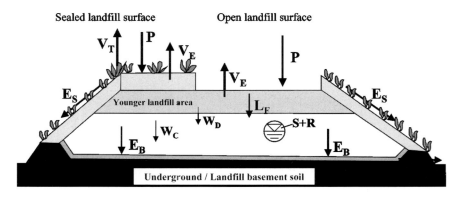

**Fig. 15.2** Factors affecting the water balance in a landfill (Hupe et al., 2003).

### 15.2.3.3 Landfill Age

Variations in leachate composition and in quantity of pollutants removed from waste are often attributed to landfill age, defined as the time from the deposition of waste or the time from the first appearance of leachate. Landfill age obviously plays an important role in the determination of leachate characteristics governed by the type of waste stabilization processes. We should emphasize that variations in composition of leachate do not depend exclusively on landfill age but also on the degree of waste stabilization and the volume of water that infiltrates the landfill. The pollutant load in leachate generally reaches maximum values during the first years of operation of a landfill (2–3 years) and then gradually decreases in the following years. This trend is generally applicable to organic components, the main indicators of organic pollution (COD, $BOD_5$, TOC), microorganism population, and the main inorganic ions (heavy metals, Cl, $SO_4$, etc.).

## 15.3
## Sanitary Landfilling and Leachate Control Strategies

### 15.3.1
### Leachate Problems in Landfills

The most typical detrimental effect of leachate discharge into the environment is that of groundwater pollution. To prevent this, the first step in landfill design development was to site the landfill far from the groundwater table and/or far from groundwater extraction wells. Thus more attention was focused on studying the hydrogeology of the area so as to identify the best siting of the landfill.

A further step in landfill technology was to site the landfill in low-permeability soil and/or to engineer impermeable liners to contain wastes and leachate. Containment, however, poses the problem of leachate treatment.

Discharge of leachate into the environment is today considered under more restrictive views. The reasons for this are:
- many severe instances of groundwater pollution by landfills
- the greater hazard posed by the size of landfill (larger than in the past)
- the need to comply with more and more restrictive legislation regarding quality standards for wastewater discharges

The leachate problem accompanies a landfill from its beginning to many decades after its closure. Therefore, leachate management facilities should also last and their effectiveness be ensured over a long period of time.

### 15.3.2
### Sanitary Landfilling and Legal Requirements

In view of all the aspects mentioned above, leachate control strategies involve the input (waste and water), the reactor (landfill), and the output (leachate and gas). This is one of the reasons why the German Landfill Ordinance (DEPV, 2002), the Technical Instructions on Waste from Human Settlements (TASI, 1993), laid down standards for disposal, including the collection, treatment, storage, and landfilling of wastes from human settlements. State-of-the-art technology is required and, with the so-called multibarrier concept, it is the waste to be dumped itself that forms the most important barrier. The other barriers are the geological barrier of the landfill site, base sealing with an effective drainage system, and surface sealing after a landfill section has been completely filled. Considering these major aspects the Instructions on Waste from Human Settlements define two classes of landfills:

- landfill class I
  - particularly high standards for mineralization levels of the waste to be dumped
  - relatively low standards for landfill deposit sites and landfill sealing (base and surface sealing)
- landfill class II
  - lower standards for mineralization levels of the waste to be dumped (mainly for municipal solid wastes)
  - considerably higher standards for landfill deposit sites and landfill sealing than apply to landfill class I

### 15.3.3
### Control of Waste Input and Pretreatment before Deposition

The first step in the waste input control strategy should be that of minimizing the amount of waste to be landfilled. This can be achieved by waste avoidance, separate collection activities, recycling centers for recyclables, incineration, and mechanical–biological pretreatment of residual municipal solid waste (MSW) or composting of biowaste.

With regard to its properties, the waste can be separated into different fractions: a light fraction of high calorific value, in some instances also a mineral fraction and a fraction rich in organics can be gained. Some of these fractions also have reutilization potential. The mechanical–biological pretreatment of MSW can be used within a waste management concept as a sole process or in combination with thermal pretreatment (Fig. 15.3).

After the mechanical–biological pretreatment, significant reductions in the remaining emissions, in the range 70%–90% in terms of loads and concentrations, are possible. Figure 15.4 shows as an example the low leachate concentration of a landfill section filled with MBP residues.

### 15.3.4
### Control of Water Input and Surface Sealing Systems

The strategy for water input control is strictly related to the quality of the waste to be landfilled. For nonbiodegradable waste and according to its potential hazard to the environment, prevention of water infiltration can be used as the main option (usually by means of top sealing). In contrast, for biodegradable waste, water input must

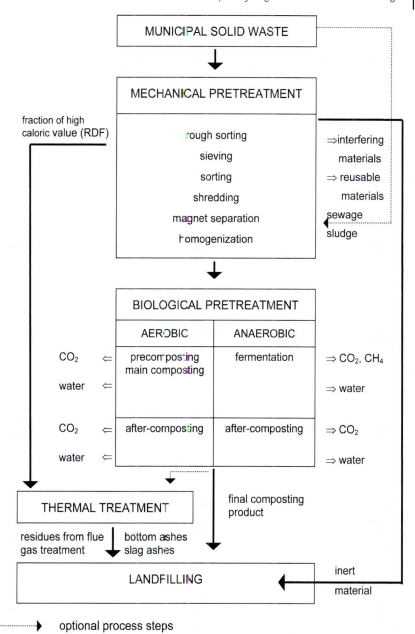

**Fig. 15.3** Scheme of waste pretreatment before landfilling.

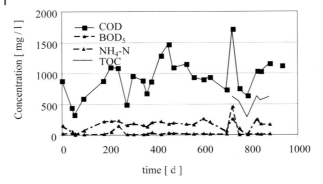

**Fig. 15.4** Effects of mechanical–biological treatment on leachate concentrations in an MBP landfill section.

be assured until a high degree of biostabilization is achieved. Here, water input should be limited to the strictly necessary amount and minimization techniques should be used. The most important parameters in this regard are (Stegmann et al., 1992):
- siting of landfills in low precipitation areas, if possible
- planting the topsoil with species characterized by high evapotranspiration
- surface lining in critical hydrological conditions
- surface water drainage and diversion
- high compaction of the refuse in place
- measures to prevent risks of cracking owing to differential settlement

Furthermore, utilization of intermediate covers in the landfill operation area could represent a useful leachate minimization technique.

The two final surface sealing systems that must be used, according to the regulations of the TASI (TASI, 1993), are shown in Figure 15.5.

### 15.3.5
#### Control of Leachate Discharge into the Environment and Base Sealing Systems

As mentioned above, uncontrolled leachate discharge into the environment is the main pollution risk of waste deposition. For this reason, regulations are more restrictive today (DepV, 2002, TASI, 1993). The following tools are used:
- The lining system should be based on the multibarrier effect (double or triple liners). Quality control of materials and construction should be improved to ensure higher safety and durability.
- A rational drainage and collection system is important to avoid accumulation of leachate inside the landfill. The main problems of drainage systems are proper choice of materials, clogging, durability, and maintenance.

In Figure 15.6 the base sealing systems for class I and II landfills, according to the requirements of the TA Siedlungsabfall (TASI, 1993), are shown. The sealing

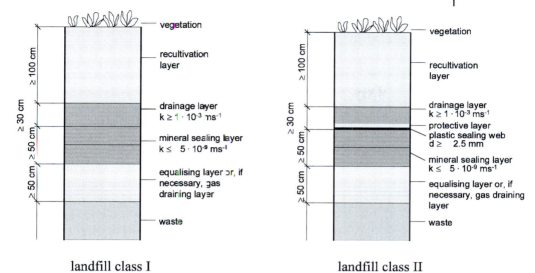

landfill class I           landfill class II

**Fig. 15.5** Landfill surface sealing systems (TASI, 1993).

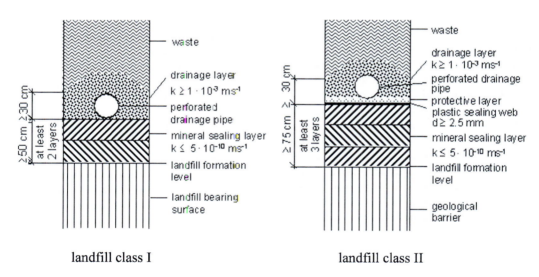

landfill class I           landfill class II

**Fig. 15.6** Landfill base sealing systems (TASI, 1993).

systems must be arranged on the landfill base and on the sloping areas. For landfill class II, the geological barrier should comprise naturally arranged, slightly permeable bedrock several meters thick. If this requirement is not met, a homogenous equalizing layer at least 3 m thick and having a permeability of $k \leq 1 \times 10^{-7}$ ms$^{-1}$ has to be installed.

The landfill base must be at least 1 m above the highest expected groundwater level, once settling under the load of the landfill has come to an end.

The surface of the base sealing system must be formed in the manner of a roof profile. Once the sealing bearing surface has finished settling, the surface of the sealing layer must exhibit a transverse gradient of ≥3% and a longitudinal gradient of ≥1%. Perforated pipes (collectors), additionally capable of being rinsed and monitored, must be provided for the collection and discharge of leachate. The leachate must be channeled by means of free flow into drainage shafts that must be installed outside the dumping area.

## 15.3.6
### Leachate Treatment

Leachate has always been considered a problematic wastewater from the point of view of treatment, because it is highly polluted and its quality and quantity are modified with time in the same landfill. Today, according to the increasingly restrictive limits for wastewater discharge, complicated and costly treatment facilities are required.

Table 15.2 shows the current German requirements for leachate quality before discharge. Several parameters are of great importance for the treatment technology that has to be applied, mainly COD and AOX, but also nitrogen and $BOD_5$. The first

**Table 15.2** Limiting concentrations for the discharge of treated leachate according to German standards (51. Anhang Rahmen-AbwasserV, 1996).

| Parameter | Limiting Concentration mg $L^{-1}$ |
|---|---|
| COD | 200 |
| $BOD_5$ | 20 |
| Nitrogen, total (sum of $NH_4 + NO_2 + NO_3$) | 70 |
| Phosphorus, total | 3 |
| Hydrocarbons | 10 |
| Nitrite-nitrogen | 2 |
| AOX | 0.5 |
| Mercury | 0.05 |
| Cadmium | 0.1 |
| Chromium | 0.5 |
| Chromium (VI) | 0.1 |
| Nickel | 1 |
| Lead | 0.5 |
| Copper | 0.5 |
| Zinc | 2 |
| Cyanide, easy liberatable | 0.2 |
| Sulfide | 1 |

two parameters require a more comprehensive treatment technology or a combination of different treatment methods.

Current treatment facilities for the treatment of leachate mainly consist of several treatment methods to meet the limiting concentrations for the effluent. Typical combinations are shown in Figure 15.7.

## 15.3.7
### Environmental Monitoring

Environmental monitoring is of extreme importance for the evaluation of landfill operational efficiency and for the observation of environmental effects on a long-term basis. The following monitoring facilities must generally be provided and checked at regular intervals for proper operation (TASI, 1993):

- groundwater monitoring system with at least one measuring station in the inflowing current of ground water and a sufficient number of measuring stations in the current of ground water flowing out of the landfill area
- measuring facilities for monitoring settlement and deformations in the landfill body and the landfill sealing systems

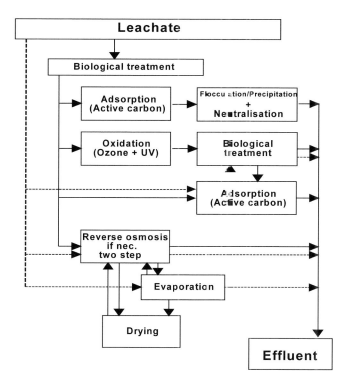

**Fig. 15.7** Methods and combinations of methods often used for leachate treatment (Ehrig et al., 1998).

- measuring facilities for recording meteorological data such as precipitation, temperature, wind, evaporation
- measuring facilities for collecting the quantities of leachate and water that are needed for analyzing the water balance
- measuring facilities for recording the quality of leachate and other waters

In addition, if the generation of landfill gas is expected, it is necessary to provide facilities for measuring landfill gas and to install gas level indicators for the purpose of emission control.

## 15.4
### Long-term Problems with Leachate

The characteristics of landfill leachate are relatively well known, at least for the first 20–30 years of life of the landfill, the period from which actual data are available. On the other hand, little is known about the leachate composition of later phases of the landfill, and the basis for making good estimates is rather weak.

For this reason, several landfills were investigated in a German joint research project 'Landfill Body' (Ehrig et al., 1997). The purpose was to describe the present stage of stability of landfills of different ages, their corresponding emissions, and the future development of emissions.

The main focus of the research program included long-term experiments in test lysimeters that were carried out to predict emissions that the solid waste in old landfills will release in the future.

### 15.4.1
### Lysimeter Tests in Landfill Simulation Reactors (LSR)

To describe the effects of future biological and thermal pretreatment on leachate emissions, landfill simulation experiments were carried out under anaerobic conditions. The test system ensured that the typical landfill phases, such as the acid phase and the stable methane phase, took place in sequence in the reactor. Choosing appropriate milieu conditions enabled the researchers to achieve an enhanced biologic degradation process. By this means, the maximum emission potential represented by gas production and leachate load were determined within reasonable periods of time (Heyer et al., 1997; Stegmann, 1981).

The range of emissions in the water phase that can occur in the landfill in the future was based on experiments in four LSR lysimeters. Figure 15.8 shows concentrations of COD and TKN on a logarithmic scale during a test period of more than 1000 d:

- The COD concentrations showed very similar qualitative declines that followed a gradual asymptotic course, which can be described as a function of dilution and mobilization. The $BOD_5$ concentrations were very low as expected, because almost all waste samples were in the stabilized methane phase when the tests began. The $BOD_5/COD$ ratio was lower than 0.1.

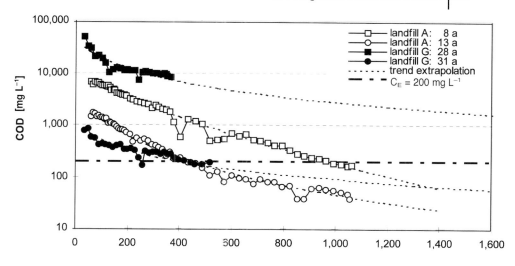

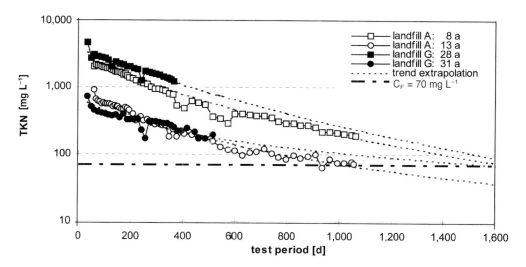

**Fig. 15.8** COD and TKN: LSR leachate concentrations in waste samples from two landfills (Heyer et al., 1998).

- There were striking differences in the magnitude of concentrations, depending on the age of deposition and the conditions within the landfill body before the sampling and because of varying waste compositions.
- The time course of nitrogen emission was comparable to that of the organic parameters. However, the decline of nitrogen in the leachate occurred more slowly, because a higher portion of organic compounds became hydrolyzed during the LSR test period. More than 90% of the TKN nitrogen was emitted as ammonia.

## 15.4.2
## Prognosis of Periods of the Long-Term Time Course of Emissions

The time course of leachate emissions depends mainly on:
- the potential substances that can be mobilized
- the water balance in the landfill, mainly the water flux
- the mobilization behavior

These LSR tests did not allow final, generalizing predictions for the development of emissions development with time. One reason was the specific conditions of each landfill site, e.g., the climate, the surface cover, the waste composition, and inhibition effects. Another reason was the high water exchange rate in the LSR tests that was used to simulate accelerated conversion, mobilization. and dilution in the landfill. However, the possible future development of emissions is discussed below.

The time course of emissions can be described with an exponential function. With the idealized conditions in the LSR test devices and the setting of a water balance, which was approximately 100 times higher than at the landfill, periods $T_E$ can be estimated, until a limiting value $C_E$ is reached. The estimations are based on the following assumptions:
- constant climatic leachate generation of 250 mm per year (this means no impermeable surface sealing, only a permeable soil cover)
- a standard height of 20 m
- the dry densities in the LSR tests are similar to those in the landfill, approximately 0.75 Mg TS m$^{-3}$
- uniform percolation through the landfill body

**Table 15.3** Estimations of periods $T_E$ for reaching the limiting values $C_E$ (Heyer, 2003).

| Parameter | $C_E$ Limiting Value | $C_0$ Concentration at Test Start [mg L$^{-1}$] | $T_{1/2}$ Half Life [a] | $T_E$ Periods [a] |
|---|---|---|---|---|
| COD | $C_E = 200$ mg $O_2$ L$^{-1}$ mean | 500–12 700 5100 | 10–40 28 | 80–360 140 |
| TKN | $C_E = 70$ mg L$^{-1a}$ mean | 200–2100 1200 | 15–57 43 | 120–450 220 |
| Cl | $C_E = 100$ mg L$^{-1}$ mean | 340–2950 1200 | 15–43 33 | 90–250 140 |
| AOX | $C_E = 500$ µg L$^{-1}$ mean | 390–2380 µg L$^{-1}$ 1600 µg L$^{-1}$ | 14–42 22 | 30–210 80 |

[a] Total amount of nitrogen = sum of ammonia, nitrite, and nitrate.

The periods $T_E$ are compiled in Table 15.3 together with the concentrations $C_0$ at the beginning of the LSR tests and the half-life values $T_{1/2}$.

According to German standards for COD in the leachate, the estimation results in a period of 80–360 years, with a mean of 140 years, until the limiting concentration of 200 mg $O_2$ $L^{-1}$ is reached. Chloride shows similar periods. All investigations and tests point to nitrogen as the component with the longest period of release of relevant concentrations into the leachate phase: 220 years on average may be necessary until a concentration of 70 mg $L^{-1}$ is reached (Heyer, 2003).

The plausibility of these estimates is difficult to judge. Each landfill has a different water balance, which can vary with the seasons or change because of surface covers, lining systems, or damage to these technical barriers.

## 15.5
## Controlled Reduction of Leachate Emissions

### 15.5.1
### In Situ Stabilization for the Closing and Aftercare of Landfills

The main option for controlling leachate quality by controlling the landfill reactor is to enhance the biochemical processes (when biodegradable wastes are deposited). Then the question arises of how to positively influence the emission behavior of municipal solid waste deposits in such a way that the duration and extent of aftercare measures can be reduced. For this purpose, two principal in situ stabilization methods can be used, depending on the boundary conditions of landfills and old deposits (Fig. 15.9):

- Humidification and irrigation methods, e.g., for younger waste deposits equipped with surface sealing and having a higher proportion of bioavailable organic material for the intensification of anaerobic degradation processes (Hupe et al., 2003).
- Aeration methods, e.g., for older waste deposits or for deposits showing a lower proportion of bioavailable organic substances and decreasing landfill gas production.

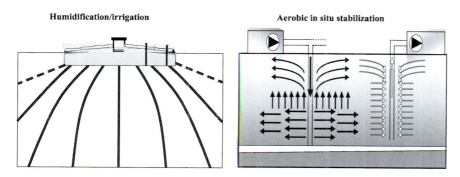

**Fig. 15.9** Methods for in situ stabilization with the goal of reducing aftercare.

## 15.5.2
### Technical Methods for Water Infiltration and Effects on Leachate

Technical methods for water infiltration must be planned so that controlled, even moisture penetration of the landfill body is guaranteed. Likewise, short circuit currents and preferred seepage paths must be avoided by using suitable measures.

Various technical methods are available for use with regard to water infiltration into the landfill body. At the same time, the effect of the infiltration plants on existing surface sealing systems or on systems that are yet to be installed is of great importance. The choice of infiltration system is additionally determined by the quality of the infiltration medium (see above) and the quantity to be infiltrated.

Depending on the landfill boundary conditions and the targets of infiltration, the following infiltration methods may be used:
- horizontal infiltration systems below the surface sealing system
  - two-dimensional infiltration methods
  - linear infiltration methods
- vertical infiltration systems
  - use of existing vertical gas collectors
  - vertical deep wells
  - infiltration injectors in short screen distances

Combinations of the individual infiltration systems can be used. The experience gained so far with regard to controlled infiltration can be summarized as follows:
- positive experiences
  - enhancement of gas production: up to three times higher
  - longer economic life of the gas
  - accelerated stabilization of the waste body and reduced leachate contamination
- negative experiences
  - blockage and incrustation in the infiltration system
  - shearing, rupture, and buckling of pipes
  - uneven water introduction or no water introduction in subareas

## 15.5.3
### Aerobic In Situ Stabilization and Effects on Leachate Contamination

Aeration processes for aerobic in situ stabilization are being used in several German landfills and old deposits with success. Common to all sites is the ultimate target: the controlled reduction of emissions and of the resultant risk potential of leachate within a relatively short period of time with the goal of economical site closure, aftercare, and securing measures (Heyer et al., 2003).

The basic technical concept of aeration consists of a system of gas wells, through which atmospheric oxygen is fed into the landfill body via active aeration in such a way that aerobic stabilization of deposited waste is accelerated. Simultaneously, the low-contaminated waste gas is collected and treated in a controlled manner by means of other gas wells.

In principle, the following processes occur during aeration in the landfill body:

- A change from anaerobic to aerobic conditions takes place, resulting in an accelerated and, in part, broader degradation of the bioavailable waste components. The increased carbon conversion during in situ aeration therefore leads to faster stabilization of organic substances.
- At the end of the stabilization process, organic compounds consist of only nearly or completely nondegradable compounds with very low residual gas potential.
- In the leachate path, accelerated decrease in the parameters COD and, above all, $BOD_5$, as well as in nitrogen (TKN or $NH_4$-N) occurs with the aerobic degradation of organic compounds and their release into the gas phase (mainly as carbon dioxide) as a result of aeration.
- Compared with strictly anaerobic conditions, the aftercare periods for the leachate emission path are reduced by at least several decades by in situ aeration. The aftercare phase is not considered complete after aeration has been terminated, but aftercare expenses are significantly reduced, since costly leachate purification measures can be terminated earlier. If leachate percolates directly into the underground, as sometimes occurs in old deposits lacking sealing and drainage systems for the collection of leachate, the polluting effects would be considerably lower.

In situ aeration is planned to operate for a period of 2–4 years under average landfill conditions. Meanwhile, results and experience concerning the operation of stabilization measures are available for several sites for a period of 2–3 years (Heyer et al., 2003). Figure 15.10 shows, as an example, the development of nitrogen contamination in the leachate and groundwater off-flow from an old landfill. From the outset of aeration in April 2001, a considerable decrease in the nitrogen contamination can be seen after one year of stabilization, in spite of several deviations.

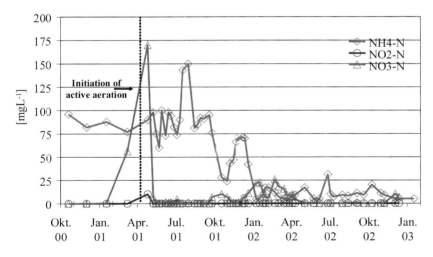

**Fig. 15.10** Changes in leachate contamination within the off-flow area during aerobic in situ stabilization of the old Kuhstedt waste deposit (Heyer et al., 2003).

## References

ABFABLV, Abfallablagerungsverordnung (German Waste Disposal Regulation). Verordnung über die umweltverträgliche Ablagerung von Siedlungsabfällen (Regulation for the Environmentally Compatible Disposal of MSW). March 2001

ABWV, Abwasserverordnung (German Waste Water Ordinance, 1997), Verordnung über Anforderungen an das Einleiten von Abwasser in Gewässer vom 21. März 1997

Andreottola, G., Chemical and biological characteristics of landfill leachate, in: *Landfilling of Waste: Leachate* (Christensen, T.H., Cossu, R., Stegmann, R. eds.), pp. 65–88, London **1992**: Elsevier.

Barber, C., *Behaviour of Wastes in Landfills, Review of Processes of Decomposition of Solid Wastes with Particular Reference to Microbiological Changes and Gas Production*, Stevenage **1979**: Water Research Centre, Stevenage Laboratory Report LR 1059.

Belevi, H., Baccini, P., Long-term behaviour of municipal solid waste landfills, *Waste Manag. Res.* **1989**, *7*, 483–499.

Christensen, T.H., Kjeldsen, P., Basic biochemical processes in landfills, in: *Sanitary Landfilling: Process, Technology and Environmental Impact* (Christensen, T.H., Cossu, R., Stegmann, R. eds.), pp. 29–49, London **1989**: Academic Press.

DepV, Deponieverordnung (German Landfill Ordinance, 2002), Verordnung über Deponien und Langzeitlager: in der Fassung des Beschlusses des Bundeskabinetts vom 24. Juli 2002, BGBl I 2002, 2807

European Council, Council Directive 1999/31/EC of 26 April 1999 on the Landfill of Waste, European Council, *Official Journal of the European Communities* L 182 (**1999**)

Ehrig, H.-J., Qualität und Quantität von Deponiesickerwasser, *Entsorgungsprax. Spez.* **1990**, *1*, 100–105

Ehrig, H.-J., Einführung in das Verbundvorhaben Deponiekörper, in: *Verbundvorhaben Deponiekörper*, Proceedings of 2. Statusseminar, Wuppertal **1997**, pp. 1–5, Umweltbundesamt, Projektträgerschaft Abfallwirtschaft und Altlastensanierung des BMBF.

Ehrig, H.-J., Stegmann, R., Biological processes, in: *Landfilling of Waste: Leachate* (Christensen, T.H., Cossu, R., Stegmann, R. eds.), pp. 185–202, London **1992**: Elsevier.

Heyer K.-U., Emissionsreduzierung in der Deponienachsorge. *Hamburger Berichte, Band 21*, Stegmann (ed), Stuttgart **2003**: Verlag Abfall aktuell.

Heyer, K.-U., Stegmann, R., Untersuchungen zum langfristigen Stabilisierungsverlauf von Siedlungsabfalldeponien, in: *Verbundvorhaben Deponiekörper*, Proceedings of 2. Statusseminar, Wuppertal **1997**, pp. 46–78, Umweltbundesamt, Projektträgerschaft Abfallwirtschaft und Altlastensanierung des BMBF.

Heyer K.-U., Hupe K., Koop A., Ritzkowski M. & Stegmann R., The low pressure aeration of landfills: experiences, operation, costs. *Proceedings Sardinia 2003*, (Christensen, T.H., Cossu, R., Stegmann, R. eds), CISA, Cagliari, Italy **2003**

Hupe K., Heyer K.-U., Stegmann R., Water infiltration for enhanced in situ stabilization. *Proceedings Sardinia Symposium 2003*, (Christensen, T.H., Cossu, R., Stegmann, R. eds), CISA, Cagliari, Italy **2003**

Kruse, K., Langfristiges Emissionsgeschehen von Siedlungsabfalldeponien, Heft 54 der Veröffentlichungen des Instituts für Siedlungswasserwirtschaft, Braunschweig **1994**: Technische Universität.

Robinson, H.D., Development of methanogenic conditions within landfill, *Proceedings 2nd International Landfill Symposium Sardinia '89*, Porto Conte **1989**, October 9–13.

Stegmann, R., Beschreibung eines Verfahrens zur Untersuchung anaerober Umsetzungsprozesse von festen Abfallstoffen im Deponiekörper, *Müll Abfall* **1981**, 2.

Stegmann, R., Spendlin, H.H., Enhancement of degradation: German experiences, in: *Sanitary Landfilling: Process, Technology and Environmental Impact*, (Christensen, T.H., Cossu, R., Stegmann, R. eds.), pp. 61–82, London **1989**: Academic Press.

Stegmann, R., Christensen, T.H., Cossu, R., Landfill leachate: an introduction, in: *Landfilling of Waste: Leachate* (Christensen, T.H., Cossu, R., Stegmann, R. eds.), pp. 3–14, London **1992**: Elsevier.

TASI, Technical Instructions on Waste from Human Settlements (TA Siedlungsabfall), Dritte Allgemeine Verwaltungsvorschrift zum Abfallgesetz vom 14. Mai 1993, Technische Anleitung zur Verwertung, Behandlung und sonstigen Entsorgung von Siedlungsabfällen, Bundesanzeiger Nr. 99a, **1993**

# 16
# Sanitary Landfills: Long-term Stability and Environmental Implications

Michael S. Switzenbaum

## 16.1
## Introduction

Our society generates significant quantities of municipal solid waste. According to Tchobanoglous et al. (1991), solid wastes comprise all the wastes arising from human and animal activities that are normally solid and that are discarded or unwanted. Municipal solid waste is usually assumed to include all community wastes (residential, commercial, institutional, construction and demolition, and municipal services) and does not include industrial and agricultural sources. Although municipal solid waste is only a relatively small fraction of the total amount of solid waste generated, proper management is essential for the control of disease vectors and for protection of the environment.

In the United States, about $1.9 \times 10^{-1}$ kg ($210 \times 10^6$ t) of municipal solid waste are generated per year. The per capita generation rate is about 2 kg (4.4 lb) per person per day (U. S. EPA, 1997b). These rates have vastly increased over the past 30–40 years, but are now starting to level off or even slightly decrease (Fig. 16.1). Generation rates in the United States are considerably higher than generation rates in European countries.

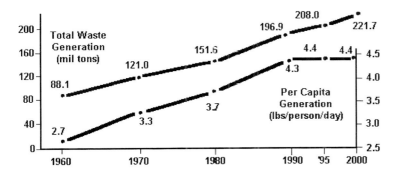

Fig. 16.1  Municipal solid waste generation rates, 1960–2000 (U.S. EPA, 1997a).

*Environmental Biotechnology. Concepts and Applications.* Edited by H.-J. Jördening and J. Winter
Copyright © 2005 WILEY-VCH Verlag GmbH & Co. KGaA, Weinheim
ISBN: 3-527-30585-8

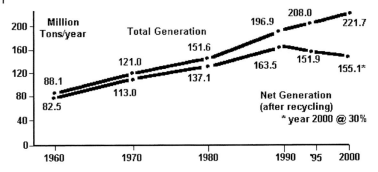

**Fig. 16.2**  Total and net waste generation, 1960–2000 (U.S. EPA, 1997a).

Because of recycling efforts, the net generation rate has been decreasing (Fig. 16.2). Net generation is the amount of solid waste remaining from the total amount generated after materials such as newspaper, glass, and aluminum have been recovered by recycling activities.

Municipal solid waste is a heterogeneous mixture of the materials that society uses. General categories include food wastes, paper, cardboard, plastics, textiles, rubber, leather, yard wastes, wood, glass, 'tin' cans, and ferrous metals.

The standard unit for reporting waste generation is mass. However, mass data are of limited value for certain applications such as landfill design, since average landfill requirements depend on the volume of waste. In addition to the original density of any materials making up the solid waste mixture, the volume also depends on how much the waste has been compacted (Table 16.1).

The comparative percentages of various components of the waste stream are variable. Composition varies geographically and is also influenced by efforts in source reduction and recycling (Table 16.2).

In summary, large amounts of municipal solid waste are generated, and the waste is diverse in nature. These wastes must be managed for the control of disease vectors and for protection of the environment.

**Table 16.1**  Density of municipal solid waste as influenced by compaction (according to Tchobanoglous et al., 1991).

| Component | Density (kg m$^{-3}$) |
| --- | --- |
| Residential | 130 |
| In compactor truck | 300 |
| In landfill (normal) | 450 |
| In landfill (well compacted) | 590 |
| Baled | 700 |

**Table 16.2** Comparison data: solid waste stream composition (% by mass) (according to O'Leary and Walsh, 1992).

| Solid Waste Component | Medford, Wisconsin 1988 Study | Franklin 1990 Estimate | Illinois Department of Natural Resources Guidelines | Cal Recovery 1988 Study | GBB Metro Des Moines |
|---|---|---|---|---|---|
| Food waste | 22.80 | 8.39 | 8.00 | 8.80 | 10.00 |
| Yard waste | 11.80 | 19.80 | 20.00 | 28.20 | 11.00 |
| Other organic | 16.20 | 7.92 | 11.00 | 8.90 | 7.80 |
| Subtotal | 50.80 | 36.11 | 39.00 | 45.90 | 28.80 |
| Newsprint | 3.10 | 5.15 | | 7.30 | 5.20 |
| Corrugated | 6.20 | 7.31 | | 6.70 | 10.80 |
| Mixed paper | 5.60 | 24.39 | | 19.10 | 32.70 |
| Subtotal | 14.90 | 36.85 | 37.00 | 33.10 | 48.70 |
| Ferrous metal | 5.20 | 7.45 | | 3.20 | 4.90 |
| Aluminum | 1.60 | 1.34 | | 1.00 | 0.30 |
| Other metal | 0.20 | 0.20 | | 0.70 | 0.20 |
| Subtotal | 7.00 | 8.99 | 6.00 | 4.90 | 5.40 |
| Plastics | 7.10 | 7.92 | 7.00 | 7.30 | 9.40 |
| Glass | 8.00 | 8.25 | 5.00 | 4.90 | 2.40 |
| Inorganics | 12.20 | 1.88 | 6.00 | 3.90 | 5.30 |
| Subtotal | 27.30 | 18.05 | 18.00 | 16.10 | 17.10 |
| Total | 100.00 | 100.00 | 100.00 | 100.00 | 100.00 |

## 16.2
## Integrated Waste Management

Tchobanoglous et al. (1991) defined solid waste management as the discipline associated with control of the generation, storage, collection, transfer and transport, processing, and disposal of solid wastes in a manner that accords with the best principles of public health, economics, engineering, conservation, aesthetics and other environmental considerations, and that is also responsive to public attitudes. Integrated solid waste management, according to Tchobanoglous et al. (1991) involves the selection and application of suitable techniques, technologies and management programs to achieve specific waste management objectives and goals. Integrated solid waste management refers to the complimentary use of a variety of waste management practices to safely and effectively handle the municipal solid waste stream with the least adverse impact on human health and the environment.

The United States Environmental Protection Agency has adapted a hierarchy in waste management, which can be used to rank management actions (U. S. EPA, 1988). The hierarchy (Fig. 16.3) is in order of preference. Source reduction is at the highest level, and landfilling is at the lowest level.

An integrated approach contains some or all of the following components:
- source reduction: reduction in amount of waste generated
- recycling: separation and collection of waste materials

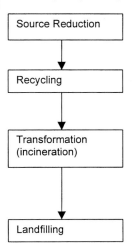

**Fig. 16.3** Waste management hierarchy (U.S. EPA, 1988).

- transformation (incineration): alterations to recover energy or other products (such as compost)
- landfilling: land disposal of wastes, those not able to be recovered, combusted or otherwise transformed

Note that, although land disposal of wastes is ranked lowest in the hierarchy, it is still a widely used waste management strategy. Certain materials cannot be reduced at the source, recycled or transformed and are therefore landfilled.

**16.3**
**Land Disposal**

Although efforts have been made in source reduction, recycling and incineration, most of the solid waste generated in the United States still ends up in landfills (56.9% in 1995). However, the percentage is decreasing (Table 16.3). The data reflect progress made in achieving the goals of the EPA's hierarchy.

**Table 16.3** Waste management practices, 1960–2000 (as percentage of waste generated) (after U.S. EPA, 1997b).

| Year: | 1960 | 1970 | 1980 | 1990 | 1995 | 2000 |
|---|---|---|---|---|---|---|
| Generated | 100 | 100 | 100 | 100 | 100 | 100 |
| Recovered for recycling/composting | 6.4 | 6.6 | 9.6 | 17.2 | 27.0 | 30 |
| Discarded after recovery | 93.6 | 93.4 | 90.4 | 82.8 | 73.0 | 70 |
| Combusted | 30.3 | 20.7 | 9 | 16.2 | 16.1 | 16.2 |
| Discarded to landfill | 60.3 | 72.6 | 81.4 | 66.7 | 56.9 | 53.7 |

Land disposal was promoted as a 'modern' means of solid waste disposal in the mid 1960s, in response to the air pollution problems associated with uncontrolled combustion of municipal solid waste. Unfortunately, land disposal was often not properly conducted. Rather than disposal into well-engineered and operated landfills, solid waste was merely buried in uncontrolled dumps. In addition, there was no distinction between municipal solid waste and hazardous waste – both were placed in these dumps. This often resulted in an adverse impact on the environment. In fact, many of the sites listed on the National Priority List of the United States Environmental Protection Agency are abandoned dumps (or orphaned landfills). A diagram of the potential impacts of improperly constructed dumps (Fig. 16.4) illustrates several detrimental effects, including groundwater contamination from leachate, surface water contamination from runoff and gas migration (which can damage plants, potentially cause explosions in confined areas and can lead to atmospheric contamination).

As of result of the problems from improperly constructed dumps, new federal regulations were promulgated. Subsequent landfill disposal criteria are listed in the Resource Conservation and Recovery Act (RCRA) under Subtitle D (Federal Register, 1991). A summary of the U.S. EPA regulations for MSW landfills is shown in Table 16.4.

The cost of landfilling has greatly increased because of the new federal regulations (Fig. 16.5). Average tipping fees increased greatly after passage of the RCRA.

At the same time, many old landfills have been closed, often because they were causing environmental damage. Figure 16.6 shows the trends in terms of the number of landfills in the United States. Although the number of landfills has decreased, the trend is to build larger landfills (often regional), which are often owned by private businesses (Denison et al., 1994).

In landfills, solid waste is disposed of in thin layers that are compacted to minimize volume and then covered (usually daily) with a thin layer of material (usually soil) to minimize environmental problems.

Intermediate cover is usually placed on the top and exposed sides of the compacted solid waste. The covered compacted material is called a cell. A series of adjacent

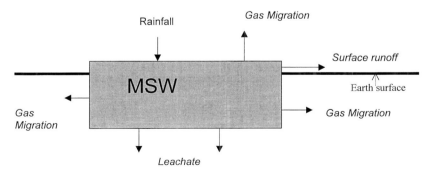

**Fig. 16.4** Environmental impacts of dumps.

**Table 16.4** Summary of U.S. EPA regulations for municipal solid waste landfills (from Tchobanoglous and O'Leary, 1994).

| Item | Requirement |
|---|---|
| Applicability | – all active landfills that receive MSW after October 9, 1993<br>– certain requirements also apply to landfills that received MSW after October 9, 1991, but were closed within 2 years<br>– certain exemptions for very small landfills<br>– some requirements are waived for existing landfills<br>– new landfills and landfill cells must comply with all requirements |
| Location requirement | – airport separation distances of 1.5 km, 3 km, and in some instances more than 3 km are required<br>– landfills located on floodplains can operate only if flood flow is not restricted<br>– construction and filling on wetlands is restricted<br>– landfills over earthquake faults require special analysis and possibly construction practices<br>– landfills in seismic impact zones require special analysis and possibly construction practices<br>– landfills in unstable soil zones require special analysis and possibly construction practices |
| Operating criteria | – landfill operators must conduct a random load checking program to ensure exclusion of hazardous wastes<br>– daily covering with 0.1524 m of soil or other suitable materials is required<br>– disease vector control is required<br>– permanent monitoring probes are required<br>– probes must be tested every 3 months<br>– methane concentrations in occupied structures cannot exceed 1.25%<br>– methane migration offsite must not exceed 5% at the property line<br>– Clean Air Act criteria must be met<br>– access must be limited by fences or other structures<br>– surface water drainage run-on to the landfill and runoff from the working face must be controlled for 24-year rainfall events<br>– appropriate permits must be obtained for surface water discharges<br>– liquid wastes or wastes containing free liquids cannot be landfilled<br>– extensive landfill operating records must be maintained |
| Liner design criteria | – geomembrane and soil liners or equivalents are required under most new landfill cells<br>– groundwater standards may be allowed as the basis for liner design in some areas |
| Groundwater monitoring | – groundwater monitoring wells must be installed at many landfills<br>– groundwater monitoring wells must be sampled at least twice a year<br>– a corrective action program must be initiated where groundwater contamination is detected |
| Closure and post-closure care | – landfill final cover must be placed within 6 months of closure<br>– the type of cover is soil or geomembrane and must be less permeable than the landfill liner<br>– postclosure care and monitoring of the landfill must continue for 30 years |
| Financial assurance | – sufficient financial reserves must be established during the site operating period to pay for closure and postclosure care |

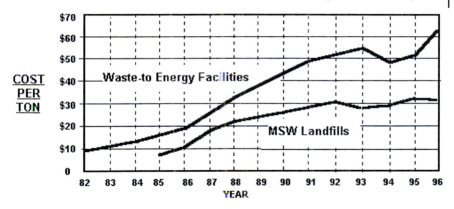

**Fig. 16.5** Landfill tipping fees, national averages in the United States (U.S. EPA, 1997a).

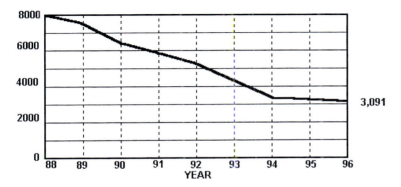

**Fig. 16.6** Number of landfills in the United States (U.S. EPA, 1997a).

cells is called a lift. A lift is usually 3–3 5 m tall. The bottom lift is placed on the bottom liner (usually a composite of clay and a flexible membrane) and a leachate collection and removal system. A landfill eventually consists of several vertical lifts and is 15–30 m tall. The landfill cap is placed on top of the top lift and is a combination of soil and synthetic materials. A schematic diagram of selected engineering features of a modern sanitary landfill is shown in Figure 16.7.

The design of a modern sanitary landfill focuses on the prevention of leachate and gas migration. Vents and collection systems ensure that any gas produced is captured and then recovered, flared, or dissipated in the atmosphere. The landfill cap is designed to prevent precipitation from entering the landfill. Liners prevent contamination of the subsurface. Leachate collection and removal systems prevent a buildup of hydraulic head on the liner.

Note, however, that the strategy of complete entombment of municipal solid waste is somewhat controversial. Many believe that, even though landfills may be carefully designed and operated, all landfills will eventually leak. Subtitle D require-

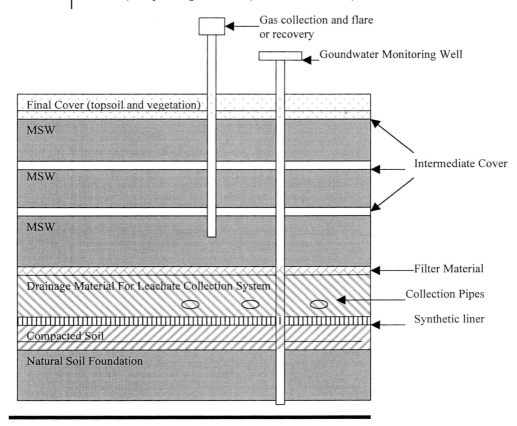

**Fig. 16.7** Profile of landfill engineered features (not to scale).

ments of RCRA can only delay, but not prevent, the generation and migration of lea-chate and methane (Denison et al., 1994). In this regard, several investigators have promoted the adoption of 'wet-cell design and operation' (i.e., leachate recycling) to accelerate decay of the organic fraction of landfilled waste and thus avoid the prob-lem of releases after the post-closure period (Pohland, 1980). Wet-cell design and op-eration also increased the methane yield in earlier years, which made gas recovery more feasible and presented the potential for reuse of the physical property of the landfill (Dension et al., 1994).

The three common configurations for landfills are area, ramp, and trench. The ar-ea configuration involves constructing a landfill above ground, and the trench con-figuration involves excavation. The ramp configuration is a variant of the area con-figuration, as it is built above ground, but on a slope.

## 16.4
## Leachate and Gas Management

The RCRA landfill design criteria include provisions for leachate and gas management to prevent groundwater contamination and to avoid any potential problems from gas migration, including explosions and fires, vegetation damage and atmospheric contamination. The rate and extent of gas generation are influenced by numerous factors but are controlled primarily by the products of microbial reactors in the landfills. Typical constituents of MSW landfill gas are shown in Table 16.5.

Methane and carbon dioxide are the principal gaseous products. Smaller amounts of nitrogen, oxygen, hydrogen and trace compounds (many of which are volatile organic compounds; VOC) can also be found in landfill gas. Some of the trace gases, although present in small amounts, can be toxic and may present risks to public health (Tchobanoglous et al., 1991). The occurrence of significant quantities of VOC is associated with older landfills, which accepted industrial and commercial wastes containing VOC. In newer landfills, the disposal of hazardous wastes is not allowed, and thus the concentrations of VOC in the landfill gas are low. Gas migration is controlled by landfill caps, vents, and recovery systems or flaring systems.

Leachate is formed when water passes through the landfilled waste materials. Leachate is a mixture of organic and inorganic, soluble and colloidal solids. It includes the products from decomposition of materials, as well as soluble constituents leached from the landfilled materials. Leachate generation rates are determined from the amount of water contained in the original material and the amount of precipitation that enters the landfill. Other factors include climate, site topography, final cover material, vegetative cover, site operating procedures and type of waste in the landfill.

Landfill leachate characteristics vary widely from season to season and from year to year, as well as from one landfill to another. Typical data on the composition of leachate are shown in Table 16.6. Particle size, degree of compaction, waste composition, landfilling technique, cover configuration, site hydrology, climate and age of the tip are among the factors that affect leachate quantity and quality (Pohland and

**Table 16.5** Typical constituents of MSW landfill gas (according to Tchobanoglous et al., 1991).

| Component | Percent (dry volume basis) |
| --- | --- |
| Methane | 45–60 |
| Carbon dioxide | 40–60 |
| Nitrogen | 2–5 |
| Oxygen | 0.1–1.0 |
| Sulfides and S compounds | 0–1.0 |
| Ammonia | 0.1–1.0 |
| Hydrogen | 0–0.2 |
| Carbon monoxide | 0–0.2 |
| Trace constituents | 0.01–0.6 |

Harper, 1986). Leachate characteristics are directly related to the natural processes occurring inside the landfill.

The effects of climate on leachate production are fairly well understood. Rainwater provides both the mode of transport of contaminants out of the landfill and the moisture required for microbial activity. Depending on the porosity of the cover material (as well as the integrity of the cap, if one exists), high levels of rainfall often result in the production of large quantities of leachate. Large amounts of rainwater effectively reduce the contaminant concentration by dilution of the leachate. In more arid regions, evapotranspiration can reduce the expected leachate volume.

Pohland and Harper (1986) have described five relatively well defined stages in landfill stabilization, which occur as landfill wastes are degraded (Table 16.7).

Under favorable conditions, generally dictated by the presence of sufficient moisture to support microbial activity, landfills function as large-scale anaerobic reactors. Complex organic materials are converted to methane gas through the biological methane fermentation process. As a result, leachates produced in younger landfills are generally characterized by the presence of substantial amounts of volatile fatty acids (VFA), which are precursors of methane. These VFA make up the majority of the COD of young leachates.

**Table 16.6** Typical compositions of leachate from new and mature landfills (according to Tchobanoglous et al., 1991).

| Constituent | Value (mg L$^{-1}$) New Landfill (<2 years) | | Mature Landfill (>10 years) |
|---|---|---|---|
| | Range | Typical | |
| BOD$_5$ | 2000–30 000 | 10 000 | 100–200 |
| TOC | 1500–20 000 | 6000 | 80–160 |
| COD | 3000–60 000 | 18 000 | 100–500 |
| Total suspended solids | 200–2000 | 500 | 100–400 |
| Organic nitrogen | 10–800 | 200 | 80–120 |
| Ammonia-N | 10–800 | 200 | 20–40 |
| Nitrate | 5–40 | 25 | 5–10 |
| Total P | 5–100 | 30 | 5–10 |
| Ortho P | 4–80 | 20 | 4–8 |
| Alkalinity as CaCO$_3$ | 1000–10 000 | 3000 | 200–1000 |
| pH | 4.5–7.5 | 6 | 6.6–7.5 |
| Total hardness as CaCO$_3$ | 300–10 000 | 3500 | 200–500 |
| Calcium | 200–3000 | 1000 | 100–400 |
| Magnesium | 50–1500 | 250 | 50–200 |
| Potassium | 200–1000 | 300 | 50–400 |
| Sodium | 200–2500 | 500 | 100–200 |
| Chloride | 200–3000 | 500 | 100–400 |
| Sulfate | 50–1000 | 300 | 20–50 |
| Total iron | 50–1200 | 60 | 20–200 |

**Table 16.7** Summary of stages of landfill stabilization (according to Pohland and Harper, 1986).

| Phase | Description | Characteristics |
|---|---|---|
| I. | Initial adjustment | – initial waste placement and moisture accumulation<br>– closure of each fill section and initial subsidence<br>– environmental changes are first detected, including onset of stabilization process |
| II. | Transition | – field capacity is exceeded and leachate is formed<br>– anaerobic microbial activity replaces aerobic<br>– nitrate and sulfate replace oxygen as primary electron acceptors<br>– a trend toward reducing conditions is established<br>– measurable intermediates (VFA) appear in leachate |
| III. | Acid formation | – intermediate VFA become predominant with the continued hydrolysis and fermentation of waste organics<br>– pH decrease, causing mobilization and possible complexation of heavy metals<br>– nutrients such as N and P are released and assimilated<br>– hydrogen produced from anaerobic oxidations may control intermediate fermentation products |
| IV. | Methane fermentation | – acetic acid, carbon dioxide, and hydrogen produced during acid formation are converted to methane<br>– pH increases from buffer level controlled by VFA to one more characteristic of the bicarbonate system<br>– redox potential is at a minimum<br>– leachate organic strength decreases dramatically due to conversion of organic matter to methane gas |
| V. | Final maturation | – relative dormancy following active biostabilization<br>– measurable gas production all but ceases<br>– organics more resistant to microbial degradation may be converted, with possible production of humic-like substance capable of complexing and remobilizing heavy metals |

Along with a high organic content, leachates can also contain considerable amounts of heavy metals (Chian and DeWalle, 1977). The oxidation–reduction potential and pH inside a biologically active landfill favor the solubilization of metals. High concentrations of iron, calcium, manganese and magnesium have been reported by several researchers (Pohland and Harper, 1986; Lema et al., 1988; Kennedy et al., 1988). Other potentially more toxic metals have also been measured in significant concentrations in leachates (Pohland and Harper, 1986).

Leachate migration is controlled by a combination of landfill cap, bottom liner, and leachate collection and removal.

The nature of leachate contaminants generally mandates that some type of treatment process be employed prior to ultimate discharge of the wastewater. Numerous studies have been performed to evaluate the performance of biological and/or physicochemical treatment processes for removing organic and inorganic leachate contaminants. In general, leachates produced in younger landfills are most effectively treated by biological processes, because the wastewater is mostly composed of volatile fatty acids, which are readily amenable to biological treatment. Both aerobic and anaerobic processes have been used for the treatment of landfill leachate (Pohland and Harper, 1986; Iza et al., 1992).

In contrast, contaminants in older leachates tend to be more refractory and inorganic in nature, making physicochemical treatment more applicable. In some situations, biological and physicochemical treatments are combined.

As mentioned previously, a particularly attractive alternative for leachate treatment is leachate recycling, which has been shown to be an effective in situ treatment (Pohland, 1980). More recently, Pohland (1998) has noted that bioreactor landfills with in situ leachate recycling and treatment prior to discharge provide accelerated waste stabilization in a more predictable and shorter time period than is achieved in a conventional landfill.

## 16.5
## Summary and Conclusions

Landfilling of municipal solid waste developed approximately 35 years ago as the principal means of solid waste management. It emerged as an alternative to incineration, which became less popular due to air quality concerns. Unfortunately, landfills were not designed and/or operated in a proper manner and therefore, had significant impact on the environment.

In response to this situation, new laws (RCRA) were passed and strict regulations were established for solid waste landfill design and operation. These include provisions to control gas and leachate migration. In addition, the United States Environmental Protection Agency developed a hierarchy that relegated landfilling to a low priority. Higher emphasis is now placed on source reduction, recycling, and transformation processes.

As a result of these new regulations and policies, the amount of municipal solid waste being landfilled has decreased. The number of landfills have also decreased (but newer landfills are generally larger). However, landfilling will remain a significant solid waste management tool, because certain components in the solid waste stream cannot be economically recovered or transformed.

Although the new regulations focus on controlling gas and leachate migrations and subsequent environmental impacts, there are concerns about the long-term stability of landfill and the integrity of the liner systems. As a result, several investigators have promoted leachate recycling or the wet-cell concept as a means of accelerating the rate of solid waste decomposition.

# References

Chian, E. S. K., DeWalle, F. B., *Evaluation of Leachate Treatment, Vol. 1. Characterization of Leachate*, U.S. EPA Report No. EAP-600/2-77-186a. Cincinnati, OH **1977**.

Denison, R. A., Ruston, J., Tyrens, J., Diedrich, R., Environmental prospectives, in: *Handbook of Solid Waste Management* (Kreith, F., ed.). New York **1994**: McGraw-Hill.

Federal Register, Solid waste disposal facility criteria: final rule, *40CFR Parts 257 and 259* (October 9, **1991**).

Iza, J. M., Keenan, P. J., Switzenbaum, M. S., Anaerobic treatment of municipal solid waste landfill leachate: operation of a pilot scale hybrid UASB/AF reactor, *Water Sci. Technol.* **1992**, *25*, 225.

Kennedy, K. J., Hamoda, M. F., Guiot, S. G., Anaerobic treatment of leachate using fixed film and sludge bed systems, *J. Water Pollut. Control Fed.* **1988**, *60*, 1675–1683.

Lema, J., Mendez, R., Blazquez, R., Characteristics of landfill leachates and alternatives for their treatment: a review, *Water Air Soil Pollut.* **1988**, *40*, 223–250.

O'Leary, P. R., Walsh, P. W., *Solid Waste Landfills*. Madison, WI **1992**: University of Wisconsin-Madison, Dept of Engineering Professional Development.

Pohland, F. G., Leachate recycle as landfill management option, *J. Environ. Eng. (ASCE)* **1980**, *106*, 1057–1069.

Pohland, F. G., In situ anaerobic treatment of leachate in landfill bioreactors, *Proc. 5th Latin American Seminar on Anaerobic Wastewater Treatment*, October **1998**, Viña del Mar, Chile.

Pohland, F. G., Harper, S. R., *Critical Review and Summary of Leachate and Gas Production from Landfills*, U.S. EPA Report no. EPA 600/2-86/073. Cincinnati, OH **1986**.

Tchobanoglous, G., Theissen, H., Vigil, S., *Integrated Solid Waste Management: Engineering Principles and Management Issues*. New York **1991**: McGraw-Hill.

Tchobanoglous, G., O'Leary, P., Landfilling, in: *Handbook of Solid Waste Management* (Kreith, F., ed.). New York **1994**: McGraw-Hill.

U. S. EPA, *The EPA Municipal Solid Waste Factbook*, Version 4.0 software. Washington DC **1997a**.

U. S. EPA, *Characterization of Municipal Solid Waste in the United States, 1996 Update*. Washington DC **1997b**: OSW Publications Distribution Center.

U. S. EPA, *The Solid Waste Dilemma: An Agenda for Action*, U.S. EPA Report no. EPA/530-SW-88.054A. Washington DC **1988**.

# 17

# Process Engineering of Biological Waste Gas Purification

Muthumbi Waweru, Veerle Herrygers, Herman Van Langenhove, and Willy Verstraete

## 17.1
### Introduction

Process engineering of biological waste gas purification aims at the selection and operation of biological waste gas purification technologies with the ultimate aim of assuring mass transfer and biodegradation of one or more pollutants in a waste gas stream. Biodegradation of the pollutants occurs when microorganisms use the pollutants as a carbon source or an electron donor. In some special situations, microorganisms using a particular substrate such as glucose, ethanol, etc., can also oxidize another pollutant. This is due to unspecific metabolism by the enzymes of organisms and is called cometabolism (Alexander, 1981). The extent to which biological waste gas purification can occur is determined mainly by the physicochemical characteristics of the pollutant(s), the intrinsic capabilities of the microbial physiology and ecology, and the operating and environmental conditions.

When selecting the bioreactor technology, focus is placed on the operational and control requirements needed to ensure an optimal chemical and physical environment for mass transfer and biodegradation so as to achieve a high and constant removal efficiency of the pollutant.

## 17.2
### Biological Waste Gas Purification Technology

### 17.2.1
#### General Characteristics

Biological waste gas purification technology currently includes bioreactors known as biofilters, biotrickling filters, bioscrubbers, and membrane bioreactors. The mode of operation of all these reactors is similar. Air containing volatile compounds is passed through the bioreactor, where the volatile compounds are transferred from the gas phase into the liquid phase. Microorganisms, such as bacteria or fungi, grow

*Environmental Biotechnology. Concepts and Applications.* Edited by H.-J. Jördening and J. Winter
Copyright © 2005 WILEY-VCH Verlag GmbH & Co. KGaA, Weinheim
ISBN: 3-527-30585-8

in this liquid phase and are involved in removing the compounds acquired from the air. The microorganisms responsible for the biodegradation usually grow as a mixture of organisms. Such a mixture of different bacteria, fungi and protozoa depends on a number of interactions and is often referred to as a microbial community. Microorganisms are generally organized in thin layers called biofilms. The pollutants in the air (such as toluene, methane, dichloromethane, ethanol, carboxylic acids, esters, aldehydes, etc.; Tolvanen et al., 1998) usually act as a source of carbon and energy for growth and maintenance of the microorganisms. Some waste gases, such as those produced during composting, are composed of many (often up to several hundred) different chemicals, such as alcohols, carbonyl compounds, terpenes, esters, organosulfur compounds, ethers, ammonia, hydrogen sulfide, etc. (Tolvanen et al., 1998; Smet et al., 1999). The remarkable aspect of the microbial community is that it generally develops to a composition so that all these different chemicals are removed and metabolized simultaneously. Microorganisms also require essential nutrients and growth factors to function and produce new cells. These include nitrogen, phosphorous, sulfur, vitamins and trace elements. Most often these nutrients and growth factors are not present in the waste gas and have to be supplied externally.

There are fundamental differences between the four types of reactors mentioned above. They range from whether the microorganisms are immobilized or dispersed to the state of the aqueous phase in the reactor (mobile or stationary). The aqueous phase significantly influences the mass transfer characteristics of the system. A short description of each of the four types of bioreactors for biological waste gas purification currently in use is given below (also see Figure 17.1).

## 17.2.2
## Technology Types

### 17.2.2.1 Biofilter
In a biofilter, the air is passed through a bed packed with organic carrier materials, e.g., compost, soil or wood bark. The compounds in the air are transferred to a biofilm that grows on the filter materials. The nutrients necessary for growth of the microorganisms are supplied by the organic matter. On top of the biofilm is a thin liquid layer. An important control parameter is the moisture content of the overall carrier matrix, which must be between 40% and 60% (w/w). To avoid dehydration, the air is generally humidified before entering the biofilter. If the waste gas contains high levels of solid particles (i.e., the waste gas is an aerosol), an aerosol removal filter can be installed before the humidification chamber. This prevents clogging of the biofilter by the particles.

### 17.2.2.2 Biotrickling Filter
A biotrickling filter is similar to a biofilter. Here, pollutants are also transferred from the gas phase to a biofilm that grows on a packing material. However, the packing materials are made of chemically inert materials, such as plastic rings. Because nu-

trients are not available in these materials, they have to be supplied to the microorganisms by recirculating a liquid phase through the reactor in co- or countercurrent flow.

### 17.2.2.3  **Bioscrubber**

A bioscrubber consists of two reactors  The first part is an absorption tower, where pollutants are absorbed by a liquid phase. This liquid phase goes to a second reactor, which is a kind of activated sludge unit, where microorganisms growing in suspended flocs in the water degrade the pollutants. The effluent of this unit is recirculated over the absorption tower in a co- or countercurrent direction to the flow of the waste gas.

### 17.2.2.4  **Membrane Bioreactor**

In a membrane bioreactor, the waste gas stream is separated from the biofilm by a membrane that is selectively permeable to the pollutants. One side of the membrane is in contact with a liquid phase supplemented with nutrients, and the other side is in contact with the waste gas stream. The nutrient-rich liquid phase is inoculated with microorganisms capable of degrading the pollutant. These microorganisms organize themselves and form a biofilm attached onto the membrane. As the pollutants migrate through the selectively permeable membrane, they enter the nutrient-rich liquid phase and are degraded. The liquid phase is maintained in a reservoir where the nutrients are refreshed, oxygen is supplied and the pH and temperature are controlled. Different types of membranes can be used, such as polar or hydrophobic membranes. They can be installed as tubular or flat sheets. Figure 17.1 shows a close-up of the site of biological activity in the four types of bioreactors. Note that in a bioscrubber the microorganisms are in the second reactor and are fully suspended as flocs or granules in the liquid.

## 17.3
## Performance Parameters

Different biological waste gas purification technologies can be compared based on performance by using a set of parameters. These parameters include
- empty bed contact time (s)
- surface loading rate ($m^3\, m^{-2}\, h^{-1}$)
- mass loading rate ($g\, m^{-3}\, h^{-1}$)
- volumetric loading rate ($m^3\, m^{-3}\, h^{-1}$)
- elimination capacity ($g\, m^{-3}\, h^{-1}$)
- removal efficiency (%)

The discussion below particularly relates to the biofilter type of reactor.

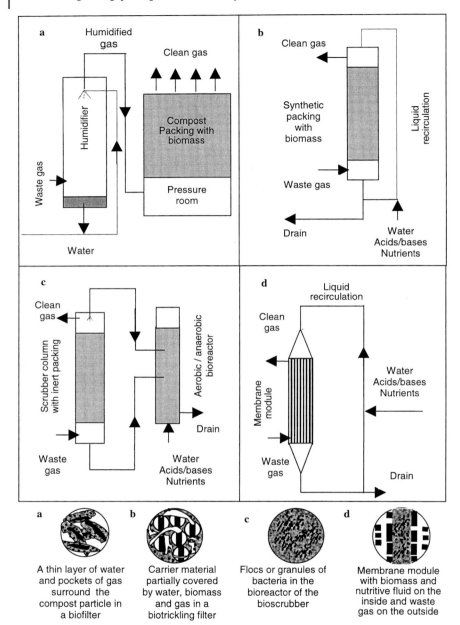

**Fig. 17.1** Schematic representation of four types of bioreactors used in biological waste gas purification and close-up view of their respective microbial configurations: (a) biofilter, (b) biotrickling filter, (c) bioscrubber, (d) membrane bioreactor.

## 17.3.1
### Empty Bed Contact Time or True Contact Time

The residence time of the gas in a bioreactor can be calculated in two different ways;

- Superficial residence time or empty bed residence time, based on the total volume of the reactor and referred to as empty bed contact time (EBCT):

$$EBCT = \frac{V\,3600}{Q} \qquad (1)$$

where $V$ = volume of filter material in the reactor (m$^3$), and $Q$ = waste gas flow rate (m$^3$ h$^{-1}$).
- True residence time $\tau$, which is based on the free space in the reactor and defined as

$$\tau = \frac{\varepsilon V\,3600}{Q} \qquad (2)$$

where $\varepsilon$ = porosity of the packing materials (dimensionless).

Often the exact porosity needed to calculate the true residence time is not known. Hence, most often the empty bed contact time is used. The EBCT is typically used for comparison of gas residence times in different reactor technologies or under different loading conditions. However, one has to remember that this gives an overestimation of the true residence time. Due to preferential currents through the larger voids in the packing, the actual residence time can differ considerably from the calculated residence time.

The residence time in the reactor is useful as an indicator of the time available for mass transfer of the pollutant from the gas phase to the liquid phase through the biofilm, which is often the factor limiting microbial degradation.

## 17.3.2
### Surface Loading Rate ($B_A$)

The surface loading rate indicates the amount of air that is passed through the bioreactor per unit surface area per unit time:

$$B_A = \frac{Q}{A} \qquad (3)$$

where $A$ = total surface area of the packing or filter material in the bioreactor (m$^2$).

One can also express the velocity of the gas (m h$^{-1}$) through the empty reactor. However, the reactor is normally filled with packing materials, which results in a velocity of gas higher than the surface loading rate.

### 17.3.3
**Mass Loading Rate ($B_V$)**

The mass loading rate gives the amount of pollutant that is introduced into the bioreactor per unit volume and per unit time:

$$B_V = \frac{QC_{g-in}}{V} \tag{4}$$

where $C_{g-in}$ = concentration of the pollutant in the inlet waste gas stream (g m$^{-3}$).

### 17.3.4
**Volumetric Loading Rate ($v_s$)**

The volumetric loading rate is the amount of waste gas passed through the reactor per unit reactor volume:

$$v_s = \frac{Q}{V} \tag{5}$$

### 17.3.5
**Elimination Capacity (EC)**

The elimination capacity EC gives the amount of pollutant removed per volume bioreactor per unit time. The overall elimination capacity is defined by Eq. (6):

$$EC = \frac{Q(C_{g-in} - C_{g-out})}{V} \tag{6}$$

where $C_{g-out}$ = concentration of the pollutant in the effluent waste gas (g m$^{-3}$).

### 17.3.6
**Removal Efficiency (RE)**

Removal efficiency is the fraction of the pollutant removed in the bioreactor expressed as a percentage. It is defined as

$$RE = \frac{(C_{g-in} - C_{g-out})}{C_{g-in}} \, 100 \tag{7}$$

We should note that the various parameters are interdependent. There are only four independent design parameters: reactor height, volumetric loading rate, and gas phase concentrations at the inlet ($C_{g-in}$) and outlet ($C_{g-out}$).

## 17.4
## Characteristics of the Waste Gas Stream

Several characteristics of the waste gas stream have to be known when considering the implementation of biological waste gas purification technologies. Table 17.1 lists the characteristics of the waste gas stream that are essential for correctly designing a biological purification system.

Physical parameters such as relative humidity and temperature are important, because they have considerable influence on microbial degradation of the pollutant. Different microorganisms have different optimal ranges of temperature and relative humidity for growth. Temperature also affects the partitioning of the pollutant between the gas and liquid phases. The waste gas flow rate influences the volumetric loading rate of pollutant on the biologically active phase and, hence, the elimination capacity. Equally important are the identity and concentration of the pollutant and/or odor units in the waste gas stream, because they influence the overall efficiency of the biological waste gas system.

It is also important to establish the chemical composition of the waste gas stream before starting to design the treatment system. The microbial degradability of the pollutant in the waste gas stream is largely dependent on its chemical identity. The pollutant can be organic or inorganic. Typical organic pollutants that are often encountered in waste gas streams are ethers, ketones, fatty acids, alcohols, hydrocarbons, amines, and organosulfur compounds. Valuable information about the biodegradability of chemicals can be obtained, e.g., from Van Agteren et al. (1998). Waste gases can also contain inorganic compounds such as $NH_3$, $NO_2$, $NO$, $H_2S$, and $SO_2$. Some of these compounds may be present at toxic levels or they may reduce the degradation capacity by, e.g., acidifying the biofilter material. Therefore, either these compounds have to be eliminated before the waste gas stream enters the bioreactor, or a means of controlling the pH has to be installed. Different compounds can also

Table 17.1 Important characteristics of a waste gas stream.

| Parameter | Unit |
| --- | --- |
| Relative humidity | % |
| Temperature | °C |
| Waste gas flow rate | $m^3\,h^{-1}$ |
| Pollutant identity (chemistry) | |
| Pollutant concentration | $g\,m^{-3}$ |
| Odor concentration | $ou\,m^{-3}$ |

Odor unit (ou) is the amount of (a mixture of) odorous compounds present in 1 $m^3$ of odorless gas (under standard conditions) at the panel threshold (CEN, 1998).

affect each other's degradation without being toxic to the microorganisms. Smet et al. (1997) reported that isobutyraldehyde (IB) had to be removed by a first layer of the biofilter before a *Hypohomicrobium*-based microbial community in a subsequent layer could develop and metabolize the dimethyl sulfide (DMS) present in the waste gas. In separate batch experiments they showed that the same *Hypohomicrobium* sp. switched its metabolism from using IB to consuming DMS when the IB concentrations decreased.

When bioreactors are used for odor abatement, the concentration of odiferous compounds in the waste gas has to be determined as well. The odor concentration in odor units per cubic meter (ou m$^{-3}$) corresponds to the number of times a waste gas sample has to be diluted with reference air before the odor of the diluted sample can be distinguished from the reference air by 50% of the members of a standard panel. In this respect, the European Committee for Standardisation (CEN) is currently involved in standardizing the determination of odor compounds by dynamic olfactometry. This will improve the reproducibility of olfactometric measurements, basically by using panels standardized with respect to 1-butanol (detection threshold of 40 ppbv) (CEN, 1998). Although the evaluation of bioreactor performance aimed at odor reduction has to be based on olfactometric measurements, design and optimization always require chemical characterization of the overall process.

The concentration of the pollutants and/or odor units largely depends on the source of the waste gas stream. Waste gas streams from, e.g., hexane oil extraction processes have a pollutant concentration in the range of a few g m$^{-3}$. On the other hand, for waste gas streams polluted with offensive odors, concentrations of the odorous compounds can be in the range of mg m$^{-3}$ or less (Smet et al., 1998).

The characteristics of the waste gas stream determine to a large extent the type of bioreactor system that can be used. Table 17.2 gives a first indication of the suitability of bioreactors for waste gas purification in relation to the characteristics of the waste gas stream. Note the preponderant importance of the Henry coefficient. Chemicals that dissolve easily in water (hydrophilic substances) can be retained efficiently by scrubbing with water. Chemicals that are poorly water soluble (high Henry coefficient) are better dealt with by means of a biofilter. In Table 17.2, the mem-

**Table 17.2** Pollutant concentrations, Henry coefficients, and concomitant operating parameters of biofilters, biotrickling filters, and bioscrubbers (after van Groenestijn and Hesselink, 1993).

|  | *Biofilter* | *Biotrickling Filter* | *Bioscrubber* |
|---|---|---|---|
| Pollutant concentration (g m$^{-3}$) | <1 | <0.5 | <5 |
| Henry coefficient (dimensionless) | <10 | <1 | <0.01 |
| Surface loading rate (m$^3$ m$^{-3}$ h$^{-1}$) | 50–200 | 100–1000 | 100–1000 |
| Mass loading rate (g m$^{-3}$ h$^{-1}$) | 10–160 | <500 | <500 |
| Empty bed contact time (s) | 15– 60 | 30–60 | 30– 60 |
| Volumetric loading rate (m$^3$ m$^{-3}$ h$^{-1}$) | 100–200 |  | 250–580 |
| Elimination capacity (g m$^{-3}$ h$^{-1}$) | 10–160 |  |  |
| Removal efficiency (%) | 95– 99 |  | 85– 95 |

brane reactor is not mentioned – depending on the nature of the membranes, it can
be suited to handle a range of compounds (Stern, 1994).

## 17.5
## Process Principles

Several processes take place in biological waste gas cleaning systems. They include
partitioning of the pollutant from the gaseous to the liquid phase, followed by its diffusion from the bulk liquid to the biofilm. Microbial degradation of the pollutant
takes place in the biofilm, and the end products diffuse back into the bulk liquid.
Mass transfer is the combined migration of compounds from the gaseous to the liquid phase and from the bulk liquid to the biofilm (Fig. 17.2).

### 17.5.1
### Equilibrium Partitioning of the Pollutant

The first step toward microbial degradation of the pollutant is partitioning of the
gaseous pollutants to the liquid phase. In a bioscrubber and biotrickling filter this is
obvious, but also in a biofilter a small water layer is normally present on top of the
microbial biofilm.

In describing gas–liquid mass transfer, the interfacial resistance between the liquid and the gas is often neglected. For practical reasons, it is usually assumed that
straightforward partitioning of the pollutant between the two phases occurs, and the
resulting concentration in both the gas and the liquid phase is at equilibrium. Equilibrium partitioning largely depends on the Henry constant of the pollutant.

The concentrations of the pollutant in the gas and liquid phases are related by Eq.
(8) (Sander, 1999).

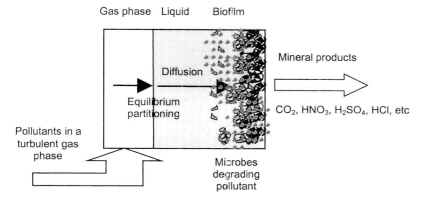

**Fig. 17.2** Schematic view of the sequence of processes leading to microbial degradation of pollutants in a biofilter.

$$K_H = \frac{C_g}{C_1} \tag{8}$$

where: $K_H$ = dimensionless Henry constant, $C_g$ = gas phase concentration (mol m$^{-3}$ or g m$^{-3}$), and $C_1$ = liquid phase concentration (mol m$^{-3}$ or g m$^{-3}$).

For pollutants with high Henry constants, partitioning of the pollutant to the liquid phase is very poor. In Table 17.3 some Henry constants are compared for different kinds of pollutants. The Henry constant varies with temperature and with salinity of the water. Dewulf et al. (1995) carried out several measurements of Henry constants and found that, in general, the Henry constant increases with a decrease in temperature and increases with an increase in salinity, as expressed by Eq. (9):

$$\ln K_H = a\,\frac{1}{T} + bZ + c \tag{9}$$

where: $K_H$ = dimensionless Henry constant, $a, b, c$ = constants, $T$ = absolute temperature (K), and $Z$ = salt concentration (g L$^{-1}$).

In a biofilter, there is a low water content (40%–60%), and therefore the gas–liquid mass transfer takes place with less interfacial resistance than in a biotrickling filter or a bioscrubber, where the water content is higher. In a membrane bioreactor there is no gas–liquid interface. Therefore, this reactor can be very suitable for treating pollutants with high Henry coefficients, provided the membrane is quite apolar such as are, e.g., silicone membranes (De Smul and Verstraete, 1999). However, we should note that in a membrane bioreactor, two mass transfers have to be dealt with: gas–membrane and membrane–biofilm. For the gas–membrane mass transfer, an equation similar to the Henry equation can be used:

$$S = \frac{C_m}{C_g} \tag{10}$$

where: $S$ = solubility ratio (dimensionless), $C_g$ = gas phase concentration (mol m$^{-3}$ or g m$^{-3}$), and $C_m$ = pollutant concentration in the membrane (mol m$^{-3}$ or g m$^{-3}$).

Methods to enhance gas–liquid mass transfer have been explored. Addition of a surface-active reagent, as, e.g., silicone oil, to the liquid phase gives good results in laboratory-scale biofilters (Budwill and Coleman, 1997). In a biotrickling filter, intermittent circulation can be used to enhance the transfer of poorly water soluble pollutants into the biofilm. De Heyder et al. (1994) used this approach to remove ethene from air and obtained an increase in removal of ethene by a factor of 2.25.

**Table 17.3** Henry constants (dimensionless) for pollutants treatable with a biotechnological waste gas treatment system (Howard and Meylan, 1997).

| Compound: | Ethanol | Butanone | Isobuteral-dehyde | Dimethyl-sulfide | Trichloro-ethene | Limonene | Hexane |
|---|---|---|---|---|---|---|---|
| $K_H$ (25 °C) | 0.00021 | 0.0023 | 0.0074 | 0.0658 | 0.403 | 0.82 | 74 |

17.5.2
**Diffusion**

Migration of the pollutant from the bulk liquid to the biologically active phase (biofilm) occurs by diffusion, which can be described by Fick's law:

$$J = -D \frac{dC_1}{dx} \tag{11}$$

where: $J$ = mass flux (mol m$^{-2}$ s$^{-1}$ or g m$^{-2}$ s$^{-1}$), $D$ = diffusion coefficient (m$^2$ s$^{-1}$), $C_1$ = liquid concentration (mol m$^{-3}$ or g m$^{-3}$), and $x$ = distance within the biofilm (m).

The value of the effective diffusion coefficient $D$ varies over some orders of magnitude, depending on the medium (Table 17.4).

Diffusion is much slower in water than in air and even slower in a membrane than in water. In porous membranes, diffusion occurs through the fluid in the pores, and in dense membranes it occurs through the membrane material itself. When choosing a membrane, a study should be made to determine an appropriate material with high diffusion characteristics for the given pollutant.

In bioreactors without membranes, the pollutant has to pass through the water phase before it reaches the biofilm. The limit between the water phase and the biofilm is as yet vaguely defined. The diffusion coefficient varies between its value in water and in the biofilm (Devinny et al., 1999). Roughly, diffusion in a biofilm is about 0.5–0.7 of that in water. Most important in the whole concept is that the concentration gradient needed for diffusion flux is maintained by a constant input from the gas phase and removal of pollutants by the microbial degradation in the biologically active phase. Figure 17.3 illustrates the concentration gradient that exists between the gas phase, through the liquid phase, to the biofilm. Although microorganisms have high affinity for most biodegradable substrates, they can have difficulty in consuming substances at concentrations below 1 µg L$^{-1}$ of water (Verstraete and Top, 1992). This so-called lower microbial threshold means that, for substances with high Henry constants, the corresponding levels ($C_{thr}$ H) remain in the gas phase. Various approaches have been described to derive appropriate kinetic parameters from such performance curves (De Heyder et al., 1997b).

**Table 17.4** Diffusion coefficients of some compounds in air, water, and membrane materials (after Reid et al., 1987).

| Compound | $D_{air}$ (m$^2$ s$^{-1}$) | $D_{water}$ (m$^2$ s$^{-1}$) | Membrane Material | $D_{membrane}$ (m$^2$ s$^{-1}$) |
|---|---|---|---|---|
| Oxygen | | | natural rubber | $2.5 \times 10^{-10}$ |
| Oxygen (25 °C) | $1.40 \times 10^{-5}$ | $2.50 \times 10^{-9}$ | polydimethyl siloxane (35 °C) | $4.0 \times 10^{-9}$ |
| Ethanol | $1.24 \times 10^{-5}$ | $1.13 \times 10^{-9}$ | poly(vinyl acetate) | $1.5 \times 10^{-13}$ |
| CO$_2$ | $1.64 \times 10^{-5}$ | $2.00 \times 10^{-9}$ | PMDA-MDA (20 °C) | $9.0 \times 10^{-13}$ |
| Benzene | $1.20 \times 10^{-5}$ | $1.30 \times 10^{-9}$ | poly(vinyl acetate) | $4.8 \times 10^{-17}$ |

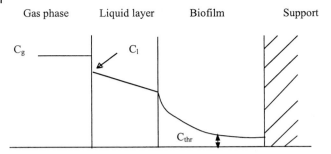

**Fig. 17.3** Concentration gradient of a pollutant between the gas phase, through the liquid phase, to the biofilm. $C_g$: concentration in the gas phase; $C_l$: concentration in the liquid phase; $C_{thr}$: lower threshold level.

17.5.3
**Microbial Degradation of the Pollutant**

Microbial metabolism of pollutants readily occurs when the pollutants are used as a source of energy. For instance, toluene is used as an electron and carbon donor by several organotrophic bacteria; they use oxygen as an electron acceptor. Ammonium is used as an electron donor by lithotrophic nitrifying bacteria; they use oxygen as an electron acceptor and carbon dioxide as a carbon source to build cell biomass (Focht and Verstraete, 1977). Sometimes the pollutant can be a cosubstrate. For instance, trichloroethene can be metabolized together with toluene (Mu and Scow, 1994).

Sufficient availability of nutrients such as minerals, vitamins and growth factors is essential for proper growth of the microbial community. Hence, the microbial biomass acts as a kind of biocatalyst that constantly maintains itself (e.g., by dying off and regrowing).

The energy released during degradation of the pollutants is used for maintenance metabolism and for growth of the microorganisms, according to the modified Monod equation:

$$\frac{dC_1}{dt} = \left(\frac{\mu}{Y_{XS}}\right) X \tag{12}$$

where: $C_1$ = concentration of the substrate dissolved in the liquid (g m$^{-3}$), $\mu$ = growth rate (g g$^{-1}$ h$^{-1}$), $Y_{XS}$ = yield of dry cell weight per mass of substrate metabolized (g g$^{-1}$), $m$ = maintenance energy consumption [g substrate (g cell dry weight)$^{-1}$ h$^{-1}$], and $X$ = dry weight of biomass in the biofilm or suspension (g m$^{-3}$).

In practice, however, it is difficult to obtain values for these parameters. Therefore, the design of reactor performance should be based on pilot experiments.

In a waste gas treatment reactor, growth of the bacteria is often minimal. This means that all the released energy serves mainly for maintenance metabolism of the bacteria. The advantage is that there is little or no waste sludge, in contrast to, e.g., wastewater treatment systems. Often one deliberately tries to minimize the growth of excess biomass, e.g., by limiting the supply of mineral nutrients such as phosphate (Wubker and Friedrich, 1996) or by seeding the reactor with protozoa that graze on the bacteria (Cox and Deshusses 1999). Microbial degradation of the pollutant depends on a multitude of factors. The most important external factors are temperature, nutrient availability and toxicity of the gaseous components. There are also a series of internal factors directly related to the way microorganisms develop and work. For instance, different species, each with different capabilities, can cooperate and achieve very effective pollutant removal. De Heyder et al. (1997a) described the stimulation of ethene removal by *Mycobacterium* sp. in the presence of an active nitrifying population. Veiga et al. (1999) identified two different bacterial species (*Bacillus* and *Pseudomonas*) and a fungus (*Trichosporon*) functioning as cooperative agents in a biofilter treating alkylbenzene gases.

The microorganisms (biomass) can be introduced into the bioreactor in several ways. In some situations, natural sources such as manure, aquatic sediments or sludge from wastewater treatment plants are used as the inoculum. Moreover, in a biofilter, the carrier material (compost, wood bark, etc.) itself has a naturally occurring microbial community. In other instances, specific bacteria or mixtures of isolated strains of naturally occurring bacteria that can metabolize the pollutant in the waste gas stream are introduced into the bioreactor (Kennes and Thalasso, 1998). Such seeding is referred to as bioaugmentation. To accelerate the removal of a particular recalcitrant pollutant, one could make use of genetically modified microorganisms having improved degradation capacities. For bioscrubbers in which the removal occurs in an activated sludge reactor system, bioaugmentation as described for wastewater systems by Van Limbergen et al. (1998) could be applied.

Generally, inoculation of the reactor with appropriate bacteria significantly decreases the startup period for these reactors (Smet et al., 1996). However, the operating conditions and the prevailing environmental factors generally exert selective pressure on the microorganisms present in the bioreactor, resulting in the development of a specific microbial community. This community can be quite different from the enrichment culture that was introduced into the reactor (Bendinger, 1992). The development of the structure and function of the microbial community in the bioreactor affects the microbial degradation rate and, hence, the extent of pollutant removal. The microorganisms are surrounded by an extracellular organic layer, which normally has a negative charge and serves many functions, including adhesion, protection, carbon storage, and ion exchange (Bishop and Kinner, 1986).

Most often, biological waste gas treatment results in non-reusable end products. For $SO_2$ scrubbing, however, a special approach has been developed (De Vegt and Buisman, 1995; Verstraete et al., 1997). As schematized in Figure 17.4, a sequence of biotechnological reactors enables $SO_2$ to be recovered as sulfur powder.

## 17.6
### Reactor Performance

The overall performance of a bioreactor is mainly determined by mass transfer, as governed by equilibrium partitioning at the gas–liquid phase and diffusion from the

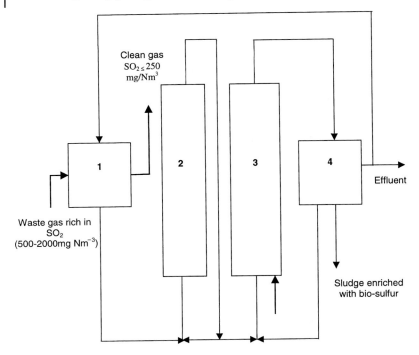

**Fig. 17.4** Conversion of $SO_2$ by means of sulfate reduction and subsequent sulfide oxidation reactors to biologically formed elemental sulfur (after Grootaerd et al., 1977). 1: absorption of $SO_2$ gas; 2: sulfate reduction; 3: partial oxidation of hydrogen sulfide; 4: separation of sludge enriched with biosulfur.

bulk liquid to the bioactive phase (biofilm), combined with microbial degradation of the pollutant. As mentioned before, this is expressed in removal efficiency (%) or elimination capacity ($g\ m^{-3}\ h^{-1}$). Figure 17.5 shows a typical elimination capacity as a function of the mass loading rate. Such a performance diagram can be experimentally determined by changing the concentration in the influent gas flow and measuring the resulting elimination capacity. In most studies these curves are determined in short-term experiments. This means that no significant growth of biomass is allowed to occur during the experiment. In Figure 17.5, two main regions can be distinguished. At the lower mass loading rate the elimination capacity is equal to the mass loading, and the removal efficiency is 100%. In this range the reactor kinetics are first-order; the microbial metabolism normally represented by the Monod equation follows a simple equation:

$$-\frac{dC_1}{dt} = \frac{KC_1}{K_S + C_1} \tag{13}$$

where: $C_1$ = concentration in the liquid ($mol\ m^{-3}$ or $g\ m^{-3}$) substrate, $K$ = maximum conversion rate ($g$ substrate ($g$ cell dry weight)$^{-1}\ h^{-1}$), and $K_S$ = substrate level at which the biomass works at half-maximum velocity ($mol\ m^{-3}$ or $g\ m^{-3}$).

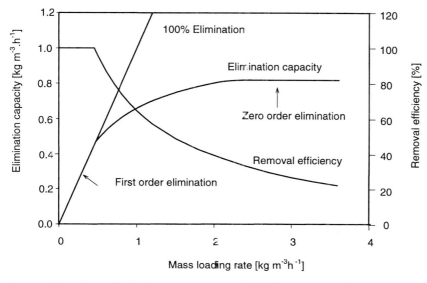

**Fig. 17.5** Typical curve for the elimination capacity of a biofilter vs. its mass loading rate.

When $K_S \geq C_1$, Eq. (13) becomes

$$-\frac{dC_1}{dt} = kC_1 \tag{14}$$

All pollutants that are fed to the biofilter are removed from the air. Low mass loading can be achieved by a low concentration of the pollutants in the gas phase and by a low gas flow rate. When the mass loading increases, complete removal of the pollutants is no longer possible. At even higher mass loading, the elimination capacity does not increase further and remains at a steady value. This is the region of zero-order kinetics: the elimination capacity is independent of the mass loading rate. At this stage the removal efficiency (%) decreases as the mass loading rate is increased further. Two phenomena explain the incomplete removal of the pollutants: diffusion and reaction limitation. In diffusion limitation, not all the pollutants diffuse into the biofilm and not all microorganisms can take part in degradation of the pollutant. When the diffusion rate is slower than the degradation rate, e.g., for pollutants with a high Henry coefficient, the concentration in the liquid phase is lower than in the gas phase. In reaction limitation, the pollutants diffuse into the complete biofilm, but the pollutants are not removed rapidly and sufficiently enough by the biocatalyst. Indeed, microbial metabolism can be hampered by other limiting factors such as shortage of nutrients, presence of toxins, etc.

## 17.7
## Reactor Control

The environmental factors prevailing in the bioreactor, such as pH, temperature, oxygen level and water content, affect the ability of the microorganism to metabolize the pollutant. Prevailing environmental factors largely determine the composition of the microbial community in the bioreactor. Most species of microorganisms exhibit optimal growth over a certain pH range (Devinny et al., 1999). A pH range of about 6–8 is suitable for most microorganisms, but some species can tolerate lower or higher pH values. Microbial activity is strongly influenced by temperature. Some microorganisms operate optimally in the mesophilic temperature range (15–40 °C), and others do so in the thermophilic temperature range (40–60 °C).

The microbial activity and the mass transfer of the pollutant from the gas phase to the biofilm are related to the water content in the bioreactor. This is especially true for biofilters that operate optimally at 40%–60% relative humidity (Devinny et al., 1999). Excess water in a biofilter may lead to loss of nutrient supplements (Smet et al., 1996). Moreover, wet pockets may be formed, in which diffusion of both pollutant and oxygen used as an electron acceptor for the microorganisms becomes limiting.

In practice, it is difficult to implement control and mitigation actions in biofilter systems. In contrast, because of the circulation of a liquid in the other reactor types, it generally is possible to optimize the latter reactor systems for temperature, pH, and nutrient supply. Eventually, if required, one can also supplement with a cosubstrate in accord with the needs of the bacteria.

## 17.8
## Perspectives

Biological waste gas purification processes have strong competitors, such as activated carbon sorption and incineration. Also other physicochemical techniques, with a small footprint, are more frequently marketed for odor treatment of air, e.g., ozonization and UV treatment. The main advantages of biocatalytic removal of pollutants are the low investment and operation costs. The main disadvantages are the often slow startup and the limited reliability, e.g., as a result of changing environmental conditions and autointoxication.

The major efforts in the near future for bioprocess engineers should therefore be directed to the development of reactors with a controllable microbial biomass. Adequate natural or even possibly genetically modified organisms should be available as ready-to-use industrial biocatalysts. They should, as occurs for activated carbon, be immediately operational upon introduction in the reactor. Moreover, the overall environment and performance in the reactor should be monitored online and, if necessary, adjusted. These aspects, mass production of biocatalysts with a guaranteed quality and implementation of process control, are crucial for the future of biological waste gas treatment.

## Acknowledgments

This paper was in part supported by a grant of the Fonds Wetenschappelijk Onderzoek, Belgium.

## References

Alexander, M., Biodegradation of chemicals of environmental concern, *Science* **1981**, *11*, 132–138.

Bendinger, B., Microbiology of biofilters for the treatment of animal-rendering plant emissions: occurrence, identification, and properties of isolated coryneform bacteria, Thesis, University of Osnabrück, Germany **1992**.

Bishop, P. L., Kinner, E. N., Aerobic fixed film processes, in: *Biotechnology*, 1st edit. Vol. 8 (Rehm, H.-J., Reed, G., eds.), pp. 113–176. Weinheim **1986**: VCH.

Budwill, K., Coleman, R. N. Effect of silicone oil on biofiltration of *n*-hexane vapors. *Forum for Applied Biotechnology, Mededelingen Faculteit Landbouwkundige en Toegepaste Biologische Wetenschappen* **1997**, *62*, 1521–1527.

CEN (Comité Européen de Normalisation), *Odors: Air Quality-Determination of Odor Concentration by Means of Dynamic Olfactometry*, Doc. Nr. Cen/Tc 292/WG 2 (**1998**).

Cox, H., Deshusses, M., Biomass control in waste air biotrickling filters by protozoan predation, *Biotechnol. Bioeng.* **1999**, *62*, 216–224.

De Heyder, B., Overmeire, A., Van Langenhove, H., Verstraete, W., Ethene removal from a synthetic waste gas using a dry biobed, *Biotechnol. Bioeng.* **1994**, *44*, 642–648.

De Heyder, B., Van Elst, T., Van Langenhove, H., Verstraete, W., Enhancement of ethene removal from waste gas by stimulating nitrification, *Biodegradation* **1997a**, *8*, 21–30.

De Heyder, B., Vanrolleghem, P., Van Langenhove, H., Verstraete, W., Kinetic characterization of mass transfer limited biodegradation of a low water soluble gas in batch experiments: necessity for multiresponse fitting, *Biotechnol. Bioeng.* **1997b**, *55*, 511–519.

De Smul, A., Verstraete, W., The phenomenology and the mathematical modeling of the silicone-supported chemical oxidation of aqueous sulfide to elemental sulfur by ferric sulfate, *J. Chem. Technol. Biotechnol.* **1999**, *74*, 456–466.

De Vegt, A. L., Buisman, C. J. N., Thiopaq bioscrubber: An innovative technology to remove hydrogen sulfide from air and gaseous streams, *Proc. GRI Sulfur Recovery Conf.* Austin, TX, Sept. 24–27, **1995**.

Devinny, J. S., Deshusses, M. A., Webster, T. S., *Biofiltration for Air Pollution Control*. New York **1999**: Lewis.

Dewulf, J., Drijvers, D., Van Langenhove, H., Measurement of Henry's law constant as function of temperature and salinity for the low temperature range, *Atmospher. Environ.* **1995**, *29*, 323–331.

Focht, D., Verstraete, W., Biochemical ecology of nitrification and denitrification, *Adv. Microb. Ecol.* **1977**, *1*, 135–214.

Grootaerd, H., De Smul, A., Dries, J., Goethals, L., Verstraete, W., Epuration biologique des eaux de lavage des fumées riches en $SO_2$, *Procédés d'une journée d'étude organisée au Faculté Polytechnique de Mons*, 13–15 May **1997**.

Howard, P. H., Meylan, W. M., *Handbook of Physical Properties of Organic Chemicals*. London **1997**: Lewis.

Kennes, C., Thalasso, F., Waste gas biotreatment technology, *J. Chem. Technol. Biotechnol.* **1998**, *72*, 303–319.

Mu, Y. D., Scow, K. M., Effect of trichloroethylene (TCE) and toluene concentrations on TCE and toluene biodegradation and the population density of TCE and toluene degraders in soil, *Appl. Environ. Microbiol.* **1994**, *60*, 2662–2665.

Reid, R. C., Prausnitz, J. M., Poling, B. E., *The Properties of Gases and Liquids*. Singapore **1987**: McGraw-Hill.

Sander, R., Compilation of Henry's law constants for inorganic and organic species of potential importance in environmental chemistry. http://www.mpch-mainz-mpg.de/~sander/res/henry.html (**1999**).

Smet, E., Chasaya, G., Van Langenhove, H., Verstraete, W., The effect of inoculation and the type of carrier material used on the biofiltration of methyl sulfides, *Appl. Microbiol. Biotechnol.* **1996**, *45*, 293–298.

Smet, E., Van Langenhove, H., Verstraete, W., Isobutyraldehyde as a competitor of the dimethyl sulfide degrading activity in biofilters, *Biodegradation* **1997**, *8*, 53–59.

Smet, E., Lens, P., Van Langenhove, H., Treatment of waste gases contaminated with odorous sulfur compounds, *Crit. Rev. Environ. Sci. Technology* **1998**, *28*, 89–117.

Smet, E., Van Langenhove, H., De Bo, I., The emission of volatile compounds during the aerobic and the combined anaerobic/aerobic composting of biowaste, *Atmospher. Environ.* **1999**, *33*, 1295–1303.

Stern, S. A., Polymers for gas separations: the next decade, *J. Membrane Sci. 94*, 1–65.

Tolvanen, O. K., Hanninen, K. I., Veijanen, A., Villberg, K., (**1998**), Occupational hygiene in biowaste composting, *Waste Managem. Res.* **1994**, *16*, 525–540.

Van Agteren, M. H., Keuning, S., Janssen, D. B., *Handbook on Biodegradation and Biological Treatment of Hazardous Organic Compounds*. Dordrecht **1998**: Kluwer.

van Groenstijn, J. W., Hesselink, P. G. M., Biotechniques for air pollution control, *Biodegradation* **1993**, *4*, 283–301.

Van Limbergen, H., Top, E. M., Verstraete, W., Bioaugmentation in activated sludge: current features and future perspectives, *Appl. Microbiol. Biotechnol.* **1998**, *50*, 16–23.

Veiga, M. C., Fraga, L. A., Amor, L., Kennes, C., Biofilter performance and characterization of a biocatalyst degrading alkylbenzene gases, *Biodegradation* **1999**, *10*, 169–176.

Verstraete, W., Top, E., Holistic environmental biotechnology, in: *Microbial Control of Pollution* (Fry, J. C., Gadd, G. M., Herbert, R. A., Jones, C. W., Watson-Craik, I. A., eds.). Society for General Microbiology Symposium Vol. 48. Cambridge **1992**: Cambridge University Press.

Verstraete, W., Tanghe, T., De Smul, A., Grootaerd, H., Anaerobic biotechnology for sustainable waste treatment, in: *Biotechnology in the Sustainable Environment* (Sayler, G. S., Sanseverion, J., Davis, K. L., eds.), Proc. Conf. Biotechnology in the Sustainable Environment, April 14–17, 1996, Knoxville, TN. pp. 343–359. New York **1997**: Plenum.

Wubker, S., Friedrich, C., Reduction of biomass in a bioscrubber for waste gas treatment by limited supply of phosphate and potassium ions, *Appl. Microbiol. Biotechnol.* **1996**, *46*, 475–480.

# 18
# Commercial Applications of Biological Waste Gas Purification

Derek E. Chitwood and Joseph S. Devinny

## 18.1
## Background

### 18.1.1
### Needs

Concerns over industrial releases of odors, toxic compounds and smog-forming chemicals have grown steadily. Increased public awareness of health impacts, declining tolerance for industrial offensiveness, expanding industrial activity and greater prosperity have led to insistence that air discharges be cleaned up even when dischargers complain that costs are high. Regulations have forced some companies to move to less-affected areas and have made others close down. All are interested in seeking to meet regulatory requirements at the lowest possible cost.

Contaminants have commonly been removed by thermal treatment or activated carbon adsorption. Thermal treatment, either direct flaring or catalytic oxidation at lower temperature, is effective when contaminant concentrations are high enough to provide a significant portion of the energy required. However, when concentrations are low and additional fuel is required, a great deal of energy is wasted, because the entire airflow must be heated to oxidize a small amount of contaminant. Activated carbon acts by surface adsorption and can produce very clean effluent air. However, the amount of contaminant adsorbed per unit weight of activated carbon is a function of the airborne concentration. When concentrations are low, the adsorptivity of activated carbon is often reduced. The amount of activated carbon to be regenerated or replaced for each kilogram of contaminant removed becomes large, and the treatment is again impractical.

Biological treatment fills the need for an economical means of treating low concentration contaminants. It has been most widely used in odor control and has gained greatest acceptance in Europe where odor regulations are strict and fuel costs are high. In recent years, however, it has been increasingly used for the control of toxic compounds and those that contribute to the formation of smog.

*Environmental Biotechnology. Concepts and Applications.* Edited by H.-J. Jördening and J. Winter
Copyright © 2005 WILEY-VCH Verlag GmbH & Co. KGaA, Weinheim
ISBN: 3-527-30585-8

18.1.2
**Biological Treatment**

The concept of biological air treatment at first seems contradictory: organisms suspended in the air cannot be biologically active. Air can be treated, however, if it is brought in close contact with a water phase that contains active organisms. Contaminants dissolve in the water where microbial degradation keeps concentrations low, driving further dissolution.

The needs for interphase contaminant transfer and for maintaining the water phase microbial ecosystem dictate the basic form of biological air treatment systems. A reactor must provide a large water surface area for efficient phase transfer and support for a substantial biomass. The three general plans for biological treatment systems are biofilters, biotrickling filters, and bioscrubbers (Ottengraf, 1986). In biofilters, the porous medium is kept damp by maintaining the humidity of the incoming air and by occasional sprinkling (Fig. 18.1). In biotrickling filters, water flows steadily over the porous bed and is continuously collected and recirculated. In a bioscrubber, phase transfer occurs in one container as the water is sprayed through the air or trickled over a packed bed, then the water is transferred to a liquid phase bioreactor where the contaminant is degraded. Bioscrubbers are less commonly used and are not further discussed.

There are some secondary considerations for design. An important operating cost is the power for pumping air. The system must be designed so that the pressure required to drive the air through the system, called the head loss, is small. It is also necessary to have a means for removing biomass so that it does not clog the system. In biotrickling filters, the flowing water may carry away biomass as rapidly as it grows, maintaining a steady state microbial ecosystem. However, heavily loaded biotrickling filters often have problems with clogging. Lightly or moderately loaded bio-

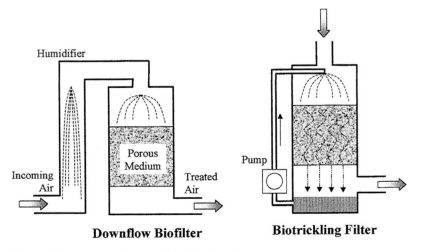

**Fig. 18.1** Two common forms of biological air treatment.

filters may reach steady state through the establishment of a culture of predatory microbes that consume the decomposers, but again, heavy contaminant loads can cause rapid growth and pore clogging. Research on means to control or remove biomass remains active.

Water is readily available in biotrickling filters, but biofilters must be provided with humidification or irrigation systems to maintain the water content of the biofilm. All the systems require a source of nutrients such as nitrates and phosphates.

Each biological system must have an initial seed culture of microorganisms. Most often, a diverse mixed culture comes with the medium or is applied separately. It is presumed that those species capable of degrading the contaminant will grow rapidly, dominate the system, and create a dense culture ideal for efficient treatment.

A major design characteristic of a biological treatment system is the detention time. This is the average time that a parcel of air is held in the system. For a given flow, the minimum detention time determines the size of the reactor and so is a prime factor in determining the costs of the system. Because true porosity is often hard to measure, empty bed residence time (EBRT), calculated using the total volume of the bed, is often used to describe systems.

## 18.1.3
## Biofilters

Biofilters are the most common form of biological air treatment system. The earliest and simplest were made by digging trenches, installing perforated pipe to serve as an air distribution system, and refilling the trenches with permeable soil. The Bohn Biofilter Corporation has installed many successful biofilters that are only slightly more complex, consisting of soil placed over a distribution system and contained within concrete traffic barriers (Bohn and Bohn, 1998). Soil biofilters are inexpensive. Soil is cheap and the construction methods are simple. Soil does not decompose significantly and so does not compact It is easily rewetted if it is inadvertently allowed to dry. However, the pores are small, so that head losses are high when flow rates are high, and high contaminant loads cause clogging. These problems are avoided by making soil biofilters large, typically with detention times of a few minutes.

More complex biofilters are a pit or box filled with sieved compost or wood chips (Fig. 18.2). Once again, air is introduced through a distribution system at the bottom, typically perforated plastic pipes embedded in a layer of gravel or a layer of perforated concrete blocks. The compost or wood chips are placed above this, commonly in a layer about 1 m thick. A sprinkler system is provided to keep the medium moist, and drainage is provided to remove collected water in the event of rain or overwatering. Although these biofilters are somewhat more elaborate than soil biofilters, they can still be made cheaply. Because the particle size can be controlled, the organic media provide large pores to reduce head loss. They also provide nutrients as they slowly decompose. The diverse and active culture of microorganisms on compost allows it to start working immediately. The reduced head loss and ideal microbial environment mean that biofilters can be made smaller.

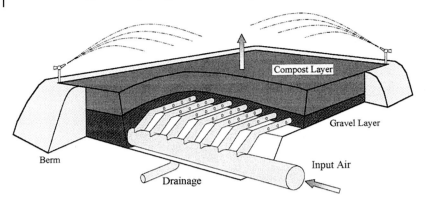

**Fig. 18.2** Schematic of an open bed compost biofilter.

Compost does have disadvantages, however. As it decomposes, small particles are generated, and the material softens and compacts. Compaction along the vertical axis tends to close the pores, increasing head loss. Horizontal shrinkage creates fissures in the medium and pulls it away from the walls of the vessel. Air is diverted from flowing through the medium to flowing through open channels, and treatment fails.

Compost also shrinks if it dries. Ideally, humidification and irrigation systems should prevent drying, but these occasionally fail. Once the compost has dried, it may become hydrophobic, making rewetting difficult.

The need to keep biofilters moist is a significant design consideration. Large amounts of air are passed through biofilters, so that air with a relative humidity even slightly lower than saturation causes a slow but steady drying that eventually disrupts operation. Relative humidity is a function of air temperature, and even modest warming also causes drying (Van Lith et al., 1997). Such warming can occur because of the pressure increase caused by a blower or because of heat generated in the biofilter by microbial oxidation of contaminants. To prevent problems, incoming air is often passed through a humidifier, typically a packed trickling bed. However, any warming of the air that occurs downstream (which always occurs for metabolic warming) again can reduce the relative humidity.

The water content of the porous medium can also be supplemented by direct irrigation, usually through sprinkling the top of the bed. This provides direct control, but it is sometimes difficult to ensure homogeneity. It is surprisingly difficult to arrange sprinklers to water uniformly, and partial clogging of the sprinkler nozzles commonly disrupts the distribution. Fine mist generators are better, because the mist droplets move with the air. Even when water is sprinkled uniformly on the surface of the bed, downward flow tends to 'finger', concentrating in streams in some areas and bypassing others. Biofilters are now often operated with both humidifiers and irrigation systems, so that each system can compensate for the shortcomings of the other.

Porous-medium biofilters can be operated with either upward or downward airflow. (In a few special designs using other supports such as rotating disks, the air flows horizontally.) Construction costs are lower for upward flow, because the biofilter can be uncovered. The bottom of the container that supports the medium also serves to contain the air, so no additional structure is needed. If downward flow is chosen, a cover is necessary to contain the air. Because biofilters are often large, this can add substantially to costs. However, there is a second constraint: irrigation water must be added from the top so that it drains downward to wet the interior of the biofilter. Drying is most severe where the air first enters the medium. In an upflow biofilter, this means that the bottom of the medium is most prone to drying, and that irrigation water reaches this critical zone only after trickling down through 1 m (typically) of porous medium. Trickling is often nonuniform, so maintaining a homogeneous water distribution is difficult. The irrigation water also occupies pore space and flows counter to the air current. In some systems, the resulting backpressure can significantly decrease the air flow during irrigation. If the air and water are added at the top, the incoming air immediately contacts a layer of wet medium, raising its humidity and helping to maintain dampness throughout the bed, and little or no additional head loss is caused during irrigation.

The difficulties with compost and the desire to produce higher specific activities have led to the use of mixed or inorganic media. Monsanto (McGrath and Nieuwland, 1998) used a proprietary mixture of foam plastic beads and organic matter. Several biofilters have been built using lava rock, which is a gravel with large pores. Biotrickling filters are often made with plastic shapes like those used in cooling towers, and biofilters have been filled with large-pore polyurethane foam. The inorganic materials solve the problems of compaction and can provide a highly uniform large pore medium for low head loss. The formed media include far less dead space within the particles, and may be much lighter. However, inorganic media provide no nutrients or seed culture. Although they are easy to rewet if they are inadvertently dried, their low water holding capacity means that they dry more easily. In all, the inorganic media hold the potential for high degradation rates and correspondingly smaller biofilters, but the systems must be managed more carefully. An appropriate seed culture must be found and applied uniformly to the medium. Nutrients must be added regularly, and the water content must be carefully controlled.

The most advanced biofilters are thus contained systems with permanent inorganic media and highly capable control systems. Sensors measure moisture in the air and on the medium, input and output contaminant concentrations, leachate pH, head loss, air flow rates, and other parameters (Devinny, 1998). A programmable logic controller can monitor biofilter operation, adjust operating parameters, and shut down the system in response to anomalous conditions. These most advanced systems hold the promise for high treatment rates and smaller, better controlled biofilters, but there are substantial increases in costs for equipment, maintenance, and operation. The choice between cheap low-tech systems and more expensive advanced biofilters is still a difficult one.

18.1.4
**Biotrickling Filters**

Providing a continuous flow of water over the surface of the porous medium precludes drying and provides a means for precise control of medium pH and nutrient content. As the water is pumped back to the top of the reactor, acids, bases, or nutrients can be added. The amount of water present in the reactor at any time is substantially larger than in a biofilter, allowing larger amounts of soluble contaminants to dissolve. These factors mean that biotrickling filters are often capable of higher specific rates of treatment: more contaminant is degraded per unit volume and the reactors can be smaller. However, the greater mechanical complexity requires more capital investment and maintenance. Biotrickling filters are more likely to be used where contaminant concentrations are at high levels that would cause clogging in biofilters.

18.1.5
**Applications for Biological Systems**

Biological treatment is certainly not appropriate for all applications. Most obviously, microorganism able to transform the contaminant to a harmless product must be available. Commonly, organic contaminants are converted to carbon dioxide and water, but other processes are possible. Many biological systems are used to convert hydrogen sulfide, a strongly odorous compound, to highly soluble sulfate ions that are captured in the water. Mineral dusts cannot be treated, however, and accumulate to eventually clog the biofilter. Some chlorinated hydrocarbons degrade only very slowly, and biological systems have not been successful in practical application to these vapors.

Contaminants that are very poorly soluble in water are more difficult to treat. A low water–air partition coefficient means that concentrations in the water phase, and thus biodegradation rates, are low (Hodge and Devinny, 1995). This restriction applies only to very low solubilities: benzene and toluene, e.g., are commonly thought of as not very soluble in water, but they are readily treated by biofiltration. However, some compounds, like isopentane and chlorinated solvents, which combine low solubility and some resistance to biodegradation, are poor candidates for biological treatment.

Biological systems are generally larger than alternative systems and cannot be used where space is limited. Simple compost biofilters are restricted to layers of medium about 1 m thick because greater depths cause compaction. A high-volume biofilter must therefore have a large footprint or bear the structural costs of constructing multiple layers.

Biological systems will be most successful for low concentrations of readily degradable contaminants in large volumes of air.

## 18.2
## Applications

There are now many hundreds of biofilters in operation worldwide. The following examples are chosen to represent the range of complexity, from the simplest to the most elaborate (Dragt and van Ham, 1991).

### 18.2.1
### Soil Bed Biofilters (Bohn and Bohn, 1998)

Sunshine Plastics of Montebello, California, was required to control VOC emissions or face pollution penalties. The facilities emit 170 $m^3$ $min^{-1}$ of waste air containing a mixture of propanol, ethanol, and acetone at a total VOC concentration between 100 and 1000 ppm. The air permit granted by the South Coast Air Quality Management District required a minimum 70% reduction in VOC emissions from the facilities.

A 486 $m^3$ soil bed biofilter was installed for VOC control by Bohn Biofilter Corporation of Tucson, Arizona (Devinny et al., 1999). The biofilter is 90 cm deep and provides an EBRT of 2.5 min. The biofilter was constructed above ground, adjacent to the facility in the parking lot. Concrete traffic dividers were used for the walls of the biofilter. The inlet air to the biofilter is humidified by fogger nozzles in the main air pipe, and additional water is added to the bed through sprinklers controlled by timers. The schedule of plant operation is variable. The biofilter operates only while the presses are working, and the blower is on about 80 h per week. Despite the use of a soil bed, the head loss across the bed is relatively low (5 cm) because of the long air residence time. The inlet and outlet concentrations are continuously monitored. Results indicate 95% destruction of VOC in the biofilter, indicating much better removal than required. The biofilter cost approximately $78000. Overall, it is typical of a simple soil bed that is cheap and effective, but large.

### 18.2.2
### Open Compost Biofilter for Treating Odors from a Livestock Facility
### (Nicolai and Janni, 1998)

Odors from livestock facilities are a significant problem in some communities. Sources of odors include buildings, manure storage, and manure land applications. A set of simple and low cost biofilters was installed at a pig farm to control odors from sow gestation and farrowing (birthing) barns. The odors are caused chiefly by hydrogen sulfide and ammonia

Readily available equipment and materials were used in construction of the biofilter. The dimensions of the biofilter were based on available space and acceptable pressure drop across the medium. Standard agricultural ventilation fans were used for the air blowers. They are not designed to operate at a static pressure greater than 62 Pa, and tests determined that a bed depth of 28 cm caused no more than 50 Pa of pressure loss across the medium at the maximum flow rate. Ventilation rates vary throughout the year, depending on the temperature in the barns. The total for the three biofilters ranged from 641 m³ min⁻¹ in winter to 4634 m³ min⁻¹ in summer. The biofilters were sized to have only a 5-s average EBRT, but varied from 18 s in winter to 3 s in summer.

Odors originate from manure that collects in a pit beneath the barn floor. Odorous air is drawn out of the pit by fans located below the floor level. In the summer months, additional wall fans are activated to increase the air replacement rate in the building. Air is transferred through plywood ducts leading to air plenums below the medium. The plenums are constructed of shipping pallets covered with a plastic mesh to prevent the medium from dropping through the pallet openings and clogging the plenum. The medium used is an equal weight of compost (unspecified) and chipped brush. The total surface area of the biofilters was 189 m².

The biofilters are effective at removing odors, hydrogen sulfide and ammonia. Odors were reduced by an average of 82% during the first 10 months of operation. Hydrogen sulfide concentrations were decreased by approximately 80%, and ammonia concentrations were lowered by 53%. Despite the high flow rates and extremely short detention time in the summer months, the effluent hydrogen sulfide concentration was consistently less than 100 ppb.

The use of available materials made the system very inexpensive. The biofilter cost less than $10 000, or approximately $0.13 per m³ s⁻¹ of air treated. Operational costs, including rodent control, are estimated to be approximately $400 per year.

## 18.2.3
### Open Bed Compost Biofilter for Wastewater Plant Odor Control (Chitwood, 1999)

The Ojai Valley Sanitary District in California operates a 7.9 m³ min⁻¹ wastewater treatment plant that includes a 280 m² below grade, open bed biofilter. The biofilter was installed to control odors, chiefly hydrogen sulfide. The builders were particularly concerned about potential complaints from users of an adjacent bicycle path.

The biofilter uses wood chips from lumber waste as the medium, in the form of strips 3–30 cm long. The biofilter is designed to treat 225 m³ min⁻¹ of waste air removed from the plant's headworks, grit chamber, and grit classifier. Air is driven by a centrifugal blower designed to deliver 225 m³ min⁻¹ at a pressure of 1250 Pa. Air is humidified by passing it through a spray chamber and is then distributed through fourteen 25-cm diameter schedule 80 PVC laterals beneath the wood chips (Fig. 18.2). Each lateral is 15 m long and has a pair of 1.6-cm diameter holes drilled every 15 cm along its length. Pairs of holes were drilled at 90° from each other, and the pipe was laid so that each hole faces 45° from the center bottom. The air is thus directed outward and downward from the pipe. The laterals are 1.2 m apart and are covered with 15 cm of 2-cm-diameter, acid-resistant, smooth river rock. The depth of the medium above the rock is 0.9 m, and 15 cm of chipped bark was added for beautification. Six sprinklers controlled by a timer provide irrigation of the biofilter once daily. No nutrients were added because it was presumed that the organic medium would provide necessary nutrients, and the biofilter was not inoculated. Results indicate greater than 99% reduction in hydrogen sulfide with an average inlet concentration of 4.5 ppm. A 70% reduction in total VOC was observed with an average inlet concentration of 5 ppm measured as methane equivalents.

## 18.2.4
### Inorganic Biofilter for Odor Control at a Wastewater Treatment Facility (Dechant et al., 1999)

The Cedar Rapids Water Pollution Control Facility in Iowa is designed to treat 2.45 m³ s⁻¹ of municipal and industrial wastewater. The treatment facility serves approximately 145 000 people, as well as wet corn milling, pulp paper, and other industries. In the early to mid 1990s, the organic loading rate of the facility was substantially increased. Concurrent with this change, odor complaints increased markedly. Studies indicated that the wastewater trickling filters were the major source of odors, with a typical hydrogen sulfide concentration between 50 ppm and 350 ppm. The primary clarifiers and air floatation thickeners also produced odors. In an effort to reduce complaints, several air pollution control technologies were evaluated.

Biofiltration was chosen because the combined capital cost and operating cost (present worth) were significantly less than for chemical scrubbers. Lava rock was chosen as the biofilter medium for its resistance to low pH. Operating costs are expected to be less than those for a compost biofilter, because the lava rock is presumed to be permanent. The system includes two biofilters in parallel. Each is 11 m wide and 21 m long. The depth of the medium is 1.8 m. They are designed to treat 4238 m³ min⁻¹ of waste air from the trickling filter, primary clarifiers, and air floatation thickeners. The resulting design EBRT is 15 s.

The biofilters are housed in a concrete structure coated with a polyvinyl chloride liner. The medium is supported by fiberglass-reinforced plastic. Air is not humidified before entering the biofilters but the medium is irrigated with 3.78 m³ min⁻¹ of water from the secondary effluent for 10 min every hour. This also serves to wash out excess acid, maintain a bed pH of between 2 and 2. 5, and add necessary nutrients for the microbial growth. Since the beginning of operation, the average inlet concentrations have been about 200 ppm of hydrogen sulfide, with removal efficiencies consistently between 90% and 95%. Measurements show an almost 50% removal of VOC. In the first 6 months of operation there have been 90% fewer odor complaints.

## 18.2.5
### Biofilter Treating Gasoline Vapor at a Soil Vapor Extraction Site (Wright et al., 1997)

Weathered gasoline vapors from a soil vapor extraction system were treated using a biofilter system in Hayward, California. The system comprised four small biofilters initially operated in parallel and later operated as two sets of in-series biofilter systems. Each of the biofilters was 1.2 m long by 1.2 m wide with a bed depth of 95 cm. The medium was a mixture of compost from sludge solids and wood products mixed with an equal portion of perlite as a bulking agent. Crushed oyster shells were used as a buffer. The EBRT was approximately 2 min during the parallel stage of the remediation and 1 min when operated in series. The relative humidity of the inlet air was increased by passing the air through a humidification chamber fitted with fogger nozzles. Excess water was recycled. As is typical of a soil vapor extraction system, the inlet concentration was highly variable. Initially, the inlet concentration was 2.7 g m$^{-3}$ as total petroleum hydrocarbon (TPH), but it decreased to 0.9 g m$^{-3}$ by day 22 of the project. The final inlet concentration during the course of the study was 0.4 g m$^{-3}$.

Bed drying was a notable problem. However, after operators thoroughly mixed the bed, rewetted the medium, and initiated regular direct irrigation, biofilter performance increased. The authors suggested that an improved inoculation would have helped to shorten the adaptation time and that humidification of the air should be supplemented with periodic direct application of water to the bed by either soaker hoses or a sprinkler system.

Sustained removal efficiency of 90% of the total petroleum hydrocarbons at an EBRT of 1.8 min and an inlet concentration of 0.4 g m$^{-3}$ was observed. Removal efficiencies for the BTEX compounds were even better. However, some compounds (believed to be methyl-substituted alkanes and cycloalkanes) were more poorly removed, probably because of their more recalcitrant nature.

### 18.2.6
#### Biofilter Treating VOC Emissions from an Optical Lens Manufacturer (Standefer et al., 1999)

A major optical lens manufacturer in Massachusetts emits a proprietary mixture of VOC containing alcohols and ketones from its coating process. The facility releases 127.5 m$^3$ min$^{-1}$ of waste air with a peak concentration of approximately 0.2 g m$^{-3}$ (1.55 kg h$^{-1}$). The purpose of VOC reduction was to assure compliance with the Massachusetts Department of Environmental Protection requirements and to maintain their minor source designation under the Clean Air Act. To meet these requirements, a 90% removal efficiency goal was set.

A biofilter was chosen over other technologies because of its relatively low capital and operating costs and restrictions on water consumption and wastewater disposal. A 3.4 m wide by 13.4 m long (45 m$^2$) biofilter was designed, constructed, and installed by PPC Biofilter. To prevent drying of the medium bed, the relative humidity of the air is raised from 6% to >95% by a 2.8-m$^3$ counter-current packed tower filled with 5-cm pall rings. The biofilter is a downflow, induced-draft (negative pressure) system, with a proprietary medium 1.5 m in depth allowing a 27-s EBRT at an approach velocity of 6 cm s$^{-1}$. PPC Biofilter research showed that typically 50%–70% of overall removal occurs in the first 35–45 cm of a biofilter bed. Because oxidation of the VOC compounds is an exothermic reaction, downflow allows application of irrigation water where the highest consumption of water occurs.

A programmable logic controller (PLC) managed supplemental water addition. Water was added when the weight of the medium as measured by load cells was below a set level. The PLC activates a solenoid valve that allows irrigation of the bed by 27 fine mist nozzles. The same PLC allows all pertinent data to be logged on a desktop computer.

The biofilter has met the removal efficiency goals. Independent results demonstrated that the biofilter removed 93% of the alcohols and 82% of the ketones with a total VOC removal of more than 91% in the first two years of operation. The hazardous air pollutant of concern has had an average removal efficiency of 97%.

## 18.2.7
### Advanced Biofilter for Controlling Styrene Emissions
### (Punti, personal communication; Thissen, 1997)

The Synergy Biofilter system, manufactured by Otto Umwelttechnik in Germany and supplied by Biorem Technologies, Inc. in the United States, is an example of a technologically advanced system combining a biofilter and an adsorber. The system treats styrene emissions from a manufacturer in Germany. During working hours, the biofilter treats 84 000 $m^3$ $h^{-1}$ of air containing up to 160 mg $m^3$ of styrene and reduces concentrations to less than 30 mg $m^3$, a removal efficiency of 81%. Considerable amounts of acetone are also present. The system includes humidification of the air stream and direct irrigation of the biofilter beds with find mist (Fig. 18.3). The microbial support medium is pellets of carbon-coated concrete, which are adsorbent and essentially permanent.

An adsorber was installed downstream of the biofilter bed. This eased requirements for treatment efficiency in the biofilter, allowing use of a smaller bed. The surface load is 371 $m^3$ $m^{-3}$ $h^{-1}$, for an empty bed detention time of only about 10 s. During nonworking hours, a smaller flow of air is heated and passed through the adsorber to remove accumulated styrene, then returned to the biofilter. Some makeup air from the facility is continuously added to cool the air and to remove residual styrene vapors from the factory work area. This approach also supplies small amounts of styrene to the biomass in the bed to reduce loss of activity during nonworking hours.

The system includes monitors for input and output concentrations, temperature, pH, airflow, and several other system parameters. The sprinkler system is activated automatically on a predetermined schedule.

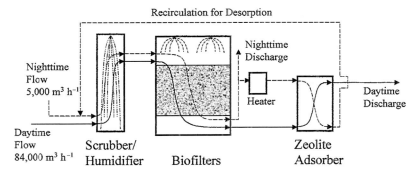

**Fig. 18.3** Biofilter system with adsorber.

## References

Bohn, H.-L., Bohn, K.-H., Accurate monitoring of open biofilters, *Proc. 1998 Conf. Biofiltration (an Air Pollution Control Technology)*, The Reynolds Group, Tustin, CA, October 22–23 (**1998**), pp. 9–14.

Chitwood, D. E., Two-stage biofiltration for treatment of POTW off-gases, Thesis, University of Southern California, Los Angeles, CA (**1999**).

Dechant, D., Ball, P., Hatch, C., Full-scale validation of emerging bioscrubber technology for odor control, in: *Proc. Water Environ. Fed. 72nd Ann. Conf. Exposition*, New Orleans (**1999**).

Devinny, J. S., Monitoring biofilters used for air pollution control. Practice Periodical of Hazardous, Toxic, and Radioactive Waste Management, *Am. Soc. Civil Eng.* **1998**, *2*, 78–85.

Devinny, J. S., Deshusses, M. A., Webster, T. S., *Biofiltration for Air Pollution Control*. Boca Raton, FL **1999**: CRC Lewis.

Dragt, A. J., van Ham, J. (eds.), Biotechniques for air pollution abatement and odor control policies, *Proc. Int. Symp.*, Maastricht, The Netherlands, 27–29 October. Amsterdam **1991**: Elsevier.

Hodge, D. S., Devinny, J. S., Modeling removal of air contaminants by biofiltration, *J. Environ. Eng.* **1995**, *121*, 21–32.

McGrath, M. S., Nieuwland, J. C., Case study: effectively treating high levels of VOCs using biofiltration, in: *Proc. 1998 Conf. Biofiltration (an Air Pollution Control Technology)*, The Reynolds Group, Tustin, CA (**1998**).

Nicolai, R., Janni, K., Biofiltration: adaptation to livestock facilities, in: *Proc. 1998 Conf. Biofiltration (an Air Pollution Control Technology)*, The Reynolds Group, Tustin, CA (**1998**).

Ottengraf, S. P. P., Exhaust gas purification, in: *Biotechnology 1st Edit.*, Vol. 8 (Rehm, H.-J., Reed, G., eds.), pp. 425–452. Weinheim **1986**: VCH.

Standefer, S., Willingham, R., Dahlstrom, R., Commercial biofilter applied to an optic lens manufacturer to abate VOCs, in: *Proc. 92nd Ann. Meeting Exhibition Air Waste Management Assoc.*, St. Louis, MO (**1999**).

Thissen, I. N., Biological treatment of exhaust air: practical experience with combined processes for treatment of organic solvents, *Fachmagazin für Kreislaufwirtschaft, Abwassertechnik und Luftreinigung* (**1997**).

Van Lith, C., Leson, G., Michelsen, R., Evaluating design options for biofilters, *J. Air Waste Manage. Assoc.* **1997**, *47*, 37.

Wright, W. F., Schroeder, E. D., Chang, P. Y., Romstad, K., Performance of a pilot-scale compost biofilter treating gasoline vapor, *J. Environ. Eng.* **1997**, *123*, 547–555.

# 19
# Perspectives of Wastewater, Waste, Off-gas and Soil Treatment

Claudia Gallert and Josef Winter

## 19.1
## Introduction

After the Second World War an expanding industry was essential to restore and improve the standard of living and stimulate the economy in Europe. The cost of this boom in industrial development (the so-called Wirtschaftswunder), however, was severe air pollution and the production of huge masses of domestic and industrial wastes and wastewater. Because atmospheric pollution was rapidly diluted and dislocated and solid wastes could be deposited within defined, spatially limited sanitary landfill areas, in the beginning of the industrial boom these pollutions were not recognized as being serious as the deterioration of surface waters, e.g., rivers and lakes, by pollutants from wastewater. The extent of pollution often exceeded the natural self-purification capacity of aquatic ecosystems, and severe environmental harm was visible to everybody. Epidemic mortality of whole populations of fish or other aquatic organisms by depletion of oxygen or the presence of toxins in the water demanded counteractions by legislative authorities. Technically achievable and environmentally acceptable atmospheric, aquatic, and terrestrial pollution limits had to be defined, fixed by state laws, and controlled by administrative officers. The standards for wastewater, waste, and off-gas treatment, as well as for drinking water preparation were defined and progressively strengthened with improving technological treatment and purification possibilities.

## 19.2
## Wastewater Handling

Domestic and industrial wastewater has to be purified by application of standardized, widely experienced state-of-the-art treatment technologies to meet the quality standards of environmental legislation. For this purpose, the wastewater must be collected, transported in public, industrial or private sewer systems, and treated in domestic sewage or industrial wastewater treatment plants to remove organic and

*Environmental Biotechnology. Concepts and Applications* Edited by H.-J. Jördening and J. Winter
Copyright © 2005 WILEY-VCH Verlag GmbH & Co. KGaA, Weinheim
ISBN: 3-527-30585-8

inorganic pollutants, as required by environmental laws and enforced by state control agencies of the respective countries. The concentration limits in the purified wastewater for residual carbon (measured as biological oxygen demand, BOD, or chemical oxygen demand, COD), nitrogen (total nitrogen or ammonia nitrogen), and phosphorous (in particular, soluble orthophosphate) that had to be met for disposal into surface waters became more stringent with time. Improvements and the development of new processes for wastewater purification were stimulated by pressure exerted by legislation.

Due to the complexity of the pollutants in different wastewater types or even in a certain wastewater of defined composition, combined multistage processes for physical, chemical, and biological removal of organic pollutants, nitrogen, and phosphorous were required. Generally, the process development for wastewater purification was always slightly ahead of an exact knowledge of the biological reactions or reaction sequences behind it. A lack of detailed biochemical knowledge on even major metabolic pathways or, in particular, on single reactions within the complex ecosystem called wastewater was always the bottleneck for specific improvements in wastewater purification techniques and treatment efficiencies, favoring a trial-and-error approach by civil engineers. Another nuisance was an apparently deep gap between the scientific approaches of civil engineers (practical process designers and operators) and life scientists (basic researchers). Even today, finding the bottleneck reactions is still one of the obstacles to improving wastewater treatment.

## 19.2.1
### Domestic Wastewater

In the 1920s treatment of domestic wastewater began in big cities, with the construction of sewer systems and large treatment units. Later, many small wastewater treatment plants, often considered less efficient, were built in smaller settlements to serve single towns or villages in less densely populated areas. With time, development was directed more and more toward centralization of wastewater treatment, with huge treatment units serving whole regions. These units were supplied with wastewater from several settlements, often via pumping stations to transport the wastewater over long distances. This development was subsidized by the government and favored by the inspecting offices, since the treatment efficiency in these plants was considered more reliable (or easier to control) than that in a large number of scattered, local, small wastewater treatment plants. However, except for accelerating costs for additional pumping stations and the construction of new central sewage treatment plants or the extension of the capacity of existing plants, another possible source of environmental pollution was or still is created: thousands of miles of main sewers becoming leaky with age and allowing wastewater to seep into soil and groundwater. This must be a major concern, since groundwater in many regions of the world is the main source of potable water, so its pollution must be prevented.

Since all wastewater sources in settlements, including rainwater from roofs of buildings and streets, often flow abundantly into mixed-water sewer systems, the wastewater reaching the treatment plant during or after rainy weather is highly diluted. All wastewater treatment facilities must be designed to cope with such unfavorable conditions with respect to the hydraulics and the chemical and biological reactions. On the one hand, dilution of wastewater is counterproductive for efficient chemical or biological treatment; but on the other hand, rainwater is periodically required to flush the channels free of sediments, due to the construction of the sewer systems having little slope. In some communities or newly developing suburbs, a dual channel system for separately collecting sewage and rainwater is available.

For testing alternative wastewater handling approaches, in some new residential areas the general strategy of collecting all wastewater types into a common sewer system for treatment in a central sewage plant has meanwhile been reversed. Less polluted rainwater, e.g., is collected in natural or artificial depressions in the ground, where it seeps into the underground, with the top soil layers serving as a natural biologically active filter. Only limited paved areas are allowed, to favor the seepage of as much rainwater as possible.

Grey water from single households can be purified biologically in special soil filters planted with *Phragmites australis*, *Typha angustifolia* or other plants that develop an aerenchym. After removal of most of the pollutants by biofiltration, the purified wastewater seeps into the underground. The natural self-purification capacity of the top soil layers for wastewater components is extended into deeper layers of the soil by improving the oxygen supply via the aerenchym of planted vegetation.

To reduce the amounts of waste and wastewater, separation toilets have been developed that handle feces and urine differently. After utilization, these toilets are flushed with very little water; the solids are separated and composted in-house, and a concentrated mixture of urine, some suspended matter, and the flushing water flows through a sewer system to a nearby biogas plant for wastewater stabilization and biogas production. The ammonia of the fermented 'yellow water' should be recoverable for use as fertilizer.

Compact treatment units for human excrements with solids separation and long-term hydrolysis, preanoxic zone fixed-bed denitrification, and fixed-bed activated sludge treatment for carbon removal and nitrogen oxidation, followed by pasteurization for direct disposal (if disposal standards are less strict) or microfiltration and UV irradiation for reutilization of the purified wastewater have been developed on a small scale for railway passenger cars or small ships (total volume 600–900 L). On a larger scale, such units are available for single buildings to serve up to 20 inhabitant equivalents (e.g., Fa. AKW, Protec GmbH, Hirschau, Germany). Several of these units have been operated successfully for years. If recontamination of the purified and ultrafiltered wastewater can be prevented, it can be recycled for reuse in toilet flushing. If the grey water from showers and bathrooms of houses is also purified in these minisewage treatment plants, much more water than necessary for flushing toilets is generated, and no drinking water at all has to be wasted for toilets. Only little maintenance, usually once a year, is required.

At present, some pilot projects are under investigation in new residential areas in Germany and other European countries to test zero-emission concepts. These are, however, realistic only for single houses set on several acres of land or in settlements with sufficient area for seepage of ponded water into the underground. A double piping system to supply the kitchen, bathroom, and laundry room with high-quality drinking water and toilets with less pure, purified rain and grey water could help to save drinking water resources. Purified rainwater from the roofs of the houses might even be used for laundry and thus further reduce the drinking water demand.

In the future, increasing costs for processing and supplying drinking water and for wastewater treatment might lead to a decrease in the drinking water consumption of single households. A separation of costs for drinking water supply and wastewater treatment might favor individual on-site wastewater treatment systems for single households or small communities. Rain and grey water purification for reutilization within the household, e.g., for laundering, toilet flushing, or watering the garden, may still sound futuristic (and to some people not acceptable), but may become necessary in the future, especially when not enough fresh water is accessible during dry seasons. First indications of a shortage of fresh water became apparent in the very dry year 2003 even in Germany, when the water level in Lake Constanz, the main drinking water reservoir in southern Germany, unusually fell several meters.

## 19.2.2
### Industrial Wastewater

Biological wastewater treatment in industry has focused on three approaches: Some companies favor a central treatment plant for all production units, requiring the enrichment of an 'omnipotent' microbial population at suboptimal loading. Other companies favor smaller treatment units for each production process, requiring the enrichment of specialized bacteria in each plant, which could then be operated at maximal loading. Mainly known from the paper recycling industry are partial process water cleaning procedures for process water recycling. Only a small part of the total process water is released for final wastewater treatment. Whereas the mechanically purified recycling water should be biologically inactive (which in practice is achieved by adding biocides), the disposed wastewater must be biodegradable by microorganisms, which demands nontoxic biocide concentrations.

Wastewater purification for disposal into a natural water source always requires a technically sophisticated combination of mechanical, chemical, and biological processes essential to purify the multiple components contained in waste fluids from production processes.

If the wastewater from certain production processes contains xenobiotic substances, and these substances cannot be adsorbed, precipitated, or biologically degraded, environmental protection has to go one step further and intervene in the production process. The process must be altered to prevent nondegradable xenobiotics from appearing in the wastewater or, if this is not possible, their concentrations must be reduced to the minimum and single wastewater streams should be recycled internally to avoid environmental pollution.

19.2.3
**Effluent Quality and Future Improvements**

Every wastewater that has been purified according to present treatment standards still contains some residual pollutants, consisting of a small proportion of BOD (biodegradable, but residing organic substances) and a higher proportion of nondegradable COD (organic substances that resist rapid biological degradation and require more drastic conditions for chemical oxidation), as well as a certain salt load. Further elimination of some or most of these compounds and of the suspended residual bacteria can be achieved by modern membrane technologies, such as reversed osmosis or ultrafiltration. These are, however, not yet widely used for wastewater treatment, and experience with their long-term performances is still rather scarce. They must, however, be used if the pathogen content of treated wastewater – a potential source of infections or epidemics – must be reduced to much lower concentrations than obtained by conventional sedimentation or soil filtration.

The residual BOD and part of the COD of the purified wastewater are degraded in the receiving water and some of the salt components may be precipitated. However, even if the biological self-purification capacity of receiving waters is not exceeded, traces of nondegradable wastewater components, such as detergents, household chemicals, antibiotics, pharmaceuticals, pesticides, and fungicides, are washed into the groundwater. Some of these substances have hormonal activity and act as endocrine disruptors. They influence the reproduction of wild fauna and may damage the flora in the receiving lakes and rivers or exert a negative effect on human health. If the raw water for the preparation of drinking water contains such contaminants, they must be carefully separated, e.g., by adsorption onto charcoal. Alternatively, membrane technologies can be used to separate large molecules of trace pollutants from the bulk of the water.

Future wastewater handling must more and more begin at the sources of waste and get away from an almost exclusive end-of-pipe-treatment, as has been practiced to date. Two main goals have to be envisioned:

- reduction of the total amount of wastewater by water-saving and water-recycling procedures
- reduction or avoidance of biologically nondegradable chemicals as water polluting ingredients

The first goal requires strict in-house or in-factory water-saving regimes and recycling techniques, and the second goal, a change in production processes and human habits.

Complete closure of the water cycle is already achieved in some industries, e.g., the paper recycling industry. However, new problems appear with recycling of production water. Massive germination of the water during interim storage and reutilization requires permanent application of biocides. Formation of fatty acids from carbohydrates by contaminating microorganisms has meant that only low-quality papers could be produced from the recycled cellulose material. If these papers are moistened, an unpleasant odor develops; in addition, their acidity favors rapid deterioration.

To promote environmentally sound production processes in industry, the so-called Ökoaudit system was introduced by environment controlling authorities in Germany. Input material, products, and wastes are analyzed, and improvements in working procedures or production processes are proceeded by the management of the company. With their participation in the Ökoaudit evaluation, a company agrees to reduce pollution year by year by a certain percentage compared to the present state. To make consumers aware that a company uses an environmentally friendly production process (which may justify a somewhat higher retail price), the participating companies are allowed to print a symbol of environmentally friendly production on their products.

## 19.3
## Solid Waste Handling

Until only recently, solid domestic wastes and residues from industrial production have been collected and simply deposited into sanitary landfills located outside residential areas. Except for sorting out most of the recyclable material and a certain degree of homogenization, no other treatment was considered necessary. Only highly poisonous industrial wastes were deposited underground in abandoned salt mines, mining caves, or tunnels.

Most waste pretreatment or treatment procedures, other than just deposition in sanitary landfills, were developed during the last two to four decades. Recently, a new deposition guideline of the European Community (EU-Deponierichtlinie 1999/31/EG, European Commission, 26 April 1999) was introduced that defines the construction, operational, and aftercare requirements of the three exclusively allowed sanitary landfill classes of the future for either:
• dangerous industrial wastes
• nondangerous wastes such as domestic refuse
• inert monowaste material without chemical or biological reactivity

The EU guideline prescribes a leak-proof construction of bottom and top seals and focuses on deposition techniques, whereas the German technical instruction for the handling of domestic wastes (TA Siedlungabfall and Abfallablagerungsverordnung AbfAblV 2001) is much more stringent. In addition to comparably detailed prescriptions regarding the construction of sanitary landfills, the TA Siedlungsabfall defines a maximum organic dry matter content of 3% or 5% of the wastes for deposition into class I or class II sanitary landfills, to guarantee an inert or quasi-inert behavior after deposition. Class III landfills are monodeposits for certain inert, nondangerous waste materials.

The German federal waste recycling law (Kreislaufwirtschaftsgesetz-KrW/AbfG 1996) defines three top priorities:

- avoidance of wastes: to reduce the amount of waste material to a minimum
- recycling of wastes:
  - as secondary raw material (substance recycling)
  - as a source of energy (energy recycling)
- deposition of wastes

Deposition of untreated wastes is restricted until 2005. After then, only deposition of fully inert wastes such as incineration slag or ash will be allowed according to the environmental law. However, due to a shortage of incineration capacity, mechanically and biologically pretreated wastes with low residual respiratory activity or methane production capacity and a high lignin-to-carbohydrate coefficient may be deposited until 2025, according to an approved amendment to the German environmental legislation.

Whereas avoidance of wastes, especially of packaging wastes, still seems to have a real potential for improvement, recycling of waste materials from other waste types has reached a high overall level. For example, glass recycling is characterized by high recovery rates, presumably because the quality of glass products made from recycled raw materials is almost the same as that of virgin glass.

Paper recycling by standardized procedures is also well introduced and accepted. However, due to breakage of fibers with every round of recycling, paper recycling is more a process of downcycling. A certain percentage of fresh fiber material must be added to achieve constant product quality. Paper from recycled raw material has to compete with paper made from low-quality wood or waste wood, which is available in high quantities in the Northern, wood-producing countries of the world.

Whether plastic material should be recycled for the production of new plastic goods is a matter of discussion. To achieve high qualities of recycled-plastic products, plastic wastes have to be cleaned of contaminants. Then the mixed plastic material must be separated into the different polymer fractions (which apparently is not efficient with the procedures now available) for specific recycling of each polymer class. It may be more favorable to use mixed plastic wastes for energy recycling, e.g., using mixed waste plastic as fuel in the production of new plastic materials from the fossil petroleum saved by not using the latter as fuel.

To achieve the low carbon content of 5% or even 3% of organic dry matter content, as required for deposition, the nonrecyclable fractions of municipal or industrial wastes must be incinerated. Pyrolysis alone does not suffice. Waste incineration leads to two main residual products: gas and slag. Both are highly polluted with toxic materials, but their concentrations of toxins are not as high as in fly ash. Whereas purification of the off-gas from waste incineration can be considered a state-of-the-art process enforced by various state laws, e.g., the 17th German Federal Ordinance on Protection from Emissions (17. BImSchV 1990), disposal or proper utilization of the ash and slag is still a matter of controversy. Incineration slag has been used to construct traffic noise protection barriers along highways. Long-term reactions of the heavy metal oxides might, however, lead to their remobilization and cause environmental harm.

After removal of the powder fraction of slag by sieving, the granular fraction can be used as a raw material in the construction industry. However, there seem to be gaseous organic inclusions in the slag which slowly diffuse out of concrete walls of houses and harm the inhabitants. In addition, some residual toxic organics remain in the ash, even when the waste incineration efficiency is high.

To obtain less toxic incineration residues in the future, detoxification of slag and fly ash is considered not only a desirable option, but a requirement – even if it seems too expensive now. In Switzerland, a new technique for separating toxic heavy metals in highly toxic filter ash from waste incineration plants from nontoxic mineral products was tested on a laboratory scale (*Chemische Rundschau* No. 15, 1998). The heavy metal ions react with hydrochloric acid to form their chlorides, which can be evaporated at 900 °C. This would avoid deposition of toxic fly ash in mining shafts or solidification of the toxic material with concrete. To save energy, this treatment should ideally start with still-hot ashes.

## 19.4
## Off-gas Purification

For removal of organic and inorganic pollutants from huge quantities of highly polluted waste gas streams from, e.g., the lime and cement industry, coal-fired power plants, or waste incineration plants, technical procedures are available for the separation of fly ash by gas cyclones, particle filtration or electrofiltration; the removal of acid and alkaline gas impurities by washing procedures; and the removal of neutral trace-gas components by washing/adsorption/gas filtration. Off-gases from composting plants, pork and chicken breeding facilities, etc., which are mainly polluted with volatile organic compounds, can be purified by gas washing and aerobic/anaerobic treatment of the washing water in biofilters. Natural and synthetic filter materials have been used as support materials for the development of active biofilms on biogas or off-gas filter units, which in principle resemble the trickling filters used in sewage treatment. To maintain a permanent high adsorption and degradation efficiency, the moisture content of the filter material must be kept high enough to keep the biofilm in an active state, and addition of trace elements may eventually be required to support optimal growth and metabolic activity of the microorganisms forming the biofilm.

## 19.5
## Soil Remediation

Due to an almost unlimited number of pollutants and to different soil and underground structures, no general guideline for soil remediation is possible. Since soil is an agglomerate of mineral compounds, including small particulate materials such as clay, gravel, and stones, and – at least in the upper layers – of organic materials (e.g., plant residues, organic fertilizers, humic substances) and its adsorption capacity for toxicants changes at different moisture contents or water conductivities, the retention of hydrophilic, water-soluble and hydrophobic, water-insoluble contaminants in soils varies. In particular, water-insoluble light compounds (NAPL = non-aqueous phase liquids) tend to accumulate on top of the groundwater level.

Besides safeguarding contaminated sites by, e.g., inertization and encapsulation, several methods for decontamination of highly polluted soils and groundwater at former industrial sites have been practiced in the past, including ex situ soil treatment (incineration, soil washing and soil extraction procedures, land farming, and biological cleaning) and in situ remediation (bioventing or biosparging, natural and enhanced natural soil remediation). Ex situ treatment is easier to control but expensive, whereas in situ remediation is less easy to control and becomes less efficient with time but is much cheaper.

In locally restricted soil compartments that have been contaminated with highly toxic chemicals or metals (e.g., mercury), excavation and thermal or chemical treatment (incineration or chemical extraction of the pollutants) may be necessary to restrain spreading of the toxic contaminants or of toxic metabolites during sanitation. By excavation and, e.g., thermal soil treatment, not only is the original soil texture destroyed, but also the redox state is shifted toward oxidized sinter products. However, a high decontamination efficiency even of micropores can be obtained within a short treatment time. If the contaminants are highly volatile, on-site decontamination should be favored, since otherwise, extensive precautions for transport in closed containers are necessary, as is true for soil contaminated, e.g., with poisonous solvents or leaded antiknock agents.

Except for those sites with highly toxic contaminants requiring ex situ treatment procedures, the more economical but eventually less quantitative in situ procedures could be used in many contaminated sites. The procedures include soil stripping with solvents or water and groundwater recirculation after purification of the stripped liquid (hydraulic procedures) or biosparging (gas venting through groundwater and the unsaturated soil), purification of the ventilating gasses from the contaminants, and reintroduction of the purified gases. If a light liquid phase is floating on the surface of the groundwater, the bioslurping procedure may be the most appropriate means for remediation; bioslurping is a combination of stimulation of biological processes and contaminant removal. Direct in situ bioremediation techniques are suitable for soils and groundwater with high liquid and gas conductivity, if biodegradation of the contaminants can be achieved by air injection, mineral or carbon addition, or groundwater recirculation. For in situ bioremediation, until now only a limited number of possible techniques have been developed for practical application, mainly because the efficiency of biological in situ removal of contaminants is limited by several factors concerning the contaminants (e.g., low solubility, strong sorption onto the soil matrix, diffusion into macropores of soils and sediments) and concerning the transport of nutrients and electron acceptors for microbial activity (e.g., permeability and porosity of the soil, its ion exchange capacity, pH, and redox potential).

The least invasive in situ remediation approach is based on the identification of intrinsic bioremediation factors, to show that under suitable environmental conditions an indigenous microbial population exists and that degradation has already occurred and is continuing. To support or enhance this natural self-curing ability, bioaugmentation and biostimulation technologies are available and sometimes suitable. In all situations, degradation rates, degradation efficiency, and groundwater flow rates should be carefully monitored, and the remediation area should be planned far enough downstream in the groundwater to avoid transportation of contaminants or metabolites out of the remediation field. By construction of funnel-and-gate systems (reactive walls) at the downstream end of the remediation field, nondegraded contaminants and residual metabolites can be adsorbed onto activated carbon or other suitable materials and be prevented from migrating into noncontaminated areas.

## 19.6
### Drinking Water Preparation

If the groundwater contains toxic substances as a consequence of soil pollution by leachates from sanitary landfills, production residues, spillages from industry, over-fertilization, or insecticide and pesticide application in agriculture, these substances must be separated quantitatively during water processing for drinking water preparation. Separation, filtration, and sanitizing procedures have been developed and have reached high technological levels. Since contamination of groundwater is still increasing and many contaminants remain for decades, water purification procedures must have high priority now and in the future, especially since drinking water resources are limited.

In arid countries with access to saline seawater, the water must be desalted by membrane-based seawater desalting processes to obtain salt-free process water or drinking water.

Although techniques for complete purification of wastewater are in principle available, the application of these multistep procedures to drinking water preparation is not likely in the near future, because of the very high water processing costs and the availability of less polluted water sources.

## 19.7
### Future Strategies to Reduce Pollution and Conserve a Natural, Healthy Environment

In industrialized countries the main strategy for handling domestic and industrial wastewater seems to be set for years or even decades, due to high investments in sewer systems and in what is considered modern wastewater treatment facilities. High-efficiency removal of carbon, nitrogen, and phosphate was intended in the past as a way to avoid damage to the receiving ecosystems.

In Germany, centralized treatment centers fed with hundreds of miles of sewers and many pumping stations have been almost completed for domestic wastewater treatment. 'Spot solutions' for new residential areas, single houses, or small villages should be promoted to gain experience with the new small-scale process alternatives.

The real challenge for wastewater purification arises in many developing countries lacking sewer systems and often lacking any wastewater treatment. Central treatment units are unaffordable, and even if they existed the sewer systems would not be capable of handling the masses of rainwater during the rainy season. This is why decentralized wastewater and waste treatment should be favored. 'Decentralized' in this context should range from single-house solutions, neighborhood solutions for a few houses, wastewater treatment solutions for a residential area or municipality, to solutions for a certain geologically defined area of human settlement. If the whole infrastructure for wastewater collection and treatment must be designed, the best solution would be the one with the shortest overall sewer length. Due to the still-unreliable electric supply outside the megacities, small-scale treatment systems should be reduced to the basic components as a starting technology, requiring little or no electricity and no skilled personnel for maintenance. Decentralized wastewater management should be favored, not only because imitating the systems of industrialized countries would not be affordable, but because the wastewater resources could be better used. Domestic wastewater and wastes, if properly collected and treated, can be upgraded to yield valuable nitrogen- and phosphate-rich fertilizers and thus save money otherwise spent on mineral fertilizers. By decentralized treatment, more nontoxic wastewater, sewage sludge, or waste compost as a source of nitrogen and phosphate is available for treating local farmland, and transport distances are short.

A process development that goes hand in hand with investigations on the respective microbiology is very important for the future development of wastewater treatment. Microbial reaction rates are higher in equatorial countries due to the high average annual temperatures.

Future microbial investigations for wastewater treatment should start with the complex ecophysiology and end with tracing and optimizing single microbial bottleneck reactions. As recently experienced with the Anammox (anaerobic ammonia oxidation) process, microbiology often seems to lag behind technical verification. Other new procedures for the removal of nitrogen from domestic or industrial wastewater are at the stage of pilot- or technical-scale testing. In parallel, microbiologists are elucidating the biochemical basis of the relevant reactions.

Although in some branches of the food and feed industries, starter cultures or even enzymes are now essential tools for production, the advantage of a broad application of starter cultures to wastewater treatment (bioaugmentation) in order to improve purification efficiencies or to degrade trace compounds is not yet widely recognized. Most reports refer to laboratory-scale experiments; only a few full-scale tests have been reported. Starter cultures containing genetically engineered specialists for metabolizing certain xenobiotics that periodically appear in more than trace concentrations may, however, help to introduce or stabilize the required metabolic capabilities. Starter cultures containing an 'omnipotent' population may be seeded only after complete process failure due to the presence of toxins, to reestablish the microbial degradation potential more quickly in wastewater treatment plants receiving wastewater having little indigenous population. Biostimulation of the autochthonic population by addition of extra substrates or electron acceptors is an alternative to be considered.

A major problem at present and in the future is the handling of surplus sludge from wastewater treatment. Dewatering procedures must be improved, and new and better sludge disintegration methods must be developed. Although in some examples the microbiological basis for the formation of bulking sludge is understood, reliable microbiological counteractions to prevent bulking are not yet available. For sludge disintegration, enzyme engineering should in the future create new, stable, and powerful lytic enzymes.

For water management in new residential areas, developments might go in the direction of dual water supply, on-spot treatment of slightly polluted wastewater, and seepage of purified wastewater in especially designed ecosystems. Concentrated wastewater streams should also be treated near to where they are generated. New residential areas must be planned with few paved areas (or existing paved areas should be depaved), so as to retain most of the rainwater for replenishing the groundwater.

Industrial production processes with better product-to-wastes ratios have to be developed by applying new production processes or by more efficient utilization of the water, e.g., by internal water cycling. Tailor-made treatment systems for every wastewater stream should be optimized, with emphasis on production procedures and on microbiological capabilities, including the use of starter cultures (bioaugmentation).

The slogan 'the waste of one company is the raw material of another' should be promoted worldwide and may be facilitated by creation of appropriate databanks. Retail prices for all goods, including those imported from developing countries, should include the full, real, or fictive costs for wastewater and residue treatment.

The potential to reduce the total amount of solid wastes in the future must be fully exploited, in particular by the packaging industry. Improvement of distribution logistics may help to prevent one-way single-product packaging, pallet-level packing and another layer of packaging for transport of larger package units.

Since incineration is the most expensive waste destruction system, it should be reserved only for those fractions that cannot be recycled or reutilized. Biowaste composting and biowaste methanation are options for organic waste fractions having a high content of naturally occurring organics. Cofermentation of biowaste fractions with sewage sludge may also be taken into consideration, if excess digester volume is available and the sewage sludge is free of toxins. Combined mechanical and biological waste inertization could be an alternative to incineration, but cannot achieve the low carbon content required by the deposition guideline of the EU.

In developing countries, direct reutilization of wastewater or wastes or product recycling seems to be more distributed than in highly industrialized countries, due to a shortage of raw materials or to restricted production or affordability. This is especially true for, e.g., plastic bottles and containers, which are often one-way articles in industrialized countries, but are reutilized several times in developing countries.

In industrialized countries, drinking water management must in the future take care of trace pollutants that have unknown effects on human health. New methods to analyze and separate residual agricultural and household chemicals and their metabolites must be developed.

Due to the high number of contaminated areas in almost every country and due to limited budgets for soil remediation, such areas should be ranked according to environmental risk. Then soil remediation techniques should be chosen that will prevent further migration of the contaminants or their possibly toxic reaction products. In addition to the common techniques for groundwater treatment (pump-and-treat, funnel-and-gate systems and reactive barriers) increasingly have to be used. For natural attenuation of contaminated areas, gen probe methods must be developed to analyze the biological or biochemical potential in-situ or from in-situ samples. In the U.S. the Environmental Protection Agency requires proof of the degradative capability of the in-situ population.

For treatment of sites with low contaminant concentrations, phytoremediation approaches for metals and organics, e.g., nitro compounds and polycyclic aromatic hydrocarbons, increasingly have to be tested. Together with other near-natural processes and the monitored natural attenuation procedures, sustainable strategies have to be developed to overcome the problems of contaminated sites. Furthermore, a variety of bacterial species and enzymes have been the target of genetic engineering to improve the performance of biodegradation, control degradation processes, and detect chemical pollutants and their bioavailability. Avoidance of environmental contamination is the future challenge for which suitable and sustainable strategies can be achieved only by an interdisciplinary collaboration between all protagonists in research and industry. The wide-ranging experience accumulated with respect to the contamination of soils and groundwater must provide a special impetus for testing the environmental impact of new chemical products before they are introduced, thus preventing subsequent contamination and undesirable reactions, such as the endocrine disruption suspected to be caused by Bisphenyl A. A benign 'design chemistry' would, therefore, have to concentrate research on identifying forms of bonding that facilitate the development of biodegradable and environmentally sound chemical products.

The supply of good-quality drinking water must especially be improved in developing countries, to reduce mortality, especially in children. Groundwater pumping from deep wells often exceeds the amount of newly formed groundwater, so wells are drilled deeper and deeper. In coastal regions this may lead to salt water infiltration, which contaminates the sweet water reserves.

Wastewater seepage and groundwater pumping often occur close together, too close to maintain a sufficient purification distance for complete degradation and sufficient sanitization. Contamination of well water with pathogenic microorganisms is favored by this mismanagement and in warm climates causes epidemics.

Off-gas purification by biological means has seen much-increased use in the past. For biological off-gas purification, existing gas ventilation, washing, and filtration techniques and the appropriate technical equipment must be improved further.

# Subject Index

*Environmental Biotechnology. Concepts and Applications.* Edited by H.-J. Jördening and J. Winter
Copyright © 2005 WILEY-VCH Verlag GmbH & Co. KGaA, Weinheim
ISBN: 3-527-30585-8